# Metallofoldamers

# Metallofoldamers

## Supramolecular Architectures from Helicates to Biomimetics

Edited By

**GALIA MAAYAN**

*Schulich Faculty of Chemistry, Technion - Israel Institute of Technology, Israel*

and

**MARKUS ALBRECHT**

*Institut für Organische Chemie, RWTH Aachen, Germany*

A John Wiley & Sons, Ltd., Publication

This edition first published 2013
© 2013 John Wiley & Sons, Ltd.

*Registered office*
John Wiley & Sons Ltd, The Atrium, Southern Gate, Chichester, West Sussex, PO19 8SQ, United Kingdom

For details of our global editorial offices, for customer services and for information about how to apply for permission to reuse the copyright material in this book please see our website at www.wiley.com.

*Library of Congress Cataloging-in-Publication Data applied for.*

A catalogue record for this book is available from the British Library.

ISBN: 9780470973233

Set in 10/12pt, Times by Thomson Digital, Noida, India
Printed and bound in Singapore by Markono Print Media Pte Ltd

# Contents

*List of Contributors*                                                    **xi**
*Foreword*                                                                **xiii**
*Preface*                                                                 **xv**


**1  Metalloproteins and Metallopeptides – Natural Metallofoldamers**        **1**
*Vasiliki Lykourinou and Li-June Ming*

    1.1   Introduction                                                    1
    1.2   Metalloproteins                                                 2
          1.2.1   Metalloproteins are Nature's "Metallofoldamers!"     2
          1.2.2   Metal-Triggered Conformational Change of Proteins    3
          1.2.3   Conformational Change of Metalloproteins Caused by
                  Ligand Binding                                       7
          1.2.4   Protein Misfolding: Causes and Implications – Cu,
                  Zn-Superoxide Dismutase                              10
    1.3   Metallopeptides                                                 12
          1.3.1   Antibiotic Metallopeptides                           13
          1.3.2   Metallopeptides in Neurodegenerative Diseases        20
          1.3.3   Other Metallopeptides                                24
    1.4   Conclusion and Perspectives                                     28
    Acknowledgements                                                            30
    References                                                                  30


**2  Introduction to Unnatural Foldamers**                                   **51**
*Claudia Tomasini and Nicola Castellucci*

    2.1   General Definition of Foldamers                                 51
    2.2   Biotic Foldamers                                                53
          2.2.1   Homogeneous Foldamers                                53
          2.2.2   β-Peptides                                           53
          2.2.3   γ-Peptides                                           59
          2.2.4   Hybrid Foldamers                                     60
          2.2.5   Aliphatic Urea Foldamers                             63
          2.2.6   Foldamers of α-Aminoxy Acids                         64
          2.2.7   Foldamers Containing Amido Groups                    65
    2.3   Abiotic Foldamers                                               70

2.4    Organization Induced by External Agents                          72
          2.4.1    Organization Induced by Solvents                     72
          2.4.2    Organization Induced by Anions                       73
2.5    Applications                                                     78
2.6    Conclusions and Outlook                                         81
References                                                             81

**3  Self-Assembly Principles of Helicates**                           **91**
*Josef Hamacek*

3.1    Introduction                                                    91
3.2    Thermodynamic Considerations in Self-Assembly                   93
          3.2.1    Mononuclear Coordination Complexes                  93
          3.2.2    Extension to Polynuclear Edifices                   96
3.3    Cooperativity in Self-Assembly                                 100
          3.3.1    Allosteric Cooperativity                           101
          3.3.2    Chelate Cooperativity                              102
          3.3.3    Interannular Cooperativity                         104
3.4    Kinetic Aspects of Multicomponent Organization                104
3.5    Understanding Self-Assembly Processes                         108
          3.5.1    Assessment of Cooperativity                        108
          3.5.2    Thermodynamic Modelling                            110
          3.5.3    Solvation Energies and Electrostatic Interactions  115
3.6    Secondary Structure and Stabilizing Interactions              118
3.7    Conclusions                                                    118
References                                                             120

**4  Structural Aspects of Helicates**                                 **125**
*Martin Berg and Arne Lützen*

4.1    Introduction                                                   125
4.2    Structural Dynamics                                            127
4.3    Template Effects                                               129
4.4    Sequence Selectivity                                           130
4.5    Self-Sorting Effects in Helicate Formation                     135
4.6    Diastereoselectivity I – "*Meso*"-Helicate versus Helicate Formation   138
4.7    Diastereoselectivity II – Enantiomerically Pure Helicates from Chiral
          Ligands                                                     139
          4.7.1    2,2′-Bipyridine Ligands                            140
          4.7.2    2,2′:6′,2″-Terpyridine and 2,2′:6′,2″:6″,2-Quaterpyridine Ligands   143
          4.7.3    2-Pyridylimine Ligands                             144
          4.7.4    Further Hexadentate N-Donor Ligands                144
          4.7.5    Oxazoline Ligands                                  144
          4.7.6    P-Donor Ligands                                    145
          4.7.7    Hydroxamic Acid Ligands                            147

| | | |
|---|---|---|
| 4.7.8 | β-Diketonate Ligands | 147 |
| 4.7.9 | Catecholate Ligands and Other Dianionic Ligand Units | 148 |
| 4.7.10 | Non-Covalently Assembled Ligand Strands | 150 |
| 4.8 | Summary and Outlook | 150 |
| | References | 151 |

**5 Helical Structures Featuring Thiolato Donors**     **159**
*F. Ekkehardt Hahn and Dennis Lewing*

| | | |
|---|---|---|
| 5.1 | Introduction | 159 |
| 5.2 | Coordination Chemistry of Bis- and Tris(Benzene-o-Dithiolato) Ligands | 162 |
| | 5.2.1 Mononuclear Chelate Complexes | 162 |
| | 5.2.2 Dinuclear Double-Stranded Complexes | 165 |
| | 5.2.3 Dinuclear Triple-Stranded Complexes | 167 |
| | 5.2.4 Coordination Chemistry of Tripodal Tris(Benzene-o-Dithiolato) Ligands | 172 |
| 5.3 | Coordination Chemistry of Mixed Bis(Benzene-*o*-Dithiol)/Catechol Ligands | 176 |
| | 5.3.1 Dinuclear Double-Stranded Complexes | 176 |
| | 5.3.2 Dinuclear Triple-Stranded Complexes | 178 |
| 5.4 | Subcomponent Self-Assembly Reactions | 181 |
| 5.5 | Summary and Outlook | 186 |
| | References | 186 |

**6 Photophysical Properties and Applications of Lanthanoid Helicates**   **193**
*Jean-Claude G. Bünzli*

| | | |
|---|---|---|
| | List of Acronyms and Abbreviations | 193 |
| 6.1 | Introduction | 194 |
| 6.2 | Homometallic Lanthanoid Helicates | 197 |
| | 6.2.1 Influence of the Triplet-State Energy on Quantum Yields | 198 |
| | 6.2.2 Radiative Lifetime and Nephelauxetic Effect | 203 |
| | 6.2.3 Site-Symmetry Analysis | 206 |
| | 6.2.4 Energy Transfer between Lanthanoid Ions | 208 |
| | 6.2.5 Lanthanoid Luminescent Bioprobes | 210 |
| | 6.2.6 Other Investigated Helicates | 219 |
| 6.3 | Heterometallic d-f Helicates | 223 |
| | 6.3.1 Basic Photophysical Properties | 223 |
| | 6.3.2 $Eu^{III}$-to-$Cr^{III}$ Energy Transfer | 227 |
| | 6.3.3 Control of f-Metal Ion Properties by d-Transition Metal Ions | 228 |
| | 6.3.4 Sensitizing NIR-Emitting Lanthanoid Ions | 235 |
| 6.4 | Chiral Helicates | 236 |
| 6.5 | Extended Helical Structures | 239 |
| 6.6 | Perspectives | 240 |
| | Acknowledgements | 241 |
| | References | 241 |

**7   Design of Supramolecular Materials: Liquid-Crystalline Helicates**   **249**
*Raymond Ziessel*

| | | |
|---|---|---|
| 7.1 | Introduction | 249 |
| 7.2 | Imino-Bipyridine and Imino-Phenanthroline Helicates | 252 |
| | 7.2.1   Liquid Crystals from Imino-Polypyridine Based Helicates | 257 |
| 7.3 | Conclusions | 266 |
| 7.4 | Outlook and Perspectives | 267 |
| | Acknowledgements | 268 |
| | References | 268 |

**8   Helicates, Peptide-Helicates and Metal-Assisted Stabilization
of Peptide Microstructures**   **275**
*Markus Albrecht*

| | | |
|---|---|---|
| 8.1 | Introduction | 275 |
| 8.2 | Selected Examples of Metal Peptide Conjugates | 276 |
| 8.3 | Helicates and Peptide-Helicates | 279 |
| | 8.3.1   Helicates | 279 |
| | 8.3.2   Peptide-Helicates | 281 |
| 8.4 | Metal-Assisted Stabilization of Peptide Microstructures | 288 |
| | 8.4.1   Loops and Turns | 288 |
| | 8.4.2   α-Helices | 292 |
| | 8.4.3   β-Sheets | 297 |
| 8.5 | Conclusion | 298 |
| | References | 300 |

**9   Artificial DNA Directed toward Synthetic Metallofoldamers**   **303**
*Guido H. Clever and Mitsuhiko Shionoya*

| | | |
|---|---|---|
| 9.1 | Introduction | 303 |
| | 9.1.1   Oligonucleotides are Natural Foldamers | 303 |
| | 9.1.2   Biological Functions and Beyond | 305 |
| | 9.1.3   DNA Nanotechnology | 306 |
| | 9.1.4   Interactions of DNA with Metal Ions | 308 |
| 9.2 | The Quest for Alternative Base Pairing Systems | 309 |
| | 9.2.1   Modifications of the Hydrogen Bonding Pattern | 310 |
| | 9.2.2   Shape Complementarity | 310 |
| | 9.2.3   Metal Coordination | 310 |
| 9.3 | Design and Synthesis of Metal Base Pairs | 311 |
| | 9.3.1   Rational Design of Metal Base Pairs | 311 |
| | 9.3.2   Model Studies | 312 |
| | 9.3.3   Synthesis of Modified Nucleosides | 312 |
| | 9.3.4   Automated Oligonucleotide Synthesis | 314 |
| | 9.3.5   Enzymatic Oligonucleotides Synthesis | 315 |
| 9.4 | Assembly and Analysis of Metal Base Pairs Inside the DNA Double Helix | 315 |
| | 9.4.1   Strategies for Metal Incorporation | 315 |
| | 9.4.2   Analytical Characterization in Solution | 316 |
| | 9.4.3   X-Ray Structure Determination | 317 |

| | | |
|---|---|---|
| 9.5 | Artificial DNA for Synthetic Metallofoldamers | 318 |
| | 9.5.1 Overview | 318 |
| | 9.5.2 The Hydroxypyridone Base Pair | 320 |
| | 9.5.3 The Salen Base Pair | 320 |
| | 9.5.4 The Imidazole, Triazole and 1-Deazaadenine-Thymine Base Pairs | 323 |
| 9.6 | Functions, Applications and Future Directions | 324 |
| | 9.6.1 Duplex Stabilization and Conformational Switching | 324 |
| | 9.6.2 Sensor Applications | 325 |
| | 9.6.3 Magnetism and Electrical Conductance | 325 |
| | 9.6.4 Future Directions | 326 |
| References | | 327 |

**10 Metal Complexes as Alternative Base Pairs or Triplets in Natural and Synthetic Nucleic Acid Structures**    **333**
*Arnie De Leon, Jing Kong, and Catalina Achim*

| | | |
|---|---|---|
| 10.1 | Introduction | 333 |
| 10.2 | Brief Overview of Synthetic Analogues of DNA: PNA, LNA, UNA, and GNA | 338 |
| 10.3 | Metal-Containing, Ligand-Modified Nucleic Acid Duplexes | 340 |
| | 10.3.1 Design Strategy | 341 |
| | 10.3.2 Duplexes Containing One Alternative Metal–Ligand Base Pair with Identical Ligands | 342 |
| | 10.3.3 Duplexes Containing One Alternative Metal–Ligand Base Pair with Different Ligands | 359 |
| 10.4 | Duplexes Containing Multiple Metal Complexes | 361 |
| 10.5 | Metal-Containing, Ligand-Modified Nucleic Acid Triplexes | 367 |
| 10.6 | Summary and Outlook | 367 |
| Acknowledgement | | 369 |
| Abbreviations | | 369 |
| References | | 370 |

**11 Interaction of Biomimetic Oligomers with Metal Ions**    **379**
*Galia Maayan*

| | | |
|---|---|---|
| 11.1 | Introduction | 380 |
| 11.2 | Single-Stranded Oligomers in Which Metal Coordination Templates, or Templates and Nucleates the Formation of an Abiotic Helix | 381 |
| 11.3 | Folded Oligomers in Which Metal Coordination Nucleates the Formation of an Abiotic Single-Stranded Helix | 384 |
| 11.4 | Folded Oligomers in Which Metal Coordination Enhances Secondary Structure and Leads to Higher-Order Architectures | 393 |
| | 11.4.1 Metal Coordination in Folded Aromatic Amide Oligomers | 394 |
| | 11.4.2 Metal Coordination in Peptidomimetic Foldamers | 396 |
| 11.5 | Concluding Remarks | 402 |
| References | | 402 |

**12  Applications of Metallofoldamers**                                          **407**
*Yan Zhao*

12.1  Introduction                                                               407
12.2  Metallofoldamers in Molecular Recognition                                  409
12.3  Metallofoldamers as Sensors for Metal Ions                                 414
12.4  Metallofoldamers as Dynamic Materials                                      419
12.5  Conclusions and Outlook                                                    429
References                                                                       430

***Index***                                                                      **433**

# List of Contributors

**Catalina Achim**, Department of Chemistry, Carnegie Mellon University, USA

**Markus Albrecht**, Institut für Organische Chemie, RWTH Aachen University, Germany

**Martin Berg**, Kekulé Institute of Organic Chemistry and Biochemistry, University of Bonn, Germany

**Jean-Claude G. Bünzli**, Institute of Chemical Sciences and Engineering, École Polytechnique Fédérale de Lausanne, Switzerland
Photovoltaic Materials, Department of Materials Chemistry, Korea University, Sejong Campus

**Nicola Castellucci**, Dipartimento di Chimica "G. Ciamician", Università di Bologna, Italy

**Guido H. Clever**, Institute for Inorganic Chemistry, Georg-August University Göttingen, Germany

**Arnie De Leon**, Department of Chemistry, Carnegie Mellon University, USA

**F. Ekkehardt Hahn**, Institut für Anorganische und Analytische Chemie, Westfälische Wilhelms-Universität Münster, Germany

**Josef Hamacek**, Department of Inorganic and Analytical Chemistry, University of Geneva, Switzerland

**Jing Kong**, Department of Chemistry, Carnegie Mellon University, USA

**Dennis Lewing**, Institut für Anorganische und Analytische Chemie, Westfälische Wilhelms-Universität Münster, Germany

**Arne Lützen**, Kekulé Institute of Organic Chemistry and Biochemistry, University of Bonn, Germany

**Vasiliki Lykourinou**, Department of Chemistry, University of South Florida, USA

**Galia Maayan**, Schulich Faculty of Chemistry, Technion – Israel Institute of Technology, Israel

**Li-June Ming**, Department of Chemistry, University of South Florida, USA

**Mitsuhiko Shionoya**, Department of Chemistry, University of Tokyo, Japan

**Claudia Tomasini**, Dipartimento di Chimica "G. Ciamician", Università di Bologna, Italy

**Yan Zhao**,  Department of Chemistry, Iowa State University, USA

**Raymond Ziessel**, Laboratoire de Chimie Organique et Spectroscopie Avancées (LCOSA)

Ecole de Chimie, Polymères, Matériaux Université de Strasbourg France

# Foreword

Since their initial presentation 25 years ago, helicates, helical metallosupramolecular architectures, have received wide attention and been subject to numerous investigations. Much of the fascination they exerted and continue to exert on chemists results from their helical structure, and in particular the relationship of double helices with the emblematic entity, double-stranded DNA. Although this feature was part of our initial motivation, the goal and the impact were much wider. Indeed, as pointed out earlier, the generation of double helicates from designed ligands and suitable metal ions had for us a much broader significance as it marked the start of the implementation of self-organization processes in supramolecular chemistry via metalloarchitectures. It represented an important first step in the formulation of supramolecular self-organization as the outcome of programmed chemical systems, whereby information stored in the structural features of the ligand molecules is processed by the metal cations via their coordination algorithm. In a further extension, the self-sorting processes occurring in mixtures of equilibrating helicates and the generation of anion-dependent circular helicates became the starting points of our studies that were to evolve into constitutional dynamic chemistry.

From double-stranded to multiple stranded, from two centers to multiple centers, from identical metal ions to different ones, the field of helicates developed extensively, exploring structural features, physical properties and coding schemes, and reaching beyond towards a great variety of other metallosupramolecular architectures. Its expansion demonstrated once again that, without claiming to imitate what biology has so powerfully achieved in the DNA molecule, chemistry is exploring a much wider scene, generating entities entirely imagined and fabricated by chemists, exerting the power of chemistry over the expressions of matter.

The present volume displays a selection of the various aspects of the chemistry of helicates and related entities. It takes stock of achievements, but also suggests numerous exciting developments to the imagination of chemists, at the triple interface of organic, inorganic and biological chemistry, as well as, from a more general perspective, towards the precise control of the self-organization of chemical entities through appropriate programming.

The editors have to be congratulated for assembling a remarkable roster of active players in the field, who deserve our warmest thanks for their expert contributions.

*Jean-Marie Lehn*

# Preface

Nature's examples of functional molecular entities are highly versatile. The living world as we know it only exists due to the specific properties and selective interactions of molecules. Hereby, the observed broad functionality of biological systems is mainly based on the structure of biomacromolecules (proteins or DNA), which strongly relies on an intra- or intermolecular connection between polymer strands. This occurs either in a covalent (e.g., –S-S– bridges) or in a noncovalent fashion. In the latter case H-bonding is probably the most prominent bonding motif. In many cases, however, metal ions have a significant role in the structure and function of biopolymers, being able to bind them and consequently to control their overall structural and thus functional features. It is now well established that the identity of the metal-binding ligands and their coordination mode, as well as side chain interactions, have a crucial role in governing metal binding and selectivity in proteins. The coordination of metal ions to proteins and peptides results in conformational changes, which lead to important functions such as catalysis, interactions with specific receptors or target proteins and the inhibition of biochemical or biophysical processes.

The effect of metal ions on conformational changes in proteins and peptides and thus on their function, which is widely discussed in Chapter 1, has inspired the design and synthesis of "metallofoldamers" – synthetic oligomers that fold upon interactions with metal ions to give various stable architectures in solution. Metallofoldamers are one class of "foldamers", which are unnatural oligomers that fold into well defined three-dimensional structures in solution via noncovalent interactions. The design, synthesis, structures and applications of foldamers, as thoroughly described in several reviews and a book, are summarized in Chapter 2. We distinguish between three types of metallofoldamers: helicates (Chapters 3–8), metallo-nucleic acids (Chapters 9–10) and metallo-peptidomimetics (Chapter 11).

The term "helicate" was introduced 25 years ago by J.-M. Lehn and describes a rigid oligomer that spontaneously folds into helical structure upon binding of metal ions to its backbone. The self-assembly principles of helicates are presented in Chapter 3. Due to their simplicity, helicates are the ideal "supramolecular *Drosophila*" to be investigated. Indeed, in the years following their introduction, the chemistry of helicates was intensely studied. This included looking into their structural aspects (Chapter 4) and exploring new ways to construct helicates with various structures and features using unique oligomer backbone chelators (Chapter 5) or nontransition metal ions (Chapter 6). Although the majority of studies were based on the helicate model character, some functional helicates were developed. Thus, liquid-crystalline helicates were designed as supramolecular materials (Chapter 7). In the quest for the creation of more biomimetic helicates, a new generation of helicates is being developed, which includes the use of biopolymers such as peptides and DNA as backbones (Chapters 8–10). This new direction in the field of helicates leads the way to new trends in the field of foldamers – the generation of synthetic oligomers which are peptide and nucleic acid mimics and can fold upon metal coordination to produce various three-dimensional structures stable in solution (Chapters 9–11). It is important to note that, while in helicates the metal ions template a helical structure via self-assembly, in biomimetic oligomers the metal ions nucleate the formation of a three-dimensional structure in a controlled manner. These biomimetic metallofoldamers are therefore macromolecular coordination compounds in which the overall structure of the macromolecule is influenced by its coordination to metal ions (Chapter 11). This copies nature's example of metallaproteins and opens the door to developing novel functional materials (Chapter 12).

The present book covers the whole field of structure control in oligomeric, polymeric, biomimetic and biological systems spanning the bridge from the simple helicates to polymers and natural or artificial peptides or DNA. This directly leads to prospective applications. We believe that the presented aspects are of interest and we hope that many scientists will be inspired and motivated as we are to do research in this fascinating field of chemistry.

Finally, we would like to deeply acknowledge the authors of the individual chapters for their singular and extraordinary contributions. We are also thankful to our coworkers and colleagues for introducing us to this fascinating field of chemistry, for fruitful discussions and for sharing our interest and passion. We also want to express our gratitude to the Wiley team, especially Paul Deards, who initiated the production of this book, as well as Sarah Tilley and Rebecca Ralf for their professional assistance during the editing and publishing process.

August 2012
*Galia Maayan*
*Markus Albrecht*

# 1

# Metalloproteins and Metallopeptides – Natural Metallofoldamers

*Vasiliki Lykourinou and Li-June Ming*
*Department of Chemistry, University of South Florida, USA*

## 1.1 Introduction

By a combination of amino acids with various properties, it is possible to obtain the natural polymers, peptides and proteins, capable of catalyzing key transformations, sustaining energy flow, and maintaining life in a narrow range of allowable conditions and starting materials. The folding, flexibility, and structural changes of peptides and proteins are integral to their functions within living systems. Several factors affect their folding and conformation, including primary sequence, subunit composition, nature and type of subunit interactions, presence of cofactors, and compartment and sequence during folding in a well controlled environment. Emphasis must also be placed on the conformational flexibility of peptides and proteins necessary in the context of their folding and function. In the case of enzymes, such conformational flexibility renders the induced fit process possible during enzymatic catalysis, as found in the large domain movements in hexokinase, HIV protease, and many others upon substrate or inhibitor binding.

Protein conformation, structure, and function are often determined or modulated by metal ions. Therefore, it is instructive to discuss the effect of metal ions or cofactors on the conformational changes of proteins prior to and during their actions. Furthermore, by comparing the folding of metallopeptides and metalloproteins to those whose conformational changes take place in the absence of metal ions can provide further structural and functional information on the metal ions in the former molecules. Such discussion is

*Metallofoldamers: Supramolecular Architectures from Helicates to Biomimetics*, First Edition.
Edited by Galia Maayan and Markus Albrecht.
© 2013 John Wiley & Sons, Ltd. Published 2013 by John Wiley & Sons, Ltd.

important in this context, as many proteins require metal ion(s) for their optimal structure and function.

## 1.2 Metalloproteins

### 1.2.1 Metalloproteins are Nature's "Metallofoldamers!"

The term "foldamer" was defined as "polymers with a strong tendency to adopt a specific compact conformation" (Gellman) or "oligomers that fold into a conformationally ordered state in solution, the structures of which are stabilized by a collection of noncovalent interactions between nonadjacent monomer units" (Moore) [1]. In this respect foldamers share common characteristic with proteins and thus the term is adopted to differentiate synthetic oligomers and polymers from "nature's foldamers" such as peptides, proteins, and nucleic acids. The term "metallofoldamers" thus is used to describe foldamers that adopt conformationally ordered states in the presence of metal ions or complexes [2]. Therein, the metal center(s) plays a key structural role in the formation of the specific conformation of the corresponding foldamer.

Metal ions may play a key role in the conformational changes of proteins or peptides crucial in their functions. Given the high occurrence of metal ions as cofactors and their integral role in the function of proteins in carrying out catalytic transformations, it is important to review how they are incorporated in the structure of proteins. The mechanism of metal incorporation ranges from a controlled manner through the use of metallochaperones to the direct incorporation from the cellular pool. In the former case this specific class of proteins binds metal ions and mediates the delivery into target enzymes through protein–protein interactions. In the case of iron transport and heme incorporation, transferrin transports iron into cells and hemopexin delivers apo-heme to the same compartment. For cytochrome-c, the heme must be attached before proper folding occurs [3], whereas the assembly of Fe–S clusters and incorporation into proteins and the folding of the F–S proteins require machinery encoded in the iron–sulfur cluster operon [4].

Nevertheless, metal ion incorporation is not always well controlled in such a manner. There are still many metalloproteins without a known chaperone counterpart for metal delivery and folding. Therefore, metal incorporation has also been suggested to be controlled by the choice of compartment in which the metal incorporation takes place. From studies in cyanobacteria, it becomes apparent that many copper and manganese containing proteins fold in different compartments where metal insertion can be controlled in a way that the Irvin–Williams series of stability can be circumvented [5].

A classic example of the structural role of metal ions that affects the function is the zinc finger domains, wherein the metal ions crosslink α–β domains and thus play a central role in the formation of the defined structures. The metal binding as well as the packing of a hydrophobic core drive the folding process. The zinc coordinates to the $Cys_2His_2$ motif and drives the folding process, while its removal causes the disruption of the proper folding. These zinc finger domains were first found in the transcription factor TFIIIA, which represents the most common nucleic acid binding motif in transcription factors [6]. Upon binding to DNA, the Zn finger domains undergo further conformational change in order to fit

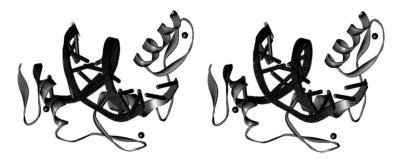

***Figure 1.1*** *Crystal structure of zinc finger–DNA complex (of Zif268). The Zn(II) ions (black spheres) in the three "fingers" (cyan) are bound to the protein through two Cys and two His residues and hold the α-helical/β-sheet structural motifs together.*

into the major groove of DNA (Figure 1.1; PDB 1ZAA) [7], representing one typical example of metal- and ligand-mediated conformational change of a natural metallofoldamer.

## 1.2.2   Metal-Triggered Conformational Change of Proteins

### 1.2.2.1   Cytochrome c and Heme Binding

The cytochrome c family of electron-transfer proteins has a high α-helix content and a heme cofactor that is postranslationally modified to covalently attach to the protein by two thioether bonds between the vinyl group of heme and two cysteine residues within the motif Cys-Xaa-Xaa-Cys-His [8]. The folding of cytochrome c has been studied by proton to deuterium exchange equilibrium monitored with NMR [9]. The folding sequences involve the interaction of the two terminal helices (N- and C-, respectively) first, followed by joining of the 60s helix. The heme center locks in the designated pocket and this is followed by final coordination of the Met80 in the sixth position. These three helices containing the axial methionine appear to be the minimal structural requirement in cytochrome c folding (Figure 1.2) [10]. The axial heme ligand is not conserved among all the proteins and can be found to be histidine, asparagine, or it can be absent [11]. The cytochrome c family has also been implicated in the apoptotic mechanism of cell death

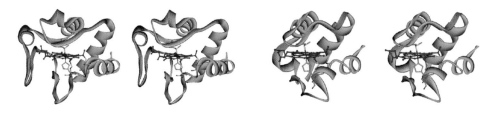

***Figure 1.2*** *Comparison of the three-dimensional structures of two different cytochromes c. Left: crystal structure of tuna cytochrome c (PDB ID 3CYT). Right: solution structure of oxidized cytochrome c from* Bacillus pasteurii *determined by NMR. The three helices are in cyan, the extra loop in the tuna enzyme is shown in lavender, the heme is shown in red.*

via a reactive oxygen species mechanism, scavenging of hydrogen peroxide, and in the assembly of cytochrome c oxidase [11].

The absence of the heme from cytochrome c causes complete destabilization of the protein due to a decrease of hydrophobic contacts as the heme resides at the hydrophobic core essential for the folding of the protein, an effect similar to guanidinium chloride denaturation [12]. The denaturant action is related to its competition with water molecules in the protein binding, resulting in an unfolding of the protein structure. Simulations of protein folding have shown that correct folding requires full heme contacts at the folding transition state in addition to the hydrophobic interactions as a critical "folding nucleus" [13]. The computational results are in agreement with H–D exchange NMR results, suggesting the initiation of the folding by the terminal helices followed by the 60s helix for the Met80 loop and β-sheets to build onto. The last helix to form is the 40s loop, concluding the significant folding role of the heme for providing a hydrophobic core to stabilize the protein and coordinating to the His and Met residues which results in further lowering the entropy en route to native folding. The heme center can also complicate the folding process as other residues or small molecules can compete for heme binding, which causes what is termed "chemical frustration" [14]. Folding of cytochrome c can therefore be modulated by choosing solvent conditions that favor one set of heme ligands over others. No crystal structures of the apo protein could be obtained for this reason [10], indicating the significance of the heme cofactor in forming and maintaining the folding this natural metallofoldamer cytochrome c.

### 1.2.2.2 α-Lactalbumin and its $Ca^{2+}$ Binding and Molten Globule

Calcium binding can cause significant conformational changes, which in turn may mediate a signaling cascade. The "EF-hand" folding is a major Ca-binding motif composed of a helix-loop-helix sequence as found in the multifunctional messenger calmodulin and S100. The assembly of the Ca binding can then be propagated to a protein partner with which the Ca-binding protein is interacting with. In the S100-type of proteins (which regulate cell cycles, cell growth, and differentiation), Ca binding influences protein folding to aid in their dimerization and further interaction with other partner proteins. Likewise, Ca binds α-lactalbumin at a domain containing a helix-loop-helix bend – close to the EF-hand domain – dubbed a Ca-binding "elbow" [8].

α-Lactalbumin is a main protein component of milk, which has been the target for investigation of calcium binding to proteins besides the EF-hand group of proteins and is used as a model for the study of protein stability. It is the regulatory subunit of lactose synthase for the synthesis of lactose from UDP-galactose and glucose in the lactating mammary gland. The protein possesses a single strong $Ca^{2+}$-binding site, which can also bind $Mg^{2+}$, $Mn^{2+}$, $Na^+$, and $K^+$, and a few distinct $Zn^{2+}$-binding sites. In bovine α-lactalbumin, $Ca^{2+}$ binds to the "elbow" region (Figure 1.3, lavender) via three carboxylates of Asp82, Asp87, and Asp88, two carbonyl groups of Lys79 and Asp84, and two water molecules in a distorted pentagonal bipyramid coordination sphere with the two carbonyl groups at the axial positions. The binding of cations to the $Ca^{2+}$ site increases the stability of α-lactalbumin against heat and various denaturing agents and proteases, while the binding of $Zn^{2+}$ to $Ca^{2+}$-saturated protein decreases the stability and causes aggregation.

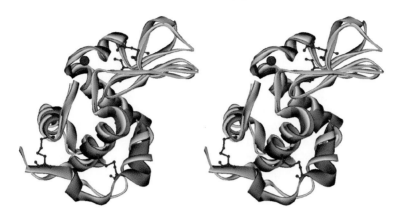

**Figure 1.3**   *Stereo view of crystal structure (PDB 1F6S) of bovine holo α-lactalbumin, showing the "Ca$^{2+}$ elbow" in lavender. The structure of hen egg white lysozyme (3LZT; yellow) is superimposed onto the α-lactalbumin structure, showing the significant similarity in folding.*

There are four disulfide bonds in α-lactalbumin (Figure 1.3, red), but none in the EF-hand proteins, which dramatically stabilize the protein conformation [8]. The correct folding of the protein, which relies on the correct formation of the proper disulfide bonds, is promoted by the high Ca$^{2+}$-binding affinity to the protein. Ca$^{2+}$ binding however does not change the secondary structure based on circular dichroism (CD) and fluorescence studies [15,16] which can effectively afford the molten globular state of the protein in the absence of Ca$^{2+}$ possessing a native-like secondary structure but a flexible tertiary structure [16]. The molten globular state of the protein can be obtained by removal of calcium, in the presence of denaturants, or in an acid denaturant state [17].

Ca$^{2+}$ does not have a significant effect on the metal-binding site, but it does affect the cleft at the opposite face (underneath the β-sheets shown in Figure 1.3) of the molecule at the joining of the α-helical (Figure 1.3, light brown) and the β-sheet (Figure 1.3, cyan) lobes by disturbing the H-bonding pattern in this region, which results in a more open conformation in apo α-lactalbumin and demonstrates the significance of Ca$^{2+}$ in the folding of α-lactalbumin [18].

Bovine α-lactalbumin (BLA) shares 38% sequence homology with hen egg white lysozyme (HEWL) with most of the differences at charged residues in BLA. However, their tertiary structures are nearly superimposable with all the four conserved disulfide bonds (Figure 1.3, yellow ribbon). Electrostatic interactions can stabilize partially unfolded conformations, which can aid in the formation of molten globule state in BLA, but not in HEWL. The thermodynamic folding barriers in the two proteins are different with a marginal barrier possible for BLA due to stabilization of partially unfolded conformations in the presence of Ca$^{2+}$, which stabilizes the fully folded state of the protein [19]. Such electrostatic interactions can be potentially engineered in artificial systems as a means to aid in structure stabilization. These studies again reveal the significance of metal ions in the proper folding and stabilization of proteins in order to afford functional natural metallofoldamers.

### 1.2.2.3  Metallothionein and Heavy Metal Regulation

The concentration of zinc must be well controlled in the cell, as high concentrations can be toxic and cause mitochondrial dysfunction. Metallothionein plays a central role in zinc homeostasis since it is the major protein important in regulation of the zinc level in the cell and its translocation [20] and it has been shown to induce an effect on brain neurons by binding to neuronal receptors and initiating pathways which cause neurite survival [21,22]. Metallothionein regulates the flow of zinc and copper in the cell and can further prevent poisoning from exposure to toxic cadmium and mercury. Its overall role is suggested to be the control of the distribution of zinc as a function of the cellular energy state and it has been implicated in the following functions [23,24]: (a) intracellular zinc transport, (b) zinc binding and exchange (e.g., with the zinc cluster protein Gal4, zinc finger transcription factors such as TFIIIA, and aconitase) as well as a zinc-specific chaperone, (c) oxidoreductive properties of cysteine bound zinc as cysteine ligands are redox-sensitive regulatory switches [25], (d) controlling cellular zinc distribution as a function of the energy state of the cell as shown by the interaction with ATP, GSH, and ROS and zinc distribution to enzymes in metabolic networks of gene expression and respiration [26], and (e) a possible role in neural activity, storing and distributing zinc for the neuronal network and protecting it against cellular damage as well as neuronal recovery through binding to neuronal receptors initiating signal transduction pathways [22].

One function of metallothionein may resemble that of the iron-storage protein ferritin in terms of its zinc-storage capability. It is a small protein rich in cysteine (20 cysteines in a total of 62 amino acids in human metallothioneine), but without aromatic amino acids such as tyrosine or histidine. The apo protein can bind a total of seven equivalents of divalent metal ions with $d^{10}$ configuration such as $Zn^{2+}$ or $Cd^{2+}$ in two noninteracting domains (Figure 1.4) [27] and up to six $Cu^+$ ions in each domain [28]. The protein can bind up to seven $Zn^{2+}$ or $Cd^{2+}$ ions in tetrahedral coordination spheres, or 12 three-coordinate $Cu^+$ ions with only cysteine residues. The binding of metal converts the random coil conformation of apo metallothionein into a folded two-domain structure (Figure 1.4). $Cd^{2+}$ binding has been extensively studied with C-13 NMR [29], wherein structural flexibility was observed. The absence of hydrophobic residues for stabilizing a folded form is compensated by the presence of the metal–thiolate core in the folding of this protein.

Apo metallothionein retains the backbone conformation imposed by the formation of the metal–thiolate clusters. REF computational studies indicate a potential H-bonding

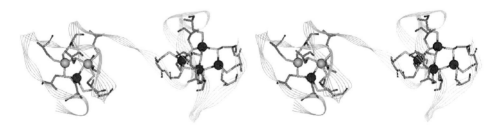

**Figure 1.4**  *Stereo view of the structure (PDB 4MT2) of Cd₅Zn₂, showing the N- (cyan) and C-(pink) domains with the coordinated Cys residue.*

network present in apo metallothionein which plays a role in the formation of a constant but flexible backbone needed to adjust to the incoming metal to ensure specific metal incorporation [30]. The metal–thiolate clusters then direct the "wrapping" of the protein into its three-dimensional structure. There are two distinct metal binding sites in the C-terminal (a-domain) and N-terminal domains (b-domain). The C-domain contains four $Zn^{2+}$ ions bound to 11 Cys residues with five serving as bridging ligands, while the N-domain binds three $Zn^{2+}$ ions via nine Cys residues and the metal binding follows a non-cooperative model up to four equivalents, as shown by $Co^{2+}$ binding [31,32]. Computational studies also support this metal-binding mechanism. MM3/MD calculations of metallothionein structure after sequential removal of metals show that the last two metals bind to independent tetrathiolate sites from terminal thiolate ligands, which further supports independent $Co^{2+}$ binding to isolated sites prior to metal cluster formation and also confirms that the C-terminal site is the first to bind metal ions [24]. Taken together, metallothionein represents an excellent example for describing the role of metal ions in the proper folding of a natural metallofoldamer.

### 1.2.3   Conformational Change of Metalloproteins Caused by Ligand Binding

*1.2.3.1   Calmodulin and $Ca^{2+}$/Ligand Binding*

The role of metal ions in stabilizing the folded states of small proteins is well established, as illustrated in zinc finger proteins [33] (Figure 1.1). Reversible binding of metal ions, where both the metal-free disordered form and the metal-bound ordered form are functional, is very widely observed among calcium-binding proteins. The coupling of the N- and C-terminal lobes in the EF-hand $Ca^{2+}$-binding calmodulin is a good example [34]. Since calcium signaling is such an important process in many metabolic systems, it is likely that this kind of reversible order–disorder equilibrium is quite common. The binding of up to four $Ca^{2+}$ ions in the two different globular ends of apo calmodulin [35] causes significant conformational changes to the molecule (Figure 1.5), including straightening of the long interdomain helix (Figure 1.5, red). Calcium binds to the sites with different affinities (i.e., a higher preference for the C-terminal binding site than the N-terminus),

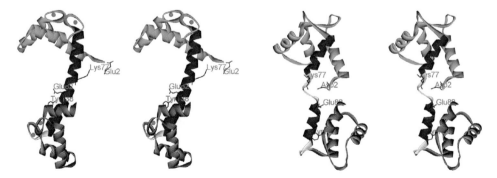

**Figure 1.5**   *Right: structure of $Ca^{2+}$-free calmodulin determined with NMR in solution, showing a kink in the middle of the interdomain helix (PDB ID 1CFD). Left: crystal structure of $Ca^{2+}$-bound calmodulin, showing dramatic conformational change upon $Ca^{2+}$ binding (PDB 1EXR).*

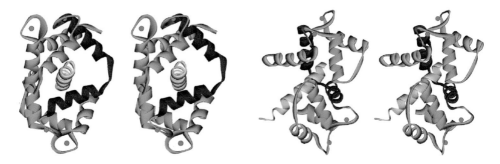

**Figure 1.6** *Two different views (rotating by ~90°) of Ca$_4$-calmodulin-bound "IQ motif" of the Cav1.1 channel (PDB ID 2VAY). The interdomain helix is colored in red, the IQ motif in yellow, and the four Ca$^{2+}$ ions as green spheres.*

resulting in conformational changes and interaction with other proteins and enzymes to perform its regulatory role (Figure 1.5) [36]. Ca$^{2+}$ binding to the C-terminal sites stabilizes the long interdomain helix via a Tyr138–Glu82 interaction, which in turn disrupts two interaction helices by breaking an Asp/Glu2–Lys77 interaction, which is followed by Ca$^{2+}$ binding to the N-terminal sites to form a binding cleft for target proteins [37].

Calmodulin can bind various molecules, including drugs, peptides, and their regulating target proteins. Calmodulin-modulated Ca$^{2+}$ signaling is thus attributed to the different responses of these target molecules to the conformational change of calmodulin upon Ca$^{2+}$ binding. The difference in Ca$^{2+}$ binding and target interactions of the two lobes also enable calmodulin to work out local and global Ca$^{2+}$ sensing and signaling through conformational change [38]. Further, the binding of target molecule to calmodulin can also influence the Ca$^{2+}$ sensitivity of calmodulin [39]. The calmodulin-binding regions in the target proteins are comprised of short helical segments of ~14–26 amino acids with a high occurrence of hydrophobic and basic residues for high affinity and specificity without the need for sequence specificity [40]. Such dramatic conformational change is illustrated in Figure 1.6 for Ca$_4$-calmodulin binding binding to the "IQ" "motif" in the α$_1$ subunit of the L-type voltage-dependent Ca$^{2+}$ channel Cav1.1, which undergoes Ca$^{2+}$- and CaM-dependent channel facilitation and inactivation [41]. The C-terminal conformation of the α$_1$ subunit is critical for channel function and has been proposed to regulate the gating machinery of the channel [42]. The binding causes a significant conformational change in calmodulin, especially the kink at positions 79–81 of the interdomain helix, which results in wrapping around the peptide (Figure 1.6). Taken together, calmodulin represents one of the best examples showing significant metal- and ligand-induced conformational changes.

### 1.2.3.2 *Carboxypeptidase A Catalytic Mechanism*

Another example of dramatic conformational changes in a metalloenzyme is well represented in the action of carboxypeptidase A, a pancreatic proteolytic enzyme. It belongs to a family of exopeptidases responsible for catalyzing the hydrolysis of peptide bonds at the C-terminus of peptides and proteins. It plays a regulatory role or complements the action

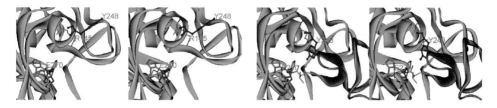

**Figure 1.7** *Stereo views of the crystal structure of (left) bovine carboxypeptidase A at 1.25 Å (PDB 1ML) with the nucleophilic water shown in red sphere and (right) the enzyme with a bound potato inhibitor (lavender; PDB 4CPA).*

of other proteolytic enzymes such as trypsin, chymotrypsin, and pepsin to aid in the production of essential amino acids [43]. Carboxypeptidase A is specific to hydrophobic C-terminal amino acid residues (such as phenylalanine, tyrosine, or tryptophan), while the B-type is specific to the charged residues Lys and Arg. Carboxypeptidase A is a monomer of 307 amino acids with a globular shape consisting of both α-helices and β-sheets (Figure 1.7). The active-site zinc ion plays a key role in the catalysis as it participates in the stabilization of the intermediate, the deprotonation of the nucleophilic coordinated water, and the electrostatic interactions critical for recognition of the terminal amino acid in the substrate peptide chain. This enzyme shares a common active-site motif of $H^{69}xxE^{72}$ for zinc binding (with a bidentate carboxylate of E72), along with a second histidine (H196) located far downstream and a water molecule to complete the coordination sphere of the metal (Figure 1.7) [44].

As in the case of hexakinase, this protein also undergoes conformational changes upon substrate binding to the active site "pocket," which closes upon substrate binding. More specifically, the negatively charged residues can interact with Arg145 residues in the active site while the hydrophobic interactions between the substrate and the hydrophobic pocket can help orient the substrate. Upon binding of potato inhibitor via its C-terminal carboxylate (Figure 1.7) [45], the bidentate Glu72 residue becomes monodentate which is accompanied by a >10 Å movement of Tyr248 to form a H-bond with the bound inhibitor along with a ~2 Å movement of the backbone of the region around Tyr248 toward the metal. Once again, conformational flexibility in metalloproteins is illustrated herein with carboxypeptidase A during its catalytic action.

### 1.2.3.3 Aminopeptidase and Alternative Catalysis

Aminopeptidases are widely distributed hydrolytic enzymes catalyzing various processes, such as peptide digestion and hormone production, some requiring metal ion(s) in the active site for full activity. The nuclearity of the active site of metallopeptidases varies from mononuclear in the case of carboxypeptidase, to dinuclear in aminopeptidases, and trinuclear in phospholipase C. Even among the dinuclear aminopeptidases, such as those isolated from bovine lens (bAP) [46], *Escherichia coli* (eAP) [47], *Aeromonas proteolytica* (aAP) [48], and *Streptomyces griseus* (sAP) [49], there is a variation in the structural and mechanistic roles of the metal ions [50]. For example, the dinuclear aAP shows selective metal binding, but mononuclear catalytic activity with the second metal playing a regulatory role [51]. In contrast, the dinuclear sAP (as well as bAP) exhibits dinuclear

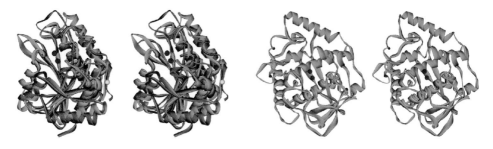

**Figure 1.8** *Left: stereo view of superimposed sAP (lavender; PDB ID 1CP7) and aAP (cyan; PDB 1AMP), with the active-site di-Zn in red sphere and Ca (in sAP) in yellow. Right: stereo view of di-Cu catechol oxidase from sweet potato, showing the very different folding from that of sAP. The two Cu ions are shown in red.*

catalysis, despite its very similar folding and active-site coordination (Asp/His on one metal, Glu/His on the other metal, and a bridging Asp) to those of aAP (Figure 1.8), which indicates that the microenvironment such as the proximal amino acid residues around the active-site coordination sphere must be more significant than the folding of the peptide backbone for proper functions and specific activities.

*Streptomyces* AP is a di-$Zn^{2+}$-containing 30-kDa enzyme which consists of a central β-sheet core surrounded by helices with the active site found within the β-sheet region (Figure 1.8), wherein the dinuclear site can selectively bind metal ions such as $Co^{2+}$ and $Mn^{2+}$ in the two metal-binding sites of sAP [52]. The various metal derivatives exhibit significant alternative catalysis toward a phosphodiester substrate, despite the fact the latter frequently serves as a transition-state inhibitor, which is not the case for aAP [53]. Moreover, the di-$Cu^{2+}$ derivative of the enzyme shows significant activity toward catechol oxidation [54], despite its protein folding and active-site coordination environment being completely different from those of catechol oxidase with three His residues bound to each Cu (Figure 1.8). The observations presented herein indicate that protein folding in not the only control for showing specific enzyme catalysis (i.e., sAP vs aAP toward alternative catalysis) and the folding and/or active-site coordination sphere does not need to be restricted to a certain pattern to exhibit a specific catalysis (i.e., di-Cu-sAP vs catechol oxidase).

### 1.2.4 Protein Misfolding: Causes and Implications – Cu, Zn-Superoxide Dismutase

The superoxide dismutase (SOD) family has four distinct groups, that is, Cu-, Zn-, Fe-, Mn-, and Ni-containing, which are responsible for catalyzing the conversion of superoxide anionic radical to $O_2$ and $H_2O_2$ to protect the cellular environment from damage by superoxides generated during respiration or through the oxidative activity of immune cells [55]. The Cu,Zn-containing SOD (SOD1) is a dimeric protein, with each monomer consisting of an eight-stranded β-barrel and electrostatic and metal-binding loops (Figure 1.9) [56]. The electrostatic loop features Arg143 for hydrogen bonding to superoxide and Thr137 in conjunction with Arg143 to limit the anions coming into the copper-active site. The catalytic site features a unique bridging His63 residue between the two metal ions at

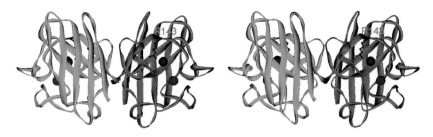

**Figure 1.9**   *Crystal structure of as-isolated dimeric human SOD1 (PDB ID 2C9U). The Cu ions are shown in blue and Zn ions in red.*

6 Å apart. The coordination site of $Cu^{2+}$ is completed by three more His residues and a water molecule in a square pyramidal geometry, whereas the $Zn^{2+}$ tetrahedral site is comprised of two more His residues and an Asp residue. The catalytic cycle starts with the binding of superoxide to $Cu^{2+}$ by displacing the coordinated water followed by electron transfer to the copper and diffusion of oxygen, which results in a trigonal planar $Cu^+$ site. A second electron transferred by another superoxide results in the regeneration of the $Cu^{2+}$ center and the release of peroxide [57].

SOD1 is properly folded through posttranslational modifications which proceed via two distinct pathways, depending on whether or not the copper chaperone CCS is required for the insertion of Cu and the formation of an intramolecular disulfide bond [58]. The formation of disulfide bridges is crucial in the oligomerization of the protein as the reduced metal free protein favors the monomeric state. Proper folding of SOD1 is important since many mutant forms of this protein have been shown to cause amyotrophic lateral sclerosis (ALS), suggested to be due to destabilized or completely unfolded structures and aggregation at room temperature [59]. More specifically, the immature reduced forms of the mutant protein without the formation of disulfide bonds [60] have been implicated in the aggregation process as they can form incorrect intermolecular disulfide crosslinks. The spinal cords of ALS transgenic mice have been found to contain significant amounts of insoluble aggregates composed of such crosslinked multimers, which however are not observed in other tissues such as the brain cortex and liver [60]. Transgenic mice expressing the human mutant G85R SOD1 protein develop paralytic ALS symptoms along with the appearance of SOD1-enriched inclusions in their neural tissues. The crystal structure of this mutant supports that metal-deficient and/or disulfide-reduced SOD1 mutants may contribute to toxicity in SOD1-linked ALS [61].

The unfolding process of SOD1 has been shown to include more than two states, involving other intermediates in the unfolding process. The irreversible inactivation process due to thermal denaturation has distinct features between apo- and holoenzymes, with the rates of inactivation showing a biphasic response as a function of temperature for the apoenzyme but a monophasic function in the case of the apoenzyme [62,66]. The role of the metal ions has been also implicated in the stabilization of the β-barrel structure of the protein fold [63]. Moreover, the unfolded state of the protein is also stabilized by metal ions [64]. The derivative Cu, E-SOD1 without a bound $Zn^{2+}$, has a lower thermal stability, supporting the primary structural role of $Zn^{2+}$ in this protein. In addition, Co(II) or Hg(II) can replace Zn(II) in order to maintain the thermal stability of the Cu(II)-free apoenzyme [65]. Overall, the apoenzyme is more sensitive to inactivating processes compared to the holoenzyme. Stability is also affected by the oxidation state of the bound copper. DSC measurements of dithionite reduced native SOD1 containing $Cu^+$ and $Zn^{2+}$ reveal one peak at 96 °C while native SOD1 containing $Cu^{2+}$ and $Zn^{2+}$ exhibits two melting transitions at 89 and 96 °C wherein the transition at 89 °C is affected by oxygen in the solution [66]. The effect of metal binding in protein stabilization is not unique to SOD1. For example the Cu-binding in *P. aeruginosa* azurin stabilizes the protein, whereas in beta-2-macroglobulin it causes native-state destabilization [67].

Taken together, the presence and identity of the metal ion bound to SOD1 and the status of the disulfide bonds in SOD1 have significant effects on the folding, stability, and catalytic efficiency of the enzyme. Destabilization and misfolding of this enzyme may result in the formation of aggregations in neural tissues and cause neurodegeneration and ALS. The above sections have briefly described the folding, structure, and function of several natural metallofoldamers–metalloproteins, which serve as a foundation for the further design and investigation of synthetic metallofoldamers.

## 1.3 Metallopeptides

Analogous to the metalloproteins discussed above which inspired the design and syntheses of metallofoldamers, a number of simple natural products such as oligopeptides and oligoketides and some antibiotics also adopt secondary or specific structures upon binding with metal ion(s) and can serve as templates for functional metallofoldamers. Metal ions play a key role in the actions of synthetic and natural metallopeptides [68,69] and are involved in specific interactions with proteins, membranes, nucleic acids, and other biomolecules. For example, Fe/Co-bleomycin binds DNA, which impairs DNA function and may also result in DNA cleavage; metallobacitracin binds the sugar-carrying undecaisoprenyl pyrophosphate to inhibit cell wall synthesis; and the specific binding of metal ions to ionophores or siderophores results in their transport through the cell membrane either causing disruption of the potential across the membrane or enabling microorganisms to acquire Fe from the environment.

In addition to the α-helical and β-sheet secondary structures, the β-turn is another important secondary structure in peptides and proteins, in which Pro frequently found at the "break point" [70] and the β-turns [71] to afford an anti-parallel β-sheet structure. In addition to Pro, Gly is also a "structure breaker" and frequently associated with Pro to form a turn [72] as observed at the G12–P13 β-turn in Cu,Zn-superoxide dismutase (Figure 1.10). Combining with His, a metal-binding site can form near the turn in metalloproteins, such as the Pro86 turn in the copper site of plastocyanin (Figure 1.10, blue). Peptides are prototypical molecules which can adopt secondary structures to exhibit broad biological activities by interacting with specific receptors or target proteins, including a large number of G protein-coupled receptors, wherein a general "turn motif" is associated with the binding [73]. Peptides are involved in many physiological regulations and bioactivities, such as the opioid peptides dynorphin, endorphin, and enkephalin, galanin (which may regulate nociception), ghrelin (which may stimulate hunger), $Ca^{2+}$-regulating calcitonin, adrenocorticotrophic hormone, and some neurotoxins. Such peptide-associated activities have triggered the design

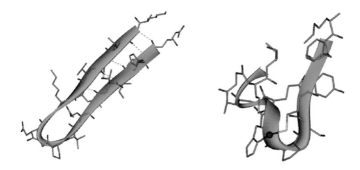

**Figure 1.10**   *Left: the tight G12-P13-β-turn in Cu,Zn-superoxide dismutase (PDB ID 1QE). Right: the metal-binding P86 turn in plastocyanin (PDB ID 1AG6).*

and synthesis of a β-turn mimetic library targeting the major recognition motifs in protein–protein and peptide–receptor interactions [74].

### 1.3.1   Antibiotic Metallopeptides

There are many antibiotic peptides of diverse structures isolated from various sources which interact with a variety of biomolecules, resulting in the inhibition of the associated biochemical or biophysical processes which frequently are associated with conformational changes of the peptides and/or the targets. A number of antibiotics need metal ions to function properly, thus dubbed "metalloantibiotics" [75], such as the peptide bacitracin (Bc) from *Bacillus* species [76] and the peptides/ketide bleomycin (Blm) from the culture medium of *Streptomyces verticullus* [77]. Metal binding to these antibiotics results in significant conformational changes, such as the ~180° twisting of the simple antibiotic streptonigrin upon metal binding [78,79]. A few prototypical antibiotic metallopeptides are discussed in this section to show conformational changes of these peptides associated with metal binding and interaction with targets.

#### 1.3.1.1   Metallo-Blm and DNA Binding

Blm is a $Cu^{2+}$-containing glycopeptidyl antibiotic excreted by *Streptomyces verticullus* [80,81] which also exhibits antiviral [82] and anticancer [83] activities and is widely used in chemotherapy. Blm contains a few uncommon amino acids, that is, β-amino-Ala, β-hydroxy-His, and methylvalerate and a peptidyl bithiazole chain for DNA binding (**1**, with potential metal-binding sites marked in red). It is the most extensively studied metalloantibiotic from various view points [75], including structures, oxidative DNA cleavage, and use as a model system to gain further insights into $O_2$ activation by nonheme Fe enzyme [84]. DNA cleavage by Fe-Blm is carried out by the active $O = Fe^{V/IV}$-Blm species via oxidation at C4′ and C2′–H proton abstraction immediately after the 5′GC and 5′GT sequences [84–87], $Fe^{2+}$-Blm can also bind and cleave RNA molecules [88], including tRNA and its precursors and rRNA [89] mainly at the junctions between double- and single-stranded regions [90], and also DNA–RNA hybrids [8b,[91]].

**(1)**

Blm can bind various metal ions [92], including $Mn^{2+}$ [93], $Fe^{2+/3+}$ [94], $Co^{2+/3+}$ [95], $Ni^{2+/3+}$ [96,97], $Cu^{+/2+}$ [98,99], $Zn^{2+}$ [98], $Cd^{2+}$ [100], $Ga^{3+}$ [101], and $Ru^{2+}$ [102] as well as the radioactive $^{105}Rh$ for radiotherapy [103], changing the conformation of the molecule from a supposedly more extended form (**1**) to a more compact form (Figure 1.11, green structure). Metallo-Blm has a distorted octahedral geometry with coordinated imidazole, pyrimidine, amines of β-aminoalamine, the amide nitrogen of β-hydroxyhistidine, and possibly the amide group of α-D-mannose, as shown by optical, NMR, EPR, and electron spin-echo envelope modulation (ESEEM) spectroscopic methods, crystallography, and chemical modeling [104–107]. This leaves an open or exchangeable site for $O_2$ or peroxo binding. Similar coordination chemistry was also suggested for $Zn^{2+}$-Blm from 2-D NMR [108], however excluding carbamoyl binding. The binding of $Co^{2+}$-Blm to DNA via the bithiazole rings does not influence the metal coordination, whereas the binding of $O_2$–Co-Blm affects the bound $O_2$ where the unpaired electron resides [109].

Low-spin diamagnetic $Co^{3+}$ complexes of Blm and analogues have structures similar to Zn-Blm based on 2-D NMR spectroscopy, except the axial ligands [110,111]. The structures of $(DNA)_2$-$Co^{3+}$-Blm and $(DNA)_2$-$(HOO)Co^{3+}$-Blm ternary complexes show further structural changes upon DNA binding (Figure 1.11) [110,112]. The metal-binding moiety is located in the minor groove of $(DNA)_2$ with the bithiazole rings intercalated into the DNA double helix, rendering the bound peroxide close to the 4'-H of the scissile ribose (<3 Å). Conversely, the bithiazole is in the minor groove in $Zn^{2+}$-Blm-DNA [113]. Nevertheless, intercalation of bithiazole may not be necessary for DNA cleavage since DNA cleavage by Fe-Blm is similarly effective compared to its derivative with the bithiazole tethered to a porous glass bead [114]. $Fe^{3+}$-Blm becomes low-spin (g = 2.41, 2.18, 1.89) with a bound hydroxide at slightly alkaline conditions [115,116]. $O_2$ binds to $Fe^{2+}$-Blm to afford a superoxide $O_2^-$–$Fe^{3+}$-Blm complex based on its $^{57}Fe$ Mössbauer spectrum [117] which can afford an active $HOO^-$-Fe-Blm complex toward DNA cleavage [11b]. The paramagnetic $Fe^{2+}$-Blm (S = 2) with distance-dependent fast relaxing

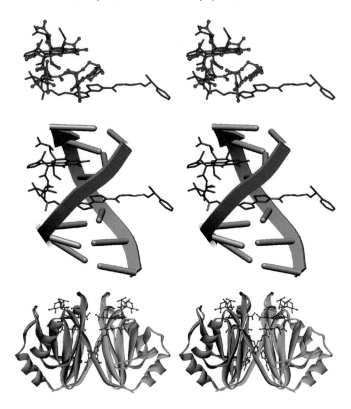

***Figure 1.11*** *Relaxed-eye stereoviews of (top) the superimposed structures of the activated "green species" HOO-Co$^{3+}$-deglycopepleomycin (green ball and stick structure; PDB 1AO2) with a more compact folding than that of the complex upon binding with d(CGTACG)$_2$ (red stick structure retrieved from PDB 1AO1) and (middle) deglycopepleomycin-Co$^{3+}$(OOH)-(CGTACG)$_2$ structure obtained from NMR studies (PDB 1AO1). Peroxide is bound to the metal from the bottom in the structure shown. Bottom: crystal structure of bleomycin-binding protein from Streptoalloteichus hindustanus complexed with a bleomycin congener of an extended conformation (PDB 2ZHP).*

isotropically shifted [118][1]H NMR signals has been studied by means of NMR [119,120]. The structural model of Fe$^{2+}$-Blm built by the use of NMR relaxation times as constraints is similar to those of Co$^{3+}$-Blm complexes. Upon binding to the Shble protein from *Streptoalloteichus hindustanus*, Cu-Blm adopts an extended conformation [121] (Figure 1.11), as opposed to that upon binding to DNA wherein bending of the bithioazole rings occurs in order to intercalate into the base pairs. Taken together, metallo-Blm represents a prototypical natural metallofoldamer which undergoes conformational change upon metal binding and binding with nucleic acid targets.

### 1.3.1.2 Bc and Cell Wall Biosynthesis

Bc is a metal-dependent peptidyl antibiotic produced by *Bacillus subtilis* and *B. licheniformis*, primarily against Gram-positive bacteria via inhibition of cell wall synthesis

[76,122]. It is widely used as a feed additive for livestock [123] and in "triple anti-biotics" (along with polymyxin and neomycin) for human external use [124]. Bc contains four D-amino acids, a thiazoline ring, and a cyclic heptapeptide structure formed via a linkage between the sidechain amine of Lys6 and the C-terminal carboxylate of Asn12 (**2**). It can inhibit metalloproteases presumably due to its metal-binding capability [125], can inhibit a membrane-bound protein disulfide isomerase (PDI) [126], and may serve as a selective inhibitor of β1 and β7 integrin following a not yet known mechanism [127]. However, it was recently shown to have only relatively minor effect on PDI activity *in vitro* [128] which requires re-evaluation of Bc as a specific inhibitor of PDI in cellular systems.

(**2**)

Bc requires a divalent metal ion for its activity [129] and can bind several divalent transition metal ions, including $Co^{2+}$, $Ni^{2+}$, $Cu^{2+}$, and $Zn^{2+}$ [130], and it triggers a slight conformational change as discussed below [131]. $Co^{2+}$-Bc binds $C_{55}$-isoprenyl (undecaisoprenyl or bactoprenyl) pyrophosphate with a formation constant of $1.05 \times 10^6 M^{-1}$ [132], which can prevent dephosphorylation of this lipid pyrophosphate to bind UDP sugars for transport of the sugars during cell wall synthesis [133]. NMR study of $Zn^{2+}$-Bc suggested that His-10 and the sulfur of thiazoline are coordinated to the metal [130]. EPR study of $Cu^{2+}$-Bc indicated a tetragonally distorted $Cu^{2+}$ center ($g_x = 2.058$, $g_y = 2.047$, $g_z = 2.261$, and $A_{z(Cu)} = 534$ MHz) with coordinated thiazoline ring nitrogen, imidazole of His10, and carboxylates of D-Glu4 and Asp11 [13b]. Extended X-ray absorption fine structure (EXAFS) study of $Zn^{2+}$-Bc in solid form revealed three Ns and one O in the first coordination sphere with a tetrahedral-like geometry [134], suggested them to be thiazoline nitrogen, His10 imidazole, D-Glu4, and possibly the N-terminal amino group. The hyperfine-shifted $^1H$ NMR spectrum of $Co^{2+}$-Bc revealed the $N_\varepsilon$ of His10, the carboxylate of D-Glu4, and the thiazoline nitrogen as the metal-binding ligands [131]. A structural model of $Co^{2+}$-Bc built with relaxation times as distance constraints revealed a hydrophobic pocket formed by the side chains of Ile5 and D-Phe9 for possible binding with the hydrocarbon chain of the sugar-carrying undecaisoprenyl pyrophosphate (Figure 1.12), whereas the structure of apo-bacitracin from 2-D NMR showed that the side chains of D-Phe9 and Ile8 are close to Leu3 [135]. Investigation of the $Co^{2+}$ complexes of Bc congeners, including the active Bc-$B_1$ and $B_2$ and the inactive stereo isomer $A_2$ and the oxidized form F, revealed that proper metal binding is essential for Bc activity. The crystal structure of a Bc-bound serine protease complex shows an extended structure [136] which would prevent metal binding. The conformational difference among Bc, its protease-bound complex, metallo-Bc, and farnasylpyrophosphate-metallo-Bc complexes reflects conformational flexibility of this peptide framework even for its seemingly rigid cyclic structure.

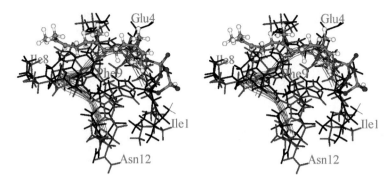

**Figure 1.12** *The structures of Bc-metal complex (pink) and a model of the ternary complex (blue) upon the binding of farnasylpyrophosphate (ball and stick) to metallo-Bc. Conformational changes due to formation of the metal complex and ligand binding are clearly seen including Phe9 and possible detachment of coordinated Glu4.*

### 1.3.1.3 Antibiotic Salivary Peptide

The histatin (Htn) family is comprised of His-rich salivary peptides in higher primates [137–140], showing antibiotic activities against *Streptococcus mutans* [141] and *S. mitis* [142], *Saccharomyces cerevisiae* [143], *Cryptococcus neoformans* [144], and *Porphyromonas gengivalis*, as well as the opportunistic pathogenic yeast *Candida albicans* [145]. Htn5 has the highest concentration among the Htns in human saliva. It is the first 24 amino acids of Htn3 (DSHAK RHHGY[10] KRKFH EKHHS[20] HRGY) and exhibits its highest activity against *C. albicans* [138]. Htn5 does not form pores in the bacterial cell membranes [146] but internalizes by binding to the heat shock protein Ssa1/2 on the cell wall [147], followed by interaction with the $K^+$ transporter TRK1 [148] which leads to apoptosis [149]. It may be internalized into mitochondria and interfere with the electron transfer processes to cause cell death [150].

Htn5 contains several potential metal-binding residues from His, Asp, Glu, and Tyr, has been shown to bind divalent metal ions in the order of $Cu^{2+} > Ni^{2+} > Zn^{2+} \gg Ca^{2+} \sim Fe^{2+}$ [151], and is suggested to bind three equivalents of $Zn^{2+}$ or $Cu^{2+}$ from a calorimetric study. $Cu^{2+}$ and $Ni^{2+}$ bind Htn5 at the N-terminal DSH with a high affinity analogous to DAH in bovine serum albumin which folds the N-terminus [152] to afford a square planar coordination geometry on the basis of electronic, NMR, and EPR results [153], and potentially also at HEXXH and two His–His sites, as found in many metallopeptides and metalloproteins. However, $Co^{2+}$ seems to bind to Htn5 first with three His residues and with two His residues in the second site [153]. Metal binding of Htn5 plays an important role in its bioactivities by fusing negatively charged vesicles ($Zn^{2+}$ binding [151a]) and exhibiting oxidative nuclease activity ($Cu^{2+}$ binding [154]). Similar to many $Cu^{2+}$ complexes, the complex $Cu_2^{2+}$–Htn5 also exhibits significant activity toward catechol oxidation [153]. Since Htn5 can effectively bind metal ions, conformation change upon each metal binding can be expected. However, it has not been revealed how all the metal centers are involved in the oxidative catalysis and in the antimicrobial activity of Htn5; and the correlation between structure, antimicrobial activities, and metallo-Htn5 is unknown.

*1.3.1.4 Ionophores and Siderophores*

Ionophores [155] are small peptides or other kinds of molecules excreted by microorganisms which can selectively bind and transport alkali or alkaline earth metal ions across cell membranes and artificial lipid bilayers, whereas siderophores [156] selectively bind and transport $Fe^{3+}$ [75]. These molecules can: (a) disturb the ionic balance across membrane, such as nactins, lasalocid, and valinomycin, (b) create pores on membranes, such as gramicidins, and (c) compete for iron in the environment, such as ferrichromes. The potential imbalance across the cell membrane may slow down cell growth or cause cell death. Consequently, the metal ions in metalloionophores serve as "magic bullets" to cause a potential imbalance and engender antibiotic activities. The mechanism of this type of antibiotic activity has been adopted in the design of channel-forming antibacterial agents [157].

Ionophores and siderophores exhibit significant conformational changes upon metal binding [155], from extended conformations to compact folded forms as in the case of nactins (e.g., nonactin, tetranactin, and dinactin) and valinomycin [158,159]. The metalloforms then bind specific receptors on the cell surface and result in the transport of metal ions into the cell. Depending on the target metal ions, the structures of the metalloforms may vary dramatically. In the case of enniatin cyclic (L-N-methyl-valine-D-hydroxy-isovalerate)$_3$, the parent ionophore has a structure very similar to its $K^+$ complex, yet quite distinct from its $Rb^+$ complex [160]. This family of antibiotics contain an O-rich metal-binding environment, including ether groups, the carbonyl group of esters and amides, and carboxylates preferably for binding with alkali and alkaline earth metal ions in different metal:ligand ratios attributed to the structures, the size of their metal-binding site, the ionic radii of the metal ions, and/or the hydration energy of the cations [75].

The gramicidins family is a peptide ionophore family produced by *Bacillus brevis* [161], of which gramicidin A is the major component. Its primary structure contains six D-form amino acids, that is, formyl-Val-Gly-Ala-(D-Leu$^4$)-Ala-(D-Val$^6$)-Val-(D-Val$^8$)-Trp-(D-Leu$^{10}$)-Trp-(D-Leu$^{12}$)-Trp-(D-Leu$^{14}$)-Trp-ethanolamine (and Trp$^{11}$ → Phe in gramicidin B and Trp$^{11}$ → Tyr in gramicidin C). This family exhibits an antibiotic mechanism different from those cation-binding ionophores described above by inserting into the lipid bilayer as a dimer and folding into a unique β-double-stranded helix [162] to create a pore of ∼4 Å (Figure 1.13) with selectivity in the order $H^+ > NH_4^+ > Cs^+ > Rb^+ > K^+ > Na^+ > Li^+ > N(CH_3)_4^+$ in 0.1 M salt [163]. However, it does not show permeability to the divalent metal ions $Mg^{2+}$, $Ca^{2+}$, $Ba^{2+}$, and $Zn^{2+}$, which bind to the entrance of the channel and prevent the transport of monovalent cations [164]. Similar configurations are observed for the peptide backbone in both solution and solid state [165] (Figure 1.13) with hydrophobic amino acid side chains facing outward for better interaction with the lipid bilayer [165–170]. However, the structures in solution determined by NMR spectroscopy exhibit a higher degree of irregularity, such as the shape of the channel. The binding of monovalent metal ions does not seem to cause a significant conformational change in gramicidins. The insertion of gramicidins into membranes does not rely on the binding of metal ions as in the case of the ionophores described above, thus it does not create a further energy barrier for metal binding and transport.

The extremely small solubility product $K_{sp}$ of ∼$10^{-38}$ for $Fe(OH)_3$ under aerobic physiological conditions makes soluble $Fe^{3+}$ in aqueous solutions very scarce. To overcome

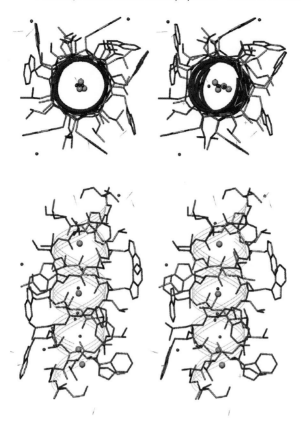

**Figure 1.13**  *Stereo relaxed eye view of the crystal structures of one channel of gramicidin A dimer (top, viewed through the channel; bottom, side view; PDB ID 1AV2). Cs$^+$ ions located in the channel are shown in green spheres, water molecules are shown in small red spheres.*

this obstacle, microorganisms excrete Fe$^{3+}$-specific siderophores which bind Fe$^{3+}$ with extraordinarily high affinity constants in the range of $\sim 10^{30}$–$10^{52}$ M$^{-1}$ and transport Fe$^{3+}$ into cells via specific receptors [156,171]. There are three families of siderophores, differing from each other by their iron-binding sites: hydroximate- (such as ferrichrome from *Penicillium* and the edible *Ustilago*), catechol-containing (e.g., enterobactin from *E. coli*), carboxylate-containing (like the simple citrate), and their combinations (such as aerobactin) [172]. Upon Fe$^{3+}$ binding, the ferrichromes fold to afford an octahedral metal coordination sphere with a more compact conformation than their metal-free apo-forms (Figure 1.14, top). The complexes are recognized and transported into the cells by species-specific receptors. For example, although ferrichrome A serves as an iron carrier for fungi, it does not in bacteria [173]. The folding is also stereo-specific. While the iron complexes of ferrioxamines B [174], D$_1$ [175], and E [176] and desferrioxamine E [177] fold into a mixture of $\Lambda$- and $\Delta$-*cis* isomers, ferrichrome complexes [178] are exclusively $\Lambda$-*cis* isomers. A few structures of the transporter protein FhuA (ferric hydroxamate

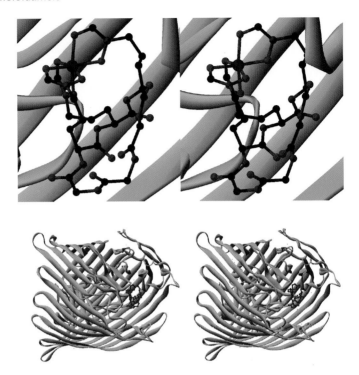

**Figure 1.14** *Bottom: stereoview of the crystal structure of FhuA with a bound $Fe^{3+}$-ferrichrome complex PDB 1FCP. The structure of the receptor with a bound Fe-albomycin complex is similar (PDB 1QKC). Top: detailed view of the complex.*

uptake A protein) with and without a bound $Fe^{3+}$-siderophore have been determined (Figure 1.14, bottom) [179]. Some $Fe^{3+}$-siderophore complexes have quite similar folding and coordination chemistry upon their binding to FhuA. The binding of $Fe^{3+}$-ferrichrome (comprised of Gly and ornithine hydroximate; Figure 1.14, top) to FhuA induces a significant conformational change at the N-terminal domain of the receptor. However, the studies could not identify how the conformational change results in the uptake and transport of the complexes through the cell membrane.

### 1.3.2 Metallopeptides in Neurodegenerative Diseases

Protein misfolding, self-assembling, and/or aggregation can be associated with neurodegenerative diseases, such as the Cu,Zn-superoxide dismutase and amyotrophic lateral sclerosis discussed above, the abnormal aggregation and accumulation of α-synuclein in Parkinson's disease, aggregation-prone mutant huntingtin in Huntington's disease, and the aggregation of prion protein in a few types of transmissible spongiform encephalopathies. In this section, the conformation and metal binding of two prototypical metallopeptides/proteins associated with neurodegenerative diseases are discussed: β-amyloid and prion, as well as their fragments.

*1.3.2.1 β-Amyloid Peptides*

Alzheimer's disease (AD) is a progressive brain-degenerative disease among elderly people and is the most common cause of dementia, which eventually leads to the loss of ability to perform daily routines. An estimate of about 2.4–4.5 million Americans have this disease, showing first symptom after age 60 in most cases [180]. Although the etiology of this disease has not been fully established, evidence for possible causes has been hypothesized and demonstrated. The aggregation of β-amyloid peptide (Aβ) in the brain as plaques and fibrils have been hypothesized to be associated with the pathogenesis of AD [181], which has gained further support from the fact that Aβ plaques are toxic to neurons and some rat models [181–183] and the overexpression of Aβ in familial AD and Down's syndrome results in early onset of the disease [184]. Nevertheless, the role of Aβ coagulation in AD has still been challenged [185]. The linkage between the structure and reactivity of Aβ and AD must be further clarified to provide possible prevention and treatment of this disease.

The Aβ peptides are generated by cleaving the ubiquitous amyloid precursor protein (APP) by α, β, and γ secretases, wherein Aβ(1–40/42) fragments (with sequence DAEFR HDSGY[10] EVHHQ KLVFF[20] AEDVG SNKGA[30] IIGLM VGGVV[40] IA) are generated by secretases β and γ while Aβ(1–16) is released by secretases α and β [186]. Aβs have several potential metal-binding residues, including Asp, His, Tyr, and Glu frequently found as ligands in metalloproteins [187]. Growing evidence has pointed out the involvement of metal ions, including $Fe^{2+/3+}$, $Cu^{2+}$, and $Zn^{2+}$, in the conformational changes of Aβ into fibrils and plaques and the role of metallo-Aβ in causing oxidative stress in the brain of AD patients [188]. The variant His[13] → Gln revealed the importance of this His for metal binding and the formation of Aβ plaques [188g]. Likewise, mouse Aβ with His[13] → Arg is much less apt to form aggregates in the presence of metal ions [188a]. Redox-active Cu- and Fe-Aβ complexes may exhibit neurotoxicity via the generation of reactive oxygen species (ROS), which can cause damage to cell membranes, proteins, nucleic acids, and other biomolecules and lead to cell death [189]. Moreover, the significance of the soluble forms of Aβ in the pathogenesis of AD has also been proposed and verified [190].

The structure and metal binding of Aβ and the chemistry associated with metallo-Aβ have emerged from studies using different physical methods [188,191]. For example: the morphology of Aβ fibrils was revealed by electron microscope; [192] Aβ fibers and the hydrophobic C-terminal fragments of the peptide were shown by X-ray diffraction [193] and solid-state NMR techniques [194] to adopt a β-sheet structure (Figure 1.15a), whereas Aβ in micelles [195] and an intermediate during Aβ fibrillogenesis [196] have α-helical structures (Figure 1.15b); an extended structure of Aβ upon binding to insulin-degrading enzyme was shown by X-ray crystallography (Figure 1.15c) [197]; the structure of the Zn-Aβ complex was determined by NMR to show a coordinated Asp11 residue along with three His residues (two $N_\delta$- and one $N_\varepsilon$-coordinated; Figure 1.15d) [198]; the structure of $Co^{2+}$-Aβ was determined on the basis of hyperfine-shifted His ring protons and molecular mechanics calculations to show three $N_\varepsilon$-coordinated His residues as well as the presence of an extended H-bonding framework (Figure 1.15e) [199]; and damage to cellular components such as membranes by metallo-Aβ-mediated ROS was determined on the basis of the reaction products [200]. From these studies, the conformational change of Aβ peptides upon

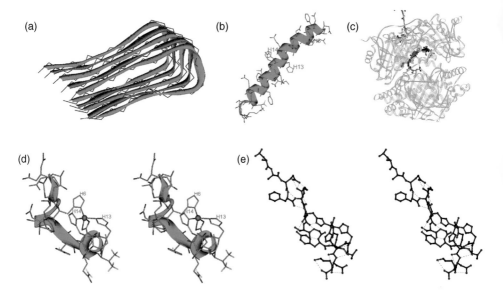

**Figure 1.15** *(a) Structure of extended β-sheet Aβ(17–42) from solid state NMR (PDB ID 2BEG). (b) Solution structure of helical Aβ(1–28) in micelles by NMR (1AMB). (c) Crystal structure of human insulin-degrading enzyme with a bound Aβ(1–42) of an extended structure, showing one monomer in cyan (2WK3). (d, e) Stereo views of (d) $Zn^{2+}$-bound Aβ (1ZE9) and (e) $Co^{2+}$-bound Aβ.*

metal binding and/or the formation of aggregates and fibrils under different conditions can be established, providing a good example of metallopeptides as natural metallofoldamers.

### 1.3.2.2 Prion Proteins and Fragments

The cellular form of prion protein $PrP^C$ (209 amino acids for the mammalian one) is a $Cu^{2+}$-binding glycoprotein which is attached to the cell surface via a glycosylphosphatidylinositol anchor [201] and contains $Cu^{2+}$ when isolated from a diseased brain [202]. The C-terminal domain of $PrP^C$ is mainly α-helical [203,204], whereas the N-terminal domain is unstructured in the absence of $Cu^{2+}$ (Figure 1.16a, b) [205,206]. $PrP^C$ is converted into an infectious form $PrP^{Sc}$ (scrapie isoform) when misfolded and aggregated (Figure 1.16c). This is responsible for transmissible spongiform encephalopathies such as bovine spongiform encephalopathy, ovine scrapie, and human Creutzfeldt–Jakob disease. Misfolding of $PrP^C$ with three α-helices and a short anti-parallel β-sheet to afford the self-assembled oligomeric β-sheet-rich $PrP^{Sc}$ is essential to the transmissible spongiform encephalopathies. Unlike the disordered protein moieties which are integrated parts of protein structure and function commonly found in signaling and regulatory proteins that can induce/adopt certain conformations for specific interactions with targets [207], the misfolded $PrP^{Sc}$ can "infect" normal $PrP^C$ and convert it into the disease-causing misfolded form. Recent studies on $PrP^C$-knockout cells and on truncated $PrP^C$ devoid of the $Cu^{2+}$-binding repeats pointed to a possible role of this protein serving as a metal transporter [208]. Moreover, the $Cu^{2+}$ centers can generate reactive oxygen species to cause oxidative stress in the brain [209] and can result in

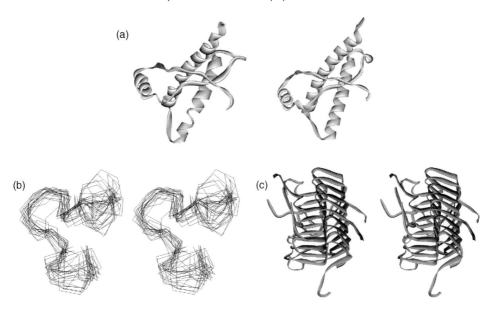

**Figure 1.16** *The NMR structures of recombinant human prion protein (globular domain extending from residues 125–228) 1QM2 (a, left) and the crystal structure of the globular domain of sheep prion protein 1UW3 (a, right). (b) Solution NMR structure of the octapeptide repeats in mammalian prion protein (1OEI). (c) Solid-state NMR structure of amyloid fibrils from the prion-forming domain of the HET-s protein (2RNM). Each monomer is highlighted in a different color for clarity.*

oxidative modification of PrP$^C$ which is also proposed to link to prion diseases [210]. However, PrP knockout mice are sensitive to Cu$^{2+}$-induced oxidative stress [211], which suggests a possible anti-oxidation role of PrP$^C$. Further research will be needed to fully establish the biological function of this structurally two-faced Janus protein.

Cu$^{2+}$ binding to PrP$^C$ triggers structural changes of the protein [212] which enhances resistance against proteases [213] and is linked to the prion diseases. The N-terminal domain of PrP$^C$ can bind up to six Cu$^{2+}$ ions at physiological pH, with the first two Cu$^{2+}$ ions binding to the amyloidogenic region (residues 90–126) and the rest to the four highly conserved octapeptidyl repeats of PHGGGWGQ in residues 58–91 (Figure 1.16b) [214]. The four Cu$^{2+}$-octarepeats are not interacting with each other as they are magnetically isolated [214a]. Synthetic octapeptide repeats show higher preference toward Cu$^{2+}$ binding than other metal ions [215]. The crystal structure of the Cu$^{2+}$ complex of the simple peptide HGGGW reveals a square pyramidal coordination sphere with the equatorial sites occupied by the His imidazole, two deprotonated Gly-amides, and a Gly-carbonyl and an axial water H-bonded to the Trp indole [214a]. The similarity of the EPR spectra between this simple complex and the Cu$^{2+}$-bound octarepeats suggests their similar coordination sphere. Zn$^{2+}$ also binds mammalian PrP with a weaker binding affinity than Cu$^{2+}$ and can affect Cu$^{2+}$ binding modes in PrP [216]. However, there is no structure solved for the entire protein or the N-terminus domain of this protein in either the native or the unfolded form. In addition to the disease-causing mammalian prions, many fungal prions are also known to undergo structural change to form amyloid filaments, which may be involved in normal or diseased states, such as the prion-forming domain at residues 218–289 of the HET-s protein from the filamentous fungus *Podospora anserine* [217] (Figure 1.16c). These fungal prions may be involved in fungal epigenetic processes [218] and also serve as good models systems for a better understanding of the structure and function of mammalian prions.

### 1.3.3 Other Metallopeptides

Peptide chains undergo dramatic conformational changes upon formation of secondary structures, which can be triggered by metal binding or interacting with target molecules, as discussed above. In addition to the natural metallopeptides discussed above, there are a number of synthetic peptides that can also undergo similar conformational changes. Such a structural property is applicable to the design of metallopeptides for further investigation of the structure and function of natural metallopeptides and metalloproteins and as therapeutic agents. A couple examples are given here.

#### 1.3.3.1 N-Terminal Binding Peptides and Ni-SOD

The N-terminal metal-binding site in serum albumin represents a typical coordination in metallopeptides and is comprised of a large number of natural and synthetic peptides, as in the case of histatin discussed above. Moreover, both the C- and N-termini of proteins are the "loose ends" in protein folding and thus can significantly contribute to protein stability when they are "tightened up" [219]. The binding of $Cu^{2+}$ to a deprotonated peptidyl amide results in the formation of a square planar metal center and a change in conformation of the peptide, as in the case of $Cu^{2+}$ binding to the octarepeats in prion discussed above [215]. A square planar geometry is also formed in several $Cu^{2+}$ complexes of Tyr-containing peptides at elevated pH through a proposed binding to the N-terminal amine and a deprotonated Gly-amide and binding to the phenolate of a Tyr side chain by showing a charge transfer transition at $\sim$400 nm, such as the monomeric $[Cu^{II}LH_{-1}]$ (L = Phe-Gly-Pro-Tyr) complex and the dimeric $[(CuLH_{-1})_2]$ (L = Tyr-Gly-Pro-Phe) complexes at $\sim$pH 8–10 on the basis of the distinct $(O^-)$Tyr to $Cu^{2+}$ charge transfer transitions at $\sim$400 nm [220]. Once again, the coordination sphere was proposed based on spectroscopic features since the structures of these complexes were not solved. $Cu^{2+}$ binding to $\alpha$-synuclein may play a role in the fibrillogenesis of Parkinson's disease [221]. $Cu^{2+}$ binds to the N-terminus Met-Asp of $\alpha$-synuclein and folds this moiety into a square planar geometry with the N-terminal amine, Asp-amido $(H_{-1})$, and Asp-carboxylate as the ligands, along with His50 which may or may not be involved [222], and a dissociation constant of 0.10 nM [222,223].

Another kind of N-terminal metal-binding site is found in Ni-containing superoxide dismutase [224] (SOD; see Section 1.2.4 for a summary of this family of enzymes). The Ni center undergoes redox cycle between $Ni^{2+}$ and $Ni^{3+}$ during the catalytic cycle, accompanied by conformational change at the N-terminal Ni coordination sphere. The Ni ion is bound to the N-terminus of the sequence **His$^1$-Cys$^2$**-Asp-Leu-Pro-**Cys$^6$** of the protein through the N-terminal amine of His1, the peptidyl amido-N of Cys2, and the thiolates of Cys2 and Cys6 in the reduced $Ni^{2+}$ state with a square planar geometry, and an

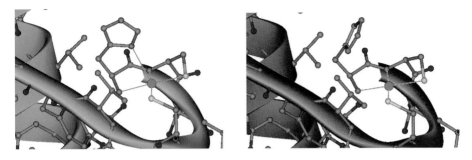

**Figure 1.17** *The N-terminal Ni-binding active site of Ni-SOD in the oxidized $Ni^{3+}$ form (left, PDB ID 1Q0D) and the reduced $Ni^{2+}$ form (right, PDB ID 1Q0K).*

additional axial ligand from His1 residue in the oxidized $Ni^{3+}$ state (Figure 1.17) [224]. Here, the axial His seems to serve as a redox and/or conformational switch of the N-terminal active-site Ni coordination of this enzyme, detaching from the metal to stabilize the $Ni^{2+}$ state via a lowering of the $d_z2$ energy level and binding to the metal to donate further electron density to $Ni^{3+}$. A large conformational change during the redox cycle may not take place so that electron transfer is not slowed down by the conformational movement. This catalytic cycle is thus a good example for demonstrating the significance of the conformation and coordination of active-site metal in retaining the proper function of "natural metallofoldamers."

The correlation between conformational change and redox activity of Ni-SOD has been modeled with the simple peptide $H_2N$-Gly-Cys-OMe, which folds to bind to $Ni^{2+}$ via the N-terminus, deprotonated amide, and Cys side chain S, as well as a couple dipeptide mimics to form complexes with a coordination sphere of $N_2S$ [225]. Various external thiolates are introduced to the complexes to complex the square-planar coordination sphere of $N_2S_2$ as in reduced Ni-SOD. The coordination sphere and reactivity of Ni-SOD is also mimicked with the $Ni^{2+}$ complex of the tripeptide Asn-Cys-Cys with a coordination sphere of $N_2S_2$ (including a deprotonated amide), as in Ni-SOD, consistent with a diamagnetic $Ni^{2+}$ in square-planar geometry [226]. This complex undergoes chiral structural transformation that is associated with its SOD activity.

### 1.3.3.2 Metal-Triggered Conformational Change in Peptides

Helical coiled-coil structures can be designed on the basis of the tendency of amino acids to form helix structures [227] and inter-stranded interactions. However, a peptide designed to form a double-stranded parallel coiled-coil structure ended up showing triple-stranded "up up down" $\alpha$-helices [228], reflecting the great conformational flexibility of peptides and their high degree of freedom in assembling into higher-order structures. Synthetic peptides can afford a helical conformation upon metal binding to metal-binding sequences such as His-x-x-x-His and Cys-x-x-x-His [229], consistent with an $\alpha$-helical $i–(i+4)$ conformation, or to moieties with unnatural metal-binding ligands [230]. Short peptides can also be triggered to form $\alpha$-helices with metal ions by the use of such a method [231], which can be as short as only one helical turn [232]. Despite their similar ligand-binding capabilities, $Ni^{2+}$ and $Cu^{2+}$ ions were found in one case to induce

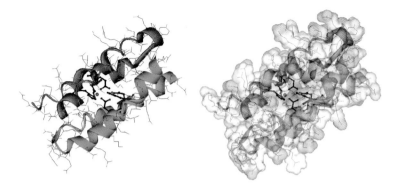

**Figure 1.18** *A dimeric four-helical bundle peptide with a dinuclear metal center, represented herein by the di-Zn form solved with NMR (PDB 2KIK). The crystal structure of this helical bundle metallopeptide has a similar structure (PDB 1EC5). The peptide folding is shown in a stereo view with the image on the right also showing the molecular surface and Gly9 labeled in red.*

conformational change differently: $Ni^{2+}$ induced the formation of a helical structure, but not $Cu^{2+}$ [233].

The role of metal ions has further been demonstrated in assisting peptides to adopt a specific conformation and then assemble into a certain tertiary or quaternary structure, such as the helical bundles [234] found in many metalloproteins, including cytochrome c (Figure 1.2). A designed helical bundle-forming metallopeptide demonstrates that hydrophobic interactions are sufficient to induce polypeptide folding, while the introduction of metal-binding sites can further "tighten up" a four-bundle helical structure [235]. Therein, the four-helix bundle is formed as a dimer of helix-turn-helix peptides and two $Zn^{2+}$-binding sites are built in to form a dinuclear center (Figure 1.18) [236]. An overall $C_2$-rotation symmetry of the dimer and the dinuclear metal sites is revealed from the crystal structure. The metal binding sites in this dimeric peptide and a few variants [237] can also accommodate various dinuclear metal centers with an overall folding and assembly analogous to the di-Zn peptide with some variations in the coordination sphere of the metal-binding site [238]. The variant L9G/L13G has a larger opening to the dimetal site than the original peptide, affording substrate accessibility and significant oxidation activities for the di-Fe complex, with $k_{cat}/K_m = 105\,M^{-1}\,s^{-1}$ for the oxidation of 3,5-dimethylcatechol and $23\,M^{-1}\,s^{-1}$ for 4-aminophenol oxidation [23b].

In addition to metal binding to the helical bundles, the prosthetic group metal–porphyrin can also be incorporated into peptides and assist the folding of the peptides and in some cases render catalysis possible. A designed peptide containing two His residues can be folded and assembled into a tetrameric helical bundle upon binding two Fe(III)-porphyrin molecules, with each one coordinated to two peptide chains via two His residues [239]. Moreover, a heterodimeric peptide with a protophorphyrin IV covalently linked to two peptide chains folds mainly to a helical structure. Here the heme-Fe is five-coordinate via binding to one His from one peptide chain, leaving an open site for possible substrate and/or peroxide binding, as in the case of heme-containing peroxidases. This heme–peptide complex indeed

(a)                                                    (b)

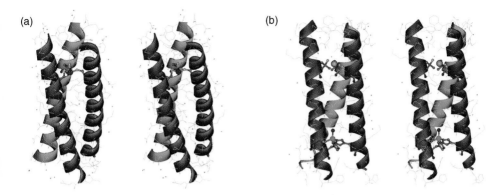

**Figure 1.19** *Stereo views of the crystal structures of (a) As(L9C)₃ (PDB 2JGO) and (b) Zn,Hg (Leu9Pen/Leu23His)₃. The metal-binding sites are shown as spheres (As, red; Hg, gray; Zn, green).*

shows a significant peroxidase activity toward 2,2′-azino-di(3-ethyl-benzothiazoline-6-sulfonic acid (ABTS; $k_{cat}/K_m = 4417 \, M^{-1} \, s^{-1}$) and 2-methoxyphenol ($870 \, mM^{-1} \, s^{-1}$) [240].

Metal ions have characteristic coordination geometries, such as the presence of bridging carboxylate and hydroxide/oxide in di-Fe centers, as illustrated above. In the case of $Hg^{2+}$, both linear and trigonal geometries are present in $Hg^{2+}$ complexes and proteins such as the $Hg^{2+}$-detoxification regulatory factor MerR [241]. The peptide Ac-G(LKA-LEEK)₄G-CONH₂ forms triple-helical bundles [242] which can be further modified with Cys at position 12 or 16 to afford mutants L12C or L16C that bind $Hg^{2+}$ in a pH-dependent manner [243], that is, a linear geometry as Hg(L12C)₂ at low pH and a trigonal planar coordination sphere as Hg(L12C)₃ at high pH [244] and as Hg(L16C)₂ at low L16C concentrations and Hg(L16C)₃ at high peptide concentrations at pH 8.5 [245]. These Cys-containing peptides also bind $Cd^{2+}$ and $As^{3+}$ to fold into triple-helical bundles [244,246]. The crystal structure of the complex As(L9C)₃ reveals parallel helical bundles and Cys-coordinated $As^{3+}$ with a tripodal coordination sphere (Figure 1.19a). This As–peptide complex mimics possible $As^{3+}$ binding to the bacterial As-responsive repressor protein ArsR which dissociates from DNA upon $As^{3+}$ binding as the regulatory control in bacterial arsenic resistance [247].

Further metal-binding properties of the peptide were pursued by the use of a double mutant L9C/L19C which binds two $Cd^{2+}$ ions sequentially, the Cys-9 site first followed by the Cys-19 site, to form a triple-helical bundle structure [248]. An analogous sulfur- and His-containing peptide L9PenL23H (Pen = penicillamine) with the sequence Ac-E WEALEKK (Pen)AALESK LQALEKK HEALEHG-NH₂ binds both $Hg^{2+}$ (at the L9Pen site) and $Zn^{2+}$ (at the L23H site) to fold into a triple-helical bundle structure (Figure 1.19b) [249]. Therein the $Zn^{2+}$ site is four-coordinate with one coordination site occupied by a water molecule analogous to the catalytic $Zn^{2+}$ site in carbonic anhydrase, rendering hydrolytic/hydration catalysis possible by this mixed-metal metallopeptide. In a different case, a much longer designed peptide of 73 amino acids is folded by $Zn^{2+}$ ($2.5 \times 10^{-8} \, M$) or $Co^{2+}$ ($1.6 \times 10^{-5} \, M$) into a four-helical bundle structure with two metal-binding Cys-x-x-x-His sequences on adjacent helices, wherein a tetrahedral $N_2S_2$

metal-binding coordination is formed that mimics the $Zn^{2+}$ finger motif [250]. A shorter Zn-binding peptide of 27 amino acids was suspected to form a helical bundle structure in the absence of $Zn^{2+}$, whereas it forms a folded helix-$\beta$-sheet structure triggered by $Zn^{2+}$ binding [251] which represents a typical example of metal switching in peptide folding.

A number of peptides are synthesized with built-in metal-binding motif(s) which include nonamino acid ligands. An amphiphilic peptide with a 2,2'-bipyridine (bpy) group attached to the N-terminus via a 4-carboxyl group, for example, (4-carboxyl-bpy)-Gly-Glu-**Leu**-Ala-Gln-Lys-**Leu**-Glu-Gln-Ala-**Leu**-Gln-Lys-**Leu**-Ala-NH$_2$, is expected to have the four Leu side chains facing to the same side once a helical structure is formed [252]. In the presence of $Ni^{2+}$, $Co^{2+}$, or $Ru^{2+}$, the peptide assembles into a 45-residue three-bundled coiled-coil structure that is confirmed by mass spectrometry of the inert $Ru^{2+}$-bound three bundle with $m/z = 5563$ [253]. The CD spectra of the metallopeptide reveal 80% $\alpha$-helicity in 150 mM NaCl solution and the formation of one diastereomer with a $\Lambda$-isomeric tris-bipyridyl-$M^{2+}$ center and a left-handed coiled-coil structure. The introduction of an additional sequence of Ala-Ala-His-Tyr to the C-terminus of the above peptide affords another three-bundled assembly in the presence of $Ru^{2+}$ with an additional tri-His binding site from the three bundles for $Cu^{2+}$ binding [254]. The use of a N-terminal monodentate metal-binding site in a peptide affords nicotinyl-$\gamma$-aminobutylic-Gly-Leu-Ala-Gln-Lys-Leu-Leu-Glu-Ala-Leu-Gln-Lys-Ala-Leu-Ala which binds $Ru^{2+}$ in a 4 : 1 ratio to assemble into an inert four-bundled coiled-coil structure, demonstrated by means of atomic absorption spectroscopy and electrospray mass spectrometry [255]. A few 3,3'-peptidyl derivatives of 2,2'-bipyridine show an extended configuration which fold to render $\beta$-sheet structures upon $Cu^{2+}$ binding based on CD spectra [256], representing another example of metal-triggered peptide folding.

## 1.4 Conclusion and Perspectives

The biological activities of proteins and peptides rely on the proper folding of their peptide chains. In some diseased stages, misfolded natural peptide chains can form organized tertiary and higher-order structures which may be further triggered by metal binding, as found in amyotrophic lateral sclerosis due to Cu,Zn-superoxide dismutase, Alzheimer's disease due to $\beta$-amyloid, and prion diseases due to the prion proteins discussed above, reflecting the great conformational flexibility of peptide chains. Metal ions can also mediate the assembly of synthetic/designed peptides to form nanoscale spheres and fibrils [257] and microflorettes [258]. In some other instances, the binding of different metal ions to a peptide chain may afford conformational changes to a certain extent and afford different bioactivities, as in the case of the various metal-dependent activities of the dinuclear aminopeptidase from *Streptomyces* discussed above. The $Ni^{2+}$- and $Fe^{2+}$-substituted forms of acireductone dioxygenase serve as another example of metalloprotein foldamers with different enzymatic reactions [259]. Future structural studies about substrate binding modes in the ES complexes and the transition state are expected to provide further insight into the mechanisms for the different catalyses by the different metal forms of each individual protein. Proteins and peptides have chiral-specific properties as their amino acids building blocks are chiral and only the L-form is incorporated into living systems. Consequently, a totally synthesized *Desuljiwibrio* iron–sulfur protein rubredoxin with only D-form amino acids binds $Fe^{3+}$ and folds as the mirror image of the L-form protein and exhibits the opposite CD spectrum [260]. Moreover, a dramatic conformational change of unfolded rubredoxin in 5 M urea upon metal binding was observed, wherein the addition of a 100-fold molar excess of $Fe^{2+}$ refolded the protein to >90% recovery with a $t_{1/2}$ of <10 ms [261]. In addition to the structures of peptides and proteins which can be dramatically affected by metal ions to afford various foldameric conformations, the structures of nucleic acids are also known to be affected by the binding of a metal center, such as the dramatic conformational change of duplex DNA upon binding with cisplatin (*cis*-diamminodichloroplatinum), a cancer chemotherapeutic agent. Here the metal center binds a DNA duplex at two $N_7$ of adjacent guanidine bases or guanidine–adenine bases in the major groove or two proximal guanidine bases of different strands in the minor groove, bending the duplex structure by 40–60° along the helix and twisting the helix by 25–32° [262]. Nucleic acids can also be designed to contain metal-binding bases to form a metallofoldermers, analogous to the case of designed peptides with a metal-binding bipyridyl group (Section 1.3.3.2), such as the dramatic conformational change from a monomeric hairpin structure to a dimeric double-stranded structure for the sequence 5′-TTAATTT-Im-Im-Im-AAATTAA upon $Ag^+$ binding (Figure 1.20) [263]. More detailed discussions about "metallo-DNA" and "metallo-PNA" can be found in this volume (Chapters 7 and 8). These studies and the several examples described in the above sections demonstrate the significant flexibility of the backbones of peptides/proteins and nucleic acids which can adopt various foldameric conformations under different conditions and/or upon metal binding. Future exploration along this direction can be expected to produce novel metalloproteins, metallopeptides, and metallo-nucleic acids with unique structures and chemical, physical, and/or biological properties for various applications.

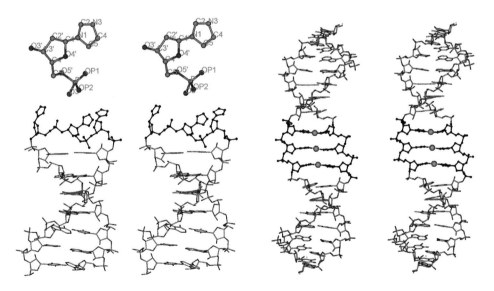

**Figure 1.20** Left: the hairpin structure of a designed nucleic acid with three imidazole-containing (Im; blue) units as shown on top. Right: the double-stranded helical structure of this nucleic acid upon $Ag^+$ (pink sphere) binding to bridge the Im pair.

## Acknowledgements

The studies of several metalloantibiotics and metallopeptides carried out in the authors' laboratory have been supported by the start-up funds, Research and Creative Scholarship Grants and the PYF Award of the University of South Florida, the Edward L. Cole Research Grant (F94USF-3) of the American Cancer Society – Florida Division, and the National Science Foundation (CHE-0718625). The authors' coworkers Drs. Xiangdong Wei, Jon Epperson, Jason Palcic, Giordano da Silva, and William Tay and Alaa Hashim, Christie Tang, and Justin Moses have made significant contributions to a better understanding of the structure–function relationship of these metallobiomolecules.

## References

1. Hecht, S. and Huc, I. (eds) (2007) Front matter, in *Foldamers: Structure Properties and Applications*, Wiley-VCH Verlag, Weinheim.
2. Maayan, G. (2009) Conformational control in metallofoldamers: design, synthesis and structural properties. *Eur. J. Org. Chem.*, **33**, 5699–5710.
3. Bertini, I., Gray, H.G., Lippard, S., and Valentine, J.S. (1994) *Bioinorganic Chemistry*, University Science Books, Mill Valley.
4. Agar, J.N., Krebs, C., Frazzon, J. *et al.* (2000) Iscu as a scaffold for Iron-sulfur cluster assembly of [2Fe-2S] and [4Fe-4S] clusters. *Biochemistry*, **39** (27), 7856–7862.
5. (a) Berks, B.C. (2008) Biochemistry: cells enforce an ion curtain. *Nature*, **455**, 1043–1044; (b) Tottey, S., Waldron, J., Firbank, S.J. *et al.* (2008) Protein-folding location can regulate manganese-binding versus copper- or zinc-binding. *Nature*, **455**, 1138–1142.
6. Cox, E.H. and McLendon, G.L. (2000) Zinc dependent protein folding. *Curr. Opin. Chem. Biol.*, **4**, 162–165.
7. Pavletich, N.P. and Pabo, C.O. (1991) Zinc finger-DNA recognition: crystal structure of a Zif268-DNA complex at 2.1 Å. *Science*, **252**, 809–817.
8. Permyakov, E. (2009) *Metalloproteomics*, John Wiley & Sons, Inc, Hoboken.
9. Maity, H., Maity, M., Krishna, M. *et al.* (2005) Protein folding: the stepwise assembly of foldon units. *Proc. Natl Acad. Sci. USA*, **102**, 4741–4746.
10. Gomez, C. and Wittung-Staffshede, P. (eds) (2011) *Protein Folding and Metal Ions: Mechanisms, Biology and Disease*, CRC, Boca Raton.
11. Bertini, I., Cavallaro, G., and Rosato, A. (2006) Cytochrome c: occurrence and functions. *Chem. Rev.*, **106**, 90–115.
12. Mande, S.C., and Sobhia, M.E. (2000) Structural characterization of protein-denaturant interactions: crystal structures of hen-egg white lysozyme in complex with DMSO and guanidinium chloride. *Prot. Eng.*, **3** (2), 133–141.
13. Maity, H.M. and Englander, S.W. (2004) How cytochrome c folds and why: submolecular foldon units and their stepwise sequential stabilization. *J. Mol. Biol.*, **343**, 223–233.
14. Jones, C.M., Henry, E.R., Hu, Y. *et al.* (1993) Fast events in protein-folding initiated by nanosecond laser photolysis. *Proc. Natl Acad. Sci. USA*, **90**, 11860–11864.
15. Anderson, P.J., Brooks, C.L., and Berliner, L.J. (1997) Functional identification of calcium binding residues in bovine a-lactalbumin. *Biochemistry*, **36**, 11648–11654.
16. Permyakov, E.A., Morozova, L.A., and Burnstein, E.A. (1985) Cation binding effects on the pH, thermal and urea denaturation transitions in α-lactalbumin. *Biophys. Chem.*, **21**, 21–31.
17. Kronman, M.J. (1989) Metal ion binding and the molecular conformational properties of alpha-Lactalbumin. *Crit. Rev. Biochem. Mol. Biol.*, **24**, 565–667.

18. Chrysina, E.D., Brew, K., and Acharya, R. (2000) Crystal structures of apo-and holo-bovine a-lactalbumin at 2.2 Å resolution reveal an effect of calcium on inter-lobe interaction. *J. Biol. Chem.*, **275**, 3701–37029.

19. Halskau, O., Perez-Jimenez, R., Ibarra-Molero, B. *et al.* (2008) Large scale modulation of thermodynamic protein folding barriers linked to elecrostatics. *Proc. Natl Acad. Sci. USA*, **105**, 8625–8630.

20. Fischer, E. and Davie, E.W. (1998) Recent excitement regarding metallothionein. *Proc. Natl Acad. Sci USA*, **95** (7), 3333–3334.

21. Henkel, G. and Krebs, B. (2004) Metallothioneins: Zinc, cadmium, mercury and copper thiolates and selenolates mimicking protein active site features-structural aspects and biological implications. *Chem. Rev.*, **104** (2), 801–824.

22. Ambjorn, M., Asmussen, J.W., Lindstam, M. *et al.* (2008) Metallothionein and a peptide modeled after metallothionein, EmtinB, induce neuronal differentiation and survival through binding to receptors of the low-density lipoprotein receptor family. *J. Neurochem.*, **104**, 21–37.

23. Maret, W. and Vallee, B.L. (1998) Thiolate ligands in metallothionein confer redox activity on zinc clusters. *Proc. Natl Acad. Sci. USA*, **95**, 3478–3482.

24. Duncan, K.R. and Stillman, M.J. (2006) Metal dependent protein folding: metallation of metallothionein. *J. Inorg. Biochem.*, **100**, 2101–2107.

25. Maret, W. (2006) Zinc coordination environments in proteins as redox sensors and signal transducers. *Antioxid. Redox Signal.*, **8**, 1419–1441.

26. Takeda, E., Taketani, Y., Sawada, N. *et al.* (2004) The regulation and function of phosphate in the human body. *Biofactors*, **21**, 345–355.

27. Braun, W., Vasak, M., Robbins, A.H. *et al.* (1992) Comparison of the NMR solution structure and the x-ray crystal structure of rat metallothionein-2. *Proc. Natl Acad. Sci. USA*, **89**, 10124–10128.

28. (a) Green, A.R., Presta, A., Gasyna, Z., and Stillman, M.J. (1994) Luminescent probe of copper-thiolate cluster formation within mammalian metallothionein. *Inorg. Chem.*, **33**, 4159–4618; (b) Stillman, M.J., Presta, A., Gui, Z., and Jiang, D.-T. (2004) Spectroscopic Studies of copper, silver and gold-metallotheionins. *Met. Based Drugs*, **1**, 375–394.

29. (a) Otvos, J.D. and Armitage, I.M. (1979) Cadmium-113 NMR of metallothionein: direct evidence for the existence of polynuclear metal binding sites. *J. Am. Chem. Soc.*, **101**, 7734–7736; (b) Boulanger, Y., Armitage, I.M., Miklossy, K.A., and Winge, D.R. (1982) [113]Cd NMR study of a metallothionein fragment. Evidence for a two-domain structure. *J. Biol. Chem.*, **257**, 3717–3719; (c) Vasak, M., Hawkes, G.E., Nicholson, J.K., and Sadler, P.J. (2005) Cadmium-113 NMR studies of reconstituted seven-cadmium metallothionein: evidence for structural flexibility. *Biochemistry*, **24**, 740–747; (d) Digilio, G., Bracco, C., Vergani, L. *et al.* (2009) The cadmium binding domains in the metallothionein isoform Cd7-MT10 from *Mytilus galloprovincialis* revealed by NMR spectroscopy. *J Biol. Inorg. Chem.*, **14**, 167–178.

30. Rigby, K.E., Chan, J., Mackie, J., and Stillman, J.M. (2006) Molecular dynamics study on the folding and metallation of the individual domains of metallothionein. *Proteins*, **62**, 159–172.

31. Bell, S.G., and Valee, B.L. (2009) The metallathionein/thionein system an oxidoreductive metabolic zinc link. *ChemBioChem*, **10**, 55–62.

32. Vasák, M., and Kägi, H.J. (1981) Metal thiolate clusters in cobalt(II)-metallothionein. *Proc. Natl Acad. Sci. USA*, **78**, 6709–6713.

33. Parraga, G., Horvath, S.J., Eisen, A. *et al.* (1988) Zn-dependent structure of a single-finger domain of yeast ADR1. *Science*, **241** (4872), 1489–1492.

34. Chen, B., Lowry, D.F., Mayer, M.U., and Squier, T.C. (2008) Helix A stabilization precedes amino terminal lobe activation upon calcium binding to calmodulin. *Biochemistry*, **47**, 9220–9226.

35. Kuboniwa, H., Tjandra, N., Grzesiek, S. *et al.* (1995) Solution structure of calcium-free calmodulin. *Nat. Struct. Biol.*, **2**, 768–776.

36. Wilson, M.A. and Brunger, A.T. (2000) The 1.0 A crystal structure of $Ca^{2+}$-bound calmodulin: an analysis of disorder and implications for functionally relevant plasticity. *J. Mol. Biol.*, **301**, 1237–1256.

37. Chen, B., Lowry, D.F., Mayer, M.U., and Squier, T.C. (2008) Helix A stabilization precedes amino terminal lobe activation upon calcium binding to calmodulin. *Biochemistry*, **47**, 9220–9226.

38. Tadross, M.R., Dick, I.E., and Yue, D.T. (2008) Mechanism of local and global $Ca^{2+}$ sensing by calmodulin in complex with a $Ca^{2+}$ channel. *Cell*, **133**, 1228–1240.

39. Halling, D.B., Georgiou, D.K., Black, D.J. *et al.* (2009) Determinants in CaV1 channels that regulate the $Ca^{2+}$ sensitivity of bound calmodulin. *J. Biol. Chem.*, **284**, 20041–20051.

40. Finn, B.E. and Forsen, S. (1995) The evolving model of calmodulin structure, function and activation. *Structure*, **3**, 7–11.

41. Zühlke, R.D., Pitt, G.S., Richard, K.D. *et al.* (1999) Calmodulin supports both inactivation and facilitation of L-type calcium channels. *Nature*, **399**, 159–162.

42. Kobrinsky, E., Darrell, E.S., Abernethy, R., and Soldatov, N.M. (2003) Voltage-gated mobility of the $Ca^{2+}$ channel cytoplasmic tails and its regulatory role. *J. Biol. Chem.*, **278**, 5021–5028.

43. Vendrell, J., Querol, E., and Aviles, F.X. (2000) Metallocarboxypeptidases and their protein inhibitors. Structure, function and biomedical properties. *Biochem. Biophys. Acta*, **1477**, 284–298.

44. Kilshtain-Vardi, A., Glick, M., Greenblatt, H.M. *et al.* (2003) Refined structure of bovine carboxypeptidase A at 1.25 Å resolution. *Acta Crystallogr. D*, **59**, 323–333.

45. Rees, D.C. and Lipscomb, W.N. (1982) Refined crystal structure of the potato inhibitor complex of carboxypeptidase A at 2.5 Å resolution. *J. Mol. Biol.*, **160**, 475–498.

46. (a) Burley, S.K., David, P.R., Taylor, A., and Lipscomb, W.N. (1990) Leucine Aminopeptidase: bestatin inhibition and a model for enzyme catalyzed peptide hydrolysis. *Proc. Natl Acad. Sci. USA*, **87**, 6878–6882; (b) Burley, S.K., David, P.R., Sweet, R.M. *et al.* (1992) Structure determination and refinement of bovine lens leucine aminopeptidase and its complex with bestatin. *J. Mol. Biol.*, **224**, 113–140; (c) Sträter, N. and Lipscomb, W.N. (1995) A bicarbonate ion as a general base in the mechanism of peptide hydrolysis by dizinc aminopeptidase. *Biochemistry*, **34**, 9200–9210; (d) Kim, H. and Lipscomb, W.N. (1994) Structure and mechanism of bovine lens leucine aminopeptidase. *Adv. Enzymol.*, **68**, 153–213.

47. Chevrier, B., Schalk, C., D'Orchymont, H. *et al.* (1994) Crystal structure of *Aeromonas proteolytica* aminopeptidase: a prototypical member of the co-catalytic zinc enzyme family. *Structure*, **2**, 283–291.

48. Greenblatt, H.M., Maras, O.A., Spungin-Bialik, A. *et al.* (1997) *Streptomyces griseus* aminopeptidase: X-ray crystallographic structure at 1.75 Å resolution. *J. Mol. Biol.*, **265**, 620–636.

49. Prescott, J.M., Wagner, F.W., Holmquist, B., and Vallee, B.L. (1985) Spectral and kinetic studies of metal-substituted Aeromonas aminopeptidase: nonidentical, interacting metal-binding sites. *Biochemistry*, **24**, 5350–5356.

50. Lowther, W.T. and Matthews, B.W. (2002) Metalloaminopeptidases: Common Functional Themes in Disparate Structural Surrounding. *Chem. Rev.*, **102**, 4581–4608.

51. (a) Van Wart, H.E. and Lin, S.H. (1981) Metal binding stoichiometry and metal ion modulation of the activity of porcine kidney leucine aminopeptidase. *Biochemistry*, **20**, 5682–5689; (b) Vallee, B.L. and Auld, D.S. (1993) Cocatalytic zinc motifs in enzyme catalysis. *Proc. Natl Acad. Sci. USA*, **90**, 2715–2718.

52. (a) Lin, L.Y., Park, H.I., and Ming, L.-J. (1997) Metal-binding and active-site structure of di-zinc Streptomyces griseus aminopeptidase. *J. Biol. Inorg. Chem.*, **2**, 744–749; (b) Ercan, A., Tay, W.M., Grossman, S.H., and Ming, L.-J. (2010) Mechanistic role of each metal ion in

Streptomyces dinuclear aminopeptidase: Peptide hydrolysis and $7 \times 10^{10}$-fold rate enhancement of phosphodiester hydrolysis. *J. Inorg. Biochem.*, **104**, 19–29.

53. (a) Park, H.I. and Ming, L.-J. (1999) A $10^{10}$ rate enhancement of phosphodiester hydrolysis by a dinuclear aminopeptidase—Transition state analogues as substrates? *Angew. Chem. Int. Ed.*, **38**, 2914–2916; (b) Ercan, A. Park, H.I., and Ming, L.-J. (2000) Enormous enhancement of the hydrolyses of phosphoesters by dinuclear centers: *Streptomyces* aminopeptidase as a "natural model system". *Chem. Commun.*, 2501–2502; (c) Ercan, A. Park, H.I., and Ming, L.-J. (2006) A 'moonlighting' di-zinc aminopeptidase from Streptomyces griseus: Mechanisms for peptide hydrolysis and the $4 \times 10^{10}$-fold acceleration of the alternative phosphodiester hydrolysis. *Biochemistry*, **45**, 13779–13793.

54. da Silva, G.F.Z. and Ming, L.-J. (2005) Catechol oxidase activity of di-$Cu^{2+}$-substituted aminopeptidase from *Streptomyces griseus*. *J. Am. Chem. Soc.*, **127**, 16380–16381.

55. Valentine, J.S., Wertz, D.L., Lyons, T.J. *et al.* (1998) The dark side of dioxygen biochemistry. *Curr. Opp. Chem. Biol.*, **2**, 253–262.

56. Strange, R.W., Antonyuk, S.V., Hough, M.A. *et al.* (2006) Variable metallation of human superoxide dismutase: atomic resolution crystal structures of Cu-Zn, Zn-Zn and as-isolated wild-type enzymes. *J. Mol. Biol.*, **356**, 1152–1162.

57. Valentine, J.S. (1994) Chapter 5, Dioxygen reactions, in *Bioionorganic Chemistry* (eds I. Bertini, H. Gray, S. Lippard, and J.S. Valentine), University Science Books, Mill Valley.

58. Leitch, J.M., Yick, P.J., and Culotta, V.C. (2009) The right to choose: Multiple pathways for activating copper,zinc superoxide dismutase. *J. Biol. Chem.*, **284**, 24679–24683.

59. Furukawa, Y. and O'Halloran, T.V. (2005) Complete loss of post-translational modifications triggers fibrilar aggregation of SOD1 in the familial form of amyotrophic lateral sclerosis. *J. Biol. Chem.*, **280**, 17266–17274.

60. Furukawa, Y., Fu, R., Deng, H.-X. *et al.* (2006) Disulfide cross-linked protein represents a significant fraction of ALS-associated Cu,Zn-superoxide dismutase aggregates in spinal cords of model mice. *Proc. Natl Acad. Sci. USA*, **103**, 7148–7153.

61. Cao, X., Antonyuk, S.V., Seetharaman, S.V. *et al.* (2008) Structures of the G85R variant of SOD1 in familial amyotrophic lateral sclerosis. *J. Biol. Chem.*, **283**, 16169–16177.

62. Lynch, S. and Colon, W. (2006) Dominant role of copper in the kinetic stability of Cu/Zn superoxide dismutase. *Biochem. Biophys. Res. Comm.*, **340**, 457–461.

63. Assfalg, M., Banci, L., Bertini, I. *et al.* (2003) Superoxide dismutase folding/unfolding pathway: role of the metal ions in modulating structural and dynamic features. *J. Mol. Biol.*, **330**, 145–158.

64. Libralesso, E., Nerinovski, K., Parigi, G., and Turano, P. (2005) $^1$H nuclear magnetic relaxation dispersion of Cu,Zn- superoxide dismutase in the native and guanidinium-induced unfolded forms. *Biochem. Biophys. Res. Comm.*, **328**, 633–639.

65. Forman, H.J. and Fridovic, I. (1973) On the stability of bovine superoxide dismutase-the effects of metals. *J. Biol. Chem.*, **248**, 2645–2649.

66. Roe, J.A., Butler, A., Scholler, D.M. *et al.* (1988) Differential scanning calorimetry of Cu, Zn-superoxide dismutase, the apoprotein, and its zinc-substituted derivatives. *Biochemistry*, **27**, 950–958.

67. Eakin, C.M., Knight, J.D., Morgan, C.J. *et al.* (2002) Formation of copper specific binding site in non-native states of beta-2-macroglobulin. *Biochemistry*, **41**, 10646–10656.

68. Ming, L.-J. (2010) Metallopeptides—from drug discovery to catalysis. *J. Chin. Chem. Soc*, **57**, 285–299.

69. (a) Kozłowski, H., Bal, W., Dyba, M., and Kowalik-Jankowska, T. (1999) Specific structure–stability relations in metallopeptides. *Coord. Chem. Rev.*, **184**, 319–346; (b) Sóvágó, I. and Ősz, K. (2006) Metal ion selectivity of oligopeptides. *Dalton Trans.*, **2006**, 3841–3854; (c) Migliorini, C., Porciatti, E., Luczkowski, M., and Valensin, D. (2012) Structural

characterization of $Cu^{2+}$, $Ni^{2+}$ and $Zn^{2+}$ binding sites of model peptides associated with neurodegenerative diseases. *Coord. Chem. Rev.*, **256**, 352–368.

70. Bataille, M., Formicka-Kozlowska, G., Kozlowski, H. *et al.* (1984) The L-proline residue as a 'break-point' in the co-ordination of metal–peptide systems. *J. Chem. Soc. Chem. Commun.*, **1984**, 231–232.

71. (a) Chou, P.Y. and Fasman, G.D. (1977) β-turns in proteins. *J. Mol. Biol.*, **115**, 135–175; (b) Chou, P.Y. and Fasman, G.D. (1979) Prediction of beta-turns. *Biophys. J.*, **26**, 367–383; (c) Fu, H., Grimsley, G.R., Razvi, A. *et al.* (2009) Increasing protein stability by improving beta-turns. *Prot. Struct. Funct. Bioinform.*, **77**, 491–498; (d) Trevino, S.R., Schaefer, S., Scholtz, J.M., and Pace, C.N. (2007) Increasing protein conformational stability by optimizing beta-turn sequence. *J. Mol. Biol.*, **373**, 211–218; (e) Bornot, A. and de Brevern, A.G. (2006) Protein beta-turn assignments. *Bioinformation*, **1**, 153–155; (f)Fuchs, P.F.J. and Alix, A.J.P. (2005) High accuracy prediction of beta-turns and their types using propensities and multiple alignments. *Prot. Struct. Funct. Bioinform.*, **59**, 828–839; (g) Fuller, A.A., Du, D., Liu, F. *et al.* (2009) Evaluating beta-turn mimics as beta-sheet folding nucleators. *Proc. Natl Acad. Sci. USA*, **106**, 11067–11072.

72. Chou, P.Y. and Fasman, G.D. (1974) Conformational parameters for amino acids in helical, β-sheet, and random coil regions calculated from proteins. *Biochemistry*, **13**, 211–222.

73. Tyndall, J.D.A., Pfeiffer, B., Abbenante, G., and Fairlie, D.P. (2005) Over one hundred peptide-activated g protein-coupled receptors recognize ligands with turn structure. *Chem. Rev.*, **105**, 793–826.

74. Whitby, L.R., Ando, Y., Setola, V. *et al.* (2011) Design, synthesis, and validation of a β-turn mimetic library targeting protein–protein and peptide–receptor interactions. *J. Am. Chem. Soc.*, **133**, 10184–10194.

75. Ming, L.-J. (2003) Structure and function of metalloantibiotics. *Med. Res. Rev.*, **23**, 697–762.

76. Ming, L.-J. and Epperson, J.D. (2002) Metal binding and structure–activity relationship of the metalloantibiotic peptide bacitracin. *J. Inorg. Biochem*, **91**, 46–58.

77. Umezawa, H., Maeda, K., Takeuchi, T., and Okami, Y. (1966) New antibiotics, bleomycin A and B. *J. Antibiot.*, **19**, 200–209.

78. Wei, X. and Ming, L.-J. (1998) NMR studies of metal complexes and DNA binding of the quinone-containing antibiotic streptonigrin. *J. Chem. Soc. Dalton Trans.*, **1998**, 2793–2798.

79. Chiu, Y. and Lipscomb, W.N. (1975) Molecular and crystal structure of streptonigrin. *J. Am. Chem. Soc.*, **97**, 2525–2530.

80. Umezawa, H., Maeda, K., Takeuchi, T., and Okami, Y. (1966) New antibiotics, bleomycin A and B. *J. Antibiot.*, **19**, 200–209.

81. Umezawa, H. and Takita, T. (1980) The bleomycins: Antitumor copper-binding antibiotics. *Struct. Bond.*, **40**, 73–99.

82. (a) Takeshita, M. Horwitz, S.B., and Grollman, A.P. (1974) Bleomycin, an inhibitor of vaccinia virus replication. *Virology*, **60**, 455–456; (b) Takeshita, M., Grollman, A.P., and Horwitz, S.B. (1976) Effect of ATP and other nucleotides on the bleomycin-induced degradation of vaccinia virus DNA. *Virology*, **69**, 453–463; (c) Takeshita, M., Horwitz, S.B., and Grollman, A.P. (1977) Mechanism of the antiviral action of bleomycin. *Ann. N.Y. Acad. Sci.*, **284**, 367–374.

83. Lazo, J.S., Sebti, S.M., and Schellens, J.H. (1996) Bleomycin, cancer chemother. *Biol. Resp. Modif.*, **16**, 39–47.

84. (a) Burger, R.M. (2000) Nature of activated bleomycin. *Struct. Bond.*, **97**, 287–303; (b) Burger, R.M. (1998) Cleavage of nucleic acids by bleomycin. *Chem. Rev.*, **98**, 1153–1170; (c) Claussen, C.A. and Long, E.C. (1999) Nucleic acid recognition by metal complexes of bleomycin. *Chem. Rev.*, **99**, 2797–2816; (d) Boger, D.L. and Cai, H. (1999) Bleomycin: synthetic and mechanistic studies. *Angew. Chem. Int. Ed.*, **38**, 449–476; (e) Stabbe, J. and

Kozarich, J.W. (1987) Mechanism of bleomycin-induced DNA degradation. *Chem. Rev.*, **87**, 1107–1136; (f) Petering, D.H. Byrnes, R.W., and Antholine, W.E. (1990) The role of redox-active metals in the mechanism of action of bleomycin. *Chem. Biol. Int.*, **73**, 133–182; (g) Hecht, S.M. (1986) The chemistry of activated bleomycin. *Acc. Chem. Res.*, **19**, 383–391.

85. Kane, S.A. and Hecht, S.M. (1994) Polynucleotide recognition and degradation by bleomycin. *Progr. Nucleic Acids Res. Mol. Biol.*, **49**, 313–352.
86. (a) Takeshita, M., Grollman, A.P., Ohtsubo, E., and Ohtsubo, H. (1978) Interaction of bleomycin with DNA. *Proc. Natl Acad. Sci. USA*, **75**, 5983–5987; (b) Cullinam, E.B., Gawron, L.S., Rustum, Y.M., and Beerman, T.A. (1991) Extrachromosomal chromatin: novel target for bleomycin cleavage in cells and solid tumors. *Biochemistry*, **30**, 3055–3061.
87. (a) Absalon, M.J., Stubbe, J., and Kozarich, J.W. (1995) Sequence-specific double-strand cleavage of DNA by Fe-bleomycin 1. The detection of sequence-specific double-strand breaks using hairpin oligonucleotides. *Biochemistry*, **34**, 2065–2075; (b) Absalon, M.J., Wu, W., Stubbe, J., and Kozarich, J.W. (1995) Sequence-specific double-strand cleavage of DNA by Fe-bleomycin. 2. Mechanism and dynamics. *Biochemistry*, **34**, 2076–2086.
88. (a) Hecht, S.M. (1994) RNA degradation by bleomycin, a naturally occurring bioconjugate. *Bioconj. Chem.*, **5**, 513–526; (b) Mascharak, P. and Hüttenhofer, A. (eds) (2001) Cleavage of RNA by Fe(II)-bleomycin, in *RNA-Binding Antibiotics* (ed. R. Schroeder and M.G. Wallis,) Landes BioScience, Rotterdam.
89. (a) Magliozzo, R.S., Peisach, J., and Ciriolo, M.R. (1989) Transfer RNA is cleaved by activated bleomycin. *Mol. Pharmacol.*, **35**, 428–432; (b) Carter, B.J., de Vroom, E., Long, E.C., van der Marel, G.A., van Boom, J.H., and Hecht, S.M. (1990) Site-specific cleavage of RNA by Fe(II)-bleomycin. *Proc. Natl Acad. Sci. USA*, **87**, 9373–9377; (c) Hüttenhofer, A., Hudson, S., Noller, H.F., and Mascharak, P.K. (1992) Cleavage of tRNA by Fe(II)-bleomycin. *J. Biol. Chem.*, **267**, 24471–24475.
90. (a) Carter, B.J. Reddy, K.S., and Hecht, S.M. (1991) Polynucleotide recognition and strand scission by Fe-bleomycin. *Tetrahedron*, **47**, 2463–2474; (b) Holmes, C.E. Carter, B.J., and Hecht, S.M. (1993) Characterization of iron(II)-bleomycin-mediated RNA stand scission. *Biochemistry*, **32**, 4293–4307.
91. (a) Morgan, M.A. and Hecht, S.M. (1994) Iron(II) bleomycin-mediated degradation of a DNA-RNA hetero-duplex. *Biochemistry*, **33**, 10286–10293; (b) Bansal, M., Lee, J.S., Stubbe, J.A., and Kozarich, J.W. (1997) Mechanistic analyses of site-specific degradation in DNA-RNA hybrids by prototypic DNA cleavers. *Nucleic Acids Res.*, **25**, 1836–1845.
92. Dabrowiak, J.C. (1980) The coordination chemistry of bleomycin: A review. *J. Inorg. Biochem.*, **13**, 317–337.
93. (a) Burger, R.M., Freedman, J.H., Horwitz, S.B., and Peisach, J. (1984) DNA-degradation by manganese(II)-bleomycin plus peroxide. *Inorg. Chem.*, **23**, 2215–2217; (b) Ehrenfeld, G.M. Murugesan, N., and Hecht, S.M. (1984) Activation of oxygen and mediation of DNA-degfradation by manganese bleomycin. *Inorg. Chem.*, **23**, 1496–1498.
94. (a) Ishida, R. and Takahashi, T. (1975) Increased DNA chain breakage by combined action of bleomycin and superoxide radical. *Biochem. Biophys. Res. Commun.*, **66**, 1432–1438; (b) Sausville, E.A., Stein, R.W., Peisach, J., and Horeitz, S.B. (1978) Properties and products of degradation of DNA by bleomycin and Iron(II). *Biochemistry*, **17**, 2746–2754; (c) Absalon, M.J., Kozarich, J.W., and Stubbe, J. (1995) Sequence-specific double-strand cleavage of DNA by Fe-bleomycin. 1. The detection of sequence-specific double-strand breaks using hairpin oligonucleotides. *Biochemistry*, **34**, 2065–2075; (d) Absalon, M.J., Wu, W., Kozarich, J.W., and Stubbe, J. (1995) Sequence-specific double-strand cleavage of DNA by Fe-bleomycin. 2. Mechanism and dynamics. *Biochemistry*, **34**, 2076–2086; (e) McGall, G.H., Rabow, L.E., Ashley, G.W., Wu, S. H., Kozarich, J.W., and Stubbe, J. (1992) New insight into the mechanism of base propenal formation during bleomycin-mediated DNA degradation. *J. Am. Chem. Soc.*, **114**, 4958–4967.

95. Sugiura, Y. (1980) Monomeric cobalt(II)-oxygen adducts of bleomycin antibiotics in aqueous solution – a new ligand type for oxygen binding and effect of axial Lewis base. *J. Am. Chem. Soc.*, **102**, 5216–5221.

96. Greenaway, F.T., Dabrowiak, J.C., Van Husen, M. *et al.* (1978) The transition metal binding properties of a 3rd generation bleomycin analogue, tallysomycin. *Biochem. Biophys. Res. Commun.*, **85**, 1407–1414.

97. (a) Sugiura, Y. (1978) Metal coordination core of bleomycin: comparison of metal complexes between bleomycin and its biosynthetic intermediate. *Biochem. Biophys. Res. Commun.*, **87**, 643–648; (b) Guan, L.L., Totsuka, R., Kuwahara, J. *et al.* (1993) Cleavage of yeast transfer RNA$^{phe}$ with Ni(III) and Co(III) complexes of bleomycin. *Biochem. Biophys. Res. Commun.*, **191**, 1338–1346; (c) Sugiura, Y. and Mino, Y. (1979) Nickel(III) complexes of histidine-containing tripeptide and bleomycin—Electron-spin resonance characteristics and effect of axial nitrogen donors. *Inorg. Chem.*, **18**, 1336–1339; (d) Guan, L.L., Kuwahara, J., Sugiura, Y. *et al.* (1993) Guanine-specific binding by bleomycin nickel(III) complex and its reactivity for guanine-quartet telomeric DNA. *Biochemistry*, **32**, 6141–6145.

98. (a) Dabrowiak, J.C., Greenaway, F.T., and Grulich, R. (1978) Transition-metal binding site of bleomycin A$_2$. A carbon-13 nuclear magnetic resonance study of the zinc(II) and copper(II) derivatives. *Biochemistry*, **17**, 4090–4096; (b) Dabrowiak, J.C., Greenaway, F.T., Longo, W.E. *et al.* (1978) A spectroscopic investigation of the metal binding site of bleomycin A$_2$. The Cu (II) and Zn(II) derivatives. *Biochim. Biophys. Acta*, **517**, 517–526; (c) Sugiura, Y. (1979) The production of hydroxyl radical from copper(I) complex systems of bleomycin and tallysomycin: comparison with copper(II) and iron(II) systems. *Biochem Biophys. Res. Commun.*, **90**, 375–383.

99. Ehrenfeld, G.M., Shipley, J.B., Heimbrook, D.C. *et al.* (1987) Copper-dependent cleavage of DNA by bleomycin. *Biochemistry*, **26**, 931–942.

100. Otvos, J.D., Antholine, W.E., Wehrl, S., and Petering, D.H. (1996) Metal coordination environment and dynamics in cadmium-113 bleomycin: Relationship to zinc bleomycin. *Biochemistry*, **35**, 1458–1465.

101. Papakyriakou, A., Mouzopoulou, B., and Katsaros, N. (2001) The detailed structural characterization of the Ga(III)-bleomycin A2 complex by NMR and molecular modeling. *J. Inorg. Biochem.*, **86**, 371 (ICBIC10 abstract).

102. Subramanian, R. and Meares, C.F. (1985) Photo-induced nicking of deoxyribonucleic acid by ruthenium(II)-bleomycin in the presence of air. *Biochem. Biophys. Res. Commun.*, **133**, 1145–1151.

103. Brooks, R.C., Carnochan, P., Vollano, J.F. *et al.* (1999) Metal complexes of bleomycin: Evaluation of [Rh-105]-bleomycin for use in targeted radiotherapy. *Nuclear Med. Biol.*, **26**, 421–430.

104. Itaka, Y., Nakamura, H., Nakatani, T. *et al.* (1978) Chemistry of bleomycin, The X-ray structure determination of P-3A Cu(II)-complex a biosynthetic intermediate of bleomycin. *J. Antibiot.*, **31**, 1070–1072.

105. Burger, R.M., Adler, A.D., Horwitz, S.B. *et al.* (1981) Demonstration of nitrogen coordination in metal–bleomycin complexes by electron spin-echo envelope spectroscopy. *Biochemistry*, **20**, 1701–1704.

106. (a) Loeb, K.E., Zaleski, J.M., Westre, T.E. *et al.* (1995) Spectroscopic dedinition of the geometric and electronic structure of the nonheme active site in iron(II) bleomycin–correlation with oxygen reactivity. *J. Am. Chem. Soc.*, **117**, 4545–4561; (b) Loeb, K.E., Zaleski, J.M., Hess, C.D. *et al.* (1998) Spectroscopic investigation of the metal ligation and reactivity of the ferrous active sites of bleomycin and bleomycin derivatives. *J. Am. Soc.*, **120**, 1249–1259.

107. (a) Oppenheimer, N.J., Chang, C., Chang, L.H. *et al.* (1982) Deglyco-bleomycin. Degradation of DNA and formation of a structurally unique Fe(II)·CO complex. *J. Biol. Chem.*, **257**,

1606–1609; (b) Akkerman, M.A.J., Neijman, E.W.J.F., Wijmenga, S.S. *et al.* (1990) Studies of the solution structure of the bleomycin $A_2$ iron(II) carbon monoxide complex by means of 2-dimensional NMR spectroscopy and distance geometry calculations. *J. Am. Chem. Soc.*, **112**, 7462–7474.

108. (a) Akkerman, M.A.J. Haasnoot, C.A., and Hilbers, C.W. (1988) Studies of the solution structure of the bleomycin-$A_2$-zinc complex by means of two-dimensional NMR spectroscopy and distance geometry calculations. *Eur. J. Biochem.*, **173**, 211–225; (b) Akkerman, M.A.J., Haasnoot, U.K., Pandit, C.W., and Hilbert, C.W. (1988) Complete assignment of the C-13 NMR spectra of bleomycin $A_2$ and its zinc complex by means of two-dimensional NMR spectroscopy. *Magn. Reson. Chem.*, **26**, 793–802; (c) Williamson, D., McLennan, I.J., Bax, A. *et al.* (1990) Two-dimensional NMR study of bleomycin and its zinc(II) complex: reassignment of $^{13}C$ resonances. *J. Biomol. Struct. Dyn.*, **8**, 375–398.

109. (a) Sugiura, Y. (1980) Monomeric cobalt(II)-oxygen adducts of bleomycin antibiotics in aqueous solution. A new ligand type for oxygen biding and effect of axial Lewis base. *J. Am. Chem. Soc.*, **102**, 5216–5221; (b) Sugiura, Y. (1978) Oxygen binding to cobalt(II)-bleomycin. *J. Antibiot.*, **31**, 1206–1208.

110. Caceres-Cortes, J., Sugiyama, H., Ikudome, K. *et al.* (1997) Interactions of deglycosylated cobalt(III)-pepleomycin (green form) with DNA based on NMR structural studies. *Biochemistry*, **36**, 9995–10005.

111. Wu, W., Vanderwall, D.E., Lui, S.M. *et al.* (1996) Studies of Co center dot Bleomycin $A_2$ green: Its detailed structural characterization by NMR and molecular modeling and its sequence-specific interaction with DNA oligonucleotides. *J. Am. Chem. Soc.*, **118**, 1268–1280.

112. (a) Vanderwall, D.E., Lui, S.M., Wu, W. *et al.* (1997) A model of the structure of HOO-Co-bleomycin bound to d(CCAGTACTGG): recognition at the d(GpT) site and implications for double-stranded DNA cleavage. *Chem. Biol.*, **4**, 373–387; (b) Hoehn, S.T., Junker, H.D., Bunt, R.C. *et al.* (2001) Solution structure of Co(III)-bleomycin-OOH bound to a phosphoglycolate lesion containing oligonucleotide: implications for bleomycin-induced double-strand DNA cleavage. *Biochemistry*, **40**, 5894–5905; (c) Wu, W., Vanderwall, D.E., Teramoto, S. *et al.* (1998) NMR studies of co center dot deglycoBleomycin $A_2$ green and its complex with d (CCAGGCCTGG). *J. Am. Chem. Soc.*, **120**, 2239–2250; (d) Lui, S.M., Vanderwall, D.E., Wu, W. *et al.* (1997) Structural characterization of Co center dot bleomycin A2 brown: Free and bound to d(CCAGGCCTGG). *J. Am. Chem. Soc.*, **119**, 9603–9613.

113. (a) Manderville, R.A., Ellena, J.F., and Hecht, S.M. (1994) Solution structure of a Zn(II)·bleomycin $A_5$-D(CGCTAGCG)$_2$ complex. *J. Am. Chem. Soc.*, **116**, 10851–10852; (b) Manderville, R.A., Ellena, J.F., and Hecht, S.M. (1995) Interaction of Zn(II)·bleomycin with d (CGCTAGCG)$_2$—A binding model based on NMR experiments and restrained molecular dynamics calculations. *J. Am. Chem. Soc.*, **117**, 7891–7903.

114. Abraham, A.T., Zhou, X., and Hecht, S.M. (1999) DNA cleavage by Fe(II)center dot bleomycin conjugated to a solid support. *J. Am. Chem. Soc.*, **121**, 1982–1983.

115. (a) Sugiura, Y. (1980) Bleomycin–iron complexes – Electron-spin resonance study, ligand effect, and implication for action mechanism. *J. Am. Chem. Soc.*, **102**, 5208–5215; (b) Burger, RM. Peisach, J., and Horwitz, SB. (1981) Activated bleomycin. A transient complex of drug, iron, and oxygen that degrades DNA. *J. Biol. Chem.*, **256**, 11636–11544; (c) Veselov, A., Sun, H.J., Sienkiewicz, A. *et al.* (1995) Iron coordination of activated bleomycin probed by Q-band and X-band ENDOR – Hyperfine coupling to activated O-17 oxygen, N-14 and exchangeable $^1$H. *J. Am Chem Soc.*, **117**, 7508–7512.

116. Takahashi, S., Sam, J.W., Peisach, J., and Rousseau, D.L. (1994) Structural characterization of iron-bleomycin by resonance Raman spectroscopy. *J. Am. Chem. Soc.*, **116**, 4408–4413.

117. Burger, R.M., Kent, T.A., Horwitz, S.B. *et al.* (1983) Mössbauer study of iron bleomycin and its activation intermediates. *J. Biol. Chem.*, **258**, 559–1564.

118. (a) La Mar, G.N.Horrocks, W.D. Jr, and Holm, R.H. (eds) (1973) *NMR of Paramagnetic Molecules*, Academic, New York, (b) Bertini, I. and Luchinat, C. (1986) *NMR of Paramagnetic Molecules in Biological Systems*, Benjamin/Cumming, Menlo Park, (c) Berliner, L.J. and Reuben, J. (eds) (1993) *NMR of Paramagnetic Molecules*, Plenum, New York, (d) La Mar, G.N. (ed.) (1995) *Nuclear Magnetic Resonance of Paramagnetic Molecules*, NATO-ASI/Kluwer, Dordrecht, (e) Ming, L.-J. (2000) Nuclear magnetic resonance of paramagnetic metal centers in proteins and synthetic complexes, in *Physical Methods in Bioinorganic Chemistry, Spectroscopy and Magnetism*, (ed L. Que Jr), University Science Books, Sausalito.

119. Pillai, R.P., Lenkinski, R.E., Sakai, T.T. *et al.* (1980) Proton NMR study of iron(II)-bleomycin: assignment of resonances by saturation transfer experiments. *Biochem. Biophys. Res. Commun.*, **96**, 341–349.

120. Lehmann, T.E., Ming, L.-J., Rosen, M.E., and Que, L. Jr (1997) NMR studies of the paramagnetic complex Fe(II)-bleomycin. *Biochemistry*, **36**, 2807–2816.

121. Miyazaki, I., Okumura, H., Simizu, S. *et al.* (2009) Structure-affinity relationship study of bleomycins and shble protein by use of a chemical array. *ChemBioChem*, **10**, 845–852.

122. Brewer, G.A., and Florey, K. (eds) (1980) *Analytical Profiles of Drug Substances*, **9**, Academic, New York, pp. 1–69.

123. Hanson, D.J. (1985) Human health effects of animal feed drugs unclear. *Chem. Eng. News*, **63**, 7–11.

124. Arky, R. (1997) *Physicians Desk Reference for Nonprescription Drugs*, 18th edn, Medical Economics, Montvale.

125. (a) Gehm, B.D. and Rosner, M.R. (1991) Regulation of insulin, epidermal growth factor, and transforming growth factor-alpha levels by growth factor-degrading enzymes. *Endocrinology*, **128**, 1603–1610; (b) Mantle, D. Lauffart, B., and Gibson, A. (1991) Purification and characterization of leucyl aminopeptidase and pyroglutamyl aminopeptidase from human skeletal muscle. *Clin. Chim. Acta. Int. J. Clin. Chem.*, **197**, 35–45; (c) Janas, J., Sitkiewicz, D., Warnawin, K., and Janas, R.M. (1994) Characterization of a novel, high molecular weight, acidic, endothelin-1 inactivating metalloendopeptidase from the rat kidney. *J. Hypertension*, **12**, 1155–1162.

126. (a) Essex, D.W., Li, M., Miller, A., and Feinman, R.D. (2001) Protein disulfide isomerase and sulfhydryl-dependent pathways in platelet activation. *Biochemistry*, **40**, 6070–6075; (b) Täger, M., Kröning, H., Thiel, U., and Ansorge, S. (1997) Membrane-bound proteindisulfide isomerase (PDI) is involved in regulation of surface expression of thiols and drug sensitivity of B-CLL cells. *Exp. Hematol.*, **25**, 601–607; (c) Clive, D.R. and Greene, J.J. (1994) Association of protein disulfide isomerase activity and the induction of contact inhibition. *Exp. Cell. Res.*, **214**, 139–144; (d) Mandel, R., Ryser, H.J., Ghani, F. *et al.* (1993) Inhibition of a reductive function of the plasma membrane by bacitracin and antibodies against protein disulfide-isomerase. *Proc. Natl Acad. Sci. USA*, **90**, 4112–4116; (d) Mizunaga, T., Katakura, Y., Miura, T., and Maruyama, Y. (1990) Purification and characterization of yeast protein disulfide isomerase. *J. Biochem.*, **108**, 846–851.

127. Mou, Y., Ni, H., and Wilkins, J.A. (1998) The selective inhibition of beta 1 and beta 7 integrin-mediated lymphocyte adhesion by bacitracin. *J. Immunol.*, **161**, 6323–6329.

128. Karala, A.-K. and Ruddock, L.W. (2010) Bacitracin is not a specific inhibitor of protein disulfide isomerase. *FEBS J.*, **277**, 2454–2462.

129. Storm, D.R. and Strominger, J.L. (1974) Binding of bacitracin to cells and protoplasts of *Micrococcus lysodeikticus*. *J. Biol. Chem.*, **249**, 1823–1827.

130. (a) Scogin, D.A., Mosberg, H.I., Storm, D.R., and Gennis, R.B. (1980) Binding of nickel and zinc ions to bacitracin A. *Biochemistry*, **19**, 3348–3342; (b) Seebauer, E.G., Duliba, E.P., Scogin, D.A. *et al.* (1983) EPR evidence on the structure of the copper(II)-bacitracin A complex. *J. Am. Chem. Soc.*, **105**, 4926–4929; (c) Mosberg, H.I., Scogin, D.A., Storm, D.R., and Gennis, R.B. (1980) Proton nuclear magnetic resonance studies on bacitracin A and its interaction with zinc ion. *Biochemistry*, **19**, 3353–3357.

131. (a) Epperson, J.D. and Ming, L.-J. (2000) Proton NMR studies of Co(II) complexes of bacitracin analogous: Insight into structure–activity relationship. *Biochemistry*, **39**, 4037–4045; (b) Tay, W.M., Epperson, J.D., da Silva, G.F.Z., and Ming, L.-J. (2010) $^1$H NMR mechanism, and mononuclear oxidative activity of the antibiotic metallopeptide bacitracin: the role of D-Glu-4, interaction with pyrophosphate moiety, DNA binding and cleavage, and bioactivity. *J. Am. Chem. Soc.*, **132**, 5652–5661.

132. (a) Storm, D.R. and Strominger, J.L. (1973) Complex formation between bacitracin peptides and isoprenyl pyrophosphates. *J. Biol. Chem.*, **248**, 3940–3945; (b) Stone, K.J. and Strominger, J.L. (1971) Mechanism of action of bacitracin: Complexation with metal ion and $C_{55}$-isoprenyl pyrophosphate. *Proc. Natl Acad. Sci. USA*, **68**, 3223–3227.

133. (a) Higashi, Y., Siewert, G., and Strominger, J.L. (1970) Biosynthesis of the peptidoglycan of bacterial cell walls. XIX. Isoprenoid alcohol phosphokinase. *J. Biol. Chem.*, **245**, 3683–3690; (b) Bhagavan, N.V. (2001) Chapter 16, in *Medical Biochemistry*, 4th edn, Harcourt, San Diego.

134. Drabløs, F., Nicholson, D.G., and Rønning, M. (1999) EXAFS study of zinc coordination in bacitracin A. *Biochim. Biophys. Acta*, **1431**, 433–442.

135. (a) Pons, M., Feliz, M., Molins, M.A., and Giralt, E. (1991) Conformational analysis of bacitracin A a naturally occurring lariat. *Biopolymers*, **31**, 605–612; (b) Kobayashi, N., Takenouchi, T., Endo, S., and Munekata, E. (1992) $^1$H NMR study of the conformation of bacitracin A in aqueous solution. *FEBS Lett.*, **305**, 105–109.

136. (a) Pfeffer, S., Hohne, W., Branner, S. *et al.* (1991) X-ray structure of the antibiotic bacitracin A. *FEBS Lett.*, **285**, 115–119; (b) Pfeffer-Hennig, S., Dauter, Z., Hennig, M. *et al.* (1996) Three dimensional structure of the antibiotic bacitracin A complexed to two different subtilisin proteases: novel mode of enzyme inhibition. *Adv. Exp. Med. Biol.*, **379**, 29–41.

137. Oppenheim, F.G., Yang, Y.-C., Diamond, R.D. *et al.* (1986) The primary structure and functional characterization of the neutral histidine-rich polypeptide from human parotid secretion. *J. Biol. Chem.*, **261**, 1177–1182.

138. Oppenheim, F.G., Xu, T., McMillian, F.M. *et al.* (1988) Anticandidal activity of major human salivary histatins. *J. Biol. Chem.*, **263**, 7472–7477.

139. Sabatini, L.M., Ota, T., and Azen, E.A. (1993) Nucleotide sequence analysis of the human salivary protein genes HIS1 and HIS2, and evolution of the STATH/HIS gene family. *Mol. Biol. Evol.*, **10**, 497–511.

140. Vanderspek, J.C., Wyandt, H.E., Skare, J.C. *et al.* (1989) Localization of the genes for histatins to human chromosome 4q13 and tissue distribution of the mRNAs. *Am. J. Hum. Genet.*, **45**, 381–387.

141. MacKay, B.J., Denepitiya, L., Iacono, V.J. *et al.* (1984) Fungistatic and fungicidal activity of human parotid salivary histidine-rich polypeptides on Candida albicans. *J. Infect. Immun.*, **44**, 695–701.

142. Murakami, Y., Nagata, H., Amano, A. *et al.* (1991) Inhibitory effects of human salivary histatins and lysozyme on coaggregation between *Porphyromonas gingivalis* and *Streptococcus mitis*. *J. Infect. Immun.*, **59**, 3284–3286.

143. Tsai, H. and Bobek, L.A. (1997) Studies of the mechanism of human salivary histatin-5 candidacidal activity with histatin-5 variants and azole-sensitive and -resistant Candida species. *Antimicrob. Agents Chemother.*, **41**, 2224–2228.

144. Tsai, H. and Bobek, L.A. (1997) Human salivary histatin-5 exerts potent fungicidal activity against Cryptococcus neoformans. *Biochim. Biophys. Acta*, **1336**, 367–369.

145. Pollock, J.J., Denepitiya, B.J., MacKay, V., and Iacono, V. (1984) Fungistatic and fungicidal activity of human parotid salivary histidine-rich polypeptides on candida albicans. *J. Infect. Immun.*, **44**, 702–707.

146. Raj, P.A., Soni, S.-D., and Levine, M.J. (1994) Membrane-induced helical conformation of an active candidacidal fragment of salivary histatins. *J. Biol. Chem.*, **269**, 9610–9619.

147. Li, X.S., Reddy, M.S., Baev, D., and Edgerton, M. (1993) Candida albicans Ssa1/2p is the cell envelope binding protein for human salivary histatin 5. *J. Biol. Chem.*, **278**, 28553–28561.

148. Baev, D., Rivetta, A., Vylkova, S. *et al.* (2004) The TRK1 potassium transporter is the critical effector for killing of Candida albicans by the cationic protein, histatin 5. *J. Biol. Chem.*, **279**, 55060–55072.

149. Koshlukova, S.E., Lloyd, T.L., Araujo, M.W.B., and Edgerton, M. (1999) Salivary histatin 5 induces non-lytic release of ATP from candida albicans leading to cell death. *J. Biol. Chem.*, **274**, 18872–18879.

150. (a) Helmerhorst, E.J. Troxler, R.F., and Oppenheim, F.G. (1991) The human salivary peptide histatin 5 exerts its antifungal activity through the formation of reactive oxygen species. *Proc. Natl Acad. Sci. USA*, **98**, 14637–14642; (b) Helmerhorst, E.J., Breeuwer, P., van't Hof, W. *et al.* (1999) The cellular rarget of histatin 5 on candida albicans is the energized mitochondrion. *J. Biol. Chem.*, **274**, 7286–7291.

151. (a) Melino, S., Rufini, S., Sette, M. *et al.* (1999) $Zn^{2+}$ ions selectively induce antimicrobial salivary peptide Histatin-5 to fuse negatively charged vesicles. Identification and characterization of a zinc-binding motif present in the functional domain. *Biochemistry*, **38**, 9626–9633; (b) Brewer, D. and Lajoie, G. (2000) Evaluation of the metal binding properties of the histidine-rich antimicrobial peptides histatins 3 and 5 by electrospray ionization mass spectrometry. *Rapid Comm. Mass Spect.*, **14**, 1736–1745; (c) Grogan, J., McKnight, C.J., Troxler, R.F., and Oppenheim, F.G. (2001) Zinc and copper bind to unique sites of histatin 5. *FEBS Lett.*, **491**, 76–80; (d) Gusman, H., Lendenmann, U., Grogan, J. *et al.* (2001) Is salivary histatin 5 a metallopeptide? *Biochim. Biophys. Acta*, **1545**, 86–95.

152. (a) Peters, T. (1960) Interaction of one mole of copper with the alpha amino group of bovine serum albumin. *Biochim. Biophys. Acta*, **39**, 546–547; (b) Peters, T. and Blumenstock, F.A. (1967) Copper-binding properties of bovine serum albumin and its amino-terminal peptide fragment. *J. Biol. Chem.*, **242**, 1574–1578; (c) Zhang, Y. and Wilcox, D.E. (2002) Thermodynamic and spectroscopic study of Cu(II) and Ni(II) binding to bovine serum albumin. *J. Biol. Inorg. Chem.*, **7**, 327–337.

153. Tay, W.M., Hanafy, A.I., Angerhofer, A., and Ming, L.-J. (2009) A plausible role of salivary copper in antimicrobial activity of histatin-5-Metal binding and oxidative activity of its copper complex. *Bioorg. Med. Chem. Lett.*, **19**, 6709–6712.

154. (a) Melino, S., Gallo, M., Trotta, E. *et al.* (2006) Metal-binding and nuclease activity of an antimicrobial peptide analogue of the salivary Histatin 5. *Biochemistry*, **45**, 15373–15383; (b) Cabras, T., Patamia, M., Melino, S. *et al.* (2007) Pro-oxidant activity of histatin 5 related Cu (II)-model peptide probed by mass spectrometry. *Biochem. Biophys. Res. Comm.*, **358**, 277–284.

155. (a) Easwaran, K.R.K. (1985) Interaction between valinomycin and metalions. *Metal Ions Biol. Syst.*, **19**, 109–137; (b) Steinrauf, L.K. (1985) Beauvericin and the other enniatins. *Metal Ions Biol. Syst.*, **19**, 139–171; (c) Nawata, Y., Ando, K., and Iitaka, Y. (1985) Nactins: Their complexes and biological properties. *Metal Ions Biol. Syst.*, **19**, 207–227; (d) Painter, G.R. and Pressman, B.C. (1985) Cation complexes of the monovalent and polyvalent carboxylic ionophores: lasalocid (X-537A), monensin, A23187 (calcimycin), and related antibiotics. *Metal Ions Biol. Syst.*, **19**, 229–294; (e) Dobler, M. (1981) *Ionophores and Their Structures*, John Wiley & Sons, Inc., New York.

156. (a) Neilands, J.B. and Valenta, J.R. (1985) Iron-containing antibiotics. *Metal Ions Biol. Syst.*, **19**, 313–333; (b) Neilands, J.B. (1983) Siderophores. *Adv. Inorg. Biochem.*, **5**, 137–199; (c) Drechsel, H. and Jung, G. (1998) Peptide siderophores. *J. Pept. Sci.*, **4**, 147–181.

157. Leevy, W.N., Donato, G.M., Ferdani, R. *et al.* (2002) Synthetic hydraphile channels of appropriate length kill *Escherichia coli*. *J. Am. Chem. Soc.*, **124**, 9022–9023.

158. (a) Dobler, M. (1972) The crystal structure of nonactin. *Helv. Chim. Acta*, **55**, 1371–1384; (b) Dobler, M. and Phizackerley, R.P. (1974) The crystal structure of the NaNCS complex of

nonactin. *Helv. Chim. Acta*, **57**, 664–674; (c) Iitaka, Y. Sakamaki, T., and Nawata, Y. (1972) Molecular structures of tetranactin and its alkali metal ion complexes. *Chem. Lett.*, **1972**, 1225–1230; (d) Nawata, Y. Sakamaki, T., and Iitaka, Y. (1974) The crystal and molecular structures of tetranactin. *Acta Crystallogr. B*, **B30**, 1047–1053.

159. (a) Hamilton, J.A., Sabesan, M.N., and Steinrauf, L.N. (1981) Crystal structure of valinomycin potassium picrate: anion effects on valinomycin cation complexes. *J. Am. Chem. Soc.*, **103**, 5880–5885; (b) Steinrauf, L.K., Hamilton, J.A., and Sabesan, M.N. (1982) Crystal structure of valinomycin-sodium picrate. Anion effects on valinomycin-cation complexes. *J. Am. Chem. Soc.*, **104**, 4085–4091; (c) Smith, G.D., Duax, W.L. *et al.* (1975) Crystal and molecular structure of the triclinic and monoclinic forms of valinomycin, $C_{54}H_{90}N_6O_{18}$. *J. Am. Chem. Soc.*, **97**, 7242–7247; (d) Devarajan, S., Vijayan, M., and Easwaran, K.R.K. (1984) Conformation of valinomycin in its barium perchlorate complex from X-ray crystallography and NMR. *Int. J. Prot. Pept. Res.*, **23**, 324–333.

160. (a) Tishchenko, G.N. and Karimov, Z. (1978) Structure of Rb complex of LDLLDL analog of enniatin B in crystals. *Crystallography (Russia)*, **23**, 729–742; (b) Dobler, M. Dunitz, J.D., and Krajewski, J. (1969) Structure of the $K^+$ complex with enniatin B, a macrocyclic antibiotic with $K^+$ transport properties. *J. Mol. Biol.*, **42**, 603–606; (c) Tishchenko, G.N. and Zhukhlistova, N.E. (2000) Crystal and molecular structure of the membrane-active antibiotic enniatin. *C. Crystallogr. Rep.*, **45**, 619–625.

161. Cross, T.A. (1997) Solid-state nuclear magnetic resonance characterization of gramicidin channel structure. *Methods Enzymol.*, **289**, 672–696.

162. Koeppe, R.E. II and Kimura, M. (1984) Computer building of beta-helical polypeptide models. *Biopolymers*, **23**, 23–38.

163. (a) Myers, V.B. and Haydon, D.A. (1972) Ion transfer across lipid membranes in the presence of gramicidin A. II. The ion selectivity. *Biochim. Biophys. Acta*, **274**, 313–322; (b) Bamberg, E., Noda, K., Gross, E., and Läuger, P. (1976) Single-channel parameters of gramicidin A, B, and C. *Biochim. Biophys. Acta*, **449**, 223–228; (c) Sawyer, D.B., Williams, L.P., Whaley, W.L. *et al.* (1990) A Gramicidins A, B, and C form structurally equivalent ion channels. *Biophys. J.*, **58**, 1207–1212.

164. (a) Hladky, S.B. and Haydon, D.A. (1972) Ion transfer across lipid membranes in the presence of gramicidin A. I. Studies of the unit conductance channel. *Biochim. Biophys. Acta*, **247**, 294–312; (b) Urry, D.W., Trapane, T.L., Walker, J.T., and Prasad, K.U. (1982) On the relative lipid membrane permeability of $Na^+$ and $Ca^{2+}$, A physical basis for the messenger role of $Ca^{2+}$. *J. Biol. Chem.*, **257**, 6659–6661.

165. Burkhart, B.M., Li, N., Langs, D.A. *et al.* (1998) The conducting form of gramicidin A is a right-handed double-stranded double helix. *Proc. Natl Acad. Sci. USA*, **95**, 12950–12955.

166. (a) Wallace, B.A. and Ravikumar, K. (1988) The gramicidin pore: crystal structure of a cesium complex. *Science*, **241**, 182–187; (b) Wallace, B.A. Hendrickson, W.A., and Ravikumar, K. (1990) The use of single-wavelength anomalous scattering to solve the crystal structure of a gramicidin A/cesium chloride complex. *Acta Crystallogr. B.*, **46**, 440–446; (c) Wallace, B.A. and Janes, R.W. (1991) Co-crystals of gramicidin A and phospholipid. A system for studying the structure of a transmembrane channel. *J. Mol. Biol.*, **217**, 625–627; (d) Doyle, D.A. and Wallace, B.A. (1997) Crystal structure of the gramicidin/potassium thiocyanate complex. *J. Mol. Biol.*, **266**, 963–977; (e) Burkhart, B.M., Li, N., Langs, D.A. *et al.* (1998) The conducting form of gramicidin A is a right-handed double-stranded double helix. *Proc. Natl Acad. Sci. USA*, **95**, 12950–12955; (f) Wallace, B.A. (1999) X-ray crystallographic structures of gramicidin and their relation to the *Streptomyces lividans* potassium channel structure. *Novartis Found. Symp.*, **225**, 23–32 and 33–37; (g) Doi, M., Fujita, S., Katsuya, Y. *et al.* (2001) Antiparallel pleated beta-sheets observed in crystal structures of N,N-bis(trichloroacetyl) and N,N-bis(m-bromobenzoyl) gramicidin S. *Arch. Biochem. Biophys.*, **395**, 85–93.

167. (a) Townsley, L.E., Tucker, W.A., Sham, S., and Hinton, J.F. (2001) Structure of gramicidins A, B, and C incorporated into sodium dodecyl sulfate micelles. *Biochemistry*, **40**, 11674–11686; (b) Separovic, F., Barker, S., Delahunty, M., and Smith, R. (1999) NMR structure of C-terminally tagged gramicidin channels. *Biochim. Biophys. Acta*, **1416**, 48–56.

168. (a) Kim, S. Quine, J.R., and Cross, T.A. (2001) Complete cross-validation and R-factor calculation of a solid-state NMR derived structure. *J. Am. Chem. Soc.*, **123**, 7292–7298; (b) Fu, R. Cotton, M., and Cross, T.A. (2000) Inter- and intramolecular distance measurements by solid-state MAS NMR: determination of gramicidin A channel dimer structure in hydrated phospholipid bilayers. *J. Biomol. NMR*, **16**, 261–268; (c) Kovacs, F. Quine, J., and Cross, T.A. (1999) Validation of the single-stranded channel conformation of gramicidin A by solid-state NMR. *Proc. Natl Acad. Sci. USA*, **96**, 7910–7915; (d) Quine, J.R. Brenneman, M.T., and Cross, T.A. (1997) Protein structural analysis from solid-state NMR-derived orientational constraints. *Biophys. J.*, **72**, 2342–2348; (e) Ketchem, R.R. Hu, W., and Cross, T.A. (1993) High-resolution conformation of gramicidin A in a lipid bilayer by solid-state NMR. *Science*, **26**, 1457–1460.

169. (a) Koeppe, R.E. II, Killian, J.A., and Greathouse, D.V. (1994) Orientations of the tryptophan 9 and 11 side chains of the gramicidin channel based on deuterium nuclear magnetic resonance spectroscopy. *Biophys. J.*, **66**, 14–24; (b) Killian, J.A., Taylor, M.J., and Koeppe, R.E. I (1992) Orientation of the valine-1 side chain of the gramicidin transmembrane channel and implications for channel functioning. A $^2$H NMR study. *Biochemistry*, **31**, 11283–11290; (c) Separovic, F., Gehrmann, J., Milne, T. *et al.* (1994) Sodium ion binding in the gramicidin A channel. Solid-state NMR studies of the tryptophan residues. *Biophys. J.*, **67**, 1495–1500; (d) Separovic, F., Hayamizu, K., Smith, R., and Cornell, B.A. (1991) C-13 chemical shift tensor of L-tryptophan and its application to polypeptide structure determination. *Chem. Phys. Lett.*, **181**, 157–162; (e) Smith, R., Thomas, D.E., Separovic, F. *et al.* (1989) Determination of the structure of a membrane-incorporated ion channel. Solid-state nuclear magnetic resonance studies of gramicidin A. *Biophys. J.*, **56**, 307–314; (c) Cornell, B.A., Separovic, F., Baldassi, A.J., and Smith, R. (1988) Conformation and orientation of gramicidin A in oriented phospholipid bilayers measured by solid state carbon-13 NMR. *Biophys. J.*, **53**67–76.

170. Burkhart, B.M., Gassman, R.M., Langs, D.A. *et al.* (1999) Gramicidin D conformation, dynamics and membrane ion transport. *Biopolymers*, **51**, 129–144.

171. Sigel, S.A. and Sigel, H. (1998) *Iron Transport and Storage in Microorganisms, Plants, and Animals*, Dekker, New York.

172. Neilands, J.B. (1995) Siderophores: Structure and Function of Microbial Iron Transport Compounds. *J. Biol. Chem.*, **270**, 26723–26726.

173. (a) Ecker, D.J. and Emery, T. (1983) Iron uptake from ferrichrome A and iron citrate in *Ustilago sphaerogena*. *J. Bacteriol.*, **155**, 616–622; (b) Wayne, R. Frick, K., and Neilands, J.B. (1976) Siderophore protection against colicins M, B, V, and $I_a$ in *Escherichia coli*. *J. Bacteriol.*, **126**, 7–12.

174. Dhungana, S., White, P.S., and Crumbliss, A.L. (2001) Crystal structure of ferrioxamine B: a comparative analysis and implications for molecular recognition. *J. Biol. Inorg. Chem.*, **6**, 810–818.

175. Hossain, M.B., Jalal, M.A.F., and van der Helm, D. (1986) The structure of ferrioxamine D1 ethanol water (1/2/1). *Acta Crystallogr. C.*, **42**, 1305–1310.

176. (a) van der Helm, D. and Poling, M. (1976) Crystal structure of ferrioxamine E. *J. Am. Chem. Soc.*, **98**, 82–86; (b) Hossain, M.B., Jalal, M.A.F., van der Helm, D. *et al.* (1998) Crystal structure of retro-isomer of the siderophore ferrioxamine E. *J. Chem. Crystallogr.*, **28**, 53–56.

177. Hossain, M.B., van der Helm, D., and Poling, M. (1983) The structure of deferriferrioxamine E (nocardamin), a cyclic trihydroxamate. *Acta Crystallogr. Sect. B*, **39**, 258–263.

178. (a) Norrestam, R. Stensland, B., and Branden, C.-I. (1975) On the conformation of cyclic iron-containing hexapeptides: the crystal and molecular structure of ferrichrysin. *J. Mol. Biol.*, **99**,

501–506; (b) van der Helm, D., Baker, J.R., Eng-Wilmot, D.L. *et al.* (1980) Crystal structure of ferrichrome and a comparison with the structure of ferrichrome A. *J. Am. Chem. Soc.*, **102**, 4224–4231; (c) Zalkin, A. Forrester, J.D., and Templeton, D.H. (1966) Ferrichrome A tetrahydrate. Determination of crystal and molecular structure. *J. Am. Chem. Soc.*, **88**, 1810–1814.

179. (a) Locher, K.P., Rees, B., Koebnik, R. *et al.* (1998) Transmembrane signaling across the ligand-gated FhuA receptor: Crystal structure of free and ferrichrome-bound states reveal allosteric changes. *Cell*, **95**, 45–56; (b) Ferguson, A.D., Hofmann, E., Coulton, J.W. *et al.* (1998) Siderophore-mediated iron transport: Crystal structure of FhuA with bound lipopolysaccharide. *Science*, **282**, 2215–2220; (c) Ferguson, A.D., Braun, V., Fiedler, H.-P. *et al.* (2000) Crystal structure of the antibiotic albomycin in complex with the outer membrane transporter FhuA. *Prot. Sci.*, **9**, 956–963.

180. NIH (2008) "*Alzheimer's Disease Fact Sheet*", NIH (Institute of Aging), Publication No. 08-6423, November 2008.

181. (a) Masters, C.L., Simms, G., Weinman, N.A. *et al.* (1985) Amyloid plaque core protein in Alzheimer disease and Down syndrome. *Proc. Natl Acad. Sci. USA*, **82**, 4245–4249; (b) Yankner, B.A., Dawes, L.R., Fisher, S. *et al.* (1989) Neurotoxicity of a fragment of the amyloid precursor associated with Alzheimer's disease. *Science*, **245**, 417–420; (c) Salkoe, D.J. (1989) Aging, amyloid, and Alzheimer's disease. *N. Engl. J. Med.*, **320**, 1484–1487; (d) Hardy, J.A. and Higgins, G.A. (1982) Alzheimer's disease: The amyloid cascade hypothesis. *Science*, **256**, 184–185.

182. (a) Katzman, R. and Saitoh, T. (1991) Advances in Alzheimer's disease. *FESEB J.*, **4**, 278–286; (b) Pike, C.J., Walenzcewicz, A.J., Glabe, C.G., and Cotman, C.W. (2001) *In vivo* aging of β-amyloid protein causes peptide aggregation and neurotoxicity. *Brain Res.*, **563**, 311–314; (c) Harris, M.E., Hensley, K., Butterfield, D.A. *et al.* (1995) Direct evidence of oxidative injury by the Alzheimer's amyloid β peptide in cultured hippocampal neurons. *Exp. Neurol.*, **131**, 193–202.

183. (a) Lambert, M.P., Barlow, A.K., Chromy, B.A. *et al.* (1998) Diffusible, nonfibrillar ligands derived from A beta(1-42) are potent central nervous system neurotoxins. *Proc. Natl Acad. Sci. USA*, **95**, 6448–6453; (b) Walsh, D.M., Klyubin, I., Fadeeva, J.V. *et al.* (2002) Naturally secreted oligomers of amyloid beta protein potently inhibit hippocampal long-term potentiation in vivo. *Nature*, **416**, 535–539; (c) Kayed, R., Head, E., Thompson, J.L. *et al.* (2003) Common structure of soluble amyloid oligomers implies common mechanism of pathogenesis. *Science*, **300**, 486–489.

184. Glenner, G.G. and Wong, C.W. (1984) Alzheimer's disease and Down's syndrome: Sharing of a unique cerebrovascular amyloid fibril protein. *Biochem. Biophys. Res. Commun.*, **122**, 1131–1135.

185. (a) Robinson, S.R. and Bishop, G.M. (2003) Aβ as a bioflocculant: Implications for the amyloid hypothesis of Alzheimer's disease. *Neurobiol. Aging*, **23**, 1051–1072; (b) Savory, J. Ghribi, O., and Herman, M.M. (2002) Is amyloid β-peptide neurotoxic or neuroprotective and what is its role in the binding of metal ions? *Neurobiol. Aging*, **23**, 1089–1092; (c) Bishop, G. M. and Robinson, S.R. (2004) The amyloid paradox: Amyloid-beta-metal complexes can be neurotoxic and neuroprotective. *Brain Pathol.*, **14**, 448–452.

186. Ling, Y., Morgan, K., and Kalsheker, N. (2003) Amyloid precursor protein (APP) and the biology of proteolytic processing: relevance to Alzheimer's disease. *Int. J. Biochem. Cell. Biol.*, **35**, 1505–1535.

187. Bertini, I., Gray, H.B.Lippard, S.J., and Valentine, J.S. (eds) (1994) *Bioinorganic Chemistry*, University Science Books, Sausalito.

188. (a) Bush, A.I., Pettingell, W.H., Multhaup, G. *et al.* (1994) Rapid induction of Alzheimer A beta amyloid formation by zinc. *Science*, **265**, 1464–1467; (b) Lovell, M.A., Robertson, J.D., Teesdale, W.J. *et al.* (1998) Copper, iron, and zinc in Alzheimer's disease senile plaques. *J. Neur. Sci.*, **158**, 47–52; (c) Moir, R.D., Atwood, C.S., Huang, X. *et al.* (1999) Mounting

evidence for the involvement of zinc and copper in Alzheimer's disease. *Eur. J. Clin. Invest.*, **29**, 569–570; (d) Atwood, C.S., Huang, X., Moir, R.D. *et al.* (1999) Role of free radicals and metal ions in the pathogenesis of Alzheimer's disease. *Met. Ions Biol. Sys.*, **36**, 309–364; (e) Atwood, C.S., Scarpa, R.C., Huang, X. *et al.* (2000) Charaterization of copper interaction with Alzheimer amyloid β peptides: Identification of an attomolar-affinity copper binding site on amyloid β1-42. *J. Neurochem.*, **75**, 1219–1233; (f) Mirura, T., Suzuki, K., Kohata, N., and Takeuchi, H. (2000) Metal binding modes of Alzheimer's amyloid β-peptide in insoluble aggregates and soluble complexes. *Biochemistry*, **39**, 7024–7031; (g) Morgan, D.M., Dong, J., Jacob, J. *et al.* (2002) Metal switch for amyloid formation: Insight into the structure of the nucleus. *J. Am. Chem. Soc.*, **124**, 12644–12645; (h) Dong, J., Atwood, C.S., Anderson, V.E. *et al.* (2003) Metal binding and oxidation of amyloid-β within isolated senile plaque cores: Raman microscopic evidence. *Biochemistry*, **42**, 2768–2773; (i)Kowalik-Jankowska, T., Ruta, M., Wiśniewska, K., and Łankiewicz, L. (2003) Coordination abilities of the 1-16 and 1-28 fragments of β-amyloid peptide toward copper(II) ions: a combined potentiometric and spectroscopic study. *J. Inorg. Biochem.*, **95**270–282.

189. (a) Markesbery, W.R. (1997) Oxidative stress hypeothesis in Alzheimer's disease. *Free Rad. Biol. Med.*, **23**, 134–147; (b) Sayre, L.M., Zagorski, M.G., Surewicz, W.K. *et al.* (1997) Mechanisms of neurotoxicity associated with amyloid β deposition and the role of free radicals in the pathogenesis of Alzheimer's disease: A critical appraisal. *Chem. Res. Toxicol.*, **10**, 518–526; (c) Bondy, S.C., Guo-Ross, S.X., and Truong, A.T. (1998) Promotion of transition metal-induced reactive oxygen species formation by β-amyloid. *Brain Res*, **799**, 91–96; (d) Behl, C. (1999) Alzheimer's disease and oxidative stress: Implications for novel therapeutic approach. *Prog. Neurobiol.*, **323**, 301–323; (e) Huang, Y., Atwood, C.S., Hartshorn, M.A. *et al.* (1999) The Aβ peptide of Alzheimer's disease directly produces hydrogen peroxide through metal ion reduction. *Biochemistry*, **38**, 7609–7616; (f) Huang, Y., Cuajungco, M.P., Atwood, C.S. *et al.* (1999) Alzheimer's Aβ interaction with Cu(II) induces neurotoxicity, radicalization, metal reduction, and cell-free hydrogen peroxide formation. *J. Biol. Chem.*, **274**, 37111–37116; (g) Yatin, S.M., Varadarajan, S., Link, C.D., and Butterfield, D.A. (1999) *In vitro* and *in vivo* oxidative stress associated with Alzheimer's amyloid β-peptide (1-42). *Neurobiol. Aging*, **20**, 325–330; (h) Praticò, D. and Delanty, N. (2000) Oxidative injury in diseases of the central nervous system: focus on Alzheimer's disease. *Physiol. Med.*, **109**, 577–585; (i) Varadarajan, S., Yatin, S., Aksenova, M., and Butterfield, D.A. (2000) Review: Alzheimer's amyloid β-peptide-associated free radical oxidative stress and neurotoxicity. *J. Struct. Biol.*, **130**, 184–208; (j) Miranda, S., Opazo, C., Larrondo, L.F. *et al.* (2000) The role of oxidative stress in the toxicity indiced by amyloid β-peptide in Alzheimer's disease. *Progr. Neurobiol.*, **62**, 633–648; (k) Rottkamp, K.A., Raina, A.K., Zhu, X. *et al.* (2001) Redox-active iron mediates amyloid-β toxicity. *Free Radical Biol. Med.*, **30**, 447–450; (l) Butterfield, D.A., Drake, J., Pocemich, C., and Castegna, A. (2001) Evidence of oxidative damage in Alzheimer's disease brain: central role for amyloid β-peptide. *Trends Mol. Med.*, **7**, 548–554; (m) Nelson, T.J. and Alkon, D.L. (2005) Oxidation of Cholesterol by Amyloid Precursor Protein and β-Amyloid Peptide. *J. Biol. Chem*, **280**7377–7387; (n) References in above papers.

190. (a) Tabaton, M. and Piccini, A. (2005) Role of water-soluble amyloid-beta in the pathogenesis of Alzheimer's disease. *Int. J. Exp. Pathol.*, **86**, 139–145. (the most recent review); (b) Gentile, M.T., Vecchione, C., Maffei, A. *et al.* (2004) Mechanisms of soluble beta-amyloid impairment of endothelial function. *J. Biol. Chem.*, **279**, 48135–48142; (c) Smith, C.C.T. Stanyer, L., and Betteridge, D.J. (2004) Soluble beta-amyloid (A beta) 40 causes attenuation or potentiation of noradrenaline-induced vasoconstriction in rats depending upon the concentration employed. *Neurosci. Lett.*, **367**, 129–132; (d) Klyubin, I., Walsh, D.M., Cullen, W.K. *et al.* (2004) Soluble Arctic amyloid beta protein inhibits hippocampal long-term potentiation *in vivo. Eur. J. Neurosci.*, **19**, 2839–2846; (e) Zerbinatti, C.V., Wozniak, D.F., Cirrito, J. *et al.*

(2004) Increased soluble amyloid-beta peptide and memory deficits in amyloid model mice overexpressing the low-density lipoprotein receptor-related protein. *Proc. Natl Acad. Sci. USA*, **101**, 1075–1080; (f) Sengupta, P., Garai, K., Sahoo, B. *et al.* (2003) The amyloid β peptide (Aβ1-40) is thermodynamically soluble at physiological concentrations. *Biochemistry*, **42**, 10506–10513; (g) Senior, K. (2003) Soluble amyloid oligomers: a common cause of neurodegeneration? *Lancet Neurol.*, **2**, 330–330; (h) Kayed, R., Head, E., Thompson, J.L. *et al.* (2003) Common structure of soluble amyloid oligomers implies common mechanism of pathogenesis. *Science*, **300**, 486–489.

191. Tjernberg, L.O., Pramanik, A., Björling, S. *et al.* (1998) Amyloid β-peptide polymerization studied using fluorescence correlation spectroscopy. *Chem. Biol.*, **6**, 53–62.

192. (a) Cohen, A.S.Shirahama, T., and Skinner, M. (eds) (1982) Electron microscopy of amyloid, in *Electron Microscopy of Protein*, vol. 3 (ed. I. Harris), Academic, New York, pp. 165–205; (b) Merz, P.A., Wisniewski, H.M., Somerville, R.A. *et al.* (1983) Ultrastructural morphology of amyloid bibrils from neuritic and amyloid plaques. *Acta Neuropathol.*, **60**, 113–124; (c) Serpell, L.C. and Smith, J.M. (2000) Direct visulization of the β-sheet structure of synthetic Alzheimer's amyloid. *J. Mol. Biol.*, **299**, 225–231.

193. (a) Eanes, E.D. and Glenner, G.G. (1968) X-ray diffraction studies on amyloid filaments. *J. Histochem. Cytochem.*, **16**, 673–677; (b) Kirschner, D., Abraham, C., and Selkoe, D. (1987) X-ray diffraction from intraneuronal paired helical filaments and extra-neuronal amyloid fibres in Alzheimer's disease indicates cross β conformation. *Proc. Natl Acad. Sci. USA*, **84**, 6953–6957; (c) Sikorski, P., Atkins, E.D.T., and Serpell, L.C. (2003) Structure and texture of fibrous crystals formed by Alzheimer's Aβ(11-25) peptide fragment. *Structure*, **11**, 915–926.

194. (a) Lansbury, P.T., Costa, P.R., Griffiths, J.M. *et al.* (1995) Structural model for the beta-amyloid fibril based on interstrand alignment of an antiparallel-sheet comprising a C-terminal peptide. *Nat. Struct. Biol.*, **2**, 990–998; (b) Benzinger, T.L.S., Gregory, D.M., Burkoth, T.S. *et al.* (1998) Propagating structure of Alzheimer's beta-amyloid(10-35) is parallel beta-sheet with residues in exact register. *Proc. Natl Acad. Sci. USA*, **95**, 13407–13412; (c) Gregory, D.M., Benzinger, T.L.S., Burkoth, T.S. *et al.* (1998) Dipolar recoupling NMR of biomolecular self-assemblies: determining inter- and intrastand distances in fibrilized Alzheimer's β-amyloid peptide. *Solid State NMR.*, **13**, 149–166; (d) Benzinger, T.L.S., Gregory, D.M., Burkoth, T.S. *et al.* (2000) Two-dimensional structure of beta-amyloid(10-35) fibrils. *Biochemistry*, **39**, 3491–3499; (e) Lynn, D.G. and Meredith, S.C. (2000) Review: Model peptides and the physicochemical approach to β-amyloids. *J. Struct. Biol.*, **130**, 153–173; (f) Mikros, E., Benaki, D., Humpfer, E. *et al.* (eds) (2001) High-resolution NMR spectroscopy of the β-amyloid(1-28) fibril typical for Alzheimer's disease. *Angew. Chem. Int. Ed.*, **40**, 3603–3605.

195. (a) Mandal, P.K. and Pettegrew, J.W. (2004) Alzheimer's disease: Soluble oligomeric A beta (1-40) peptide in membrane mimic environment from solution NMR and circular dichroism studies. *Neurochem. Res.*, **29**, 2267–2272; (b) Shao, H.Y., Jao, S.C., Ma, K., and Zagorski, M.G. (1999) Solution structures of micelle-bound amyloid beta-(1-40) and beta-(1-42) peptides of Alzheimer's disease. *J. Mol. Biol.*, **285**, 755–773; (c) Marcinowski, K.J., Shao, H., Clancy, E.L., and Zagorski, M.G. (1998) Solution structure model of residues 1-28 of the amyloid beta peptide when bound to micelles. *J. Am. Chem. Soc.*, **120**, 11082–11091; (d) Coles, M., Bicknell, W., Watson, A.A. *et al.* (1998) Solution structure of amyloid beta-peptide(1-40) in a water-micelle environment. Is the membrane-spanning domain where we think it is? *Biochemistry*, **37**, 11064–11077; (e) Fletcher, T.G. and Keire, D.A. (1997) The interaction of beta-amyloid protein fragment (12-28) with lipid environments. *Prot. Sci.*, **6**, 666–675; (f) Talafous, J., Marcinowski, K.J., Klopman, G., and Zagorski, M.G. (1994) Solution structure of residues 1-28 of the amyloid beta-peptide. *Biochemistry*, **28**, 7788–7796.

196. Kirkitadze, M.D., Condron, M.M., and Teplow, D.B. (2001) Identification and characterization of kinetic intermediates in amyloid β-protein fibrillogenesis. *J. Mol. Biol.*, **312**, 1103–1119.

197. Guo, Q., Manolopoulou, M., Bian, Y. *et al.* (2010) Molecular basis for the recognition and cleavages of IGF-II, TGF-alpha, and amylin by human insulin-degrading enzyme. *J. Mol. Biol.*, **395**, 430–443.

198. Zirah, S., Kozin, S.A., Mazur, A.K. *et al.* (2005) Structural changes of region 1-16 of the Alzheimer disease amyloid beta-peptide upon zinc binding and in vitro aging. *J. Biol. Chem.*, **281**, 2151–2161.

199. da Silva, G.F.Z., Tay, W.M., and Ming, L.-J. (2005) Catechol oxidase-like oxidation chemistry of the 1-20 and 1-16 fragments of Alzheimer's disease-related β-Amyloid peptide: their structure-activity correlation and the fate of hydrogen peroxide. *J. Biol. Chem.*, **280**, 16601–16609.

200. (a) Some recent publications: Mirzabekov, T., Lin, M.C., Yuan, W.L. *et al.* (2004) Channel formation in planar lipid bilayers by a neurotoxic fragment of the β-amyloid peptide. *Biochim. Biophys. Res. Commun.*, **202**, 1142–1148; (b) Terzi, E. Hölzemann, G., and Seeling, J. (1995) Self-association of β-amyloid peptide (1-40) in solution and binding to lipid membrane. *J. Mol. Biol.*, **252**, 633–642; (c) Müller, W.E., Koch, S., Eckert, A. *et al.* (1995) β-amyloid peptide decrease membrane fluidity. *Brain Res.*, **674**, 133–136; (d) Allen, D.D., Galdzicki, Z., Brining, S.K. *et al.* (1997) Beta-amyloid induced increase in choline flux across PC12 cell membranes. *Neurosci. Lett.*, **234**, 71–73; (e) Inoue, S. Kuroiwa, M., and Kisilevsky, R. (1999) Basement membranes, microfibrils and β amyloid fibrillogenesis in Alzheimer' disease: high resolution ultrastructural findings. *Brain Res. Rev.*, **29**, 218–231; (f) Rodrigues, C.M.P., Solá, S., Brito, M. A. *et al.* (2001) Amyloid β-peptide disrupts mitochondrial membrane lipid and protein structure: Protective role of tauroursodeoxycholate. *Biochem. Biophys. Res. Commun.*, **281**, 468–474; (g) Butterfield, D.A. and Lauderback, C.M. (2002) Lipid peroxidation and protein oxidation in Alzheimer's disease brain: Potential causes and consequences involving amyloid β-peptide-associated free radical oxidative stress. *Free Radical Biol. Med.*, **32**, 1050–1060; (h) Kiuchi, Y., Isobe, Y., Fukushima, K., and Kimura, M. (2002) Disassembly of amyloid β-protein fibril by basement membrane components. *Life Sci.*, **70**, 2421–2431; (i) Mremer, J.J. and Murphy, R.M. (2003) Kinetics of adsorption of β-amyloid peptide Aβ(1-40) to lipid bilayers. *J. Biochem. Biophys. Met.*, **57**, 159–169; (j) Mingeot-Leclercq, M.P., Lins, L., Bensliman, M. *et al.* (2003) Piracetam inhibits the lipid-destabilising effect of the amyloid peptide Aβ C-terminal fragment. *Biochim. Biophys. Acta*, **1609**, 28–38; (k) Wood, W.G., Eckert, G.P., Igbavboa, U., and Müller, W.E. (2003) Amyloid beta-protein interactions with membranes and cholesterol: causes or casualties of Alzheimer's disease. *Biochim. Biophys. Acta*, **1610**, 281–290.

201. (a) Prusiner, S.B. (1998) Prions. *Proc. Natl Acad. Sci. USA*, **95**, 13363–13383; (b) Brown, D. R., Qin, K.F., Herms, J.W. *et al.* (1997) The cellular prion protein binds copper *in vivo*. *Nature*, **390**, 684–687.

202. Wadsworth, J.D., Hill, A.F., Joiner, S. *et al.* (1999) Strain-specific prion-protein conformation determined by metal ions. *Nat. Cell Biol.*, **1**, 55–59.

203. (a) Riek, R., Hornemann, S., Wider, G. *et al.* (1996) NMR structure of the mouse prion protein domain PrP(121-321). *Nature*, **382**, 180–182; (b) Zahn, R., Liu, A., Lührs, T. *et al.* (2000) NMR solution structure of the human prion protein. *Proc. Natl Acad. Sci. USA*, **97**, 145–150.

204. Haire, L.F., Whyte, S.M., Vasisht, N. *et al.* (2004) The crystal structure of the globular domain of sheep prion protein. *J. Mol. Biol.*, **336**, 1175–1183.

205. Donne, D.G., Viles, J.H., Groth, D. *et al.* (1997) Structure of the recombinant full-length hamster prion protein PrP(29-231): the N terminus is highly flexible. *Proc. Natl Acad. Sci. USA*, **94**, 13452–13457.

206. Zahn, R. (2003) The octapeptide repeats in mammalian prion protein constitute a pH-dependent folding and aggregation site. *J. Mol. Biol.*, **334**, 477–88.

207. (a) Dyson, H.J. and Wright, P.E. (2005) Intrinsically unstructured protein and their functions. *Nat. Rev. Mol. Cell Biol.*, **6**, 197–208; (b) Schlessinger, A., Schaefer, C., Vicedo, E. *et al.*

(2011) Protein disorder – a breakthrough invention of evolution? *Curr. Opin. Struct. Biol.*, **21**, 412–418.

208. (a) Pauly, P.C. and Harris, D.A. (1998) Copper Stimulates Endocytosis of the Prion Protein. *J. Biol. Chem.*, **273**, 33107–33110; (b) Sumudhu, W., Perera, W.S., and Hooper, N.M. (2001) Ablation of the metal ion-induced endocytosis of the prion protein by disease-associated mutation of the octarepeat region. *Curr. Biol.*, **11**, 519–523; (c) Brown, D.R. (2000) PrPSc-like prion protein peptide inhibits the function of cellular prion protein. *Biochem. J.*, **352**, 511–518.

209. Nadal, R.C., Abdelraheim, S.R., Brazier, M.W. *et al.* (2007) Prion protein does not redox-silence Cu(2+), but is a sacrificial quencher of hydroxyl radicals. *Free Radical Biol. Med.*, **42**, 79–89.

210. (a) Requena, J.R., Groth, D., Legname, G. *et al.* (2001) Copper-catalyzed oxidation of the recombinant SHa(29-231) prion protein. *Proc. Natl Acad. Sci. USA*, **98**, 7170–7175; (b) Canello, T., Engelstein, R., Moshel, O. *et al.* (2008) Methionine sulfoxides on PrPSc: a prion-specific covalent signature. *Biochemistry*, **47**, 8866–8873; (c) Canello, T., Frid, K., Gabizon, R. *et al.* (2010) Oxidation of Helix-3 Methionines Precedes the Formation of PK Resistant PrP (Sc). *PLOS Pathogens*, **6**, 1553–7366.

211. Brown, D.R., Nicholas, R.S., and Canevari, L. (2002) Lack of prion protein expression results in a neuronal phenotype sensitive to stress. *J. Neurosci. Res.*, **67**, 211–224.

212. (a) Younan, N.D., Klewpatinond, M., Davies, P. *et al.* (2011) Copper(II)-Induced Secondary Structure Changes and Reduced Folding Stability of the Prion Protein. *J. Mol. Biol.*, **410**, 369–382; (b) Viles, J.H. Klewpatinond, M., and Nadal, R.C. (2008) Copper and the structural biology of the prion protein. *Biochem. Soc. Trans.*, **36**, 1288–1292; (c) Davies, P. and Brown, D.R. (2008) The chemistry of copper binding to PrP: is there sufficient evidence to elucidate a role for copper in protein function? *Biochem. J.*, **410**, 237–244; (d) Gaggelli, E., Kozlowski, H., Valensin, D., and Valensin, G. (2006) Copper homeostasis and neurodegenerative disorders (Alzheimer's, prion, and Parkinson's diseases and amyotrophic lateral sclerosis. *Chem. Rev.*, **106**, 1995–2044.

213. (a) Qin, K.F., Yang, D.S., Yang, Y. *et al.* (2000) Copper(II)-induced conformational changes and protease resistance in recombinant and cellular PrP. Effect of protein age and deamidation. *J. Biol. Chem.*, **275**, 19121–19131; (b) Kuczius, T., Buschmann, A., Zhang, W. *et al.* (2004) Cellular prion protein acquires resistance to proteolytic degradation following copper ion binding. *Biol. Chem.*, **385**, 739–747.

214. (a) Burns, C.S., Aronoff-Spencer, E., Dunham, C.M. *et al.* (2002) Molecular features of the copper binding sites in the octarepeat domain of the prion protein. *Biochemistry*, **41** (12), 3991–4001; (b) Klewpatinond, M., Davies, P., Bowen, S. *et al.* (2008) Deconvoluting the $Cu^{2+}$ binding modes of full-length prion protein. *J. Biol. Chem.*, **283**, 1870–1881.

215. (a) Hornshaw, M.P., McDermott, J.R., and Candy, J.M. (1995) Copper Binding to the N-Terminal Tandem Repeat Regions of Mammalian and Avian Prion Protein. *Biochem. Biophys. Res. Commun.*, **207**, 621–629; (b) Hornshaw, M.P., Mcdermott, J.R., Candy, J.M., and Lakey, J.H. (1995) Copper binding to the N-terminal tandem repeat region of mammalian and avian prion protein: structural studies using synthetic peptides. *Biochem. Biophys. Res. Commun.*, **214**, 993–999.

216. Walter, E.D., Stevens, D.J., Visconte, M.P., and Millhauser, G.L. (2007) The prion protein is a combined zinc and copper binding protein: $Zn^{2+}$ alters the distribution of $Cu^{2+}$ coordination modes. *J. Am. Chem. Soc.*, **129**, 15440–15441.

217. Wasmer, C., Lange, A., Van Melckebeke, H. *et al.* (2008) Amyloid fibrils of the HET-s(218-289) prion form a beta solenoid with a triangular hydrophobic core. *Science*, **319**, 1523–1526.

218. Tuite, M.F., and Serio, T.R. (2010) The prion hypothesis: from biological anomaly to basic regulatory mechanism. *Nat. Rev. Mol. Cell Biol.*, **11**, 823–833.

219. (a) Yang, L.H., Ahmed, S.A., Rhee, S., and Miles, E.W. (1997) Importance of conserved and variable C-terminal residues for the activity and thermal stability of the beta subunit of

tryptophan synthase. *J. Biol. Chem.*, **272**, 7859–7866; (b) Fuciños, P., Atanes, E., López-López, O. *et al.* (2011) Production and characterization of two N-terminal truncated esterases from *Thermus thermophilus* HB27 in a mesophilic yeast: Effect of N-terminus in thermal activity and stability. *Prot. Expr. Purif.*, **78**, 120–130.

220. Livera, C., Pettit, L.D., Bataille, M. *et al.* (1988) Copper(II) complexes with some tetrapeptides containing the 'break-point' prolyl residue in the third position. *J. Chem. Soc. Dalton Trans.*, **1988**, 1357–1360.

221. Brown, D.R. (2007) Interactions between metals and α-synuclein: Function or artifact? *FEBS J.*, **274**, 3766–3774.

222. (a) Sung, Y.H. Rospigliosi, C., and Eliezer, D. (2006) NMR mapping of copper binding sites in alpha-synuclein. *Biochim. Biophys. Acta*, **1764**, 5–12; (b) Lee, J.C., Gray, H.B., and Winkler, J.R. (2008) Copper(II) binding to alpha-synuclein, the Parkinson's protein. *J. Am. Chem. Soc.*, **130**, 6898–6899; (c) Davies, P., Wang, X., Sarell, C.J. *et al.* (2011) The Synucleins Are a Family of Redox-Active Copper Binding Proteins. *Biochemistry*, **50** (1), 37–47.

223. (a) Dudzik, C.G., Walter, E.D., and Millhauser, G.L. (2011) Coordination features and affinity of the $Cu^{2+}$ site in the α-synuclein protein of Parkinson's disease. *Biochemistry*, **50**, 1771–1777; (b) Drew, S.C., Leong, S.L., Pham, C.L. *et al.* (2008) $Cu^{2+}$ binding modes of recombinant alpha-synuclein—insights from EPR spectroscopy. *J. Am. Chem. Soc.*, **130**, 7766–7773; (c) Kowalik-Jankowska, T., Rajewska, A., Wisniewska, K. *et al.* (2005) Coordination abilities of N-terminal fragments of α-synuclein towards copper(II) ions: a combined potentiometric and spectroscopic study. *J. Inorg. Biochem.*, **99**, 2282–2291.

224. (a) Wuerges, J., Lee, J.W., Yim, Y.I. *et al.* (2004) Crystal structure of nickel-containing superoxide dismutase reveals another type of active site. *Proc. Natl Acad. Sci. USA*, **101** (23), 8569–8574; (b) Barondeau, D.P., Kassmann, C.J., Bruns, C.K. *et al.* (2004) Nickel superoxide dismutase structure and mechanism. *Biochemistry*, **43**, 8038–8047.

225. (a) Gale, E.M., Cowart, D.M., Scott, R.A., and Harrop, T.C. (2011) Dipeptide-based models of nickel superoxide dismutase: solvent effects highlight a critical role to Ni–S bonding and active site stabilization. *Inorg. Chem.*, **50**, 10460–10471; (b) Gale, E.M., Narendrapurapu, B. S., Simmonett, A.C. *et al.* (2010) Exploring the effects of H-bonding in synthetic analogues of nickel superoxide dismutase (Ni-SOD): Experimental and theoretical implications for protection of the Ni–SCys bond. *Inorg. Chem.*, **49**, 7080–7096; (c) Gale, E.M., Patra, A.K., and Harrop, T.C. (2009) Versatile methodology toward $NiN_2S_2$ complexes as nickel superoxide dismutase models: structure and proton affinity. *Inorg. Chem.*, **48**, 5620–5622.

226. (a) Krause, M.E., Glass, A.M., Jackson, T.A., and Laurence, J.S. (2010) Novel tripeptide model of nickel superoxide dismutase. *Inorg. Chem.*, **49**, 362–364; (b) Krause, M.E., Glass, A.M., Jackson, T.A., and Laurence, J.S. (2011) MAPping the chiral inversion and structural transformation of a metal-tripeptide complex having Ni-superoxide dismutase activity. *Inorg. Chem.*, **50**, 2479–2487.

227. Padmanabhan, S., Marqusee, S., Ridgeway, T. *et al.* (1990) Relative helix-forming tendencies of nonpolar amino acids. *Nature*, **344**, 268–270.

228. Lovejoy, B., Choe, S., Cascio, D. *et al.* (1993) Crystal structure of a synthetic triple-stranded alpha-helical bundle. *Science*, **259**, 1288–1293.

229. (a) Reza Ghadiri, M. and Choi, C. (1990) Secondary structure nucleation in peptides. Transition metal ion stabilized α-helices. *J. Am. Chem. Soc.*, **112**, 1630–1632; (b) Reza Ghadiri, M. and Fernholz, A.K. (1990) Peptide architecture. Design of stable α-helical metallopeptides via a novel exchange-inert ruthenium(III) complex. *J. Am. Chem. Soc.*, **112**, 9633–9635.

230. Ruan, F., Chen, Y., and Hopkins, P.B. (1990) Metal ion-enhanced helicity in synthetic peptides containing unnatural, metal-ligating residues. *J. Am. Chem. Soc.*, **112**, 9403–9404.

231. Kelso, M.J., Beyer, R.L., Hoang, H.N. *et al.* (2004) α-Turn mimetics: short peptide α-helices composed of cyclic metallopentapeptide modules. *J. Am. Chem. Soc.*, **126**, 4828–4842.

232. Kelso, M.J., Hoang, H.N., Appleton, T.G., and Fairlie, D.P. (2000) The first solution stucture of a single α-Helical turn. A pentapeptide α-Helix stabilized by a metal clip. *J. Am. Chem. Soc.*, **122**, 10488–10489.

233. Impellizzeri, G., Pappalardo, G., Purrello, R. *et al.* (1998) Synthesis, spectroscopic characterisation, and metal ion interaction of a new α-Helical peptide. *Chem. Eur. J.*, **4**, 1791–1798.

234. Hill, R.B., Raleigh, D.P., Lombardi, A., and DeGrado, W.F. (2000) *De Novo* design of helical bundles as models for understanding protein folding and function. *Acc. Chem. Res.*, **33**, 745–754.

235. Handel, T.M., Williams, S.A., and DeGrado, W.F. (1993) Metal ion-dependent modulation of the dynamics of a designed protein. *Science*, **261**, 879–885.

236. (a) Lombardi, A., Summa, C.M., Geremia, S. *et al.* (2000) Retrostructural analysis of metalloproteins: Application to the design of a minimal model for diiron proteins. *Proc. Natl Acad. Sci. USA*, **97**, 6298–6305; (b) Faiella, M., Andreozzi, C., de Rosales, R.T.M. *et al.* (2009) An artificial di-iron oxo-protein with phenol oxidase activity. *Nat. Chem. Biol.*, **5**, 882–884.

237. (a) Di Costanzo, L., Wade, H., Geremia, S. *et al.* (2001) Toward the de novo design of a catalytically active helix bundle: a substrate-accessible carboxylate-bridged dinuclear metal center. *J. Am. Chem. Soc.*, **123**, 12749–12757; (b) DeGrado, W.F., Di Costanzo, L., Geremia, S. *et al.* (2003) Sliding helix and change of coordination geometry in a model Di-Mn$^{II}$ protein. *Angew. Chem. Int. Ed.*, **42**, 417–420.

238. (a) Korendovych, I.V., Senes, A., Kim, Y.H. *et al.* (2010) De Novo design and molecular assembly of a transmembrane diporphyrin-binding protein complex. *J. Am. Chem. Soc.*, **132**, 15516–15518; (b) Nanda, V., Rosenblatt, M.M., Osyczka, A. *et al.* (2005) De Novo design of a redox-active minimal rubredoxin mimic. *J. Am. Chem. Soc.*, **127**, 5804–5805.

239. (a) Cochran, F.V., Wu, S.P., Wang, W. *et al.* (2005) Computational De Novo design and characterization of a four-helix bundle protein that selectively binds a nonbiological cofactor. *J. Am. Chem. Soc.*, **127**, 1346–1347; (b) Korendovych, I.V., Senes, A., Ho Kim, Y. *et al.* (2010) De Novo design and molecular assembly of a transmembrane diporphyrin-binding protein complex. *J. Am. Chem. Soc.*, **132**, 15516–15518.

240. Nastri, F., Lista, L., Ringhieri, P. *et al.* (2011) A heme-peptide metalloenzyme mimetic with natural peroxidase-like activity. *Chem. Eur. J.*, **17**, 4444–4453.

241. (a) Utschig, L.M. Bryson, J.W., and O'Halloran, T.V. (1995) Hg-199 NMR of the metal receptor-site in mer R and its protein-DNA complex. *Science*, **268**, 380–385; (b) Utschig, L.M. Wright, J.G., and O'Halloran, T.V. (1993) Biochemical and spectroscopic probes of mercury (II) coordination environments in proteins. *Met. Enzymol.*, **226**, 71–97.

242. Chakraborty, S., Touw, D.S., Peacock, A.F. *et al.* (2010) Structural comparisons of apo- and metalated three-stranded coiled coils clarify metal binding determinants in thiolate containing designed peptides. *J. Am. Chem. Soc.*, **132**, 13240–13250.

243. Dieckmann, G.R., McRorie, D.K., Lear, J.D. *et al.* (1998) The role of protonation and metal chelation preferences in defining the properties of mercury-binding coiled coils. *J. Mol. Biol.*, **280**, 897–912.

244. Matzapetakis, M., Farrer, B.T., Weng, T.-C. *et al.* (2002) Comparison of the binding of Cadmium(II), Mercury(II), and Arsenic(III) to the de Novo designed peptides TRI L12C and TRI L16C. *J. Am. Chem. Soc.*, **124**, 8042–8054.

245. Dieckmann, G., McRorie, D., Tierney, D. *et al.* (1997) De Novo design of mercury-binding two- and three-helical bundles. *J. Am. Chem. Soc.*, **119**, 6195–6196.

246. Touw, D.S., Nordman, C.E., Stuckey, J.A., and Pecoraro, V.L. (2007) Identifying important structural characteristics of arsenic resistance proteins by using designed three-stranded coiled coils. *Proc. Natl Acad. Sci.USA*, **104**, 11969–11974.

247. (a) Zhou, T., Radaev, S., Rosen, B.P., and Gatti, D.L. (2000) Structure of the ArsA ATPase: the catalytic subunit of a heavy metal resistance pump. *EMBO J*, **19**, 4838–4845; (b) Shi, W.,

Dong, J., Scott, R.A. *et al.* (1996) The role of arsenic-thiol interactions in metalloregulation of the ars operon. *J. Biol. Chem.*, **271**, 9291–9297.

248. Matzapetakis, M. and Pecoraro, V.L. (2005) Site-selective metal binding by designed α-Helical peptides. *J. Am. Chem. Soc.*, **127**, 18229–18233.

249. Zastrow, M.L., Peacock, A.F.A., Stuckey, J.A., and Pecoraro, V.L. (2012) Hydrolytic catalysis and structural stabilization in a designed metalloprotein. *Nat. Chem.*, **4**, 118–123.

250. Regan, L. and Clarke, N.D. (1990) A tetrahedral zinc(II)-binding site introduced into a designed protein. *Biochemistry*, **29**, 10878–10883.

251. Cerasoli, E., Sharpe, B.K., and Woolfson, D.N. (2005) ZiCo: A peptide designed to switch folded state upon binding Zinc. *J. Am. Chem. Soc.*, **127**, 15008–15009.

252. Doerr, A.J. and McLendon, G.L. (2004) Design, Folding, and activities of metal-assembled coiled Coil proteins. *Inorg. Chem.*, **43**, 7916–7925.

253. Reza Ghadiri, M., Soares, C., and Choi, C. (1992) A convergent approach to protein design. Metal ion-assisted spontaneous self-assembly of a polypeptide into a triple-helix bundle protein. *J. Am. Chem. Soc.*, **114**, 825–831.

254. Reza Ghadiri, M. and Martin, A. (1993) Case de novo design of a novel heterodinuclear three-helix bundle metalloprotein. *Angew. Chem. Int. Ed. Eng.*, **32**, 1594–1597.

255. Reza Ghadiri, M., Soares, C., and Choi, C. (1992) Design of an artificial four-helix bundle metalloprotein via a novel ruthenium(II)-assisted self-assembly process *J. Am. Chem. Soc.*, **114**, 4000–4002.

256. Schneider, J.P. and Kelly, J.W. (1995) Synthesis and efficacy of square planar copper complexes designed to nucleate β-sheet structure. *J. Am. Chem. Soc.*, **117**, 2533–2546.

257. Tsurkan, M.V. and Ogawa, M.Y. (2007) Formation of peptide nanospheres and nanofibrils by metal coordination. *Biomacromolecules*, **8**, 3908–3913.

258. Pires, M.M. and Chmielewski, J. (2009) Self-assembly of collagen peptides into microflorettes via metal coordination. *J. Am. Chem. Soc.*, **131**, 2706–2712.

259. (a) Dai, Y. Pochapsky, T.C., and Abeles, R.H. (2001) Mechanistic studies of two dioxygenases in the methionine salvange pathway of *klebsiella pneumonia*. *Biochemistry*, **40**, 6379–6387; (b) Dai, Y. Wensink, P.C., and Abeles, R.H. (1999) One protein two enzymes. *J. Biol. Chem.*, **274**, 1193–1195.

260. Zawadzke, L.E. and Berg, J.M. (1992) A Racemic Protein. *J. Am. Chem. Soc.*, **114**, 4002–4003.

261. Morleo, A., Bonomi, F., Iametti, S. *et al.* (2010) Iron-nucleated folding of a metalloprotein in high urea: resolution of metal binding and protein folding events. *Biochemistry*, **49**, 6627–6634.

262. (a) Wu, Y.B., Pradhan, P., Havener, J. *et al.* (2004) NMR solution structure of an oxaliplatin 1,2-d (GG) intrastrand cross-link in a DNA dodecamer duplex. *J. Mol. Biol.*, **341**, 1251–1269; (b) Gelasco, A. and Lippard, S.J. (1998) NMR solution structure of a DNA dodecamer duplex containing a *cis*-diammineplatinum(II) d(GpG) intrastrand cross-link, the major adduct of the anticancer drug cisplatin. *Biochemistry*, **37**, 9230–9239; (c) Parkinson, J.A., Chen, Y., del Socorro Murdoch, P. *et al.* (2000) Sequence-dependent bending of DNA induced by cisplatin: NMR structures of an A. T-rich 14-mer duplex. *Chem. Eur. J.*, **6**, 3636–3644; (d) Yang, D., van Boom, S.S., Reedijk, J. *et al.* (1995) Structure and isomerization of an intrastrand cisplatin-cross-linked octamer DNA duplex by NMR analysis. *Biochemistry*, **34**, 12912–12920; (e) Spingle, B. Whittington, D.A., and Lippard, S.J. (2001) 2.4 Å crystal structure of an oxaliplatin 1,2-d(GpG) intrastrand cross-link in a DNA dodecamer duplex. *Inorg. Chem.*, **40**, 5596–5602; (f) Coste, F., Malinge, J.M., Serre, L. *et al.* (1999) Crystal structure of a double-stranded DNA containing a cisplatin interstrand cross-link at 1.63 å resolution: hydration at the platinated site. *Nucleic Acids Res.*, **27**, 1837–1846; (g) Takahara, P.M., Rosenzweig, A.C., Frederick, C.A., and Lippard, S.J. (1995) Crystal structure of double-stranded DNA containing the major adduct of the anticancer drug cisplatin. *Nature*, **377**, 649–652.

263. Johannsen, S., Megger, N., Böhme, D. *et al.* (2010) Solution structure of a DNA double helix with consecutive metal-mediated base pairs. *Nat. Chem.*, **2**, 229–234.

# 2

# Introduction to Unnatural Foldamers

*Claudia Tomasini and Nicola Castellucci*
*Dipartimento di Chimica "G. Ciamician", Università di Bologna, Italy*

## 2.1 General Definition of Foldamers

What is a foldamer?

The word "foldamers" was coined by Samuel H. Gellman in 1996 [1] to describe discrete artificial oligomers that adopt specific and stable conformations similar to those seen among proteins and nucleic acids. This neologism means "folding molecules" and refers mainly to medium size molecules (about 500–5000 amu) that fold into definite secondary structures (i.e., helices, turns and sheets), thus being able to mimic biomacromolecules despite their smaller size. The essential requirement of a foldamer is to possess a well defined, repetitive secondary structure, imparted by conformational restrictions of the monomeric unit. Before the term "foldamer" was coined, many nucleic acid analogues and peptide analogues had already been successfully designed to mimic the structures and, potentially, the biological properties of their natural counterparts. Typical examples are peptide nucleic acids (PNAs) [2] and *N*-substituted oligoglycines (peptoids) [3].

The path to creating useful foldamers involves several daunting steps.

1. One must identify new polymeric backbones with suitable folding propensities. This goal includes developing a predicatively useful understanding of the relationship between the repetitive features of monomer structure and conformational properties at the polymer level.
2. One must endow the resulting foldamers with interesting chemical functions, by design, by randomization and screening ("evolution"), or by some combination of these two approaches.

*Metallofoldamers: Supramolecular Architectures from Helicates to Biomimetics*, First Edition.
Edited by Galia Maayan and Markus Albrecht.
© 2013 John Wiley & Sons, Ltd. Published 2013 by John Wiley & Sons, Ltd.

3. For technological utility, one must be able to produce a foldamer efficiently, which will generally include preparation of the constituent monomers in a stereochemically pure form and optimization of heteropolymer synthesis.

This is Gellman's definition. Later, Moore proposed the following narrower definition: "any oligomer that folds into a conformationally ordered state in solution, the structures of which are stabilized by a collection of noncovalent interactions between nonadjacent monomer units" [4]. This definition covers both "single-stranded foldamers that only fold and multiple-stranded foldamers that both associate and fold."

The investigation on these new structural scaffolds has blossomed in many laboratories as they hold promise for addressing chemical, physico-chemical and biological problems and represent a new frontier in research. Many groups have explored oligomers with a wide backbone variety as potential foldamers. Figure 2.1 shows a graph of the research articles published in the period 1996–2010, indicating the tremendous increase of interest in the research topic.

As the final goal is the interaction of foldamers with living organisms, the research is directed towards a mimic with α-peptidic secondary structures and their biological functions without having to cope with the hydrolytic and metabolic instability typical of α-peptides. Of course the foldamers that have been developed contain various skeletons and differ both in structures and in applications. Synthetic foldamers are far too numerous to be presented or even simply mentioned here, thus in this introduction we will describe only some recent and typical examples. There is a very interesting book on foldamers which Ivan Huc and Stefan Hecht [5] published in 2007 for Wiley; and several excellent reviews on various aspects of foldamer chemistry recently appeared in several high-impact journals [6]. In contrast, metallofoldamers have never been extensively and systematically described, thus this book fills a gap on this topic.

In a recent review, Guichard and Huc divided foldamers into two large families: "biotic" foldamers and "abiotic" foldamers [7]. "Biotic" foldamers are molecules whose

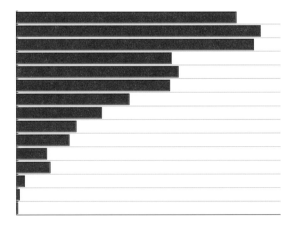

**Figure 2.1** *Number of research articles concerning foldamers published in the period 1996 (bottom) to 2010 (top)* (source: Web of Science).

synthesis has been guided by analogy to biopolymers with which they share comparable folding principles. After the seminal work of Gellman [8] and Seebach [9] on the synthesis and conformational analysis of β-peptides, which demonstrated that β-peptides adopt helical conformations more stable than those of α-peptides, this research field blossomed to include investigations of the higher γ and δ homologues [10]. It offers myriad possibilities for the construction of sophisticated folded architectures with possible applications ranging from biomedicine to material science [11]. Another possibility is the replacement of the amide bonds by, for example, imides, ureas, hydrazide or hydroxyamide functions [12].

In the other family, "abiotic" foldamers have been projected and synthesized by several research groups. They are mainly aromatic-rich sequences: (a) oligo-phenylene-ethynylenes [13], (b) sequences of alternating aromatic electron donors and acceptors [14], (c) aryl-oligomers, in particular those based on aza-heterocycles (pyridines, pyrimidines, pyridazines, etc.) [15], (d) aromatic tertiary amide, imide or urea oligomers and (e) aromatic oligoamides [16].

Another interesting topic is the study of the driving force for the folding of these very heterogeneous backbones. Some of these compounds tend to be self-organized, while others assume secondary structures only when they are in contact with external agents such as solvents, anions or metals. Thus, besides the description of some foldamers that are able to self-organize, we will show some examples of organization driven by external agents.

Finally, we will report some examples of foldamer applications for the formation of pharmacologically active compounds and discrete tertiary structures.

## 2.2 Biotic Foldamers

### 2.2.1 Homogeneous Foldamers

The first question that researchers asked themselves, when they began to think about molecules able to mimic peptides, was [17]: "what happens if one or two $CH_2$ groups are introduced into each amino acid building block in the chain of a peptide or protein, thereby providing homologues of the proteinogenic α-amino acids?" To answer this question, the most obvious kind of molecules to utilize and analyze were β-amino acids. Later, γ-amino acids and other functions, such as ureas, amino-oxy groups, imides and so on (Figure 2.2), were introduced into the peptide skeleton. Here we report some examples of foldamers containing unusual amino acids or unusual functions that replace the amide moiety.

### 2.2.2 β-Peptides

β-Peptides consisting of homologated proteinogenic amino acids were first prepared and investigated in the mid1990s [8,9]. Within 15 years, they evolved as a totally new class of unnatural peptidic oligomers with most surprising chemical and biological properties.

Before an overall description of β-peptides, it is worth introducing the synthesis of β-amino acids. In contrast to their α-counterparts, β-amino acids were not commercially

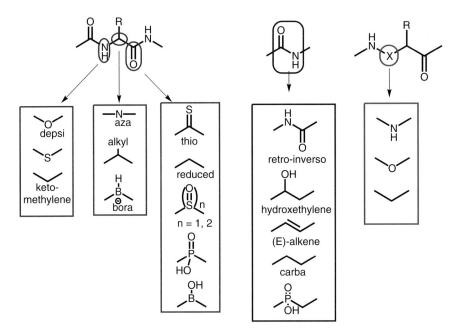

**Figure 2.2** *Some examples of chemical functions that have been introduced in the peptide skeleton for the synthesis of new foldamers.*

available till a few years ago. β-Amino acids are subdivided into $\beta^2$-, $\beta^3$- and $\beta^{2,3}$-amino acids depending on the position of the side chain at the 3-aminopropionic acid core (Figure 2.3). In addition, cyclic amino acids have the amino group integrated in a ring, as is the case in β-proline.

A general synthesis of $\beta^3 h$Xaa-type β-amino acids was obtained by Seebach via Arndt–Eistert α-amino acid homologation and now most of these compounds are commercially available [18]. The $\beta^{2,3}$-amino acids were prepared from the $\beta^3$-derivatives by enolate alkylation [19] or from enolate esters by the Davies method [20]. For the $\beta^2$-homoamino acids, the overall enantioselective Mannich reaction or benzyloxycarbonylmethylation/Curtius degradation was applied using a modified Evans auxiliary [21].

Besides the methods that utilize the chiral pool, several syntheses of β-amino acids rely on classical resolution or a stoichiometric use of chiral auxiliaries (Figure 2.4) [22]. Recently, several methods that utilize the catalytic asymmetric synthesis have been developed, including transition metal catalysis, organocatalysis, biocatalysis and important synthetic methods, such as hydrogenation, the Mannich reaction and conjugate

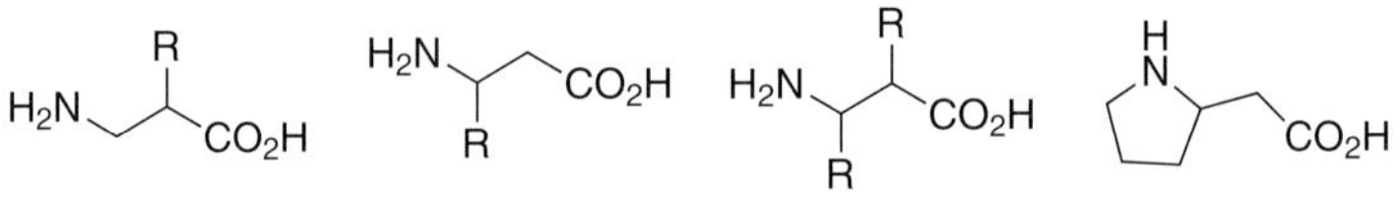

$$\beta^2- \qquad \beta^3- \qquad \beta^{2,3}- \qquad \text{β-proline}$$

**Figure 2.3** *Chemical structure of different kinds of β-amino acids.*

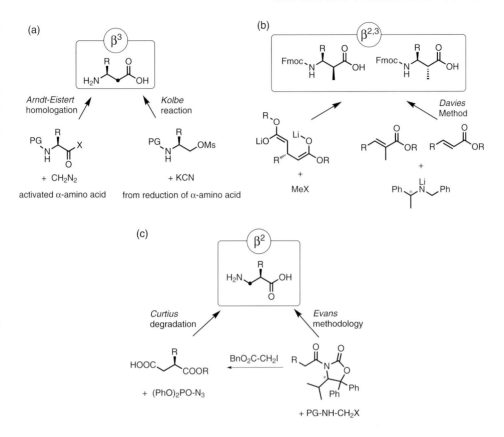

**Figure 2.4** *three generally applicable methods for the preparation of various types of β-amino acids with 21-proteinogenic side chains (R–).*

additions [23]. The synthetic methods are so many that it is impossible to provide an exhaustive number of examples to show the state of the art (some examples are shown in Figure 2.5). Many other examples are reported in a review recently published by Feringa *et al.* [24].

Cyclic β-amino acids have a broad range of use as building blocks for the preparation of modified analogues of biologically active peptides. In cyclic β-amino acids, the amino and carboxyl functions are situated on neighboring atoms, thus these compounds can exist as *R* or *S* isomers, with a total of four possible enantiomers. The availability of a vast number of stereo- and regio-isomers, together with the possibility of ring size expansion and further substitutions on the ring, significantly extends the structural diversity of β-amino acids, thereby providing enormous scope for molecular design. Several methods have been developed for the synthesis of enantiomerically pure cyclic β-amino acids [30]. Figure 2.6 shows some examples that can help us to demonstrate how many methods may be used to prepare these compounds.

β-Peptides fold to helices or hairpin-type structures, and they can be constructed such that they do not fold but are linear or assemble to pleated sheets. In contrast to their natural α-peptidic counterparts, β-peptides form such secondary structures in protic solutions

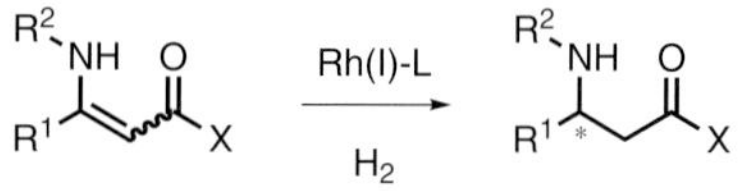

Rhodium-catalyzed hydrogenation of (*E*)- and (*Z*)-β-dehydroamino acid derivatives

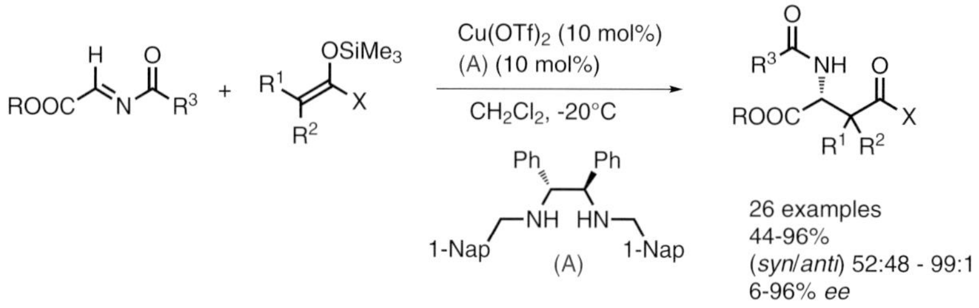

Cu-catalyzed addition of sylil enol ethers to *N*-acylimino esters

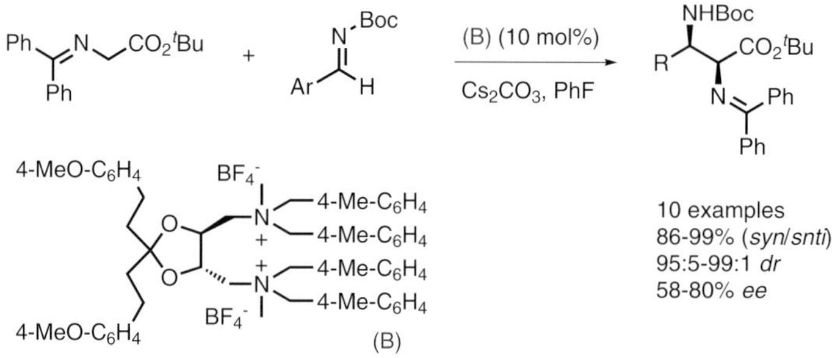

Mannich reaction catalyzed by a chiral phase transfer catalyst

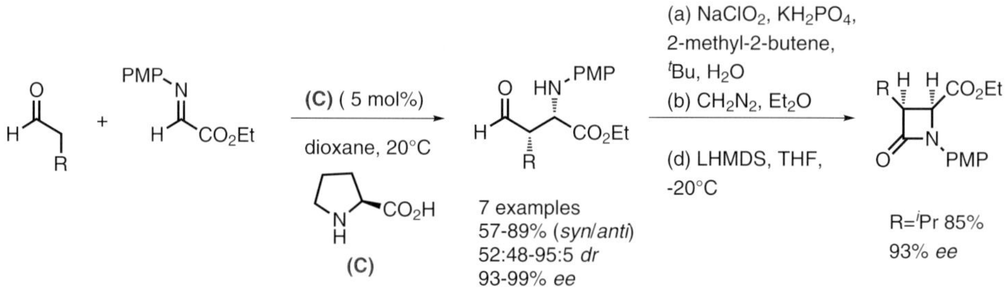

Proline-catalyzed asymmetric Mannich reaction of aldehydes and α-imido carboxylates to yield β-lactams after three consecutive steps [28]

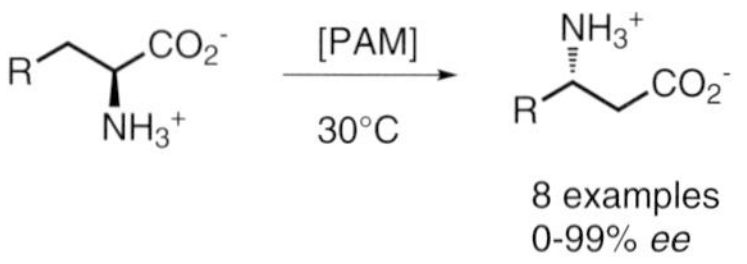

Phenylalanine amino mutase (PAM)-catalyzed synthesis of β-amino acids from α-amino acids [29]

**Figure 2.5** *Some examples of catalytic asymmetric syntheses of β-amino acids.*

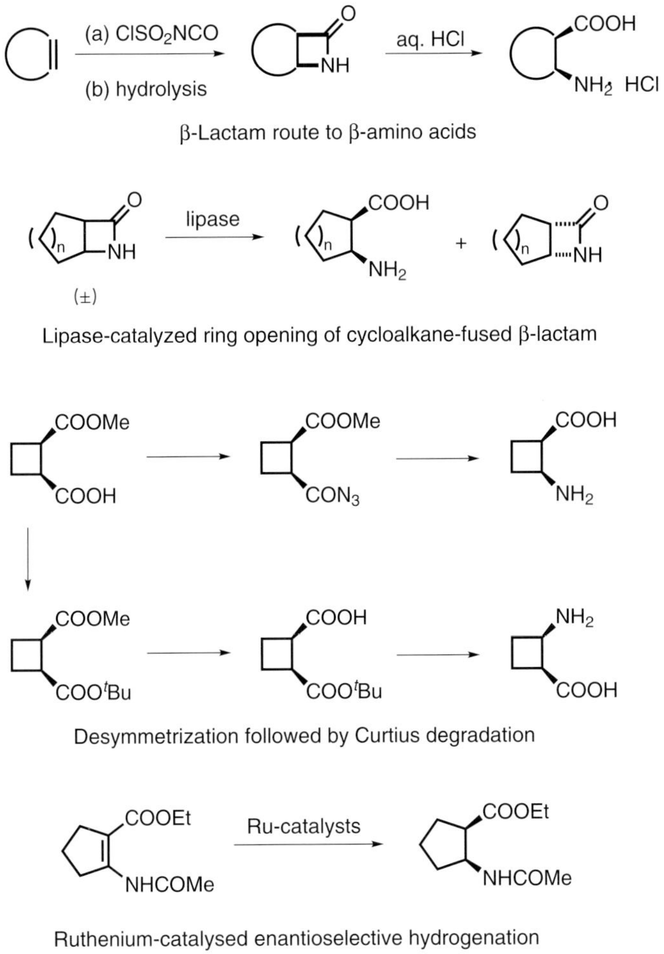
(a) ClSO2NCO
(b) hydrolysis
NH
aq. HCl
COOH
NH2 HCl
β-Lactam route to β-amino acids
lipase
NH
(±)
COOH
NH2
+
O
NH
Lipase-catalyzed ring opening of cycloalkane-fused β-lactam
COOMe
COOH
CON3
NH2
COO'Bu
Desymmetrization followed by Curtius degradation
COOEt
NHCOMe
Ru-catalysts
Ruthenium-catalysed enantioselective hydrogenation

***Figure 2.6*** *Synthetic methods for the synthesis of enantiomerically pure cyclic β-amino acids.*

(MeOH, $H_2O$) with chain lengths as short as four residues and without restricted backbone rotation, as in oligomers containing 2-amino-cyclopentane- and 2-amino-cyclohexane-carboxylic acid moieties [34].

The conformations of β-peptides can be analyzed in terms of the main chain torsional angles, which are assigned the angles ω, φ, θ and ψ (Figure 2.7) in the convention of Balaram [35].

Folded helical or turnlike conformations of β-peptides require a gauche conformation about the θ torsion angle defined by the C2–C3 bond. A *trans* rotamer leads to a fully extended conformation, provided the values of φ and ψ are appropriate. The effects of substituents on the local conformation of a β-amino acid are summarized in Figure 2.8. β-Alanine is highly flexible, analogous to Gly in the α-amino acids. Alkyl substituents at positions 2 and 3 favor a gauche conformation about the C2–C3 bond [36].

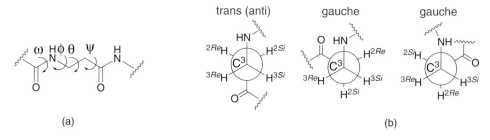

(a)                                                                 (b)

**Figure 2.7** *(a) Definition for the torsional angles in β-peptides. (b) Rotamers for β-alanine regarding the θ dihedral.*

C2,C3-Disubstituted amino acids are even more conformationally constrained and favor gauche conformers when the substituents are *anti*. Gauche-type torsion angles are even more strongly promoted when these atoms are included in a cyclohexane or cyclopentane ring [37]. The ring size determines the precise C2–C3 torsional preference, which in turn influences the β-peptide helix type [38].

Five distinct helices have so far been identified in the field of β-peptides: the 14-, a 12-, a 10-, the 12/10- and an 8-helix, which are defined by the sizes of 8-, 10-, 12-, and 14-membered hydrogen-bonded rings (Figure 2.9).

Besides helical arrangements, β-peptides, like the α-peptidic prototypes, can attain nonaggregating extended chain structures, assemble to pleated sheets, fold to hairpin turns and stack in crystals. Extended arrangements have also been identified, namely parallel and antiparallel pleated sheets, a nonaggregating linear structure and stacks. NMR spectroscopy has emerged as the most useful tool in determining the secondary structures of β-peptides, although some samples suitable for single-crystal or powder X-ray diffraction structure determination have been obtained. For an exhaustive review on helices and other secondary structures of β-peptides see Seebach *et al.* [39].

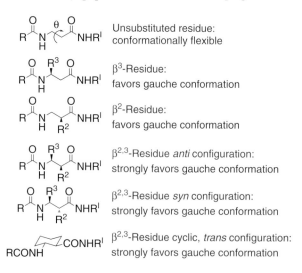

**Figure 2.8** *Effect of substituents on the torsional angle θ.*

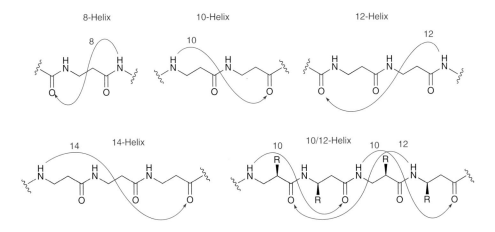

**Figure 2.9** *Nomenclature for β-peptide helices based on hydrogen-bonding patterns.*

### 2.2.3 γ-Peptides

After three decades of studies on homogeneous oligomers containing β-amino acids, molecules based on γ-amino acids were investigated. Very recently a review was published, reporting all the work that has been done in the past 10 years on γ-peptides [40].

Although this homologation reduces the number of potential hydrogen bonds for an oligomer of the same length, the γ-peptides have shown their capability to adopt various stable conformations, such as helices, sheets and turns. In 1998, Seebach and Hannessian reported simultaneously that homogeneous oligomers containing monosubstituted γ-amino acids can form stable helical conformations in solution [41].

Hofmann later performed calculations on unsubstituted and monosubstituted γ-peptides (with one methyl group on the α-, β- or γ-position), employing *ab initio* MO theory at various levels of approximation [42]. He showed that the observed 14- and 9-helix were the most stable conformations. He also claimed that, for unsubstituted and monosubstituted γ-peptides, mixed helices could also be observed. In these cases, the most stable helices are the 22/24- and the 14/12-helices. The hydrogen bonds are orientated alternately in opposite directions leading to a small helix dipole (Figure 2.10). These mixed helices should then be favored in less polar media. It is worth mentioning that $\gamma^4$-peptide helices have the same screw sense and macrodipole as α-peptide helices, whereas $\beta^3$-peptide helices have the opposite.

Several examples have been reported recently and they confirm what was predicted by Hoffman. All the different types of secondary structures have been observed, ranging from helices, to sheets, turns and extended structures. For instance a 9-helix has been observed in monosubstituted γ-peptides [43]. Disubstituted $\gamma^{2,4}$-amino acids have also been used for γ-peptide elaboration. This additional substitution reduces the number of accessible conformations for the backbone. In fact, only two of the nine possible conformers for a γ-residue do not possess unfavorable *syn*-pentane interactions. Thus, Hanessian synthesized tetramers and hexamers and concluded that they all adopt a right-handed 14-helix conformation in pyridine-$d_5$ [44].

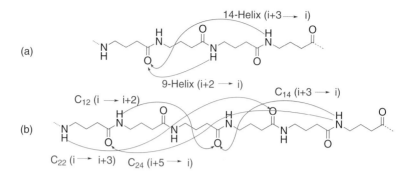

(a)

(b)

**Figure 2.10** *H-bonding in (a) 9-helix and 14-helix and (b) in mixed helices: 14/12-helix or 24/22-helix.*

Compared to β-peptides, the introduction of a supplementary carbon in the backbone can be a source of structural diversity [45]. It is likely that other new structural features or properties will emerge by the development of original amino acid building blocks. Potentially interesting results can be expected in the field of wider helices, as they were predicted by Hofmann to be very stable.

### 2.2.4 Hybrid Foldamers

Recently, several examples of foldamers have been reported, containing mixtures of α- and β-, α- and γ- or β- and γ-amino acids in alternating order [46]. These foldamers allow the formation of new kind of helices, besides those that may be obtained from α-, β- and γ-peptides, as shown in Figure 2.11.

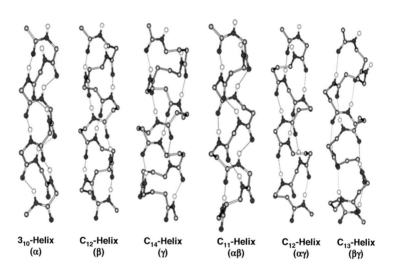

**Figure 2.11** *Helices with 4 → 1 hydrogen-bond patterns in α, β, γ and hybrid peptides.* Reprinted with permission from Ref. [46b]. Copyright 2011 American Chemical Society.

The virtues of alternating longer sequences of α- and β-amino acids, so-called α/β-peptide foldamers, were recognized only a few years ago. During this course, several new helix types were also discovered. In the year 2004, two groups independently investigated the solution properties of α/β-peptides with a 1 : 1 backbone alternation [47].

Beside helical structures with hydrogen bonds pointing only into one direction, either backward or forward along the sequence, also mixed helices (β-helices) with hydrogen bonds alternately changing in forward and backward direction were found in these regular hybrid peptides [48]. For instance, the oligomers described by the Gellman's group revealed a "split personality," meaning that there was a coexistence of an 11-helix along with a 14/15-helix, rapidly interconverting in solution. For relatively long foldamers, the 14/15-helix is favored over the 11-helix [49], analogous to the α-helix that is favored over the $3_{10}$-helix in larger α-peptide structures (Figure 2.12) [50].

Recently, Sharma et Hofmann [51] presented the concept of hybrid helices by the combination of two or more types of homologous peptides with structural and conformational diversity, resulting in well defined helical patterns. Sugar derived β- and γ-amino acids in combination with regular α-amino acids were used as building blocks for their approach. Within this work, they connected β-, α/β- and α/γ-peptide sequences within one oligomer, combining the special features of a β-12/10-helix with an α/β-11/9-helix and another α/γ-12/10-helix (Figure 2.13).

β/γ-Peptides have only recently emerged in the literature [52]. These oligomers are nevertheless of particular interest because the backbone of a β/γ-dipeptide possesses the same number of atoms as an α-tripeptide.

Hofmann performed calculations on unsubstituted hybrid β/γ-peptides [53]. He showed that the most stable conformations were the 11- or 13-helix conformation and the

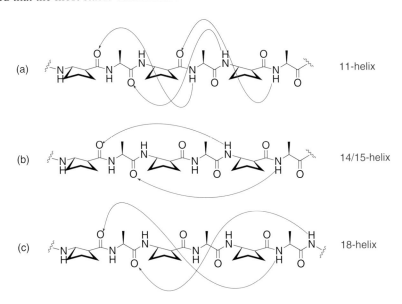

**Figure 2.12** *Hydrogen-bond patterns that define the helical secondary structures of the α/β-peptides considered here, with hydrogen bonds from carbonyl groups to amide protons in the C-terminal direction: (a) 11-helix, (b) 14/15-helix, (c) 18-helix.*

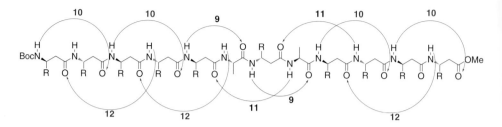

**Figure 2.13** *Example of a so-called hybrid helix, consisting of three individual parts, in which two β-peptide segments surround a short α/β-9/11-helix.*

mixed 11/13- or 20/22-helices and compared the 13-helix of the hybrid β/γ-peptides to the secondary structure of the native α-peptides. (Figure 2.14).

Gellman *et al.* recently demonstrated that β/γ-peptides containing appropriately preorganized subunits adopt the 13-helix in solution and in the solid state [54]. They combined a constrained (R,R,R)-γ-residue with (R,R)-*trans*-2-aminocyclopentanecarboxylic acid, hoping that they would favor the formation of a left-handed β/γ-peptide 13-helix, while the right-handed helix should be favored by residues with S configurations. The crystal structure of β/γ-tetrapeptide contains two molecules in the asymmetric unit. Each independent molecule forms one 13-atom H-bonded ring, involving the NH group of the second ACPC residue and the carbonyl of the N-terminal Boc group. Despite this deviation from the 13-helical H-bonding pattern, the backbone torsion angles for the β- and γ-residues generally fall in ranges predicted by Hofmann *et al.* for the α/β/γ-peptide 13-helix.

Finally, we would like to mention the α/δ-hybrid peptides that were recently studied by Sharma and Hoffman [55]. In a concerted theoretical and experimental study, they presented α/δ-hybrid peptides as a novel foldamer class. Taking advantage of the rigidity of the side chain in the design control, the new δ-amino acids, prepared from δ-glucose with a δ-xylose furanoside side chain, has been very efficiently utilized in the synthesis of

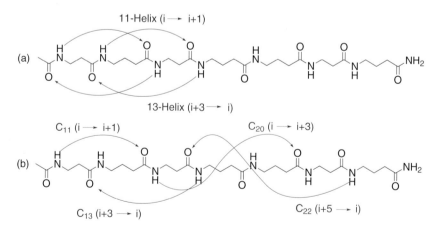

**Figure 2.14** *H-bonding (a) in 11-helix and 13-helix and (b) in mixed helices: 11/13-helix or 20/22-helix.*

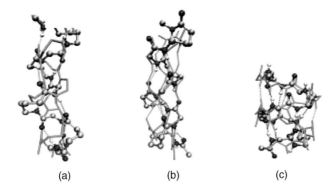

(a)                    (b)                    (c)

**Figure 2.15**  *Superimposition of (a) the 13/11-helices of α/δ- and β/γ-hybrid peptides, (b) the 13-helix of α/δ-hybrid peptides and the native α-helix and (c) the most stable mixed 20/22-helix of α/δ-hybrid peptides and the gramicidin A helix.* Reprinted with permission from Ref. [55]. Copyright 2009 WILEY-VCH Verlag GmbH & Co. KGaA, Weinheim.

these foldamers. Detailed structure analyses show the formation of a novel 13/11-motif corresponding to a mixed or β-helix. The authors have also compared the conformational properties of α/δ-hybrid peptides with those of other hybrid peptides, for instance, α/β-, α/γ- and β/γ-hybrid peptides (Figure 2.15). Remarkably, the most stable mixed 20/22-helix is overlapping with the structure of the gramicidin A membrane channel. This relationship demonstrates the considerable potential for α/δ-hybrid peptides to mimic native peptide and protein structures.

### 2.2.5  Aliphatic Urea Foldamers

Urea, thiourea and their derivatives have been playing a central role in the field of supra-molecular chemistry since the 1940s–1950s, when urea itself was recognized to form self-assembled tunnel host structures, so-called urea inclusion compounds, in the presence of appropriate guest molecules [56]. Despite receiving far less attention than their amide counterparts, oligomers with a urea-type backbone (–[RNHCONH]$_n$–) have gradually made their mark in the field [57]. The urea linkage shares a number of desirable features (i.e., rigidity, planarity, polarity and hydrogen bonding capacity) with the amide group. However, the subtle differences between the two linkages are significant when they are introduced in oligomers with self-assembling and/or folding propensity:

1. The dipole moment of ureas exceeds that of amides.
2. As a result of competitive conjugation, the double bond character of the CONH bond is significantly reduced in ureas compared to amides (rotation barriers for ureas are around 10–12 kcal mol$^{-1}$ vs 16–20 kcal mol$^{-1}$ for amides).
3. Due to the presence of one additional NH-group, *N,N′*-disubstituted ureas have a propensity to form three-centered H-bonds.

*N,N′*-Linked oligoureas are formally obtained by the substitution of NH for the α-CH$_2$ of the amino acid constituents of γ-peptides. Guichard's group was been working for several years on the synthesis and self-assembly of aromatic and aliphatic urea oligomers [58] and have demonstrated that the urea modification are compatible with the γ-peptide

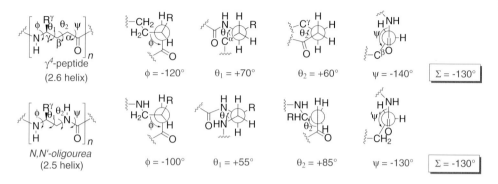

**Figure 2.16** *Comparison of main chain dihedral angles between helical N,N'-linked oligourea and γ⁴-peptide backbones.*

14-helical fold [59]. γ-Amino acid residues within the 14-helix are characterized by large ψ values ~140° (or −140°) [60]. The additional nitrogen was believed to act as a rigidifying element by fixing the pseudo ψ angle to a value close to 170–180°. Detailed NMR studies provided compelling evidence that enantiopure *N,N'*-linked oligoureas adopt a well defined and stable 2.5-helical fold, akin to the γ⁴-peptide 14-helix [61]. The helix is right-handed with a pitch of about 5.1 Å and held by H-bonds closing both 12- and 14-membered rings (12/14-helix). The structural analogy between oligourea and the γ-peptide helical backbone is evident when comparing the main backbone torsion angles (Figure 2.16).

### 2.2.6 Foldamers of α-Aminoxy Acids

α-Aminoxy acids are analogues of β-amino acids in which the β-carbon atom in the β-amino acid backbone is replaced by an oxygen atom [62]. The first synthesis of an α-aminoxy diamide was reported by Yang's group in 1996 [63]. Theoretical calculations suggest that it adopts a rigid eight-membered ring hydrogen-bonded structure (so-called αN-O turn) in its most favorable conformation (Figure 2.17). In the presence of side chains, *ab initio* calculation results suggest that chiral L-α-aminoxy acids prefer a left-handed chiral *N-O* turn.

The formation of a rigid αN-O turn structure in α-aminoxy diamides suggested that oligomers of aminoxy acids might have interesting conformational features and prompted their investigation. Thus, a series of oligomers was synthesized and subjected to conformational studies as performed previously for α-aminoxy diamides (the longest one is

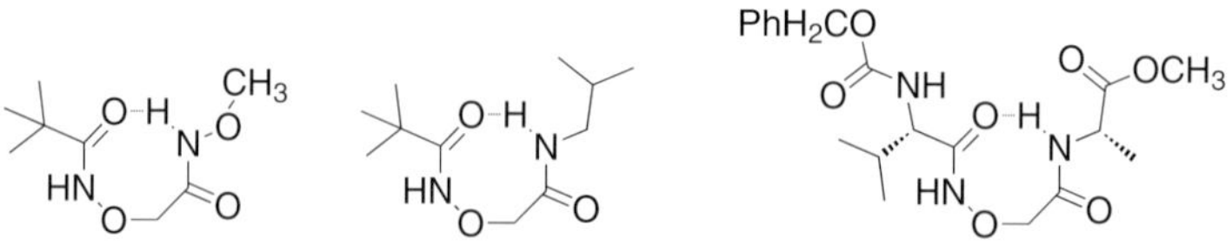

**Figure 2.17** *Some examples of dipeptides containing α-aminoxy acids (the αN-O turn is shown).*

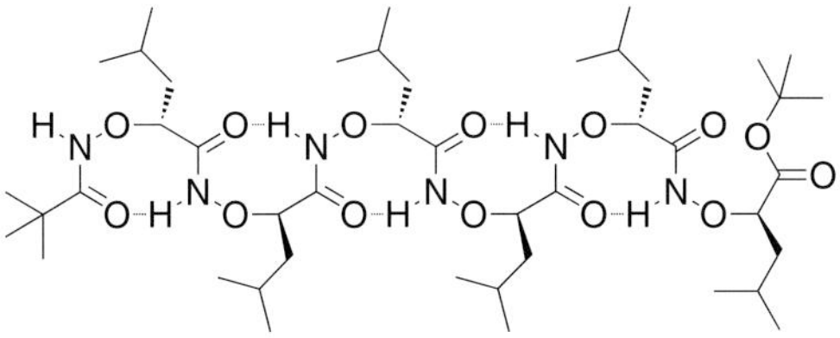

**Figure 2.18** *An oligomer of α-aminoxy acids. The αN-O turns, leading to an eight-membered-ring hydrogen-bond, are reported.*

depicted in Figure 2.18) [64]. Both FT-IR and $^1$H NMR spectroscopies showed that intramolecular hydrogen bonds are formed between the amide NH groups at the $i + 2$ position and the carbonyl oxygen atoms at the $i$ position. Furthermore, the CD curves of the oligomers in 2,2,2-trifluoroethanol were almost superimposable, indicating that their secondary structures are very similar. The same results were observed in the crystal structures.

The exploration of the conformations of hybrid peptides containing α-amino acids and α-aminoxy acids showed that, in peptides of alternating D-α-amino acids and L-α-aminoxy acids, the seven-membered-ring intramolecular hydrogen bond (i.e., γ-turn) is initiated by a succeeding α*N-O* turn [65]. It signifies a new strategy to induce a γ-turn at specific sites of short peptides by incorporating an α-aminoxy acid immediately after the particular α-amino acid of interest.

### 2.2.7 Foldamers Containing Amido Groups

Interesting new structures may be prepared, replacing the proline moieties with pseudoprolines (ψPro) [6]. This term was introduced recently to indicate synthetic proline analogues which are usually obtained by cyclocondensation of the amino acids cysteine, threonine or serine with aldehydes or ketones. Five-membered cycles that contain in the ring a nitrogen and a carboxy unit close one to the other may be ascribed to the family of pseudoprolines: all these compounds share all the same properties because the nitrogen near to a carbonyl group behaves as a rigid spacer, owing to the presence of the endocyclic carbonyl, which strictly imparts a *trans* conformation to the adjacent peptide bond. This effect is due to the tendency of the two carbonyls to lie apart one from the other (Figure 2.19) [66].

Our group has extensively studied the conformational behavior of oxazolidin-2-one homo-oligomers [67]. These compounds lead to a new type of helical structure, similar to that adopted by poly-(L-Pro)$_n$ with *trans* tertiary peptide bonds (type II). Both repeating systems generate left-handed, ternary (3$_1$ symmetry) helices, but the oxazolidin-2-one system is rigid, whereas the (L-Pro)$_n$ system is remarkably more flexible, due to the

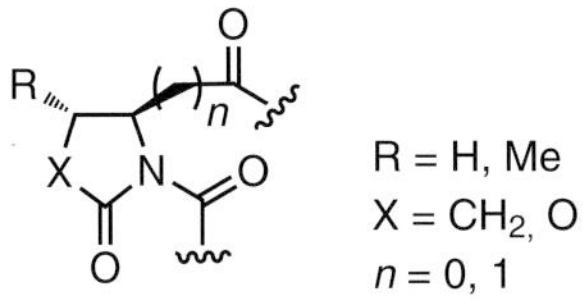

R = H, Me
X = CH$_2$, O
n = 0, 1

**Figure 2.19** *Preferential conformation of the imidic bond.*

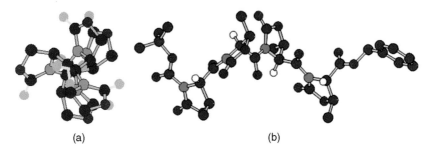

(a)                                        (b)

**Figure 2.20** *(a) Front view and (b) side view of the calculated Boc-(L-Oxd)$_5$-OBnhelix. The terminally protecting groups have been removed.* Reprinted with permission from Ref. [68c]. Copyright 2003 WILEY-VCH Verlag GmbH & Co. KGaA, Weinheim.

*cis–trans* isomerism about the tertiary amide bonds. As a consequence of this remarkable property causing a local constraint, these imido-type oligomers are forced to fold in ordered conformations, that is, in combination with other kinds of interaction (H bond, apolar interactions, etc.), lead to the formation of foldamers (Figure 2.20) [68].

Hybrid foldamers, where the Oxd moiety alternates with an α-amino acid, lead to the formation of different secondary structures. The simplest oligomers of this family that can be prepared have a general formula Boc-(Gly-L-Oxd)$_n$-OBn (Gly = glycine; Figure 2.21) [69].

In order to establish whether these molecules are foldamers, their behavior was analyzed by IR, CD and $^1$H NMR. All the techniques agree in the detection of the formation of weak hydrogen bonds and, as a result, a secondary structure. To further investigate the preferred conformations of oligomers of the Boc-(AA-Oxd)$_n$-OBn series, Gly was replaced with L-Ala, thus introducing a methyl as a side chain [70].

Again, information on the preferred conformation of the Boc-(L-Ala-D-Oxd)$_n$-OR (with $n = 1$–6 and R = Bn, H) oligomers in solution was obtained in structure-supporting solvents (methylene chloride, deuterochloroform, methanol), by FT-IR absorption, $^1$H NMR and CD techniques. The results obtained from all the techniques demonstrate that, from the tetramer level, the oligomers assume a folded structure. The most interesting evidence has been obtained from the *per*-residue CD spectra (Figure 2.22). An ellipticity increase, associated with a reversal of the Cotton effect, was observed for Boc-(L-Ala-D-Oxd)$_5$-OH and more dramatically for Boc-(L-Ala-D-Oxd)$_6$-OH, thus suggesting the formation of an ordered secondary structure. The CD spectrum of Boc-(L-Ala-D-Oxd)$_6$-OH, recorded at different concentrations (not shown), does not show any change as the

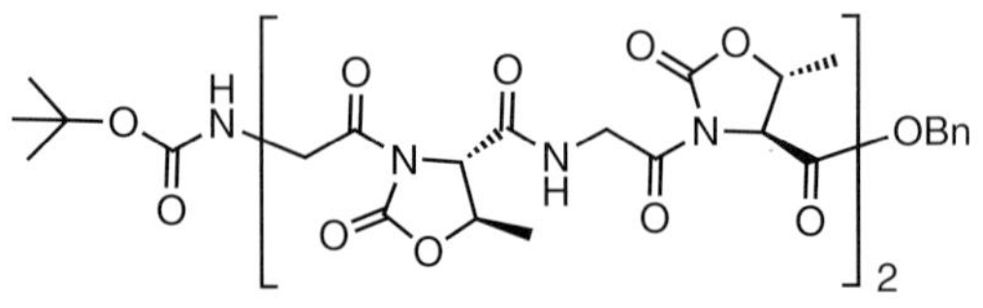

**Figure 2.21** *Chemical structure of the longest investigated Boc-(Gly-L-Oxd)$_n$-OBn oligomer.*

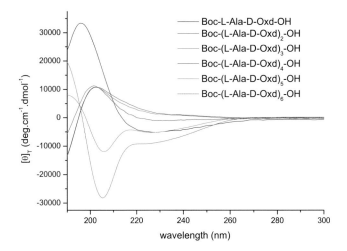

**Figure 2.22** *Normalized per-residue CD spectra of Boc-(L-Ala-D-Oxd)$_n$-OH (n = 1–6) (1 mM concentration in MeOH solution).*

concentration decreases: this finding excludes the formation of self-associated secondary structures.

From a general inspection of the spectra shown in Figure 2.22, Boc-(L-Ala-D-Oxd)$_5$-OH and Boc-(L-Ala-D-Oxd)$_6$-OH clearly fold in the $3_{10}$-helix. It may be considered a subtype of the polypeptide $3_{10}$ helix, being stabilized by alternate $1 \leftarrow 4$ intramolecular C=O . . . H–N H-bonds and having approximately the same fold of the peptide chain. This outcome may be ascribed to the cooperative effect of the rigid –CO-N(CH<)-CO– moiety, which always tend to assume a *trans* conformation and the alternate formation of C=O . . . H—N H-bonds (Figure 2.23).

The formation of fibers through self-assembly is of particular interest, as protein fibers are involved in intra- and extracellular functions. In order to understand aggregation phenomena, oligopeptides may be designed and prepared with the aim of interfering with or mimicking these processes. Indeed, the potential applications of such supramolecular assemblies exceed those of synthetic polymers since the building blocks may introduce a biological function in addition to mechanical properties.

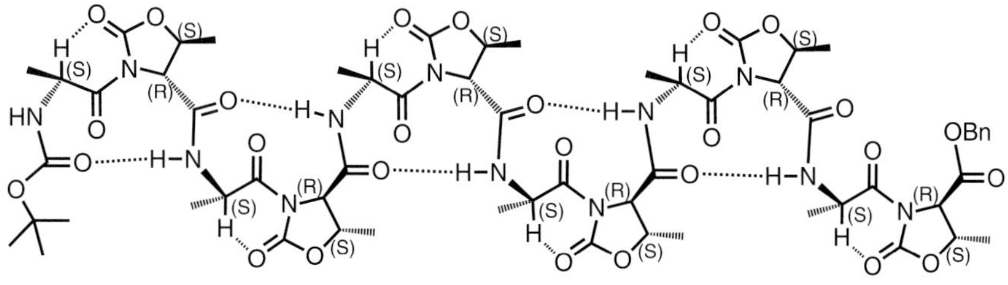

**Figure 2.23** *Conformation of Boc-(L-Ala-D-Oxd)$_6$-OBn, where the three stabilizing effects are shown: (a) the rigid –CO-N(CH<)-CO– moiety, (b) the C=O . . . H—C H-bonds, (c) the C=O . . . H—N H-bonds.*

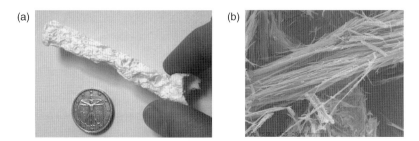

**Figure 2.24** *(a) Boc-L-Phe-D-Oxd-OBn after slow evaporation of a 1:1 mixture of cyclohexane/ethyl acetate. (b) SEM image of filament bundles of Boc-L-Phe-D-Oxd-OBn. Scale bar: 20 μm.* Reprinted with permission from Ref. [71b]. Copyright 2003 WILEY-VCH Verlag GmbH & Co. KGaA, Weinheim.

A solid fiberlike material (Figure 2.24) was obtained after slow solvent evaporation of a 20–25 mM solution of Boc-L-Phe-D-Oxd-OBn in a 1:1 mixture of cyclohexane/ethyl acetate [71]. This material shows strong birefringence and has defined edges. It is made of bundles of crystalline-shaped filaments clustered and aligned along the main bundle direction.

Crystals suitable for X-ray diffraction study were grown by slow evaporation of a solution of Boc-L-Phe-D-Oxd-OBn in diethyl ether at room temperature. Interestingly, linear chains are formed in the crystal packing of Boc-L-Phe-D-Oxd-OBn by only one intermolecular hydrogen bond between the units; thus the solid-state structure of Boc-L-Phe-D-Oxd-OBn can be considered as a borderline case of a parallel β-sheet structure, as this fiberlike material is stabilized only by single hydrogen bonds between dipeptide units.

Another interesting outcome in the formation of supramolecular materials is the self-assembly of the two epimers, Boc-L-Phe-D-Oxd-(S)-β³-hPhg-OBn and Boc-L-Phe-D-Oxd-(R)-β³-hPhg-OBn [72]. Both of these molecules contain the scaffold L-Phe-D-Oxd, but differ by the introduction of the two enantiomers of β³-homophenylglycine (Figure 2.25). The reversal of the stereogenic center contained in this moiety has important effects on the preferred molecular conformation and, as a result, on the crystal morphology.

A different aggregation in the solid state is expected for the two epimers, because one tends to form a γ-turn structure in solution, while the other does not, as demonstrated by IR and $^1$H NMR spectroscopy. Therefore an analysis of the two compounds has been carried out by means of microscopy and X-ray diffraction. The two compounds show a different morphology by optical and electronic microscopy, as reported in Figure 2.26.

By means of X-ray diffraction analysis, we could demonstrate that, in Boc-L-Phe-D-Oxd-(S)-β³-hPhg-OBn crystals, each molecule is engaged in two intermolecular hydrogen bonds with the two neighbors, but the two interactions are not equivalent since one is

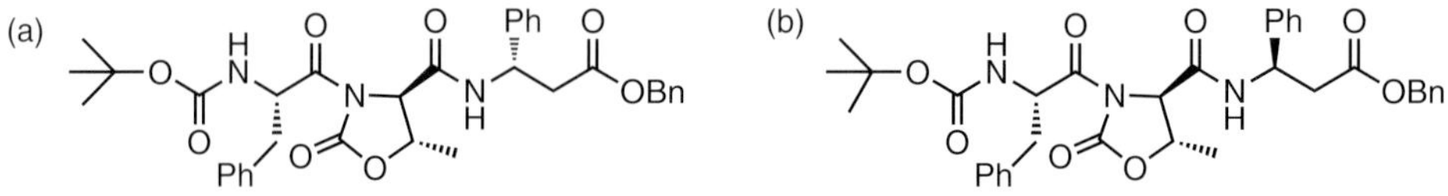

**Figure 2.25** *Chemical structure of: (a) Boc-L-Phe-D-Oxd-(S)-β³-hPhg-OBn and (b) Boc-L-Phe-D-Oxd-(R)-β³-hPhg-OBn.*

**Figure 2.26** *(a, c) Crystals of Boc-L-Phe-D-Oxd-(S)-β³-hPhg-OBn precipitated from ethanol observed by means of an optical and an electron microscope, respectively. (b, d) Crystals of Boc-L-Phe-D-Oxd-(R)-β³-hPhg-OBn precipitated from ethanol observed by means of an optical and an electron microscope, respectively.* Reprinted with permission from Ref. [72]. Copyright 2010 American Chemical Society.

between an amidic hydrogen of one side chain and a carbonyl oxygen of the other amidic moiety, while the second one is between the second amidic hydrogen and the carbonyl oxygen of the Oxd ring. The crystal packing therefore consists of parallel chains with a helical arrangement (Figure 2.27) running along the $c$ axis. The formation of a ternary helix in the solid state is in agreement with the SEM analysis as the crystals of Boc-L-Phe-D-Oxd-(*S*)-β³-*h*Phg-OBn from methanol appear elongated in one direction with a hexagonal cross-section.

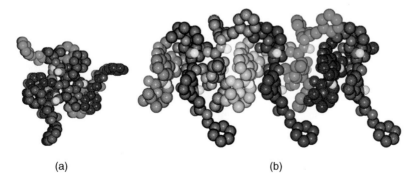

(a)                              (b)

**Figure 2.27** *Space filling model showing the helical arrangement of one of the chains of Boc-L-Phe-D-Oxd-(S)-β³-hPhg-OBn: (a) top view, (b) view along the c axis.* Reprinted with permission from Ref. [72]. Copyright 2010 American Chemical Society.

A completely different packing was observed for epimer Boc-L-Phe-D-Oxd-($R$)-$\beta^3$-$h$Phg-OBn, as the hydrogen bonding pattern in the crystal lattice shows the formation of infinite parallel β-sheets running along the *a* axis, as previously observed for Boc-L-Phe-D-Oxd-OBn. Thus, the reversal of the absolute configuration of the stereogenic center of the $h$Phg moiety ends in a dramatic variation of the preferential conformation of the two compounds, which in turn induces a different crystal packing and consequently a different crystal morphology.

The synthetic approach to supramolecular materials of foldamers containing the 4-carboxy oxazolidin-2-one unit or related molecules is highly tunable with endless variations, so, simply by changing the design and the synthesis, a wide variety of foldamers with the required properties may be prepared "on demand."

## 2.3  Abiotic Foldamers

The potential usefulness of a particular class of foldamer in the design of secondary and tertiary structural motifs depends much on the predictability of their folding. Biotic (both homogeneous and hybrid) foldamers are mainly constituted of aliphatic chains that may assume a wide variety of conformations, as we have just shown. To overcome this problem, several groups in recent years envisaged the synthesis of oligomers containing rigid scaffolds, such as aromatic rings. Following this method, predictability is associated with the structure itself, which can be designed to have no choice but to fold in a desired conformation. These foldamers containing several aromatic rings are very different from natural oligomers and were recently defined "abiotic" by Huc and Guichard [6]. These compounds are generally easy to prepare and have very stable secondary structures. This is a very important feature, as tertiary structures are hard to build using unstable secondary building blocks. Several reviews recently reported the synthesis and applications of aromatic foldamers [73]. We will present here a short overview of their general structures and explain in Section 2.4 how they get organized in the presence of external agents.

Depending on the primary driving force for folding, these unnatural aromatic oligomers can be classified into two main categories, although there are a number of cases where a combination of both applies: (a) molecules that fold because of hydrogen bonding preferences built into the repeat units and (b) those in which folding is primarily driven by solvophobic or aromatic interactions. There is no need to explore the conformational space accessible to the entire molecule to determine its most stable conformation since it primarily results from local conformational preferences. Taking as an illustration the Ramachandran plots used to map the torsion angles corresponding to stable folded conformations in peptides, the experimentally encountered values for the torsion angles in fully predictable foldamers are reduced to very small areas [5].

In hydrogen-bonded aromatic foldamers, the folding process is dependent on preorganization due to hydrogen bond donor and acceptor units that have been deliberately introduced to favor certain conformations. Folding usually requires a nonpolar solvent like chloroform, which does not disrupt this hydrogen-bonding pattern. Based on the hydrogen bond donor and acceptor functionality, these can be further divided into two categories: ones in which the aromatic rings are connected through amide bonds (i.e., oligoamides)

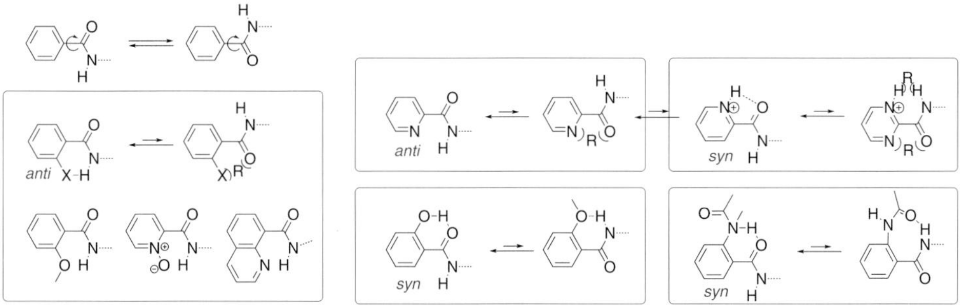

**Figure 2.28** *Conformational preferences of various NHCO-aryl linkages.*

and ones in which they are not. Figure 2.28 shows the conformational preferences of various oligoamides NH . . . OC.

One of the first examples of an arylamide oligomer folding into helical strands was published by Hamilton and coworkers [74], who demonstrated that anthranilic acid, a structural analogue of a β-amino acid, forms intramolecular hydrogen bonding in the resulting oligoamides, so that the strong tendency of the amide bond to adopt *trans* geometry causes folding into helices and sheets. The DeGrado group exploited three-centered hydrogen bond networks to design amphiphilic antibacterial foldamers. The sulfur atom of the thioether functionality in these arylamide foldamers acts as a bifurcated hydrogen bond acceptor, leading to a rigid conformation with segregation of hydrophobic and hydrophilic character [75]. Lehn and coworkers reported helical foldamers based on repeating 2,6-diaminopyridine and 2,6-pyridinedicarbonyl units [76]. These molecules fold into helical conformations because of intrastrand hydrogen bonding and π-stacking, both in chloroform and DMSO. Upon the addition of acid, the diaminopyridine rings are preferentially protonated, leading to a change in conformational preferences and a concomitant loss of helical character. This loss of helicity is completely reversible and folding is restored when excess triethylamine is added to the system. These foldamers are racemic in solution but display chirality when dissolved in chiral solvents (L- and D-diethyl tartrate). In an attempt to obtain higher order assemblies from arylamide systems, Huc and coworkers synthesized oligomers of 7-amino-8-fluoro-quinoline-2-carboxylic acid; these subunits found a quadruplex helix arrangement in crystals grown from a mixture of toluene, dichloroethane and hexane [77].

Other hydrogen-bonded foldamers have been prepared with aromatic rings connected by groups different from amides. For instance, Gong reported aromatic oligoureas as another class of intramolecular hydrogen-bonded foldamers, which are composed of aromatic rings connected through the urea functionality [78]. Meijer and coworkers reported helically folding poly(ureidophthalimide) oligomers due to hydrogen-bonding interactions between the urea hydrogens and the carbonyl group of the phthalimide moiety. The incorporation of several PEG units into the side chains imparts water solubility to these compounds [79]. The Li group reported hydrazide-based foldamers that form spherical vesicles on heating in methanol [80].

Foldamers primarily driven by solvophobic or aromatic interactions are usually composed of hydrophobic ring systems with polar side chains. Folding is driven by π-stacking

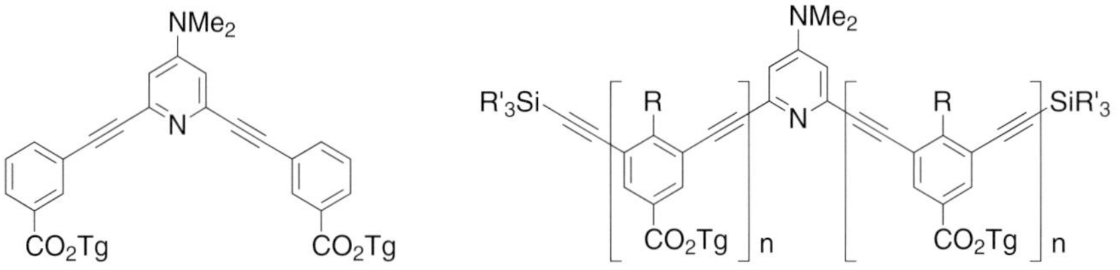

**Figure 2.29** *General structure of oligo(m-phenylene ethynylene) foldamers.*

interactions and favorable side chain–solvent interactions while minimizing destabilizing backbone–solvent contact. Unlike their hydrogen-bonded counterparts, the solvents used for these foldamers are more polar in nature (e.g., aqueous solutions).
Several examples of this class of foldamers were reported by Moore and coworkers
[81], who elaborated their oligo(*m*-phenylene ethynylene) foldamer to incorporate a
4-dimethylaminopyridine (DMAP) ring in the center (Figure 2.29). These foldamers
are capable of selecting, preferentially binding and reacting with substrates that present a better match to their size and shape. Thus, these foldamers can be likened to
enzymes, such as *t*RNA synthetases, that use multiple sieving protocols to ensure
specificity and fidelity. Hecht and coworkers modified the oligo(*m*-phenylene ethynylene) backbone pioneered by Moore and coworkers by introducing an azobenzene
ring to take advantage of its photoswitching properties [82].

## 2.4 Organization Induced by External Agents

Several authors noticed that the organization adopted by aromatic foldamers is strongly
influenced by external agents such as solvents or ions. As the organization of foldamers
driven by metal cations will be extensively discussed in the following chapters of this
book, we report here a short overview of some examples of the effect of solvents and
anions on foldamer organization.

### 2.4.1 Organization Induced by Solvents

We present here only one very interesting example, which was reported by Huc and coworkers; the authors demonstrated that pyridinedicarboxamide helices can be capped with
two quinoline-2-carboxylates on either end to form isolated capsules capable of binding
small guests like methanol and water in chloroform [83]. The host molecules give rise to
distinct NMR signals depending on whether they are empty, half-full or full, thereby facilitating the analysis. The binding event is slow and can be captured on the NMR timescale.
Capsules with larger cavities have been generated by replacing the pyridine ring with a
1,8-diaza-anthracene motif; these are capable of binding alkanediol guests [84].

Progressively increasing the size of the helical capsules by including additional monomers in the center of their sequence allows their binding properties to be tuned in a modular fashion. More guests or larger guests can be included when the capsule size is
rationally increased (Figure 2.30). Capsules with larger cavities prefer to accommodate
larger guests rather than a large number of water molecules for obvious entropic reasons.

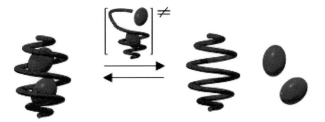

**Figure 2.30** *Encapsulation of an egg-shaped guest by partial unfolding of a helix possessing a reduced diameter at both ends.* Reprinted with permission from Ref. [83a]. Copyright 2005 WILEY-VCH Verlag GmbH & Co. KGaA, Weinheim.

### 2.4.2 Organization Induced by Anions

Anions, because of their biological and environmental relevance, are at the base of many scientific programs within the field of supramolecular chemistry [85]. The design of synthetic anion receptors with high affinity and selectivity and with a specific shape that are capable of differentiating between anions of different size and geometry could be done thanks to the directionality of the hydrogen bond [86]. Several foldamers exhibit strong tendencies to form well ordered secondary structures by encapsulating guests in a non-covalent way, within their internal cavities. As the programmed sequence of monomers affords foldamers that fold into well defined secondary structures, it is possible to achieve a certain secondary structure by designing the monomer. For instance, to obtain foldamer-based anion receptors, the monomer must have hydrogen bond donors in order to interact with anions. Some of the most used monomers contain the NH protons of amides, ureas, pyrroles and indoles and the polarized CH protons of neutral or positively charged aromatic heterocycles, such as 1,2,3-triazole. The NH proton present in the pyrrole heterocycle forms a strong hydrogen bond with anions [87]. Sessler and coworkers studied several oligopeptides containing pyrrolic moieties. The exapyrrole shown in Figure 2.31 is fully conjugated, and several experiments in the solid state and in solution show that the dihydrochloride salt exists in a planar, S-shaped conformation, where the two clefts coordinate a chloride anion due to the formation of NH . . . Cl$^-$ hydrogen bonds [88]. In a similar way, the arms of the di- and tri-pyrrolic moieties directly linked to a chromophore can adjust themselves to create a concave pocket that comprises four or six NH donor groups in a convenient manner suitable for binding an anion [89,90].

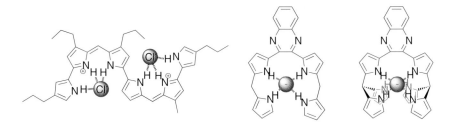

**Figure 2.31** *Proposed anion binding modes for receptor.*

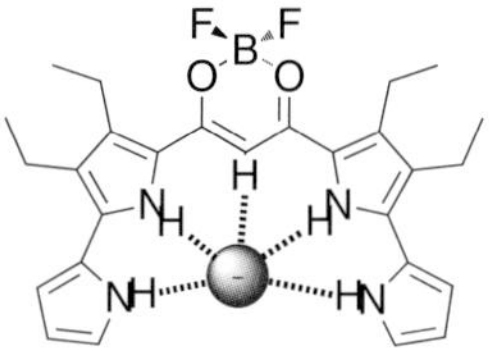

**Figure 2.32** *Dipyrrolyl diketone boron complexes are able to bind chloride, acetate and dihydrogenophosphate ions.*

Further evidence for the chelating attitude of pyrrole moieties was reported by Maeda and coworkers, where the pyrrolic arms of the oligomer (Figure 2.32) are linked to a diketo-substituted boron complex [91]. This particular bridge offers a modular way of controlling the electronic and subsequent binding properties of the oligomer, which is able to bind chloride, acetate and dihydrogenophosphate ions.

In 2005, Jeong and coworkers reported a series of oligoindoles in which monomeric units were connected sequentially by ethynyl linkers [92]. For instance the oligomer reported in Figure 2.33 adopts an expanded conformation in the absence of an anion, but folds into helical conformations in the presence of a chloride, thus encapsulating the anion within a helical conformation. The same group synthesized indolocarbazole oligomers that possessed extended π-surfaces relative to the corresponding biindole-based ones, thus possibly providing increased π-π stacking and hydrophobic interaction [93].

1,4-Disubstituted 1,2,3-triazoles are universal ligation tools [94] whose capacity for independent function has received far less attention. Recent reports, however, indicated that the size and dipole moment (around 5 D) of triazoles make them interesting candidates for amide bond surrogates, and Arora and coworkers reported the contributions of triazoles to the conformational preferences of peptidotriazole oligomers [95]. Craig and coworkers reported acyclic oligomers based on the aryl 1,2,3-triazole unit [96]. The

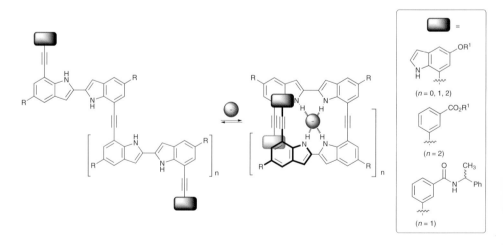

**Figure 2.33** *Oligoindoles that can fold around an anion.*

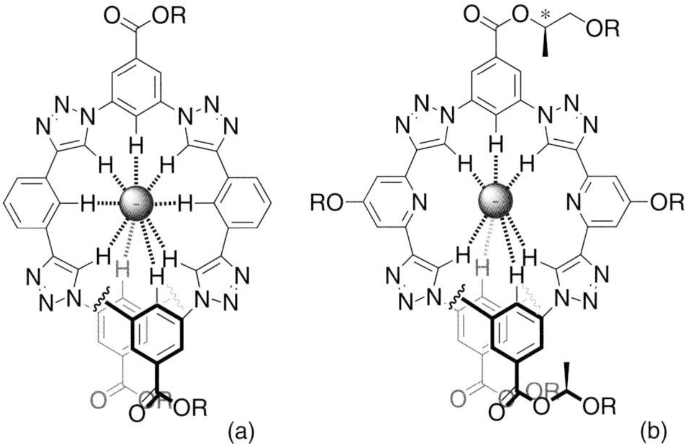

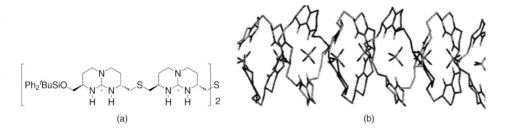

**Figure 2.34** *1,4-Diaryl-1,2,3-triazole oligomers depicted in their chloride binding conformations.*

oligomer shown in Figure 2.34a has appreciable conformational freedom only around the arene–triazole single bonds, but it is complexed in a helical fashion when it is bound with an anion. The size of the binding cavity is complementary to chloride: indeed larger anions give lower association constants, due to an improper fit within the cavity. Meudtner and Hecht independently reported oligomers obtained by alternating aryl, pyridyl and 1,2,3-triazole units [97]. In this case, the dipole–dipole repulsion between the heterocycles, together with the dipole effect of triazole, forces the oligomers to adopt a helical conformation in a water–acetonitrile mixture (Figure 2.34b). CD signals show the most significant effect when the molecule binds with fluoride, while with chloride and bromide there is a signal decrease and helicity inversion.

De Mendoza and coworkers reported the synthesis of an oligomer composed by four bicyclic guanidinium salts connected by a thioether spacer unit: in the presence of a sulfate anion, two strands of it fold into a double-helical structure around it (Figure 2.35) [98].

Kruger and Martin reported a bispyridyl ligand that forms a helical dimer in the presence of hydrogen chloride (Figure 2.36a) [99]. The chloride ions are coordinated in a pincer fashion by the two pyridinium moieties via ionic hydrogen bonds, as well as

**Figure 2.35** *(a) Tetraguanidinium salts. (b) Optimized model of a sulfate helicate from(S,S)-guanidines.* Reprinted with permission from Ref. [98]. Copyright 1996 American Chemical Society.

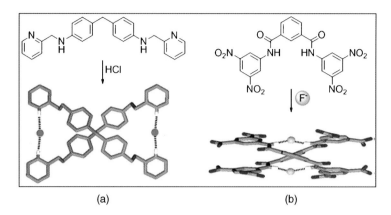

(a)                                                                      (b)

**Figure 2.36** *(a) Bispyridyl ligand forms an helical dimer in the presence of hydrogen chloride in the solid state. (b) Side view of the X-ray crystal structure of the fluoride-directed assembly of an isophthalamide cleft into a double helix.* Ref. [100]. Reproduced by permission of the Royal Society of Chemistry.

weaker interactions with methylene and aromatic hydrogens within the binding pocket. A similar observation was reported by Gale and coworkers for the fluoride-directed assembly of an isophthalamide cleft into a double helix (Figure 2.36b) [100].

Interesting results have been obtained when the folding is driven by organic anions. Li and coworkers reported the folding of a linear arylamide oligomer synthesized by coupling naphthalene-2,7-diamine with 1,3,5-benzenetricarboxylic acid segments (Figure 2.37) [101]. It adopts no compact conformation without an anionic template and it folds into a helical structure in the presence of a benzenetricarboxylate, upon hydrogen-bonding formed with NH and CH donor groups arranged in a complementary fashion to stabilize the complex. This complementarity towards tricarboxylate is crucial in generating the helical folded structure, as halide anions, nitrate, acetate or isophthalate do not induce the formation of a helical complex.

Up to now, we have focused our attention to open chained oligomers, but it is possible to program a sequence of monomers in order to obtain cyclic oligomers that display interesting dynamic conformational behavior as a consequence of their size. Generally, small cycles are relatively rigid structures with limited conformational freedom, while larger ones are flexible and can display twisting and folding behaviors, similar to their acyclic counterparts. The resulting secondary structures have the shape of a figure "eight," containing two binding pockets which are geometrically separated from each other.

Böhmer and coworkers presented a cyclic hexaurea (Figure 2.38), formed by four rigid xanthenes units and two diphenyl ether units connected with six urea moieties, displaying their NH group in the inner part of the cycle [102]. In the presence of chloride, the molecule folds into two binding cavities, due to the large flexibility induced by the ether units.

Sessler and coworkers synthesized a cyclic decapyrrole named turcasarin, which can twist to adopt a left-/right-handed enantiomeric figure of eight loops [103]. The twist observed in the crystal structure affords two binding areas, each of which binds two chloride ions hydrogen bonded to the pyrrole NH hydrogens. With this molecule it is possible to state that conjugated macrocycles are not necessarily rigid and flat structures, but can

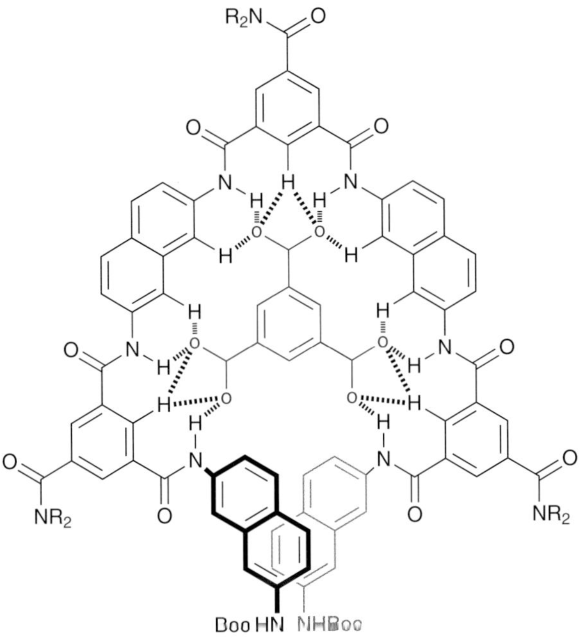

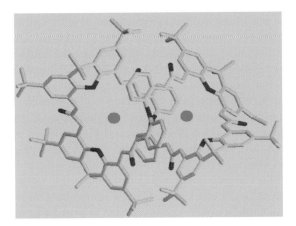

**Figure 2.37** *Proposed folded conformation of arylamide oligomer upon binding with the benzene-1,3,5-tricarboxylate anion.*

also display dynamic behavior like the twisting and folding that make them potentially chiral. Following this work, the Sessler group obtained a similar result with an oligopeptide containing bipyrroles and 2,6-diamidopyridine units, connected via enamine linkages (Figure 2.39) [104]. Anion binding studies show selective binding toward tetrahedral

**Figure 2.38** *Structure of the 1:2 complex of hexamer with tetrabutylammonium chloride.* Reprinted with permission from Ref. [102]. Copyright 2006 WILEY-VCH Verlag GmbH & Co. KGaA, Weinheim.

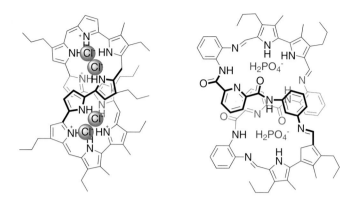

**Figure 2.39**  *Cyclic oligopyrroles twist into loops in figures of eight and encapsulate anions by hydrogen bonds in two distinct cavities.*

anions, such as dihydrogenphosphate and hydrogen sulfate. In contrast, no binding is observed with chloride, bromide and nitrate, probably due to the shape mismatch of these anions to the binding pocket generated by a tetrahedral template.

Gale and coworkers reported a polymer consisting of 3,4-dichloro-2,5-diamido-substituted pyrrole units (Figure 2.40a) [105], which, on the addition of fluoride ions, dimerizes in an orthogonal manner via NH . . . N$^-$ hydrogen bonds. Maeda and coworkers reported a polymeric network formed by chloride bridges with BF$_2$ complexes of acyclic dipyrrolyldiketones (Figure 2.40b) [106].

## 2.5  Applications

Interesting applications of foldamers include the preparation of pharmacologically active compounds, because one of the major concerns in the development of pharmaceutically potent peptides as drugs is their poor proteolytic stability and rapid degradation.

Amazingly, the introduction of single acyclic β-amino acid residues into major histo-compatibility complex (MHC) class I binding peptides led to increased stability of these peptides against enzymatic cleavage [107]. Gellman and coworkers [108] first investigated the proteolytic stability of α/β-peptides containing cyclic β-residues. Along this line, another study concluded that β-amino acids incorporated within a peptide sequence can protect the amide bonds of neighboring α-amino acids [109]. If more alterations are introduced in the backbone, more results may be obtained. For instance, α/β-peptides with a 1 : 1 backbone alternation were shown to exhibit distinct antibacterial activity [110]. Additionally, some studies of the Gellman group provided experimental support for the hypothesis that the formation of a globally amphiphilic helix is not required for host-defense peptide mimicry [111], as amphiphilic conformations of random copolymers without any helical structure were also proven to exhibit potent antibiotic activity (Figure 2.41).

Another very important goal is the formation of discrete tertiary structures.

Most foldamer research to date has focused on the formation of secondary structures, but creating foldamers with a discrete tertiary structure has long been recognized as a

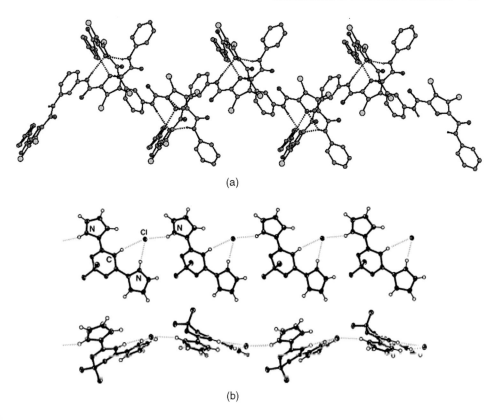

(a)

(b)

**Figure 2.40**   *(a) Amido-pyrrole oligomers form an extended supramolecular polymeric chain in the solid state upon deprotonation of the pyrrolic NH by fluoride. Ref. [91]. Reproduced by permission of the Royal Society of Chemistry. (b) Anion-bridged self-assembly (top and side view). Counter cations, $Bu_4N^+$ and solvents are omitted for clarity.* Reprinted with permission from Ref. [106]. Copyright 2005 WILEY-VCH Verlag GmbH & Co. KGaA, Weinheim.

major aim. Conformational order at this level is important not only as a fundamental structural goal but also as a prelude to developing foldamers with sophisticated functions, such as catalysis, which, among proteins, generally require discrete tertiary folding. Efforts to generate a foldamer tertiary structure have built upon the hierarchical design strategy put forward by DeGrado and coworkers for the *de novo* development of

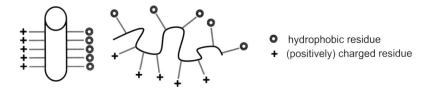

○  hydrophobic residue

+  (positively) charged residue

**Figure 2.41**   *Schematic representation of a globally amphiphilic helix and a globally amphiphilic conformation.*

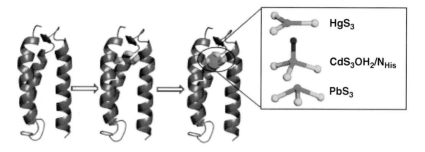

**Figure 2.42** *A well folded and stable construct that encapsulates the heavy metals Hg$^{II}$, Cd$^{II}$ and Pb$^{II}$ with high affinity and predefined coordination geometry.* Reprinted with permission from Ref. [113]. Copyright 2011 WILEY-VCH Verlag GmbH & Co. KGaA, Weinheim.

α-peptides with helix bundle tertiary structure [112]. They recently reported, in collaboration with the Pecoraro group, the preparation of a single polypeptide chain capable of binding metal ions with a high affinity and predefined coordination geometry (Figure 2.42) [113]. An understanding of the biochemistry of the binding of heavy metals to a single polypeptide chain is potentially useful for the development of peptide-based water purification systems or sensors for specific heavy metal ions.

Several groups have reported homogeneous β-peptides that accomplish the first step of the hierarchical approach, self-assembly to discrete helix bundles in aqueous solution [114]. Schepartz and coworkers recently reported that certain β$^3$-peptides self-assemble in aqueous solution into discrete bundles of unique structure and defined stoichiometry [115]. The thermodynamic stability of a β-peptide bundle can be enhanced by optimizing the length of these four interhelical salt bridges (Figure 2.43). These results provide another critical step in the "bottom-up" formation of β-peptide assemblies with defined sizes, reproducible structures and sophisticated function.

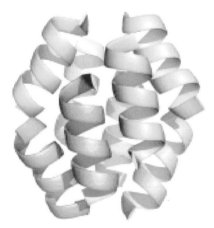

**Figure 2.43** *Ribbon diagram of the Zwit-EYYK octamer as determined by X-ray crystallography.* Reprinted with permission from Ref. 115a]. Copyright 2011 WILEY-VCH Verlag GmbH & Co. KGaA, Weinheim.

## 2.6 Conclusions and Outlook

We have shown a short overview on the amazing amount of work that has been done in the past 15 years within the newborn but very promising field of foldamers. Many new monomers have been prepared and introduced into oligomers. They may have an aliphatic or an aromatic skeleton, may mimic α-, β-, γ- or δ-amino acids or even not resemble an amino acid at all, but they all have the common characteristic of promoting a folded structure. At the moment several groups are engaged in the study of foldamers able to promote the formation of secondary, tertiary and even quaternary structures, as we previously briefly reported. These supramolecular structures may have noteworthy applications as readily tunable molecular frameworks for the recognition and inhibition of bacterial cell membranes, protein–RNA interactions, protein–protein interactions and enzymes.

Metallofoldamers have a central role in this study, as they may efficiently mimic metalloproteins. They have an impressive ability to form single-handed helical structures and other chiral architectures. Moreover, they may be applied as sensors due to their selective folding when binding to a specific metal ion or as responsive materials.

While several overviews, reviews and even books have been published in recent years in the field of foldamers, a systematic presentation of metallofoldamers has never been published till now. This book fills the gap and is an important milestone in the study of the design, preparation and application of metallofoldamers.

## References

1. Gellman, S.H. (1998) Foldamers: a manifesto. *Acc. Chem. Res.*, **31**, 173–180.
2. Nielsen, P.E., Egholm, M., Berg, R.H., and Buchardt, O. (1991) Sequence-selective recognition of DNA by strand displacement with a thymine-substituted polyamide. *Science*, **254**, 1497–1500.
3. Simon, R.J., Kania, R.S., Zuckermann, R.N. *et al.* (1992). Peptoids: a modular approach to drug discovery. *Proc. Natl Acad. Sci. USA*, **89**, 9367–9371.
4. Hill, D.J., Mio, M.J., Prince, R.B. *et al.* (2001) A field guide to foldamers. *Chem. Rev.*, **101**, 3893–4011.
5. Hecht, S. and Huc, I. (2007) *Foldamers: Structure, Properties, and Applications*, Wiley-VCH, Weinheim.
6. (a) Cubberley, M.S. and Iverson, B.L. (2001) Models of higher-order structure: foldamers and beyond. *Curr. Opin. Chem. Biol.*, **5**, 650–653; (b) Seebach, D., Beck, A.K., and Bierbaum, D. J. (2004) The world of (- and (-peptides comprised of homologated proteinogenic amino acids and other components. *Chem. Biodivers.*, **1**, 1111–1239; (c) Sanford, A.R., Yamato, K., Yang, X. *et al.* (2004) Well-defined secondary structures. *Eur. J. Biochem.*, **271**, 1416–1425; (d) Cheng, R.P. (2004) Beyond *de novo* protein design – *de novo* design of non-natural folded oligomers. *Curr. Opin. Struct. Biol.*, **14**, 512–520; (e) Balbo Block, M.A., Kaiser, C., Khan, A., and Hecht, S. (2005) Discrete organic nanotubes based on a combination of covalent and non-covalent approaches. *Top. Curr. Chem.*, **245**, 89–150; (f) Fulop, F., Martinek, T.A., and Toth, G.K. (2006) Application of alicyclic β-amino acids in peptide chemistry. *Chem. Soc. Rev.*, **35**, 323–334. (g) Goodman, C.M., Choi, S., Shandler, S., and DeGrado, W.F. (2007) Foldamers as versatile frameworks for the design and evolution of function. *Nat. Chem. Biol.*, **3**, 252–262; (h) Bautista, A.D., Craig, C.J., Harker, E.A., and Schepartz, A. (2007) Sophistication of foldamer form and function *in vitro* and *in vivo*. *Curr. Opin. Chem. Biol.*, **11**, 685–692;

(i) Smaldone, R.A., and Moore, J.S. (2008) Sophistication of foldamer form and function *in vitro* and *in vivo*. *Chem. Eur. J.*, **14**, 2650–2657.

7. Guichard, G. and Huc, I. (2011) Synthetic foldamers. *Chem. Commun.*, **47**, 5933–5941.

8. Appella, D.H., Christianson, L.A., Karle, I.L. *et al.* (1996) β-Peptide Foldamers: Robust Helix Formation in a New Family of β-Amino Acid Oligomers. *J. Am. Chem. Soc.*, **118**, 13071–13072.

9. (a) Seebach, D., Overhand, M., Kühnle, F.N.M. *et al.* (1996) β-peptides: synthesis by *Arndt-Eistert* homologation with concomitant peptide coupling. Structure determination by NMR and CD spectroscopy and by X-ray crystallography. Helical secondary structure of a β-hexapeptide in solution and its stability towards pepsin. *Helv. Chim. Acta*, **79**, 913–941; (b) Seebach, D. and Matthews, J.L. (1997) β-Peptides: a surprise at every turn. *Chem. Commun.*, **1997**, 2015–2022.

10. (a) Rueping, M., Mahajan, Y.R., Jaun, B., and Seebach, D. (2004) Design, synthesis and structural investigations of a β-peptide forming a 314-Helix stabilized by electrostatic interactions. *Chem. Eur. J.*, **10**, 1607–1615; (b) Wezenberg, S.J., Metselaar, G.A., Rowan, A.E. *et al.* (2006) Synthesis, characterization, and folding behavior of β-amino acid derived polyisocyanides. *Chem. Eur. J.*, **12**, 2778–2786; (c) Fabian, L., Kalman, A., Argay, G. *et al.* (2005) Crystal engineering with alicyclic β-amino acids: construction of hydrogen-bonded bilayers. *Crystal Growth Des.*, **5**, 773–782. (d) Martinek, T.A., Hetnyi, A., Fulop, L. *et al.* (2006) Secondary structure dependent self-assembly of β-peptides into nanosized fibrils and membranes. *Angew. Chem. Int. Ed.*, **45**, 2396–2400. (e) Hetenyi, A., Szakonyi, Z., Mandity, I. *et al.* (2009) Sculpting the β-peptide foldamer H12 helix *via* a designed side-chain shape. *Chem. Commun.*, **2009**, 177–179; (f) Martinek, T.A., Mandity, I.M., Fulop, L. *et al.* (2006) Effects of the alternating backbone configuration on the secondary structure and self-assembly of β-peptides. *J. Am. Chem. Soc.*, **128**, 13539–13544. (g) Mandity, I.M., Wéber, E., Martinek, T.A. *et al.* (2009) Design of peptidic foldamer helices: a stereochemical patterning approach. *Angew. Chem. Int. Ed*, **48**, 2171–2175.

11. Arvidsson, P.I., Ryder, N.S., Weiss, H.M. *et al.* (2005) Exploring the antibacterial and hemolytic activity of shorter- and longer-chain β-,α,β-, and γ-peptides, and of β-peptides from β2-3-Aza- and β3-2-methylidene-amino acids bearing proteinogenic side chains. *Chem. Biodivers.*, **2**, 401–419.

12. (a) Semetey, V., Rognan, D., Hemmerlin, C. *et al.* (2002) Stable helical secondary structure in short-chain *N,N′*-linked oligoureas bearing proteinogenic side chains. *Angew. Chem. Int. Ed.*, **41**, 1893–1895; (b) Salaun, A., Potel, M., Roisnel, T. *et al.* (2005) Crystal structures of Aza-β³-peptides, a new class of foldamers relying on a framework of hydrazinoturns. *J. Org. Chem.*, **70**, 6499–6502; (c) Li, X. and Yang, D. (2006) Peptides of aminoxy acids as foldamers. *Chem. Commun.*, **2006**, 3367–3379; (d) Tomasini, C., Angelici, G., and Castellucci, N. (2011) Foldamers based on oxazolidin-2-ones. *Eur. J. Org. Chem.*, **2011**, 3648–3669.

13. Nelson, J.C., Saven, J.G., Moore, J.S., and Wolynes, P.G. (1997) Solvophobically driven folding of nonbiological oligomers. *Science*, **277**, 1793–1796.

14. Lokey, R.S. and Iverson, B.L. (1995) Synthetic molecules that fold into a pleated secondary structure in solution. *Nature*, **375**, 303–305.

15. Bassani, D.M., Lehn, J.-M., Baum, G., and Fenske, D. (1997) Designed self-generation of an extended helical structure from an achiral polyheterocylic strand. *Angew. Chem. Int. Ed.*, **36**, 1845–1847.

16. (a) Hamuro, Y. Geib, S.J., and Hamilton, A.D. (1996) Oligoanthranilamides. Non-peptide subunits that show formation of specific secondary structure. *J. Am. Chem. Soc.*, **118**, 7529–7541; (b) Berl, V., Huc, I., Khoury, R. *et al.* (2000) Interconversion of single and double helices formed from synthetic molecular strands. *Nature*, **407**, 720–723; (c) Zhu, J., Parra, R.D., Zeng, H. *et al.* (2000) A new class of folding oligomers: Crescent oligoamides. *J. Am. Chem.*

*Soc.*, **122**, 4219–4220; (d) Jiang, H., Lèger, J.-M., and Huc, I. (2003) Aromatic δ-Peptides. *J. Am. Chem. Soc.*, **125**, 3448–3449.

17. Seebach, D. and Gardiner, J. (2008) β-peptidic peptidomimetics. *Acc. Chem. Res.*, **41**, 1366–1375.

18. Kirmse, W. (2002) 100 years of the wolff rearrangemen. *Eur. J. Org. Chem.*, **14**, 2193–2256.

19. Seebach, D., Abele, S., Gademann, K. *et al.* (1998) β2- and β3-Peptides with Proteinaceous Side Chains: Synthesis and solution structures of constitutional isomers, a novel helical secondary structure and the influence of solvation and hydrophobic interactions on folding. *Helv. Chim. Acta*, **81**, 932–982.

20. Davies, S.G., Garrido, N.M., Kruchinin, D. *et al.* (2006) Homochiral lithium amides for the asymmetric synthesis of β-amino acids. *Tetrahedron Asymm.*, **17**, 1793–1811.

21. Seebach, D., Schaeffer, L., Gessier, F. *et al.* (2003) Enantioselective preparation of 2-aminomethyl carboxylic acid derivatives: solving the (2-amino acid problem with the chiral auxiliary 4-Isopropyl-5,5-diphenyloxazolidin-2-one (DIOZ). Preliminary Communication. *Helv. Chim. Acta*, **86**, 1852–1861.

22. (a) Cole, D.C. (1994) Recent stereoselective synthetic approaches to β-amino acids. *Tetrahedron*, **50**, 9517–9582; (b) Juaristi, E. (1997) *Enantioselective Synthesis of β-Amino Acids*, Wiley-VCH, New York, (c) Cardillo, G. and Tomasini, C. (1996) Asymmetric synthesis of ß-amino acids and α-substituted β-amino acids. *Chem. Soc. Rev.*, **25**, 117–128; (d) Juaristi, E. and Lòpez-Ruiz, H. (1999) Recent advances in the enantioselective synthesis of beta-amino acids. *Curr. Med. Chem.*, **6**, 983–1004.

23. (a) Liu, M. and Sibi, M.P. (2002) Recent advances in the stereoselective synthesis of β-amino acids. *Tetrahedron*, **57**, 7991–7997; (b) Juaristi, E. and Soloshnok, V. (2005) *Enantioselective Synthesis of β-Amino Acids*, John Wiley & Sons, Inc., Hoboken; (c) Ma, J.-A. (2003) Recent developments in the catalytic asymmetric synthesis of α- and ß-amino acids. *Angew. Chem. Int. Ed.*, **42**, 4290–4299.

24. Weiner, B., Szymański, W., Janssen, D.B. *et al.* (2010) Recent advances in the catalytic asymmetric synthesis of β-amino acids. *Chem. Soc. Rev.*, **39**, 1656–1691.

25. (a) Bruneau, C. Renaud, J.-L., and Jerphagnon, T. (2008) Synthesis of β-aminoacid derivatives *via* enantioselective hydrogenation of β-substituted-β-(acylamino)acrylates. *Coord. Chem. Rev.*, **252**, 532–544; (b) Zhang, W., Chi, Y., and Zhang, X. (2007) Developing chiral ligands for asymmetric hydrogenation. *Acc. Chem. Res.*, **40**, 1278–1290; (c) Shimizu, H., Nagasaki, I., Matsumura, K. *et al.* (2007) Developments in asymmetric hydrogenation from an industrial perspective. *Acc. Chem. Res.*, **40**, 1385–1393; (d) Juaristi, E., Gutiérrez-Garcìa, V.M., and Lopez- Ruiz, H. (2005) *Enantioselective Synthesis of β-Amino Acids* (eds E. Juaristi and V. Soloshnok), Wiley-VCH, Hoboken, p. 159; (e) Brown, J.M. (1999) *Comprehensive Asymmetric Catalysis*, vol. **1** (eds E.N. Jacobsen, A. Pfaltz, and H. Yamamoto), Springer, Berlin, p. 121; (f) Tang, W. and Zhang, X. (2003) New chiral phosphorus ligands for enantioselective hydrogenation. *Chem. Rev.*, **103**, 3029–3070; (g) Heller, D., De Vries, A.H.M., and De Vries, J.G. (2008) *The Handbook of Homogeneous Hydrogenation* (eds J.G. de Vries and C.J. Elsevier), Wiley-VCH, Weinheim.

26. Kobayashi, S., Matsubara, R., Nakamura, Y. *et al.* (2003) Catalytic, asymmetric mannich-type reactions of *N*-acylimino esters: reactivity, diastereo- and enantioselectivity, and application to synthesis of N-acylated amino acid derivatives. *J. Am. Chem. Soc.*, **125**, 2507–2515.

27. Ooi, T., Kameda, M., Fujii, J.-I., and Maruoka, K. (2004) Catalytic asymmetric synthesis of a nitrogen analogue of dialkyl tartrate by direct mannich reaction under phase-transfer conditions. *Org. Lett.*, **6**, 2397–2399.

28. Cordova, A., Watanabe, S.-I., Tanaka, F. *et al.* (2002) A highly enantioselective route to either enantiomer of both α- and β-amino acid derivatives. *J. Am. Chem. Soc.*, **124**, 1866–1867.

29. Klettke, K.L., Sanyal, S., Mutatu, W., and Walker, K.D. (2007) β-Styryl- and β-Aryl-β-alanine products of phenylalanine aminomutase catalysis. *J. Am. Chem. Soc.*, **129**, 6988–6989.

30. Fülöp, F., Martinek, T.A., and Tóth, G.K. (2006) Application of alicyclic β-amino acids in peptide chemistry. *Chem. Soc. Rev.*, **35**, 323–334.

31. Fülöp, F. (2001) The chemistry of 2-Aminocycloalkanecarboxylic acids. *Chem. Rev.*, **101**, 2181–2204.

32. Izquierdo, S., Rua, F., Sbai, A. *et al.* (2005) (+)- and (−)-2-Aminocyclobutane-1-carboxylic Acids and their incorporation into highly rigid β-peptides: stereoselective synthesis and a structural study. *J. Org. Chem.*, **70**, 7963–7971.

33. Davies, S.G., Smith, A.D., and Price, P.D. (2005) The conjugate addition of enantiomerically pure lithium amides as homochiral ammonia equivalents: scope, limitations and synthetic applications. *Tetrahedron Asymm.*, **16**, 2833–2891.

34. Cheng, R.P., Gellman, S.H., and DeGrado, W.F. (2001) β-peptides: from structure to function. *Chem. Rev.*, **101**, 3219–3232.

35. Banerjee, A. and Balaram, P. (1997) Stereochemistry of peptides and polypeptides containing omega amino acids. *Curr. Sci.*, **73**, 1067–1077.

36. Seebach, D., Abele, S., Gademann, K. *et al.* (1998) β2- and β3-peptides with proteinaceous side chains: synthesis and solution structures of constitutional isomers, a novel helical secondary structure and the influence of solvation and hydrophobic interactions on folding. *Helv. Chim. Acta*, **81**, 932–982.

37. (a) Appella, D.H., Barchi, J.J., Durell, S.R., and Gellman, S.H. (1999) Formation of short, stable helices in aqueous solution by β-amino acid hexamers. *J. Am. Chem. Soc.*, **121**, 2309–2310; (b) Applequist, J., Bode, K.A., Appella, D.H. *et al.* (1998) Theoretical and experimental circular dichroic spectra of the novel helical foldamer Poly[(1R,2R)-*trans*-2-aminocyclopentanecarboxylic acid]. *J. Am. Chem. Soc.*, **120**, 4891–4892; (c) Wang, X., Espinosa, J.F., and Gellman, S.H. (2000) 12-Helix formation in aqueous solution with short β-peptides containing pyrrolidine-based residues. *J. Am. Chem. Soc.*, **122**, 4821–4822; (d) Seebach, D., Abele, S., Gademann, K., and Jaun, B. (1999) Pleated sheets and turns of (-peptides with proteinogenic side chains. *Angew. Chem. Int. Ed.*, **38**, 1595–1597.

38. Appella, D.H., Christianson, L.A., Klein, D.A. *et al.* (1997) Residue-based control of helix shape in β-peptide oligomers. *Nature*, **387**, 381–384.

39. Seebach, D., Hook, D.F., and Glattli, A. (2006) Helices and other secondary structures of β- and γ-peptides. *Biopolym. Pept. Sci.*, **84**, 23–37.

40. Bouillère, F., Thétiot-Laurent, S., Kouklovsky, C., and Alezra, V. (2011) Foldamers containing γ-amino acid residues or their analogues: structural features and applications. *Amino Acids*, **41**, 687–707.

41. (a) Hintermann, T., Gademann, K., Jaun, B., and Seebach, D. (1998) γ-peptides forming more stable secondary structures than α-peptides: synthesis and helical NMR-solution structure of the γ-hexapeptide analog of H-(Val-Ala-Leu)2-OH. *Helv. Chim. Acta*, **81**, 983–1002; (b) Hanessian, S., Luo, X., Schaum, R., and Michnick, S. (1998) Design of secondary structures in unnatural peptides: stable helical γ-Tetra-, Hexa-, and octapeptides and consequences of α-substitution. *J. Am. Chem. Soc.*, **120**, 8569–8570.

42. (a) Baldauf, C., Günther, R., and Hofmann, H.-J. (2003) Helix formation and folding in (-peptides and their vinylogues. *Helv. Chim. Acta*, **86**, 2573–2588; (b) Baldauf, C., Günther, R., and Hofmann, H.-J. (2004) Mixed helices—a general folding pattern in homologous peptides? *Angew. Chem. Int. Ed.*, **43**, 1594–1597.

43. (a) Sharma, G.V.M., Jayaprakash, P., Narsimulu, K. *et al.* (2006) A left-handed 9-Helix in γ-peptides: synthesis and conformational studies of oligomers with dipeptide repeats of C-linked Carbo-γ4-amino acids and γ-aminobutyric acid. *Angew. Chem. Int. Ed.*, **45**, 2944–2947; (b) Vasudev, P.G., Shamala, N., Anando, K., and Balaram, P. (2005) C9 Helices and

ribbons in γ-peptides: crystal structures of gabapentin oligomers. *Angew. Chem. Int. Ed.*, **44**, 4972–4975.

44. Hannessian, S., Luo, X., and Schaum, R. (1999) Synthesis and folding preferences of γ-amino acid oligopeptides: stereochemical control in the formation of a reverse turn and a helix. *Tetrahedron Lett.*, **40**, 4925–4929.

45. Vasudev, P.G., Chatterjee, S., Shamala, N., and Balaram, P. (2011) Structural chemistry of peptides containing backbone expanded amino acid residues: conformational features of β, γ, and hybrid peptides. *Chem. Rev.*, **111**, 657–687.

46. (a) For some very recent reviews see: Pils, L.K.A. and Reiser, O. (2011) α/β-Peptide foldamers: state of the art, b) Structural chemistry of peptides containing backbone expanded amino acid residues: conformational features of β, γ, and hybrid peptides. *Amino Acids*, **41**, 709–718; (b) Horne, W.S., and Gellman, S.H. (2008) Foldamers with heterogeneous backbones. *Acc. Chem. Res.*, **41**, 1399–1408.

47. (a) De Pol, S., Zorn, C., Klein, C.D. *et al.* (2004) Surprisingly stable helical conformations in α/β-peptides by incorporation of *cis*-β-aminocyclopropane carboxylic acids. *Angew. Chem. Int. Ed.*, **43**, 511–514; (b) Hayen, A., Schmitt, M.A., Ngassa, F.N. *et al.* (2004) Two helical conformations from a single foldamer backbone: "Split Personality" in short α/β-peptides. *Angew. Chem. Int. Ed.*, **43**, 505–510.

48. (a) Baruah, P.K., Sreedevi, N.K., Gonnade, R. *et al.* (2007) Enforcing periodic secondary structures in hybrid peptides: a novel hybrid foldamer containing periodic γ-turn motifs. *J. Org. Chem.*, **72**, 636–639; (b) Schramm, P., Sharma, G.V.M., and Hofmann, H.-J. (2009) Helix formation in beta/delta-hybrid peptides: correspondence between helices of different peptide foldamer classes. *Biopolym. Pept. Sci.*, **92**, 279–291; (c) Sharma, G.V.M., Shobar Babu, B., Ramakrishna, K.V.S. *et al.* (2009) Synthesis and structure of α/δ-hybrid peptides— access to novel helix patterns in foldamers. *Chem. Eur. J.*, **15**, 5552–5566.

49. (a) Schmitt, M.A., Choi, S.H., Guzei, I.A., and Gellman, S.H. (2005) Residue requirements for helical folding in short α/β-peptides: crystallographic characterization of the 11-Helix in an optimized sequence. *J. Am. Chem. Soc.*, **127**, 13130–13131; (b) Choi, S.H., Guzei, I.A., and Gellman, S.H. (2007) Crystallographic characterization of the α/β-peptide 14/15-Helix. *J. Am. Chem. Soc.*, **129**, 13780–13781.

50. Bolin, K.A. and Millhauser, G.L. (1999) α and 3₁₀: the split personality of polypeptide helices. *Acc. Chem. Res.*, **32**, 1027–1033.

51. Sharma, G.V.M., Chandramouli, N., Choudhary, M. *et al.* (2009) α and 3₁₀: the split personality of polypeptide helices. *J. Am. Chem. Soc.*, **131**, 17335–17344.

52. An early example was described by Karle, I.L., Pramanik, A., Banerjee, A. *et al.* (1997) ω-amino acids in peptide design. Crystal structures and solution conformations of peptide helices containing a β-alanyl-γ-aminobutyryl segment. *J. Am. Chem. Soc.*, **119**, 9087–9095.

53. Baldauf, C., Günther, R., and Hofmann, H.-J. (2006) Helix formation in α,γ- and β,γ-hybrid peptides: theoretical insights into mimicry of α- and β-peptides. *J. Org. Chem.*, **71**, 1200–1208.

54. Guo, L., Almeida, A.M., Zhang, W. *et al.* (2010) Helix formation in preorganized β/γ-peptide foldamers: hydrogen-bond analogy to the α-Helix without α-amino acid residues. *J. Am. Chem. Soc.*, **132**, 7868–7869.

55. Sharma, G.V.M., Babu, B.S., Ramakrishna, K.V.S. *et al.* (2009) Synthesis and structure of α/δ-hybrid peptides – access to novel helix patterns in foldamers. *Chem. Eur. J.*, **15**, 5552–5566.

56. (a) Harris, K.D.M. (1997) Meldola lecture: understanding the properties of urea and thiourea inclusion compounds. *Chem. Soc. Rev.*, **26**, 279–289; (b) Cram, D.J. and Cram, J.M. (1997) *Container Molecules and their Guests*, Royal Society of Chemistry, Cambridge.

57. Fischer, L. and Guichard, G. (2010) Folding and self-assembly of aromatic and aliphatic urea oligomers: Towards connecting structure and function. *Org. Biomol. Chem.*, **8**, 3101–3117.

58. (a) See for instance: Semetey, V., Didierjean, C., Briand, J.-P. *et al.* (2002) Self-assembling organic nanotubes from enantiopure cyclo-*N,N'*-linked oligoureas: design, synthesis, and crystal structure. *Angew. Chem. Int. Ed.*, **41**, 1895–1898; (b) Claudon, P., Violette, A., Lamour, K. *et al.* (2010) Consequences of isostructural main-chain modifications for the design of antimicrobial foldamers: helical mimics of host-defense peptides based on a heterogeneous amide-/urea backbone. *Angew. Chem. Int. Ed.*, **49**, 333–336; (c) Fischer, L., Claudon, P., Pendem, N. *et al.* (2010) The canonical helix of urea oligomers at atomic resolution: insights into folding-induced axial organization. *Angew. Chem. Int. Ed.*, **49**, 1067–1070.

59. (a) Semetey, V., Rognan, D., Hemmerlin, C. *et al.* (2002) Stable helical secondary structure in short-chain *N,N'*-linked oligoureas bearing proteinogenic side chains. *Angew. Chem. Int. Ed.*, **41**, 1893–1895; (b) Hemmerlin, C., Marraud, M., Rognan, D. *et al.* (2002) Helix-forming oligoureas: temperature-dependent NMR, structure determination, and circular dichroism of a nonamer with functionalized side chains. *Helv. Chim. Acta*, **85**, 3692–3711.

60. (a) Hintermann, T., Gademann, K., Jaun, B., and Seebach, D. (1998) γ-peptides forming more stable secondary structures than α-peptides: synthesis and helical NMR-solution structure of the γ-hexapeptide analog of H-(Val-Ala-Leu)2-OH. *Helv. Chim. Acta*, **81**, 983–1002; (b) Hanessian, S., Luo, X., Schaum, R., and Michnick, S. (1998) Design of secondary structures in unnatural peptides: stable helical γ-Tetra-, Hexa-, and octapeptides and consequences of α-substitution. *J. Am. Chem. Soc.*, **120**, 8569–8570; (c) Seebach, D., Brenner, M., Rueping, M., and Jaun, B. (2002) γ2-, γ3-, and γ2,3,4-amino acids, coupling to γ-hexapeptides: CD spectra, NMR solution and X-ray crystal structures of γ-peptides. *Chem. Eur. J.*, **8**, 573–584.

61. Violette, A., Averlant-Petit, M.C., Semetey, V. *et al.* (2005) N,N'-linked oligoureas as foldamers: chain length requirements for helix formation in protic solvent investigated by circular dichroism, NMR spectroscopy, and molecular dynamics. *J. Am. Chem. Soc.*, **127**, 2156–2164.

62. (a) Li, X. Wu, Y.-D., and Yang, D. (2008) α-Aminoxy acids: new possibilities from foldamers to anion receptors and channels. *Acc. Chem. Res.*, **41**, 1428–1438; (b) Yang, D., Liu, G.-J., Hao, Y. *et al.* (2010) Conformational studies on peptides of α-aminoxy acids with functionalized side-chains. *Chem. Asian J.*, **5**, 1356–1363; (c) Li, X., and Yang, D. (2006) Peptides of aminoxy acids as foldamers. *Chem. Commun.*, **32**, 3367–3379.

63. Yang, D., Ng, F.-F., Li, Z.-J. *et al.* (1996) An unusual turn structure in peptides containing α-aminoxy acids. *J. Am. Chem. Soc.*, **118**, 9794–9795.

64. (a) Yang, D., Qu, J., Li, B. *et al.* (1999) Novel turns and helices in peptides of chiral α-aminoxy acids. *J. Am. Chem. Soc.*, **121**, 589–590; (b) Yang, D., Li, B., Ng, F.F. *et al.* (2001) Synthesis and characterization of chiral N–O turns induced by α-aminoxy acids. *J. Org. Chem.*, **66**, 7303–7312.

65. Yang, D., Li, W., Qu, J. *et al.* (2003) A new strategy to induce γ-turns: peptides composed of alternating α-aminoxy acids and α-amino acids. *J. Am. Chem. Soc.*, **125**, 13018–13019.

66. Tomasini, C. and Villa, M. (2001) Pyroglutamic acid as a pseudoproline moiety: a facile method for its introduction into polypeptide chains. *Tetrahedron Lett.*, **42**, 5211–5214.

67. Tomasini, C., Angelici, G., and Castellucci, N. (2011) Foldamers based on Oxazolidin-2-one. *Eur. J. Org. Chem.*, **2011**, 3648–3669.

68. (a) Lucarini, S. and Tomasini, C. (2001) Synthesis of oligomers of trans-(4S,5R)-4-Carboxybenzyl 5-Methyl Oxazolidin-2-one: an approach to new foldamers. *J. Org. Chem.*, **66**, 727–732; (b) Bernardi, F., Garavelli, M., Scatizzi, M. *et al.* (2002) Pseudopeptide foldamers: the homo-oligomers of pyroglutamic acid. *Chem. Eur. J.*, **8**, 2516–2525; (c) Tomasini, C., Trigari, V., Lucarini, S. *et al.* (2003) Pseudopeptide foldamers–the homo-oligomers of benzyl (4S,5R)-5-Methyl-2-oxo-1,3-oxazolidine-4-carboxylate. *Eur. J. Org. Chem.*, 259–267.

69. Luppi, G., Soffrè, C., and Tomasini, C. (2004) Stabilizing effects in oxazolidin-2-ones-containing pseudopeptides. *Tetrahedron Asymm.*, **15**, 1645–1650.

70. Tomasini, C., Luppi, G., and Monari, M. (2006) Oxazolidin-2-one-containing pseudopeptides that fold into β-Bend ribbon spirals. *J. Am. Chem. Soc.*, **128**, 2410–2420.

71. (a) Angelici, G., Falini, G., Hofmann, H.-J. *et al.* (2008) A fiberlike peptide material stabilized by single intermolecular hydrogen bonds. *Angew. Chem. Int. Ed.*, **47**, 8075–8078; (b) Angelici, G., Falini, G., Hofmann, H.-J. *et al.* (2009) Nanofibers from Oxazolidi-2-one containing hybrid foldamers: what is the right molecular size? *Chem. Eur. J.*, **15**, 8037–8048.

72. Angelici, G., Castellucci, N., Falini, G. *et al.* (2010) Pseudopeptides designed to form supramolecular helixes: the role of the stereogenic centers. *Crystal Growth Des.*, **10**, 923–929.

73. (a) Huc, I. (2004) Aromatic oligoamide foldamers. *Eur. J. Org. Chem.*, **1**, 17–29; (b) Saraogi, I. and Hamilton, A.D. (2009) Recent advances in the development of aryl-based foldamers. *Chem. Soc. Rev.*, **38**, 1726–1743; (c) Sanford, A. and Gong, B. (2003) The evolution of helical foldamers. *Curr. Org. Chem.*, **7**, 1649–1659.

74. (a) Hamuro, Y., Geib, S.J., and Hamilton, A.D. (1994) Novel molecular scaffolds: formation of helical secondary structure in a family of oligoanthranilamides. *Angew. Chem. Int. Ed.*, **33**, 446–448; (b) Hamuro, Y., Geib, S.J., and Hamilton, A.D. (1996) Oligoanthranilamides. Nonpeptide subunits that show formation of specific secondary structure. *J. Am. Chem. Soc.*, **118**, 7529–7541; (c) Hamuro, Y., Geib, S.J., and Hamilton, A.D. (1997) Novel folding patterns in a family of oligoanthranilamides: non-peptide oligomers that form extended helical secondary structures. *J. Am. Chem. Soc.*, **119**, 10587–10593.

75. (a) Tew, G.N., Liu, D.H., Chen, B. *et al.* (2002) Supramolecular chemistry and self-assembly special feature: *de novo* design of biomimetic antimicrobial polymers. *Proc. Natl Acad. Sci. USA*, **99**, 5110–5114; (b) Choi, S., Clements, D.J., Pophristic, V. *et al.* (2005) The design and evaluation of heparin-binding foldamers. *Angew. Chem. Int. Ed.*, **44**, 6685–6689.

76. Kolomiets, E., Berl, V., and Lehn, J.M. (2007) Chirality induction and protonation-induced molecular motions in helical molecular strands. *Chem. Eur. J.*, **13**, 5466–5479.

77. Gan, Q., Bao, C.Y., Kauffmann, B. *et al.* (2008) Quadruple and double helices of 8-fluoroquinoline oligoamides. *Angew. Chem. Int. Ed.*, **47**, 1715–1718.

78. Zhang, A.M., Han, Y.H., Yamato, K. *et al.* (2006) Aromatic oligoureas: enforced folding and assisted cyclization. *Org. Lett.*, **8**, 803–806.

79. (a) Sinkeldam, R.W., van Houtem, M., Koeckelberghs, G. *et al.* (2006) Synthesis of 3,6-diaminophthalimides for ureidophthalimide-based foldamers. *Org. Lett.*, **8**, 383–385; (b) Sinkeldam, R.W., van Houtem, M., Pieterse, K. *et al.* (2006) Chiral Poly(ureidophthalimide) foldamers in water. *Chem. Eur. J.*, **12**, 6129–6137.

80. Cai, W., Wang, G.T., Xu, Y.X. *et al.* (2008) Vesicles and organogels from foldamers: a solvent-modulated self-assembling process. *J. Am. Chem. Soc.*, **130**, 6936–6937.

81. (a) Nelson, J.C., Saven, J.G., Moore, J.S., and Wolynes, P.G. (1997) Solvophobically driven folding of nonbiological oligomers. *Science*, **277**, 1793–1796; (b) Prince, R.B., Barnes, S.A., and Moore, J.S. (2000) Foldamer-based molecular recognition. *J. Am. Chem. Soc.*, **122**, 2758–2762; (c) Stone, M.T. and Moore, J.S. (2004) A Water-Soluble m-Phenylene ethynylene foldamer. *Org. Lett.*, **6**, 469–472; (d) Smaldone, R.A. and Moore, J.S. (2008) Sequence dependence of methylation rate enhancement in meta-phenyleneethynylene foldamers. *Chem. Commun.*, **2008**, 1011–1013; (e) Smaldone, R.A. and Moore, J.S. (2007) Foldamers as reactive sieves: reactivity as a probe of conformational flexibility. *J. Am. Chem. Soc.*, **129**, 5444–5450; (f) Smaldone, R.A. and Moore, J.S. (2008) Reactive sieving with foldamers: inspiration from nature and directions for the future. *Chem. Eur. J.*, **14**, 2650–2657.

82. Khan, A., Kaiser, C., and Hecht, S. (2006) Prototype of a photoswitchable foldamer. *Angew. Chem. Int. Ed.*, **45**, 1878–1881.

83. (a) Garric, J. Leger, J.M., and Huc, I. (2005) Molecular apple peels. *Angew. Chem. Int. Ed.*, **44**, 1954–1958; (b) Garric, J., Leger, J.M., and Huc, I. (2007) Encapsulation of small polar guests in molecular apple peels. *Chem. Eur. J.*, **13**, 8454–8462.

84. Bao, C., Kauffman, B., Gan, Q. *et al.* (2008) Converting sequences of aromatic amino acid monomers into functional three-dimensional structures: second-generation helical capsules. *Angew. Chem. Int. Ed.*, **47**, 4153–4156.

85. (a) Gale, P.A. and Gunnlaugsson., T. (2010) Preface: supramolecular chemistry of anionic species themed issue. *Chem. Soc. Rev.*, **39**, 3595–3596; (b) Juwarker, H. and Jeong, K.S. (2010) Encapsulation of small polar guests in molecular apple peels. *Chem. Soc. Rev.*, **38**, 585–605.

86. Juwarker, H. and Jeong, K.-S. (2010) Anion-controlled foldamers. *Chem. Soc. Rev.*, **39**, 3664–3674.

87. Sessler, J.L., Cyr, M.J., and Lynch, V. (1990) Synthetic and structural studies of sapphyrin, a 22-.pi.-electron pentapyrrolic "expanded porphyrin". *J. Am. Chem. Soc.*, **112**, 2810–2813.

88. Sessler, J.L. and Weghorn, S.J. (1994) Vincent Lynch and Kjell Fransson, 5,15,25-tris-nor-Hexapyrrin: the first structurally characterized linear hexapyrrin. *J. Chem. Soc. Chem. Commun.*, **1994**, 1289–1290.

89. Sessler, J.L., Maeda, H., Mizuno, T. *et al.* (2002) Quinoxaline-oligopyrroles: Improved pyrrole-based anion receptors. *Chem. Commun.*, **2002**, 862–863.

90. Sessler, J.L., Dan Pantos, G., Katayev, E., and Lynch, V.M. (2003) Pyrazine analogues of dipyrrolylquinoxalines. *Org. Lett.*, **5**, 4141–4144.

91. Maeda, H., Fujii, Y., and Mihashi, Y. (2008) Diol-substituted boron complexes of dipyrrolyl diketones as anion receptors and covalently linked 'pivotal' dimmers. *Chem. Commun.*, **2008**, 4285–4287.

92. Chang, K.-J., Kang, B.-N., Lee, M.-H., and Jeong, K.-S. (2005) Oligoindole-based foldamers with a helical conformation induced by chloride. *J. Am. Chem. Soc.*, **127**, 12214–12215.

93. Kim, U.-I., Suk, J.-m., Naidu, V.R., and Jeong, K.-S. (2008) Folding and anion-binding properties of fluorescent oligoindole foldamers. *Chem. Eur. J.*, **14**, 11406–11414.

94. Bock, V.D., Hiemstra, H., and Van Maarseveen, J.H. (2006) CuI-catalyzed Alkyne–Azide "Click" cycloadditions from a mechanistic and synthetic perspectiv. *Eur. J. Org. Chem.*, **2006**, 51–68.

95. Angelo, N.G. and Arora, P.S. (2007) Solution- and solid-phase synthesis of triazole oligomers that display protein-like functionality. *J. Org. Chem.*, **72**, 7963–7967.

96. Juwarker, H., Lenhardt, J.M., Pham, D.M., and Craig, S.L. (2008) Selective oxidation of aliphatic C—H bonds in the synthesis of complex molecules. *Angew. Chem. Int. Ed.*, **47**, 3740–3743.

97. Meudtner, R.M. and Hecht, S. (2008) Helicity inversion in responsive foldamers induced by achiral halide ion guests. *Angew. Chem. Int. Ed.*, **47**, 4926–4930.

98. Sanchez-Quesada, J., Seel, C., Prados, P. *et al.* (1996) Anion helicates: double strand helical self-assembly of chiral bicyclic guanidinium dimers and tetramers around sulfate templates. *J. Am. Chem. Soc.*, **118**, 277–278.

99. Keegan, J., Kruger, P.E., Nieuwenhuyzen, M. *et al.* (2001) Anion directed assembly of a dinuclear double helicate. *Chem. Commun.*, **2001**, 2192–2193.

100. Coles, S.J., Frey, J.G., Gale, P.A. *et al.* (2003) Anion-directed assembly: the first fluoride-directed double helix. *Chem. Commun.*, **2003**, 568–569.

101. Xu, Y.-X., Wang, G.-T., Zhao, X. *et al.* (2009) Folding of aromatic amide-based oligomers induced by Benzene-1,3,5-tricarboxylate anion in DMSO. *J. Org. Chem.*, **74**, 7267–7273.

102. Meshcheryakov, D., Böhmer, V., Bolte, M. *et al.* (2006) Two chloride ions as a template in the formation of a cyclic hexaurea. *Angew. Chem. Int. Ed.*, **45**, 1648–1652.

103. Sessler, J.L. and Siedel, D. (2003) Synthetic expanded porphyrin chemistry. *Angew. Chem. Int. Ed.*, **42**, 5134–5175.

104. Sessler, J.L., Weghorn, S.J., Lynch, V.M., and Johnson, M.R. (1994) Turcasarin, the largest expanded porphyrin to date. *Angew. Chem. Int. Ed.*, **33**, 1509–1512.

105. Gale, P.A., Navakhun, K., Camiolo, S. *et al.* (2002) Anion–anion assembly: a new class of anionic supramolecular polymer containing 3,4-Dichloro-2,5-diamido-substituted pyrrole anion dimers. *J. Am. Chem. Soc.*, **124**, 11228–11229.

106. Maeda, H. and Kusunose, Y. (2005) Dipyrrolyldiketone difluoroboron complexes: novel anion sensors with C—H⋯?X— interactions. *Chem. Eur. J.*, **11**, 5661–5666.

107. (a) Guichard, G., Zerbib, A., Leal, F.-A. *et al.* (2000) Melanoma peptide MART-1(27-35) analogues with enhanced binding capacity to the human class I histocompatibility molecule HLA-A2 by introduction of a β-amino acid residue: implications for recognition by tumor-infiltrating lymphocytes. *J. Med. Chem.*, **43**, 3803–3808; (b) Reinelt, S., Marti, M., Dedier, S. *et al.* (2001) Protein structure and folding. *J. Biol. Chem.*, **276**, 24525–24530.

108. Schmitt, M.A., Weisblum, B., and Gellman, S.H. (2007) Interplay among folding, sequence, and lipophilicity in the antibacterial and hemolytic activities of α/β-peptides. *J. Am. Chem. Soc.*, **129**, 417–428.

109. Ahmed, S. and Kaur, K. (2009) The proteolytic stability and cytotoxicity studies of L-aspartic acid and L-diaminopropionic acid derived β-peptides and a mixed α/β-peptide. *Chem. Biol. Drug. Des.*, **73**, 545–552.

110. Schmitt, M.A., Weisblum, B., and Gellman, S.H. (2004) Unexpected relationships between structure and function in α,β-peptides: antimicrobial foldamers with heterogeneous backbones. *J. Am. Chem. Soc.*, **126**, 6848–6849.

111. Mowery, B.P., Lee, S.E., Kissounko, D.A. *et al.* (2007) Mimicry of antimicrobial host-defense peptides by random copolymers. *J. Am. Chem. Soc.*, **129**, 15474–15476.

112. Bryson, J.W., Betz, S.F., Lu, H.S. *et al.* (1995) Protein design: a hierarchic approach. *Science*, **270**, 935–941.

113. Chakraborty, S., Yudenfreund Kravitz, J., Thulstrup, P.W. *et al.* (2011) Design of a three-helix bundle capable of binding heavy metals in a triscysteine environment. *Angew. Chem. Int. Ed.*, **50**, 2049–2053.

114. (a) Daniels, D.S., Petersson, E.J., Qiu, J.X., and Schepartz, A. (2007) High-resolution structure of a β-peptide bundle. *J. Am. Chem. Soc.*, **129**, 1532–1533; (b) Cheng, R.P. and DeGrado, W.F. (2002) Long-range interactions stabilize the fold of a non-natural oligomer. *J. Am. Chem. Soc.*, **124**, 11564–11565. (c) Raguse, T.L., Lai, J.R., LePlae, P.R., and Gellman, S.H. (2001) Toward β-peptide tertiary structure: self-association of an amphiphilic 14-helix in aqueous solution. *Org. Lett.*, **3**, 3963–3966.

115. (a) Craig, C.J., Goodman, J.L., and Schepartz, A. (2011) Enhancing β3-peptide bundle stability by design. *ChemBioChem*, **12**, 1035–1038; (b) Daniels, D.S., Petersson, E.J., Qiu, J.X., and Schepartz, A. (2007) High-resolution structure of a β-peptide bundle. *J. Am. Chem. Soc.*, **129**, 1532–1533; (c) Goodman, J.L., Molski, M.A., Qiu, J., and Schepartz, A. (2008) Tetrameric β3-peptide bundles. *ChemBioChem*, **9**, 1576–1578; (d) Goodman, J.L., Petersson, E.J., Daniels, D.S. *et al.* (2007) Biophysical and structural characterization of a robust octameric β-peptide bundle. *J. Am. Chem. Soc.*, **129**, 14746–14751.

# 3

# Self-Assembly Principles of Helicates

## *Physicochemical Principles of Self-Assembly Processes*

*Josef Hamacek*

*Department of Inorganic and Analytical Chemistry, University of Geneva, Switzerland*

## 3.1 Introduction

The development of supramolecular chemistry in the 1990s was closely linked with the introduction of novel and attractive semantic terms (self-organization, self-recognition, spontaneous assembly, etc.) which emerged for describing the macroscopic events resulting from a high complexity of studied systems [1,2]. The design of new supramolecular compounds was often inspired by biologically active compounds or geometrical objects. Each particular category of supramolecules is formed by metallosupramolecular compounds, which are obtained as the product of strict self-assembly, according to Lindsey's classification of self-assembly processes [3]. These architectures are built due to the formation of coordination bonds between metal ions and appropriate ligands, which are usually designed as organic receptors possessing several coordinating atoms or multidentate binding sites. Although the supramolecules are the most stable products of the reactions (the system reaches a thermodynamic minimum), the bond formation is completely kinetically reversible and allows the repair of "mistakes" arising from competitive pathways. The early description and characterization of self-assembly processes was limited to structural factors governing supramolecular reactions. The

*Metallofoldamers: Supramolecular Architectures from Helicates to Biomimetics*, First Edition.
Edited by Galia Maayan and Markus Albrecht.
© 2013 John Wiley & Sons, Ltd. Published 2013 by John Wiley & Sons, Ltd.

improvement of characterization and analytical techniques allowed aesthetically fascinating structures of great complexity to be revealed. Apparently, the coordination processes resulted in a single multicomponent product that was formed relatively easily, and some "magical" properties were often attributed to supramolecular systems. To rationalize the design of complex assemblies, the empirical laws for obtaining desired polynuclear complexes have been described by many authors. In this context, the Raymond's attempt to rationalize the designing of linear helicates and 3D structures can be mentioned [4]. Relatively recently, progress in spectroscopic methods (NMR, ESMS) and other characterization techniques has allowed a reliable exploration of self-assembly equilibria in solution. All these achievements significantly contributed to a better understanding of self-organized systems.

Artificial helicates represent archetypal metallosupramolecular compounds [5], whose design was inspired by biochemical self-assemblies such as the double-stranded DNA structure, whereby hydrogen bonding is replaced with coordination bonds. The helicates are discrete well defined assemblies usually composed of a limited number of components, that is, one ligand and one metal ion [6]. This simple system is thus ideally suited for fundamental investigations of supramolecular self-assemblies in order to better understand and describe their structure, thermodynamics and the formation mechanism behind. The NMR spectra of helicates are relatively simple due to a high molecular symmetry. Using specific structural NMR probes [6] in combination with other spectroscopic techniques allows complex speciation studies to be performed. Not surprisingly, a number of features controlling self-assembly were discovered with helicates, which gives them a special importance in a general understanding of self-organization. This progress is essential: (i) for a precise design of new functional materials with a bottom-up approach and (ii) for applying efficient predictive strategies to evaluate the structural and physico-chemical properties of desired compounds before a time-consuming synthesis [7].

The aim of this chapter is to provide a conceptual overview of basic self-assembly principles and related problems, which are discussed in order of increasing complexity. The details about physical origins and mathematical treatments can be found by experienced readers in original papers. A special focus is given to their implementation in the chemistry of helicates. In the first part, physico-chemical concepts of fundamental processes dealing with the formation of metal–ligand bonds and their supramolecular counterparts will be given for mononuclear coordination compounds. These principles will be then transposed to supramolecular helicates, highlighting the specific features accompanying this extension, especially the consideration of intramolecular interactions. A separate part is devoted to the description and assessment of cooperative interactions that may play an important role in the stabilization of self-assembled edifices. Kinetic factors are discussed in relation to the self-assembly mechanisms of dinuclear helicates. An overview of tools for assessing cooperativity in helicates is presented. The development and applications of thermodynamic modelling are described in detail in the next section. Finally, possible secondary interactions that control the structure of metallohelicates (helical conformational folding) are also qualitatively discussed. Tutorial examples of helical assemblies are given across the whole chapter. However, these examples are non-exhaustive and are selected only for illustrating the phenomena in question.

## 3.2　Thermodynamic Considerations in Self-Assembly

Spontaneous reactions tend to minimize the free energy $\Delta G$ of a chemical system at given experimental conditions. In mononuclear coordination compounds, the formation of favourable interactions between metal ions and ligands overcomes the energy-consuming desolvation processes of reagents. The structure of the resulting complexes depends on the concentration and inherent properties of the metal ions and ligands involved. When describing the self-assembly of supramolecular architectures including helicates, the same thermodynamic principles are taking place. Let us consider in detail the physico-chemical description of simple mononuclear complexes that are well experienced in coordination chemistry and that are often used as building blocks in supramolecular chemistry. These basic concepts will then be extended to the appropriate thermodynamic description of polynuclear compounds and illustrated with examples of helicates.

### 3.2.1　Mononuclear Coordination Complexes

Basic physico-chemical principles can be briefly summarized in the following three points that are closely interconnected [2,8]. To maximize the energetic gain associated with the complexation of one metal ion (i.e., a receptor), the number of its interactions with $x$-dentate ligands must be also maximal. An optimal arrangement of coordinating ligands around metal ions with respect to their intrinsic properties is known in coordination chemistry as *stereochemical matching* [1]. Ideally, a metal ion with the coordination number $CN$ may accommodate $CN/x$ ligands. The coordinating atoms must be spatially disposed in such a way as to respect the coordination geometry of the metal ion. At this point, molecular recognition takes place and selects only complementary polydentate ligands. As an example, let us take a copper cation, which is commonly employed for the assembly of double-stranded helicates. The satisfaction of its tetrahedral coordination preferences requires the coordination of two bipyridine-based chelating ligands (Figure 3.1). The mononuclear copper complex forms the maximum of possible coordination bonds provided by the bidentate ligand and reaches its maximum occupancy state with a maximal energetic gain. The copper coordination sphere is saturated with the bidentate ligand

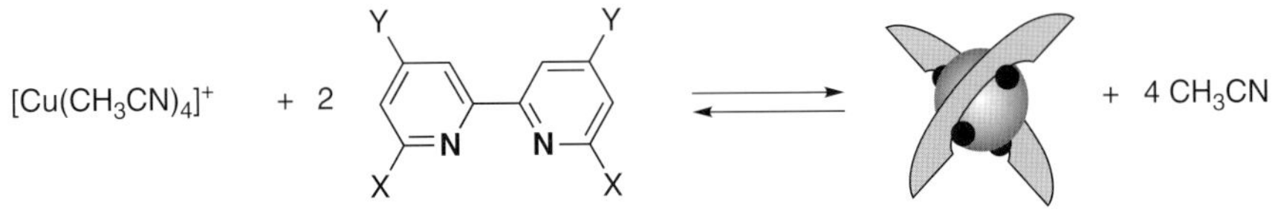

| Point group | $T_d$ | $C_{2v}$ | $D_{2v}$ | $C_{\infty h}$ |
|---|---|---|---|---|
| $\sigma^{ext}$ | 12 | 2 | 4 | 1 |
| $\sigma^{int}$ | $3^4$ | 1 | 1 | 3 |
| $\sigma^{mix}$ | 1 | 1 | 1 | 1 |

$$\omega_{1,2}^{Cu,L} = \frac{(12 \cdot 3^4 \cdot 1)\,(2 \cdot 1 \cdot 1)^2}{(4 \cdot 1 \cdot 1)(1 \cdot 3 \cdot 1)^4} = 12$$

**Figure 3.1**　*Reaction scheme for the formation of a simple copper(I) complex. The statistical factors $\omega_{m,n}^{M,L}$ are assessed with the symmetry number method [10].*

and the energetic gain cannot be further increased except by using a ligand with a stronger affinity. This classical principle was reformulated in supramolecular chemistry as the *principle of maximum occupancy* [9], implying that all receptor binding sites are involved in the assembly.

Of course, the reaction can be driven to a predicted completion only at specific exogenous conditions ($T$, pH, solvent, etc.). Among these, the stoichiometry of the reactants is probably the most important. An optimal set of external conditions can be referred as *experimental matching*.

### 3.2.1.1   Chelate Effect

Coordination reactions often occur with $x$-dentate ligands (Figure 3.1). The formation of multiple binding interactions (multivalency) is energetically favourable (decrease of $\Delta G$): (i) due to a positive entropy change in comparison with the binding of the equivalent number of monodentate ligands and (ii) due to a favourable enthalpy resulting from the preorganization of donor atoms [2]. This behaviour is well known in coordination chemistry as the chelate effect. A closer inspection of the chelate binding shows that the reaction of a multidentate ligand consists of an intermolecular binding of the first donor atom to a metal ion. This initial step is followed by one or more intramolecular connections with other donors in such a way as to incorporate the cation into the ring. A schematic illustration of the chelate effect in Figure 3.2 shows the reaction of two ligands **L**, each possessing two donor atoms, with a divalent metallic receptor **M**. The receptor and two ligands may mutually interact to give three possible complexes. The intermolecular binding of the first donor atom in **L** occurs with the microscopic affinity $\kappa$ and provides a partially bound $1:1$ open complex $o$-**ML**. The latter species may be further transformed by an intermolecular reaction with the second ligand to form a **ML**$_2$ complex, or by an intramolecular reaction to form the cyclic complex $c$-**ML** with the successive binding constant $\kappa.EM$, where $EM$ is the microscopic effective molarity and essentially parameterizes the difference between intra- and intermolecular processes with respect to the entropy and enthalpy [11–13]. Since $EM$ has a concentration unit, the corresponding free energy

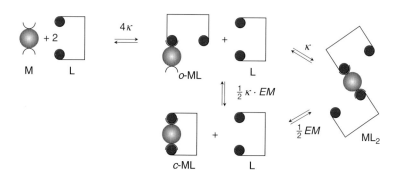

**Figure 3.2**   *Chelate effect in coordination chemistry with competing intramolecular and intermolecular reactions.* Adapted with permission from [11]. Copyright Wiley-VCH Verlag GmbH & Co. KGaA, Weinheim

change thus refers to the chosen reference state, that is, $c^{\theta} = 1$ M, and intramolecular binding is favoured ($EM > 1$ M) or disfavoured ($EM < 1$ M). In this context, a deep analysis of elementary processes in the chelate effect was recently presented by Ercolani [11,12]. The chelate effect efficiency depends on the concentration of **L** and on the magnitude of $EM$. However, the chelating donor atoms in common ligands are localized in the topological proximity and form preorganized binding sites, where the reaction with a neighbouring donor atom is strongly favoured because of a much higher effective molarity ($EM \gg 1$ M) compared with the donor atoms belonging to other ligand molecules. The favourable short-distance chelate effect thus occurs in a large domain of ligand concentrations. However, an optimal degree of freedom must be programmed in polydentate ligands that are often incorporated in segmental ligands for preparing helicates. The optimal ring size for a maximum energetic gain is five or six atoms including the coordinated metal ion [2]. The choice of donor atoms and ring sizes depends on the overall system design.

Considering the kinetic point of view, the binding of the second donor atom proceeds much more rapidly compared to the second monodentate ligand [2]. The $x$-dentate ligand connects to the receptor in a single connection point and in one reaction step, and such binding events can be globally viewed as an intermolecular connection. Moreover, a knowledge of the overall energetic effect associated with different multidentate binding sites is more suitable for a quantitative comparison between systems, without evaluating the contributions of individual donor ions. The total number of such intermolecular connections between metallic receptors and identical $x$-dentate ligands will be equal to $CN/x$.

The above simplification of the thermodynamic description found a practical utility in the interpretation and mechanistic descriptions of self-assembly processes [14] and in the thermodynamic modelling of stability constants [15]. The global free energy changes related to this complexation process are given as $\Delta G_{\text{inter}} = -RT \ln(k_i^{\text{M}}) < 0$ [13,15], where $k_i^{\text{M}}$ is the microscopic affinity constant related to the intermolecular connection of one $x$-dentate coordination site $i$ to a metallic receptor. In the case of the bipyridyl ligand containing two nitrogen donor atoms, the affinity constant can be calculated as $k_{\text{N}_2}^{\text{M}} = 2\kappa_{\text{N}}^2 . EM$, where $\kappa_{\text{N}}$ is the microscopic affinity of the pyridyl nitrogen donor atom to the metal ion. $\Delta G_{\text{inter}}$, which is associated with intermolecular connections, is negative and represents the main driving force responsible for the completion of thermodynamic assemblies.

### 3.2.1.2 Statistical Factors

The formation of self-assembled edifices may occur through different reaction pathways. The number of these possibilities is referred to as the degeneracy of the microscopic state [16]. This purely entropic driving force corresponds to the changes in rotational degeneracy related to the transformation of reactants into products. The statistical factors $\omega_{m,n}^{\text{M,L}}$ for the general complexation process (Equation 3.1) are calculated as $\omega_{m,n}^{\text{M,L}} = l/r$ [10,13], where $l$ is the number of microspecies $[\text{M}_m\text{L}_n]$ that can be formed, if all the identical atoms are labelled and $r$ corresponds to the same definition for the reverse reaction [13].

$$m[\text{M}(\text{solv})_x] + n\text{L} \quad \underset{r}{\overset{l}{\rightleftharpoons}} \quad [\text{M}_m\text{L}_n] + x\,\text{solv} \qquad \beta_{m,n}^{\text{M,L}} \qquad (3.1)$$

These factors can be conveniently calculated by using: (i) the direct counting method [10] or (ii) the symmetry number method [18], which provide equivalent results, as demonstrated by Ercolani *et al.* [10] While the direct counting method is more illustrative and useful for relatively simple systems, the symmetry number method is particularly suitable for large supramolecular assemblies. Using the latter method, the overall statistical factor for equilibrium 1 is calculated as the ratio of total symmetry numbers for reacting and produced microspecies according to Equation 3.2.

$$\omega_{m,n}^{M,L} = 1/r = \frac{(\sigma_{tot}^{M})^{m}(\sigma_{tot}^{L})^{n}}{(\sigma_{tot}^{M_{m}L_{n}})} \tag{3.2}$$

The total symmetry number $\sigma_{tot}$ of a microspecies is the product of the external $\sigma_{ext}$ and internal $\sigma_{int}$ symmetry numbers modulated by the mixing entropy $\sigma_{mix}$ produced by the number of isomeric microspecies contributing to the macrospecies. The graphical application of the symmetry numbers method for the formation of mononuclear Cu(I) complexes is given in Figure 3.1 and gives $\omega_{1,2}^{Cu,L} = 12$. The counting using the direct method gives six possibilities to arrange two bidentate ligands in the coordination tetrahedron of Cu(I). In addition, these two ligands may adopt two possible arrangements (head–head, head–tail). Since the resulting assembly is achiral, the overall statistical factor is calculated by multiplying all these possibilities, and $\omega_{1,2}^{Cu,L} = 2 \times 6 = 12$ is in agreement with the method of symmetry numbers. The statistical factor translates into the free energy contribution to the overall energetic balance with the expression $\Delta G_{stat}^{M,L} = -RT \ln(\omega_{m,n}^{M,L})$.

### 3.2.2    Extension to Polynuclear Edifices

The thermodynamic description of intermolecular connections and the chelate effect presented in the previous section apply also for large supramolecular systems [6]. However, their appropriate parameterization requires a special consideration of new factors related to these complexes. The formation of multicomponent assemblies necessarily implies the presence of two or more *x*-dentate binding sites within the ligand structure, which may become rather sophisticated. Therefore, the number of possible products increases rapidly with the number of components and with the complexity of ligands. The binding sites are interconnected with an organic linker, which plays an important role in the rational programming of desired assemblies. A good connector must exhibit a smooth balance between flexibility and rigidity in order to provide good spatial dispositions for efficient metal binding. With this difficult task in mind, a number of synthetic challenges must be resolved in order to prepare a "good" organic receptor. Moreover, the connection of different binding sites into a discrete supramolecular edifice requires intramolecular binding events (macrocyclization processes), which operate over long distances, compared with the chelate effect. Particularly in helicates, at least one intramolecular binding takes place and the magnitude of this interaction is crucial for the stability of the final edifice with respect to concurrent intermolecular reactions. Let us develop these phenomena in detail.

#### 3.2.2.1    Modelling Intramolecular Interactions

The mechanistic and thermodynamic description of long-distance intramolecular connections (ring size $\gg 6$, distinct binding sites) is, in principle, reminiscent of the short-distance chelate effect discussed for binding *x*-dentate binding sites in mononuclear complexes (Figure 3.2). The illustration of this macrocyclization reaction is given for the

**Figure 3.3** *Schematic view and the thermodynamic description of intramolecular macrocyclic connections in the self-assembly of double-stranded helicates (i.e., Ref. [19]).*

dinuclear double-stranded helicate in Figure 3.3. The initial intermolecular connection of a metal ion to the ligand coordination site $i$ occurs with the microscopic binding affinity $k_i^M$. The intramolecular ring closure follows with the stability constant $K_{intra} = k_i^M \cdot EM$, where $EM$ is the effective molarity for the macrocyclization process. The associated free energy contribution is given by the relation $\Delta G_{intra}^{M,L} = -RT \ln(k_i^M \cdot EM)$. Similarly to the chelate effect, effective molarity is obtained as the ratio of stability constants measured for the intramolecular and reference intermolecular reactions. In the absence of strains, $EM$ is a purely entropic contribution and indicates the advantage ($EM > 1\,M$) or hindrance ($EM < 1\,M$) of an intramolecular connection with respect to its intermolecular counterpart. However, the values of $EM$ in macrocyclization processes are much smaller compared to the chelate effect [15,20], and the interplay between $EM$ and the ligand concentration plays a more important role. While the ligand concentration is a driving force for disassembly and dissociation [11], the magnitude of $EM$ directly controls the balance between intra- and intermolecular interactions and measures the system preorganization.

The deliberate manipulating of $EM$ through the ligand structure represents an important tool in controlling self-assembly processes. The thought-out design of spacers between binding sites (length, spatial orientation, stacking interactions, etc.) contributes to the system preorganization for the next binding event. Using intersite linkers of differing rigidity may control the topology of the resulting assemblies. However, the formed edifice may escape the planned supramolecular design, for instance, due to too restricted degrees of freedom when connecting the binding units. For instance, the introduction of a rigid spacer between binding sites hinders the second intramolecular connection for steric reasons [21]. The most stable self-assembled complex thus contains an unoccupied metal binding site, which violates the principle of maximum occupancy. Another manipulation of $EM$ is shown for assemblies with tripodal ligands, where a deliberate programming of low effective molarity, resulting from ligand steric strains, prevents the formation of mononuclear complexes, to the benefit of tetranuclear three-dimensional helicates [22].

### 3.2.2.2 Theoretical Models for Effective Molarity

The effective molarity term $EM$, which was introduced for describing intramolecular interactions, is considered by chemists as the empirical experimental parameter. However, the theoretical concept of effective molarity is cited in literature as effective concentration $c^{eff}$ [16]. Theoretical models for estimating $EM$ in chelate and macrocyclization processes are available, but account only for the entropic part of $EM$ in the absence of enthalpic contributions ($\Delta H_{intra} = \Delta H_{inter}$). Let us briefly resume here two situations from the literature, detailed discussions of which can be found in Refs [13,15c].

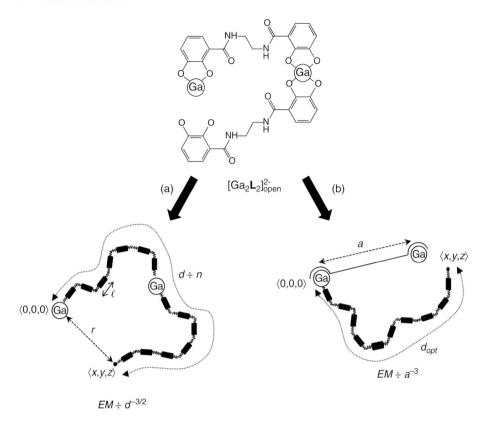

**Figure 3.4**   *Modelling effective molarity of macrocyclization in supramolecular self-assembly illustrated with a Ga(III) complex. (a) Connecting a freely jointed chain. (b) Connecting a ligand of an optimal length to preorganized binding sites separated with a [13]. Reproduced by permission of The Royal Society of Chemistry.*

First, for binding sites connected with a flexible chain, the effective concentration can be accessed by statistical theory [23] and modelled as the probability density $W(x,y,z)$ of finding one site <x, y, z> at a distance $r$ from the other site placed in the origin <0,0,0>; for a visual illustration see Figure 3.4a. $\beta$ depends on the chain length expressed as the mean square end to end distance $\langle r^2 \rangle$ (Equation 3.3).

$$W(x, y, z) = \left(\frac{\beta}{\pi^{0.5}}\right)^3 \exp(-\beta^2 r^2), \quad \beta = \left(\frac{3}{2\langle r^2 \rangle}\right)^3 \tag{3.3}$$

If the binding sites connect to the same receptor, $r$ tends to 0. It was shown that in this case the effective molarity (mol/dm$^3$) is proportional to the total length of the chain $d$ and we can write $c^{\text{eff}} \dot{=} d^{-3/2}$.

The second situation accounts for a receptor where the binding sites are held at a fixed distance $r = a$ by a rigid chain (Figure 3.4b). These sites are connected with a ditopic ligand and the derived expression for the effective concentration depends on $c^{\mathrm{eff}} \doteq a^{-3}$ for a long and flexible chain of optimized length $d_{\mathrm{opt}}$.

The present relations can be advantageously applied for estimating the entropy dependence of both processes affected with effective molarity *EM*: (i) chelate effect, which depends on the distance $d$ separating binding sites in a ligand, (ii) macrocyclic intra-molecular connection, where the receptor coordination sites are separated by the distance $a$. The enthalpic part of *EM*, which usually translates into ring strains in semirigid and rigid ligands, is more difficult to access theoretically. Therefore, related thermodynamic data extracted rigorously from real systems are required for developing satisfactory models.

### 3.2.2.3   *Number of Metal–Ligand Connections*

Supramolecular systems tend to form a maximum of connections in order to minimize $\Delta G$. It usually happens in agreement with the principle of maximum occupancy, although that can be disobeyed in some cases (e.g., lanthanide helicates in [21]). Although the number of intra- and intermolecular connections can be accessed via direct counting, it can often be convenient to use a simple mathematical calculus. We will limit ourselves to discrete binary complexes $[\mathrm{M}_m\mathrm{L}_n]$, which are formed with $m$ metal ions and $n$ ligands, that is, $N = m + n$ components. In addition, we assume that metal ions can form $CN/x$ connections and each ligand has $y$ $x$-dentate binding sites. To access correctly the total number of connections, two situations must be considered. First, if all ligand binding sites are occupied by metal ions (ligand saturation), the total number of connections in this assembly is equal to $B = n \cdot y$. This case is met more often and accounts for the majority of helicates. Second, if all metal binding sites are occupied (metal saturation), the total number of connections is equal to $B = m \cdot CN/x$. Among the total number of connections, we distinguish intermolecular (inter $= N - 1 = m + n - 1$) and intramolecular (intra $= B -$ inter) connections. The validity of the formula is illustrated for 1D, 2D and 3D helicates in Figure 3.5. The above relations can find applications in the quantitative description of self-assembly, and in thermodynamic modelling (see below).

### 3.2.2.4   *Statistical Factors in Polynuclear Systems*

The number of possibilities to arrange microspecies in polynuclear complexes significantly increases and the statistical factors become an important entropic contribution to the overall free energy. Fortunately, the symmetry numbers method introduced in Section 3.2.1 is particularly convenient and robust for calculating statistical factors. The application of Equation 3.2 is then straightforward. Since its introduction to supramolecular chemistry by Ercolani *et al.* [10], this method is now systematically applied in the thermodynamic modelling of supramolecular assemblies. Probably the most difficult task consists in determining the point groups of all compounds in equilibrium. An example of calculating for model triple-stranded helicates with Ga(III) [24] is given in Figure 3.6a [13]. For comparison, the same assembly is analysed with the direct counting method in Figure 3.6b. It is evident that applying the latter approach requires much more effort than using the first method.

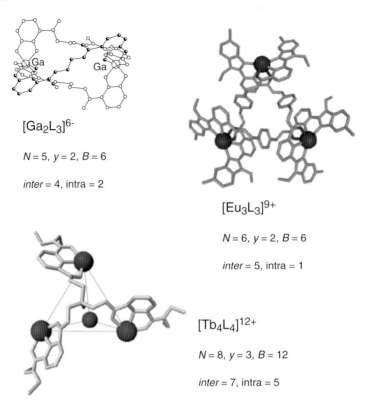

$[Ga_2L_3]^{6-}$

$N = 5, y = 2, B = 6$

$inter = 4, intra = 2$

$[Eu_3L_3]^{9+}$

$N = 6, y = 2, B = 6$

$inter = 5, intra = 1$

$[Tb_4L_4]^{12+}$

$N = 8, y = 3, B = 12$

$inter = 7, intra = 5$

**Figure 3.5**   *Calculation of the number of intra and intermolecular connections in polynuclear linear [24] and circular helicates [25] and tetrahedra [22].*

## 3.3   Cooperativity in Self-Assembly

The self-organization of molecules into helicoidal structures is often accompanied by a predominant formation of the desired products. Their thermodynamic stabilization is usually associated with the existence of cooperative interactions between components. In this context, the typical cooperative self-assembly process borrowed from biology is the "zipping up" of a double-helical structure upon the multiple binding interactions between adenine and uracyl bases [2]. The beginning of the assembly process is dominated by unfavourable entropy changes. However, further association of bases increases the enthalpic term due to the induced preorganization, which translates into the overall decrease of free energy. Although the cooperativity concept is well understood for intermolecular binding, the recognition of cooperativity in metalloorganic systems suffered from the confusion of inter- and intramolecular binding events. Ercolani shed light on these phenomena [11,12,26] and introduced a reliable classification of cooperativity interactions. If we consider intra- and intermolecular reactions as fundamentally different binding processes, three types of cooperativity can be independently accessed, which report to the following reaction pairs: intermolecular →

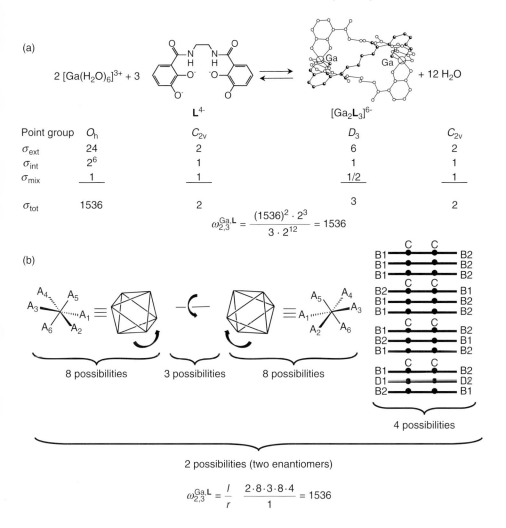

**Figure 3.6** *Application of (a) the symmetry number method and (b) the direct counting method for calculating statistical factors in* $[Ga_2L_3]^{6-}$. [13] *Reproduced by permission of The Royal Society of Chemistry.*

intermolecular, intermolecular → intramolecular, intramolecular → intramolecular. All these situations may occur in a helicate assembly and will be reviewed in detail below.

### 3.3.1 Allosteric Cooperativity

The well understood concept of allosteric cooperativity is often applied in biology for studying protein–ligand interactions strictly relying on intermolecular processes [27,28]. This type of cooperativity corresponds to the behaviour, where the initial binding of a ligand to a receptor induces allosteric effects, which changes the binding affinity of the identical receptor site toward the second ligand. The allosteric cooperativity is quantitatively described with an interaction parameter $\alpha$ that may be extracted from the ratio of statistically corrected experimental stability constants for successive binding events $\alpha = K_{n+1}/K_n$, whereby the constant $K_{n+1}$ is the examined constant and $K_n$ is taken as the reference. For non-interacting binding sites (non-cooperative), the successive constants $K_n$ and $K_{n+1}$ can be expressed with the microscopic association constant $K$ multiplied by statistical factors. Positive cooperativity is attributed to the affinity enhancement ($\alpha > 1$), while the opposite situation holds for negative cooperativity ($\alpha > 1$).

The cooperativity assessment can be clearly illustrated for the binding of a monovalent ligand **L** to a divalent receptor **R** possessing two equivalent binding sites in Equation 3.4, whereby each equilibrium is characterized with the experimental microscopic binding constant ($K_1$ and $K_2$, respectively) multiplied by the statistical factor.

$$\text{-R-} + 2L \overset{2K_1}{\rightleftharpoons} \text{L-R-} \overset{\frac{1}{2}K_2}{\rightleftharpoons} \text{L-R-L} \tag{3.4}$$

The reference microscopic affinity $K$ for the binding of L to **R** can be: (i) evaluated independently for a model system or alternatively (ii) assigned to $K_1$. Therefore, in the absence of interactions between binding sites, $K = K_1 = K_2$, and the cooperativity factor $\alpha = K_2 K_1/(K)^2 = 1$. In the presence of cooperativity, the cooperativity factor will thus differ from unity. Deviations from the statistical behaviour may find their origin in electrostatic, conformational or steric interactions, for instance. The associated free energy contribution expressed as a Boltzman factor can be obtained using the relation $\Delta E^\alpha = -RT \ln \alpha$. Allosteric cooperativity is currently encountered in classical coordination chemistry, and the tutorial example of $[\text{Ni(NH}_3)_6]^{2+}$ complexes is treated in [13,15a]. In this system, the anticooperative fixation of $NH_3$ is attributed to repulsive interligand interactions. Analogously, the presence of cooperativity can be tested for simple mononuclear complexes. Although the allosteric cooperativity can also be detected in polynuclear assemblies (pair homocomponent interactions, i.e., between coordinated ligand moieties), its reliable assessment is more complicated and requires alternative approaches, such as thermodynamic modelling (see below).

Alternatively, allosteric cooperativity may arise when metal ions bind to a receptor. This situation is often encountered in coordination chemistry and can be described with Equation 3.4, if L accounts for a metal ion and **R** for a receptor of cations. The cooperativity factor is accessed analogously with the procedure applied for the ligand binding. Accordingly, the assessment of cooperativity is described in the literature for coordination complexes with simple receptors, for instance those in Figure 3.7.

### 3.3.2 Chelate Cooperativity

While allosteric cooperativity is restricted to intermolecular interactions, chelate cooperativity operates with both inter- and intramolecular interactions. It refers not only to the chelate effect (Section 3.2.1) but also to the macrocyclization reactions described in Section 3.2.2, where the presence of cooperativity may be of high importance for the formation of supramolecular assemblies. To describe this kind of cooperativity in the correct

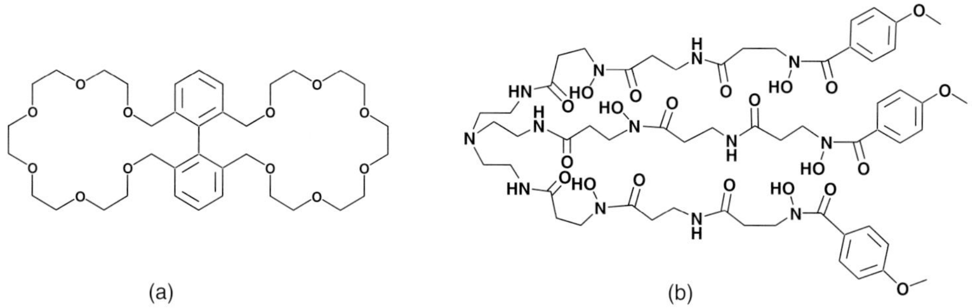

(a)                    (b)

**Figure 3.7** *Examples of ditopic organic receptors: (a) for $K^+$ [29] and (b) for $Fe^{3+}$ [30], showing allosteric cooperativity upon metal binding.*

way, we must consider two related but distinct chelate cooperativity factors which are associated with intermolecular binding processes [11]. Access to these parameters can be illustrated for the binding of a ditopic ligand L to a divalent receptor M (Figure 3.2). While the first binding event provides the opened complex *o*-ML via intermolecular binding, the second binding may occur in intramolecular or intermolecular fashion to provide *c*-ML or $ML_2$, respectively. Considering these equilibria, the apparent constant $K_{app}$ for the second binding is expressed with Equation 3.5, whereby $K$ is the reference microscopic constant for intermolecular binding.

$$K_{app} = \frac{[ML_2] + [c - ML]}{[o - ML].[L]} = K\left(1 + \frac{EM}{2[L]}\right) \tag{3.5}$$

The overall equilibrium constant for the binding process is $4K \cdot K_{app}$ in the presence of chelate cooperativity, which may compare with the overall constant $4K^2$ in the absence of cooperativity. The ratio of these constants provides the chelate cooperativity factor $\beta'$, which is always positive ($\beta' > 1$), and depends on both $EM$ and the ligand concentration and thus satisfies all the requirements for a chelate cooperativity factor.

The term $\beta = EM/2[L]$ indeed properly compares intramolecular reactions with respect to intermolecular binding. The deviation of the factor $\beta$ from unity is thus a measure of chelate cooperativity. Non-cooperative behaviour occurs for $EM/2 = [L]$ ($\beta = 1$), while positive and negative cooperativity occur for $\beta > 1$ and $\beta < 1$, respectively. The expression for $\beta$ can be generalized for the case of $x$-valent ligand B to an $x$-valent receptor $^xA$ (Equation 3.6) [11].

$$\beta = \frac{2}{x^x}\left(\frac{EM}{2[L]}\right)^{x-1} \tag{3.6}$$

The effective molarity and the ligand concentration represent two important parameters which play a key role in tuning chelate cooperativity. Moreover, since the effective molarity, and mainly its enthalpy part, depends on the ligand structure, chelate cooperativity can be manipulated via ligand design (see above).

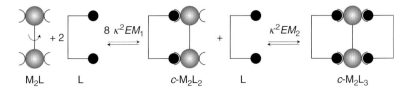

**Figure 3.8** *Schematic illustration of interannular cooperativity in self-assembly processes.* Adapted with permission from [11]. Copyright 2011 Wiley-VCH Verlag GmbH & Co. KGaA, Weinheim.

### 3.3.3 Interannular Cooperativity

In order to describe the interplay of two or more intramolecular binding processes, interannular cooperativity has been recently introduced by Ercolani [11,12]. Let us consider the ligand binding event and the intramolecular ring closure with $EM_1$, which suppress an internal rotational freedom by the freezing of torsional motion, for instance. Consequently, the binding of the second ligand molecule and the closure of the virtual second ring can be facilitated or hindered. The origin of this behaviour is attributed to the variation of $EM_2$ accompanying the second ring closure with respect to the first one ($EM_1$) and not to an increase in site binding affinity. The reference value of $EM$ can be evaluated independently for a model receptor undergoing the same interactions. The interannular cooperativity factor can be quantitatively expressed as the ratio $\gamma = EM_1 \cdot EM_2/EM^2$. This situation is shown in Figure 3.8 for a tetravalent receptor and its binding with a divalent ligand. Interannular cooperativity is clearly evidenced for porphyrin-type compounds [31], but it can likely be detected in self-assemblies of triple- or more stranded helicates.

### 3.4 Kinetic Aspects of Multicomponent Organization

With the increasing complexity of supramolecular systems (the number of interacting components, ligand branching, etc.), the probability of "good" coordination interactions decreases, and the number of possible products increases. Consequently, more complexation and decomplexation events (reversibility) are necessary for convergence to thermodynamic products. The reaction path to the final assembly can be theoretically predicted as the sequence of coordination steps within the assembly tree, which also includes unfruitful sequences [2]. Nevertheless, kinetic studies are required for identifying the role of real reaction intermediates that are preferentially formed through self-organization. This knowledge significantly contributes to a better understanding of metallosupramolecular assembly processes. However, compared with the structural characterization of supramolecular compounds, the measuring of kinetic parameters for self-assembly processes has been somewhat retarded. This is probably related to the high complexity of the systems, where the simultaneous formation of different complexes makes the interpretation of kinetic data difficult. Moreover, the instrumentation for measuring fast reactions was limited to stopped-flow techniques with spectrophotometric or fluorimetric detection.

Supramolecular complexation reactions are governed by the equilibria between solvated reactants, that is, metal ions and organic ligands. Assuming the Eigen–Wilkins mechanism (Equation 3.7, where $K_{os}$ is the equilibrium constant for the outer-sphere complex, $k_{ex}$ is the solvent exchange rate, $S$ and *solv* are solvent molecules) [32], the reaction rates for the replacement of solvent molecules, which may be considered as monodentate ligands, from the first coordination sphere of metal ions may vary extensively and depend essentially on the nature of the metal ions.

$$M(solv)_x + L(S) \xrightleftharpoons{K_{os}} M(solv)_x \ldots L + S \xrightarrow{k_{ex}} M(solv)_{x-1}L + S + solv \qquad (3.7)$$

These substitution reactions were classified by Langford and Grey in 1965 [33]. This work represents a mechanistic support for interpreting elementary reaction steps in supramolecular chemistry. To give a short reminder, the ligand substitution reactions are divided as associative, dissociative and interchange. In addition, we distinguish two kinds of intimate mechanisms: associative and dissociative activation modes [34]. These elementary reactions may take place many times along the reaction pathway and are determining for the overall reaction rates.

The strict assembly of supramolecular compounds is supposed to spontaneously provide a single compound. The first evidence of a complex formation pathway was brought by Lehn and coworkers in the study of pentanuclear helicates [35]. The proposed mechanism identifies hairpin type intermediates that slowly rearrange in the final helicates. Pioneering extensive kinetic studies (which are probably the only ones in the field of helicate self-assembly) were carried out by Albrecht-Gary and coworkers [14], who launched kinetic studies of different double- and triple-stranded helicates in order to identify the reaction intermediates and the formation mechanisms in solution.

The collaboration of Albrecht-Gary with Lehn allowed a full characterization of the formation of trinuclear double-stranded helicates [36]. The schematic representation of the proposed mechanism consists of four elemental steps (Figure 3.9). The reaction is initiated with a fast binding of two metal ions to the tritopic ligand in the terminal binding sites. This can be explained by the prevention of some unfavourable electrostatic interactions, if a copper ion would occupy the central position. Consequently, the binding of the third metal ion occurs in the central site. However, the assembly is still not complete at this point; the slow rearrangement processes (self-repairing steps) occur in the terminal phase in order to adopt the helical structure with a minimum of potential energy.

The self-assembly of triple-stranded iron(II) helicates $[Fe_2L_3]^{4+}$ with a ditopic ligand is obviously favoured in ligand excess, where the ligand coordination sites are preferentially bound to the iron(II) cation (Figure 3.10) [37]. This interesting feature found its origin in the extra stabilization of low-spin tris-bipyridine complexes. In line with this interpretation, a hairpin complex $[Fe_2L_2]^{4+}$ appears even in metal excess instead of a virtual $[Fe_2L_2]^{4+}$ helicate. The helicate assembly terminates by the metal binding to the second already preorganized site, which reduces its effective charge by polarization.

Interesting features have been revealed during the study of the self-assembly of dinuclear triple-stranded helicates containing highly charged lanthanides [38,39]. The formation of a helicate $[Eu_2L_3]$ with a neutral ligand is favoured at stoichiometric conditions, which leads to fast and quantitative self-assembly processes (Figure 3.11a). The binding

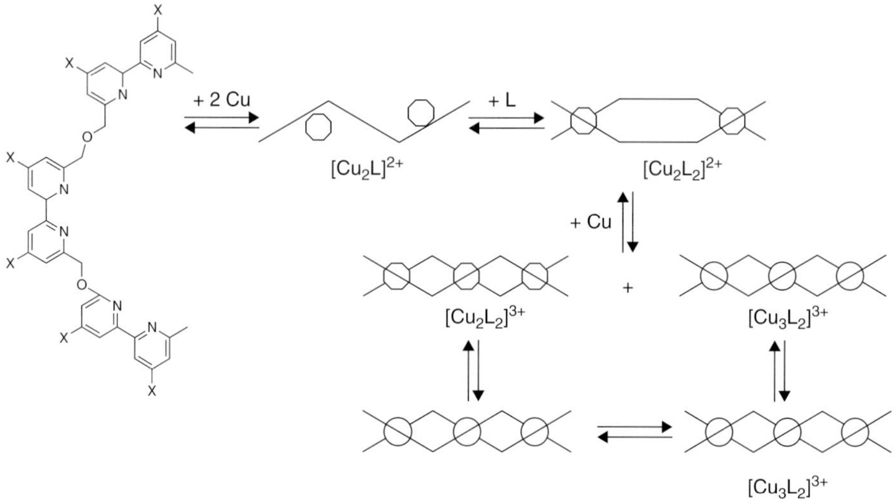

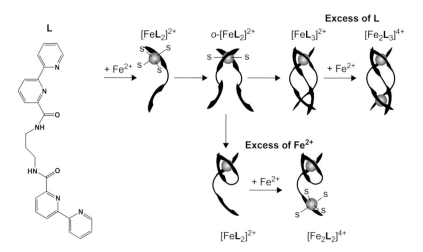

**Figure 3.9**   *Schematic representation of the self-assembly mechanism for the double-stranded helicates [Cu$_3$L$_2$]$^{3+}$*. Adapted with permission from [36]. Copyright 2001 Wiley-VCH Verlag GmbH & Co. KGaA.

of two metal ions to the same ligand does not occur because the intermetallic repulsions are apparently too strong. However, using a negatively charged analogous ligand, the metal charges are neutralized and helicates are observed even in metal excess (Figure 3.11b). In both systems, the conditions in ligand excess are unfavourable because of strong negative interligand interactions (steric hindrance, charge repulsion). Likewise, the reaction propagates through the dinuclear unsaturated species [Eu$_2$L$_2$], which appears to

**Figure 3.10**   *Self-assembly mechanism for the formation of diferrous triple-stranded helicates [Fe$_2$L$_3$]$^{4+}$ (S = solvent molecules).* Reprinted from [14] with permission from Elsevier.

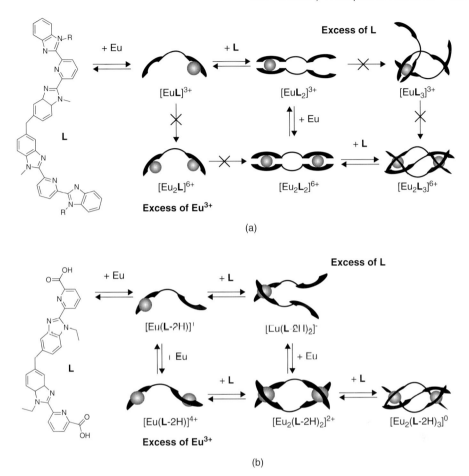

**Figure 3.11**  *Schematic representation of the self-assembly mechanisms for triple-stranded lanthanide helicates [Eu$_2$L$_3$].* (a) Adapted with permission from [38b]. Copyright 2003 American Chemical Society. (b) Adapted with permission from [ref 39b] Copyright 2011 WILEY-VCH Verlag GmbH & Co. KGaA, Weinheim.

be a key reaction intermediate in the formation of triple-stranded helicates. While [Eu$_2$L$_2$] with a neutral ligand is a labile "side by side" species, the analogous complex with a charged ligand apparently adopts a helical form, which is already well preorganized for braiding of the third ligand. This terminal step clearly occurs with positive cooperativity, and probably all kinds of cooperative interactions (see Section 3.3) contribute to the thermodynamic stabilization of the final triple-stranded helicate (Figure 3.11b).

All kinetic studies showed a stepwise formation of helicates and supramolecular compounds, which definitively disproves the hypothesis of "magic" self-assembly processes and confirms the validity of classical coordination chemistry. It has been shown that the self-assembly mechanism and the reaction rates significantly depend on stoichiometric conditions and on the magnitude of electrostatic interactions. A final slow rearrangement step is often observed and applies to self-repair and thermodynamic re-equilibration. This is also in line with a slow formation kinetics observed for complicated assemblies of pentanuclear helicates [35,40], whose completion at stoichiometric conditions requires a long equilibration time. The kinetics and reaction intermediates can then be conveniently examined with NMR or ESMS.

## 3.5    Understanding Self-Assembly Processes

### 3.5.1    Assessment of Cooperativity

#### 3.5.1.1    *Cooperativity Tests for Intermolecular Interactions*

To reveal cooperativity interactions in metallo-organic compounds, chemists borrowed established tests valid for biological systems, and the formal aspects of these methods were reviewed by Permutter-Hayman [27]. Cooperativity is usually evaluated with graphical tests such as the binding curve and Scatchard and Hill plots, which are based on the calculation of the occupancy $r$ (Equation 3.8), in other words, the average number of ligands bound per receptor.

$$r = \frac{[PL] + 2[PL_2] + \cdots + n[PL_n]}{[P] + [PL] + [PL_2] + \cdots + [PL_n]} \tag{3.8}$$

According to the curve shape, one can detect deviations from the statistically governed intermolecular binding of ligands (or metal ions) to a receptor and eventually detect positive or negative cooperativity. These methods are strictly limited and give a correct indication of allosteric cooperativity only for systems with intermolecular interactions (see Section 3.3). As an illustrative example, the Scatchard plot for $NH_3$ binding to a Ni(II) receptor is shown in Figure 3.12. The concave upward curve for this simple mononuclear

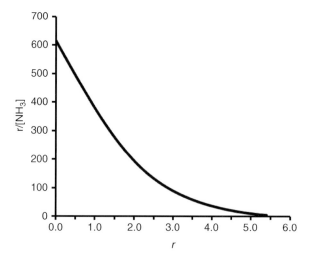

**Figure 3.12**    *Scatchard plot for the formation of $[Ni(NH_3)_x]^{2+}$, which shows a negative allosteric cooperativity indicated by a concave upward curve shape [41].*

system indicates negative cooperativity. Further, one would expect a straight line and a concave downward curve, respectively, in the absence of interactions and for positive cooperativity.

Although these tests are truly applicable for some supramolecular systems [41], their general use in supramolecular chemistry is not appropriate. In spite of this, the above graphical tests were originally applied to double- and triple-stranded helicates and revealed a strong positive cooperativity [42–44]. These confusing analyses were identified in 2003 by Ercolani [26], who showed the application of the Scatchard plot was inappropriate in self-assembled compounds, where both (virtually non-equivalent) inter- and intramolecular processes operate.

In order to reliably detect cooperativity in multicomponent assemblies, Ercolani developed the concept of statistical repetitive binding [26], where the microscopic cumulative constant of a multicomponent assembly $\beta_{mn}^{M,L}$ is obtained as the product of the degeneracy factor $\omega_{mn}^{M,L}$, the microscopic equilibrium constants $k_{inter}^{M,L}$ for inter-molecular reactions and the microscopic constants $k_{intra}^{M,L}$ associated with intramolecular reactions (Equation 3.9).

$$mM + nL \rightleftharpoons [M_mL_n] \qquad \beta_{mn}^{M,L} = \omega_{mn}^{M,L} \cdot \prod_{inter} k_{inter}^{M,L} \cdot \prod_{intra} k_{intra}^{M,L} \qquad (3.9)$$

The present description is adequate for self-assembly processes: (i) in the absence of interligand and intermetallic interactions and (ii) if the microscopic constants do not vary within related complexes. The application of this modelling to a simple dinuclear assembly with two ligands is given in Figure 3.13. The first three steps refer to strictly inter-molecular binding events between metal ions and ligand coordination sites characterized with $k_{inter}^{M,L}$. The fourth step corresponds to the intramolecular macrocyclization associated with $k_{intra}^{M,L}$. If the microscopic stability constants $k_{inter}^{M,L}$ and $k_{intra}^{M,L}$ can be independently

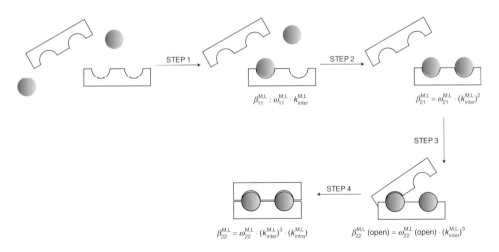

**Figure 3.13**   *Statistical repetitive binding for a dinuclear complex [M₂L₂].* [45] Adapted by permission of The Royal Society of Chemistry.

evaluated from the experimental macroscopic stability constants for simple model reactions, the macroscopic stability constant of any species $[M_mL_n]$ can be predicted with Equation 3.9. The decision about the cooperativity of interactions is based on comparing the predicted constants with the experimental ones. Ercolani initially applied his model to Lehn's double-stranded helicates [26]. The predicted constant of $[Cu_3L_2]^{3+}$ is indeed close to the experimental values, which indicates that statistical repetitive binding occurs, in contrast to positive cooperativity suggested in the original paper [43]. It is worth noting that the correct parameterization of $k_{intra}^{M,L}$ requires the formation of at least macrobicyclic (i.e., trinuclear) complexes. In this context, the trinuclear lanthanide helicates [45] fulfil this criterion, but escape Ercolani's analysis due to the lack of thermodynamic data.

The present approach has clearly shown that the classical protein/ligand cooperativity tests do not allow a reliable detection of cooperativity in supramolecular assemblies. In addition, the results of Ercolani's tests indicate that positive deviations from statistical repetitive binding are rare in metallo-organic assemblies. Although the model takes into account intra- and intermolecular reactions, it suffers from a number of limitations (e.g., identical binding affinity along the strand, no quantification of interaction parameters) and is not widely applied for sophisticated assemblies.

### 3.5.2 Thermodynamic Modelling

#### 3.5.2.1 Site-Binding Model

The difficulties in testing cooperativity stimulated coordination chemists to search for new tools that can be used to describe and analyse supramolecular compounds [45]. Not surprisingly, this innovative effort is closely connected with designing new polynuclear highly charged helicates with controlled properties [44]. Therefore, the initial development of thermodynamic models was associated with the description of intermetallic interactions in polynuclear compounds. Inspired by proton-binding models in polyelectrolytes [46], the site-binding model for metal binding to a multisite receptor was formulated in 2004 [16]. To illustrate the use of this thermodynamic modelling, let us consider a pre-assembled linear receptor containing $n$ adjacent coordination sites that can bind $n$ metal ions according to Equation 3.10, as schematically shown in Figure 3.14.

$$n\mathrm{M}^{m+} + \mathbf{R} \xrightleftharpoons{\beta_n} \mathrm{M}_n\mathbf{R} \qquad \beta_n = \omega_{n,1}^{M,R} \cdot \prod_{i=1}^{n} k_i^M \cdot \prod_{i<j}^{n} u_{ij}^{M,M} \qquad (3.10)$$

The stability constants for these equilibria are modelled by the product of: (i) statistical factors $\omega_{n,1}^{M,R}$, (ii) microscopic binding affinities $k_i^M$ and (iii) interaction parameters $u_{ij}^{M,M}$ accounting for intermetallic interactions within the receptor and expressed as the Boltzman factor $u_{ij}^{M,M} = \exp(-\Delta E_{ij}/RT)$. The associated total free energy of complexation is given in Equation 3.11.

$$\Delta G_{n,1}^{M,R} = -RT \ln\left(\omega_{n,1}^{M,R}\right) - RT \sum_{1}^{n} \ln\left(k_i^M\right) + \sum \Delta E_{ij} \qquad (3.11)$$

Receptor

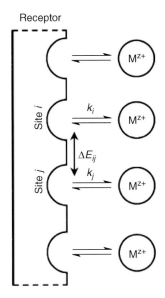

**Figure 3.14** *Site-binding model for polynuclear complexes with a linear receptor* **R**.

The application of the site-binding model is straightforward for mono-stranded helicates, where the ligand behaves as a linear receptor **R** for successively fixing metallic cations [39,47]. However, the extension of the model to double- and triple-stranded helicates requires the consideration of a virtual receptor preorganized from the ligands forming the supramolecular edifice. The metal binding to **R** thus occurs in an intermolecular fashion. This dissection of self-assembly processes for dinuclear triple-stranded helicates is graphically illustrated in Figure 3.15 together with the associated site-binding model for each species [48]. Although $\Delta G_{\text{preorg}}$ is difficult to access, it similarly influences all complexes related to the same receptor. This translation of the zero-level of the free energy of formation can be arbitrarily set to $\Delta G_{\text{preorg}} = 0$. Consequently, in the case of a ditopic preorganized receptor (Figure 3.15), the binding equilibria are defined with the stability constants $\beta_{1,1}^{\text{M,R}} = 2k_{\text{N}_6\text{O}_3}^{\text{M}}$ and $\beta_{2,1}^{\text{M,R}} = (k_{\text{N}_6\text{O}_3}^{\text{M}})^2 u_{1-2}^{\text{M,M}}$, where $k_{\text{N}_6\text{O}_3}^{\text{M}}$ is the microscopic affinity for the binding site, that is, $\text{N}_6\text{O}_3$. To extract reliable values of $k_{\text{N}_6\text{O}_3}^{\text{M}}$ and $u_{1-2}^{\text{M,M}}$, a sufficient number of experimental stability constants should be available. Alternatively, the thermodynamic stability constants for closely related receptors can be simultaneously considered. In this context, the site-binding model was elegantly applied for the analysis of homo- and heteronuclear complexes, and the fitted binding affinities allowed establishing the trends in relative affinities of metal ions along the series [48]. Moreover, the analysed data systematically show repulsive intermetallic interactions ($\Delta E_{1-2}^{\text{M,M}} > 0$, negative cooperativity). In addition to this example, the site-binding model was successfully adapted for trinuclear triple-stranded helicates possessing two different coordination sites [44,48].

Despite limiting hypotheses concerning the preorganized receptor, the site-binding model is particularly suitable for the rationalization of metal exchange processes and relative trends in the series of similar complexes. Moreover, its use is fully adequate for the

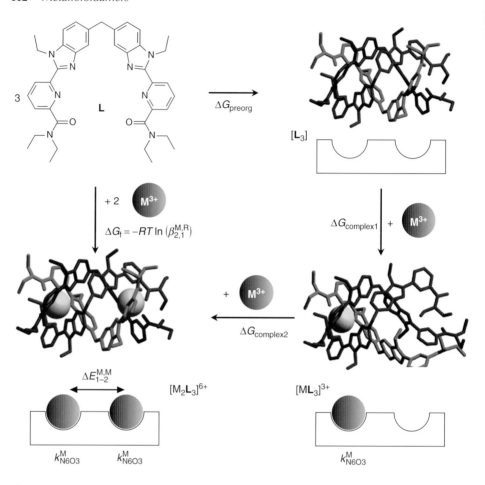

**Figure 3.15** *Self-assembly of double-stranded helicates. The preorganization of the receptor followed by successive intermolecular metal binding is associated with the site-binding models for each microspecies.* Adapted with permission from [48]. Copyright 2004 American Chemical Society.

analysis of monostranded helicates. The obtained microscopic parameters can be further applied for predictions of polynuclear topologies, where different metal ions are bound along multisite receptors. These problems concerning large assemblies can be advantageously treated using transfer matrix techniques, which allow the prediction of binding isotherms, and the receptor's microstates or occupancy [16]. Finally, the model of a virtually preassembled receptor has also been adapted for analysing cooperativity in supramolecular systems [41].

### 3.5.2.2 Thermodynamic Additive Free Energy Model

In order to escape the number of limitations connected with the site-binding model and statistical repetitive binding, a considerable effort was devoted to the development of an universally valid thermodynamic description of multicomponent self-assembly processes.

The basic principles of this approach were formulated in 2005 [15a,b] as an extension of the site-binding model. The main difference compared with the site-binding model is the explicit considerations of inter- and intramolecular reactions and homocomponent interactions. Although the complexity of modelling with the thermodynamic additive free energy model (TAFEM) increases, this thermodynamic description is applicable to the stability constant of any species formed by self-assembly.

The total free energy related to the formation of a supramolecular assembly (Equation 3.1) is obtained as the sum of the elementary energetic contributions intervening in the process, which can be expressed with the general Equation 3.12.

$$\Delta G_{m,n}^{M,L} = -RT \ln\left(\omega_{m,n}^{M,L}\right) - RT \overset{inter}{\sum_{i=1}} \ln\left(k_i^M\right) - RT \overset{intra}{\sum_{j=1}} \ln\left(k_i^M . EM_j\right) + \sum_{i<j} \Delta E_{ij}^{M,M}$$
$$+ \sum_{k<l} \Delta E_{kl}^{L,L} \tag{3.12}$$

The related global stability constant is then obtained as the product of all microscopic parameters using straightforward transformations (Equation 3.13).

$$\beta_{m,n}^{Lu,L} = \omega_{m,n}^{M,L} \cdot \overset{tot}{\prod_{i=1}}\left(k_i^M\right) \cdot \overset{intra}{\prod_{j=1}}\left(EM_j\right) \cdot \prod_{i<j}\left(u_{ij}^{M,M}\right) \prod_{k<l}\left(u_{kl}^{L,L}\right) \tag{3.13}$$

Note that all these energetic parameters [statistical factors $\omega_{m,n}^{M,L}$ (Equation 3.2), microscopic binding affinities for inter- ($k_i^M$) and intramolecular reactions ($k_i^M.EM$), homocomponent interactions $\Delta E^{M,M}$ and $\Delta E^{L,L}$] were discussed in Section 3.1 in terms of basic principles applicable to complexation processes.

The application of this innovative approach can be illustrated for the assembly of dinuclear triple-stranded helicates (i.e., with lanthanides [15c]), which represent the most frequently occurring type of helicates. Figure 3.16 schematically shows the structures of all possible microspecies in the system, their point groups (necessary for calculating symmetry numbers) and the thermodynamic model of their microscopic constants. In the present case, the expression for microconstants is identical to the macroscopic stability constants. However, in more complex systems, the expression for the macroconstant is obtained as the sum of microconstants [15].

The number of stability constants available for modelling is often limited: (i) due to the lack of reliable experimental data for existing complexes or (ii) because of the thermodynamic instability of elusive complexes. Therefore, the analysis of real systems and obtaining physically meaningful parameters often requires making reasonable simplifications and reducing the number of thermodynamic parameters by: (i) empirical estimations, (ii) theoretical predictions or (iii) by an independent evaluation of reference processes. For instance, the combination of thermodynamic data collected for the series of different but structurally related di-, tri- and tetranuclear europium helicates allowed the establishment of a reasonably simplified set of thermodynamic descriptors. The applied hypotheses are the following: (i) binding affinities and interligand and intermetallic interactions are identical in all microspecies, taking into account the same intermetallic distances, (ii) long-range intermetallic distances are modelled with Coulomb's

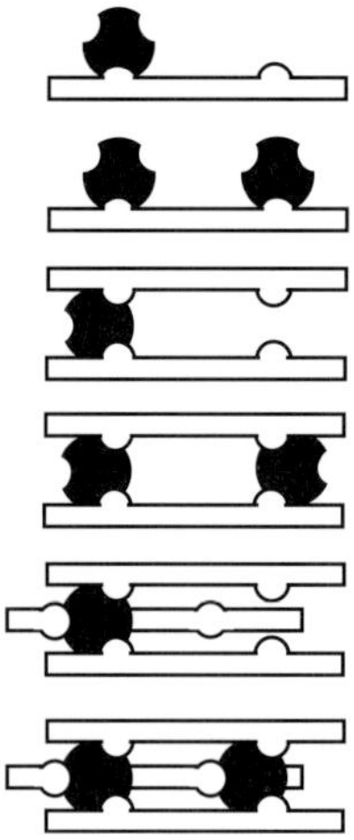

| Structures | Point groups | Microconstants |
|---|---|---|
| | $C_s$ | $\beta_{1,1}^{M,L} = 12 \cdot \left(k^M\right)$ |
| | $C_{2v}$ | $\beta_{2,1}^{M,L} = 36 \cdot \left(k^M\right)^2 \cdot \left(u_{1-2}^{L,L}\right)^6 \cdot \left(u_{1-2}^{M,M}\right)$ |
| | $C_1$ | $\beta_{1,2}^{M,L} = 48 \cdot \left(k^M\right)^2 \cdot \left(u_{1-2}^{L,L}\right)$ |
| | $C_2$ | $\beta_{2,2}^{M,L} = 144 \cdot \left(k^M\right)^4 \cdot \left(u_{1-2}^{L,L}\right)^2 \cdot \left(u_{1-2}^{M,M}\right) \cdot \left(EM\right)$ |
| | $C_3$ | $\beta_{1,3}^{M,L} = 32 \cdot \left(k^M\right)^3 \cdot \left(u_{1-2}^{L,L}\right)^3$ |
| | $D_3$ | $\beta_{2,3}^{M,L} = 96 \cdot \left(k^M\right)^6 \cdot \left(u_{1-2}^{L,L}\right)^6 \cdot \left(u_{1-2}^{M,M}\right) \cdot \left(EM\right)^2$ |

**Figure 3.16** *Schematic structure, symmetries and statistical factors for all possible microspecies formed between L and metal ions, including the dinuclear triple-stranded helicate [M$_2$L$_3$] [15c].*

law and (iii) the effective concentration of long-distance macrocyclization processes is estimated with statistical theory [23]. Finally, the system of complexes was reasonably described with five thermodynamic parameters, which were extracted by nonlinear fitting procedures [15c]. The free energy of intermolecular connections for two different binding sites in the analysed system ($\Delta G_{N_3}^{Eu} = -31(1)\,kJ/mol$, $\Delta G_{N_2O}^{Eu} = -33(1)\,kJ/mol$) are very similar. The effective molarity term decreases the energetic gain of each intramolecular reaction by $\Delta G_{corr}^{Eu} = 6(5)\,kJ/mol$. Despite this, the inter- and intramolecular connections represent the most favourable energetic contributions to the total free energy of formation. In contrast, the homocomponent interaction parameters are positive ($\Delta E_{1-2}^{Eu,Eu} = 10(4)\,kJ/mol$, $\Delta E_{1-2}^{L,L} = 6(2)\,kJ/mol$) and thus disfavourable for complex formation. Both factors contribute to the overall negative allosteric chelate cooperativity, which may be expressed as the coefficient $\alpha$ in Equation 3.14 [11,15a].

$$\alpha = \prod_k \exp\left(-\Delta E_k^{M,M}/RT\right) \cdot \prod_l \exp\left(-\Delta E_l^{L,L}/RT\right) \qquad (3.14)$$

An even more complex package of *d*- and *f*-heteronuclear helicates was analysed by Riis-Johanessen *et al.* [49]. The complex fitting procedure allowed the direct determination of apparent long-range intermetallic interactions without making any restrictive assumptions. The obtained results demonstrate a contra-intuitive alternation of attractive and repulsive interactions with the intermetallic distance, which is explained by the changes in solvations energies with respect to geometrical parameters.

Thermodynamic modelling with TAFEM was also successfully extended to 3D helicates formed with tripodal ligands and lanthanides [22]. The fitted parameters are reminiscent of the values previously obtained for the system of different linear triple-stranded

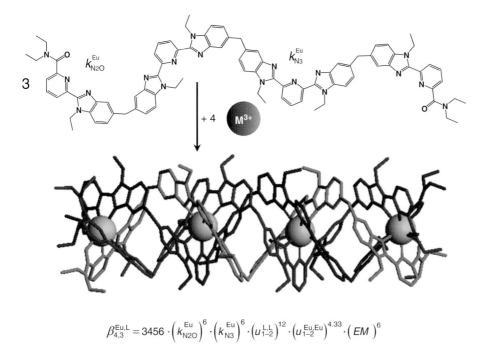

$$\beta_{4,3}^{Eu,L} = 3456 \cdot \left(k_{N2O}^{Eu}\right)^{6} \cdot \left(k_{N3}^{Eu}\right)^{6} \cdot \left(u_{1-2}^{L,L}\right)^{12} \cdot \left(u_{1-2}^{Eu,Eu}\right)^{4.33} \cdot \left(EM\right)^{6}$$

**Figure 3.17** *Prediction of tetranuclear triple-stranded helicate with TAFEM [15c,51].*

helicates [15c]; however, a limited set of stability constants compromises a more sophisticated interpretation of data. Analogously, more reliable parameters can be obtained by fitting a larger set of thermodynamic data eventually collected for tetranuclear and pentanuclear 3D complexes with helicate segments [40].

The first predictive application of TAFEM resulted in an estimation of the thermodynamic stability constant of a tetranuclear europium triple-stranded helicate (Figure 3.17). Using the microscopic parameters extracted from the set of dinuclear and trinuclear helicates, the calculated stability constant amounts to log $\beta_{4,3}^{Eu,L} = 42.5$[50] and closely agrees with the value determined using spectrophotometric titrations log $\beta_{4,3}^{Eu,L} = 43.2(1.9)$ [15c]. A calculation of system speciation with TAFEM also proved to be efficient for a non-helical system consisting of Pd(II) complexes with a bipy ligand. A knowledge of microscopic affinity constants allowed the construction of speciation maps for different concentrations and a prediction of the best conditions for the assembly of molecular squares [51].

### 3.5.3 Solvation Energies and Electrostatic Interactions

The cooperativity factors in supramolecular compounds, accessed for instance through thermodynamic modelling (Equation 3.14), are usually interpreted in terms of electrostatic interactions, steric hindrance and so on. However, the correct estimation of the free energy contribution in solution requires a consideration of solvation energies that

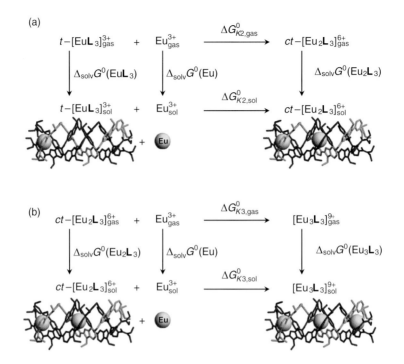

**Figure 3.18**  *Thermodynamic Born–Haber cycles for the successive complexation of Eu(III) to (a) t-[EuL$_3$]$^{3+}$ to give ct-[Eu$_2$L$_3$]$^{6+}$ and (b) ct-[Eu$_2$L$_3$]$^{6+}$ to give [Eu$_3$L$_3$]$^{9+}$.* Reprinted with permission from [52]. Copyright 2007 American Chemical Society.

compensate to some extent the above factors. Since we are dealing with metallo-organic assemblies, let us investigate the predominant effect of electrostatic interaction on the global free energy change.

For metallic cations considered as point charges, the magnitude of intermetallic interactions in the gas phase can be calculated with Coulomb's law as the electrostatic work $\Delta E^{MM}$, which depends on the intermetallic distance. The transformation of Coulombic interactions in the gas phase into free energy in solution can be done using a Born–Haber cycle, as shown for several helical complexes [19,52]. A demonstrative example of this procedure is shown in Figure 3.18. The related values of solvation energies can be estimated from the Born equation using the hydrodynamic radii of diffusing supramolecular complexes, which can be conveniently measured with diffusion ordered spectroscopy (DOSY-NMR).

The values of Coulombic repulsions and the solvation energies are usually huge, but with opposite contributions. Therefore, the overall effect of homocomponent interactions is of limited magnitude and depends on the charge distribution within the assembly, its shape and the organization in solution [13]. The competition between electrostatic interactions and solvation energies is illustrated in Figure 3.19 for dinuclear double-stranded helicates, which differ in spacer length and thus in intermetallic distances, while other

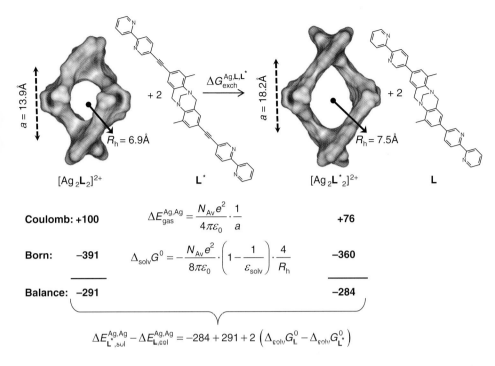

**Figure 3.19** *Quantitative analysis of the Coulomb and Born energetic contributions to the apparent intermetallic interactions in solution. (a is the intermetallic separation, $R_h$ is the pseudo-spherical hydrodynamic radius. [13] Reproduced by permission of The Royal Society of Chemistry.*

parameters remain unchanged [19]. Although the variations of Coulomb and Born energetic contributions are relatively important, the overall balance is only few kJ/mol. Moreover, the decrease of intermetallic distance with the concomitant increase of Coulombic repulsions contra-intuitively produces a more favourable solvation effect, which apparently induces positive intermetallic interactions. Therefore, it is not so surprising to experimentally find positive or negative values of homocomponent interaction energies in solution [19,49].

In summary, an efficient and thought-out manipulation of subtle homocomponent interactions (cation–cation, ligand–ligand) is challenging. Moreover, these parameters may be strongly sensitive to minor structural changes in the supramolecular structure. The compensation of Coulombic interaction by solvation energies must be taken into account in the predictive design of stabilizing interactions in self-assembly. In this context, a favourable solvation effect may be responsible for the stabilization of compact polynuclear complexes (i.e., tetranuclear 3D structures [22]) despite a large number of intermetallic repulsive interactions. Alternatively, the use of negatively charged binding units instead of neutral ones maximizes the coordination interactions, and the charge neutralization contributes to a decrease of intermetallic repulsions, which considerably favours the ultimate energetic balance [15a,39].

## 3.6    Secondary Structure and Stabilizing Interactions

Analogous to biological systems, the formation of metallo-organic helical structures is often induced by intrinsic structural modifications resulting from the interactions of "informed" components (helical conformational folding). This preorganization can be described as a self-templating process for the propagation of self-assembly. In addition, different weak interactions may favourably act in the assembly of helical structures. A common way is templating by anions, which are present in the reaction mixture and which are the key elements controlling the final helical structure. This effect is often encountered with circular helicates and can be illustrated with Lehn's pentanuclear helicate templated by $Cl^-$ (Figure 3.20a) that transforms into a hexanuclear analogue in the presence of sulfates [53]. The templating ions may also behave as switching components between different supramolecular forms. For instance, the conversion of triple-stranded helicate into 3D tetrahedral cages is induced by the presence of an alkylammonium guest [54].

In many cases, the "third" component in the self-assembly process contributes to the stabilization of helicates, and several examples can be found in the literature. In the solid-state structure of dinuclear triple-stranded helicates, a potassium cation coordinated to internal oxygen atoms weakly interacts with the double bond [55]. Although the presence of potassium is not required for the existence of this assembly in solution, this interaction indicates a possible way of stabilizing it. In this context, the presence of hydrogen bonding can also be crucial. This stabilizing effect was identified as the origin of positive cooperativity detected with thermodynamic modelling in triple-stranded dinuclear lanthanide helicates (Figure 3.11b) [15a].

A "miraculous" effect of secondary interactions is shown with the isolation of neutral amphiphilic lanthanide helicates [56]. While the usual self-assembly process gives only an insoluble mixture of oligomers instead of the expected $[Ln_2L_3]$ helicate, the addition of two equivalents of $Ag^+$ leads: (i) to a complete dissolution and (ii) to the almost quantitative formation of the dinuclear lanthanide helicate, where $Ag^+$ cations sit at its extremities and interact with the ligand chains (Figure 3.20b).

As shown in the above examples, the secondary interactions may delicately influence the self-assembly process and bring a decisive energetic contribution, which will direct the reaction toward a specific product. The stabilization may possibly come from a change of the overall solvation energy, or from partial compensation of a low effective molarity, which is commonly detected in both double- and triple-stranded helicates. The programming of secondary weak interactions would be beneficial for the designing of new helicates. However, their thermodynamic modelling was overlooked until now due to the lack of reliable experimental data, but their effect is hidden in different cooperativity factors. In this context, molecular dynamics simulations are proving to be a powerful tool for optimizing supramolecular interactions, if we refer to their recent application to different heteronuclear metallo-helicates [56].

## 3.7    Conclusions

The present chapter describes physico-chemical principles that govern coordination processes in multicomponent self-assembly. The basic concepts of coordination chemistry

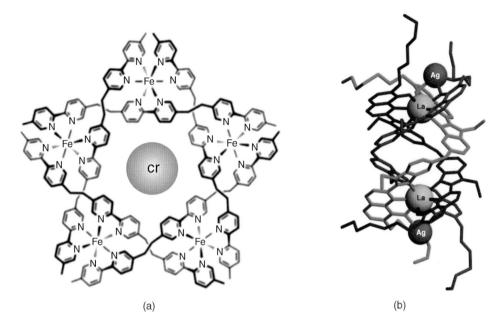

(a)                                                                (b)

**Figure 3.20** *Effect of secondary interactions on self-assembly processes. (a) Formation of the pentanuclear helicate templated with Cl⁻ anion. Reprinted with permission from [ref 55] Copyright 1996 WILEY-VCH Verlag GmbH & Co. KGaA, Weinheim. (b) Effect of stabilizing Ag⁺ interaction on the formation of triple-stranded helicates.* Reprinted with permission from [ref 56] Copyright 2011 WILEY-VCH Verlag GmbH & Co. KGaA, Weinheim.

are introduced for mononuclear coordination compounds and subsequently extended to metallo-organic supramolecular compounds, where intramolecular reactions additionally take place. The phenomenological understanding of self-assembly processes was often hindered by supramolecular terminology and several commonly used concepts have been adequately rationalized only recently. This is the case of cooperativity and intramolecular macrocyclization processes, for instance. A deeper understanding of these chemical tools was stimulated by the development of predictive strategies for designing functional edifices. The fastidious effort resulted in the elaboration of powerful thermodynamic models that quantitatively identified key energetic factors governing the self-assembly processes.

Helicate assemblies play a crucial role in the above evolution. These compounds were often used as model assemblies for structural, thermodynamic and kinetic investigations. A relative abundance of previously collected data permitted more sophisticated analyses to be carried out in order to develop thermodynamic modelling. Future initiatives in this field should be focused on acquiring physico-chemical data that are still relatively scarce for three-dimensional nanostructures. This will allow an improvement of the present thermodynamic description and its adaptation for predicting advanced supramolecular architectures. In spite of this considerable effort, some supramolecular assemblies may still escape the predictive design due to some unexpected conformational and coordination preferences, or the chemist's choice of inappropriate structural elements.

# References

1. Lehn, J.-M. (1995) *Supramolecular Chemistry: Concepts and Perspectives*, VCH, Weinheim.
2. Steed, J.W. and Atwood, J.L. (2009) *Supramolecular Chemistry*, John Wiley & Sons, Ltd., Chichester.
3. Lindsey, J.S. (1991) Self-assembly in synthetic routes to molecular devices. Biological principles and chemical perspectives: a review. *New J. Chem.*, **15**, 153–180.
4. Caulder, D.L. and Raymond, K.N. (1999) Supermolecules by design. *Acc. Chem. Res.*, **32**, 975–982.
5. Machado, V.G., Baxter, P.N.W., and Lehn, J.-M. (2001) Self-assembly in self-organized inorganic systems: A view of programmed metallosupramolecular architectures. *J. Braz. Chem. Soc.*, **12** (4), 431–462.
6. Piguet, C., Bernardinelli, G., and Hopfgartner, G. (1997) Helicates as versatile supramolecular complexes. *Chem. Rev.*, **97**, 2005–2062.
7. Piguet, C., Borkovec, M., Hamacek, J., and Zeckert, K. (2005) Strict self-assembly of polymetallic helicates: the concepts behind the semantics. *Coord. Chem. Rev.*, **249**, 705–726.
8. Piguet, C. and Bünzli, J.-C. (2010) Self-assembled lanthanide helicates: from basic thermodynamics to applications, in *Handbook on the Physics and Chemistry of Rare Earths* (eds K.J. Gschneidner, J.-C., Bünzli, and V.K. Pecharsky), Elsevier, New York.
9. (a) Krämer, R., Lehn, J.-M., and Marquis-Rigault, A. (1993) Self-recognition in helicate self-assembly: Spontaneous formation of helical complexes from mixture of ligands and metal ions. *Proc. Natl Acad. Sci. USA*, **90**, 5394–5398; (b) Lehn, J.-M. and Eliseev, A.V. (2001) Dynamic combinatorial chemistry. *Science*, **291**, 2331–2333.
10. Ercolani, G., Piguet, C., Borkovec, M., and Hamacek, J. (2007) Symmetry numbers and statistical factors in self-assembly and multivalency. *J. Phys. Chem. B*, **111** (42), 12195–12203.
11. Ercolani, G. and Schiaffino, L. (2011) Allosteric, chelate and interannular cooperativity: a mise au point. *Angew. Chem. Int. Ed.*, **50**, 1762–1768.
12. Ercolani, G. (2006) Thermodynamics of metal-mediated assemblies of porphyrines. *Struct. Bond.*, **121**, 167–215.
13. Piguet, C. (2010) Five thermodynamic describers for addressing serendipity in the self-assembly of polynuclear complexes in solution. *Chem. Commun.*, **46** (34), 6209–6231.
14. Elhabiri, M. and Albrecht-Gary, A.-M. (2008) Supramolecular edifices and switches based on metals. *Coord. Chem. Rev.*, **252**, 1079–1092.
15. (a) Hamacek, J., Borkovec, M., and Piguet, C. (2005) A simple thermodynamic model for quantitatively addressing cooperativity in multicomponent self-assembly processes—Part 1: Theoretical concepts and application to monometallic coordination complexes and bimetallic helicates possessing identical binding sites. *Chem. Eur. J.*, **11**, 5217–5226; (b) Hamacek, J., Borkovec, M., and Piguet, C. (2005) A simple thermodynamic model for quantitatively addressing cooperativity in multicomponent self-assembly processes - Part 2: Extension to multimetallic helicates possessing different binding sites. *Chem. Eur. J.*, **11**, 5227–5237; (c) Dalla-Favera, N., Hamacek, J., Borkovec, M. *et al.* (2008) Linear polynuclear helicates as a link between discrete supramolecular complexes and programmed infinite polymetallic chains. *Chem. Eur. J.*, **14**, 2994–3005.
16. Borkovec, M., Hamacek, J., and Piguet, C. (2004) Statistical mechanical approach to competitive binding of metal ions to multi-center receptors. *J. Chem. Soc. Dalton Trans.*, **2004**, 4096–4105.
17. Bishop, D.-M. and Laidler, K.J. (1965) Symmetry numbers and statistical factors in rate theory. *J. Chem. Phys.*, **42**, 1688.
18. (a) Benson, S.W. (1958) Statistical factors in the correlation of rate constants and equilibrium constants. *J. Am. Chem. Soc.*, **80**, 5151–5154; (b) Bailey, W.F. and Monahan, A.S. (1978)

Statistical effects and the evaluation of entropy differences in equilibrium processes. Symmetry corrections and entropy of mixing. *J. Chem. Educ.*, **55**, 489–493.

19. Dalla Favera, N., Kiehne, U., Bunzen, J. *et al.* (2010) Intermetallic interactions within solvated polynuclear complexes: a misunderstood concept. *Angew. Chem. Int. Ed.*, **49** (1), 125–128.

20. Canard, G., Koeller, S., Bernardinelli, G., and Piguet, C. (2008) Effective concentration as a tool for quantitatively addressing preorganisation in multicomponent assemblies: Application to the selective complexation of lanthanide cations. *J. Am. Chem. Soc.*, **130**, 1025–1040.

21. Lemonnier, J.-F., Guénée, L., Bernardinelli, G. *et al.* (2010) Planned failures from the principle of maximum site occupancy in lanthanide helicates. *Inorg. Chem.*, **49** (3), 1252–1265.

22. (a) Hamacek, J., Besnard, C., Penhouet, T., and Morgantini, P.-Y. (2011) Thermodynamics, structure and properties of polynuclear lanthanide complexes with a tripodal ligand: Insight into their self-assembly. *Chem. Eur. J.*, **17**, 6757–6764; (b)Hamacek, J., Bernardinelli, G., and Filinchuk, Y. (2008) Tetrahedral assembly with lanthanides: Toward discrete polynuclear complexes. *Eur. J. Inorg. Chem.*, **2008**, 3419–3422.

23. (a) Kuhn, W. (1934) The shape of fibrous molecules in solutions. *Kolloid-Z.*, **68**, 2–15; (b) Jacobson, H. and Stockmayer, W.H. (1950) Intramolecular reaction in polycondensations. I. Theory of linear systems. *J. Chem. Phys.*, **18**, 1600–1606; (c) Flory, P.J., Suter, U.W., and Mutter, M. (1976) Macrocyclization equilibriums. 1. Theory. *J. Am. Chem. Soc.*, **98**, 5733–5739; (d)Winnik, M.A. (1981) Cyclisation and the conformation of hydrocarbon chains. *Chem. Rev.*, **81**, 491–524; (e) Jencks, W.P. (1981) On the attribution and additivity of binding energies. *Proc. Natl Acad. Sci. USA*, **78**, 4046–4050; (f) Ercolani, G. (1998) Physical basis of self-assembly macrocyclizations. *J. Phys. Chem. B*, **102**, 5699 5703; (g) Galli, C. and Mandolini, L. (2000) The role of ring strain on the ease of ring closure of bifunctional chain molecules. *Eur. J. Org. Chem.*, **2000**, 3117; (h) Gargano, J.M., Ngo, T., Kim, J.Y. *et al.* (2001) Multivalent inhibition of AB5 toxins. *J. Am. Chem. Soc.*, **123**, 12909.

24. Enemark, E.J. and Stack, T.D.P. (1998) Stereospecificity and self-selectivity in the generation of a chiral molecular tetrahedron by metal-assisted self-assembly. *Angew. Chem. Int. Ed.*, **37**, 932–935.

25. Senegas, J.-M., Koeller, S., and Piguet, C. (2005) Isolation and characterization of the first circular single-stranded polymetallic lanthanide-containing helicate. *Chem. Commun.*, **2005**, 2235–2237.

26. Ercolani, G. (2003) Assessment of cooperativity in self-assembly. *J. Am. Chem. Soc.*, **125**, 16097–16103.

27. Perlmutter-Hayman, B. (1986) Cooperative binding to macromolecules. A formal approach. *Acc. Chem. Res.*, **19**, 90–96.

28. (a) Van Holde, K.E. (1985) *Physical Biochemistry*, 2nd edn, Prentice Hall, Englewood Cliffs, (b) Ben-Naim, A. (1998) Cooperativity in binding of proteins to DNA. II. Binding of bacteriophage λ repressor to the left and right operators. *J. Chem. Phys.*, **108**, 6937–6946.

29. Rebek, J.J., Costello, T., Marshall, L. *et al.* (1985) Allosteric effect in organic chemistry: Binding cooperativity in a model for subunit Interactions. *J. Am. Chem. Soc.*, **107**, 7481–7487.

30. Blanc, S., Yakirevich, P., Leize, E. *et al.* (1997) Allosteric effects in polynuclear triple-stranded ferric complexes. *J. Am. Chem. Soc.*, **119**, 4934–4944.

31. (a) Shinkai, S., Sugasaki, A., Ikeda, M., and Takeuchi, M. (2001) Positive allosteric systems designed on dynamic supramolecular scaffolds: Toward switching and amplification of guest affinity and selectivity. *Acc. Chem. Res.*, **34**, 494–503; (b) Takeuchi, M., Sugasaki, A., Ikeda, M., and Shinkai, S. (2001) Molecular design of artificial molecular and ion recognition systems with allosteric guest responses. *Acc. Chem. Res.*, **34**, 865–873; (c) Ercolani, G. (2005) The Origin of cooperativity in double-wheel receptors. freezing of internal rotation or ligand-induced torsional strain? *Org. Lett.*, **7**, 803–805; (d) Wilson, G.S. and Anderson, H.L. (1999) A conjugated triple strand porphyrin array. *Chem. Commun.*, **1999**, 1539–1540.

32. (a) Eigen, M. and Wilkins, R.G. (1965) The kinetics and mechanism of formation of metal complexes. *Adv. Chem. Ser.*, **49**, 55–67; (b) Eigen, M. and Tamm, K. (1962) Sound absorption in electrolytes as a consequence of chemical relaxation. I. Relaxation theory of stepwise dissociation. *Z. Electrochem.*, **66**, 93–107; (c) Wilkins, R.G. (1970) Mechanisms of ligand replacement in octahedral nickel(II) complexes. *Acc. Chem. Res.*, **3**, 408–416.

33. Langford, C.H. and Gray, H.B. (1965) *Ligand Substitution Processes*, Benjamin, New York.

34. Helm, L. and Merbach, A.E. (2005) Inorganic and bioinorganic solvent exchange mechanisms. *Chem. Rev.*, **105**, 1923–1959.

35. Marquis-Rigault, A., Dupont-Gervais, A., Van Dorsselaer, A., and Lehn, J.-M. (1996) Investigation of the self-assembly pathway of pentanuclear helicates by electrospray mass spectrometry. *Chem. Eur. J.*, **2**, 1395–1398.

36. Fatin-Rouge, N., Blanc, S., Pfeil, A. *et al.* (2001) Self-assembly of tricuprous double helicates: thermodynamics, kinetics and mechanism. *Helv. Chim. Acta*, **84**, 1694–1711.

37. Fatin-Rouge, N., Blanc, S., Leize, E. *et al.* (2000) Self-assembly of a diferrous triple-stranded helicate with bis(2,2'-bipyridine) ligands: Thermodynamic and kinetic intermediates. *Inorg. Chem.*, **39**, 5771–5778.

38. (a) Piguet, C., Bünzli, J.-C.G., Bernardinelli, G. *et al.* (1993) Self-assembly and photophysical properties of lanthanide dinuclear triple-helical complexes. *J. Am. Chem. Soc.*, **115**, 8197–8206; (b) Hamacek, J., Blanc, S., Elhabiri, M. *et al.* (2003) Self-assembly mechanism of a bimetallic europium triple-stranded helicate. *J. Am. Chem. Soc.*, **125**, 1541–1550.

39. (a) Elhabiri, M., Scopelliti, R., Bünzli, J.-C.G., and Piguet, C. (1998) *Chem. Commun.*, **1998**, 2347–2348; (b) Elhabiri, M., Hamacek, J., Bünzli, J.-C.G., and Albrecht-Gary, A.M. (2004) *Eur. J. Inorg. Chem.*, **2004**, 51–62.

40. El Aroussi, B., Zebret, S., Besnard, C. *et al.* (2011) Rational design of a ternary supramolecular system:self-assembly of pentanuclear Lanthanide Helicates. *J. Am. Chem. Soc.*, **133**, 10764–10767.

41. Hamacek, J. and Piguet, C. (2006) How to adapt Scatchard plot for graphically addressing cooperativity in multicomponent self-assemblies. *J. Phys. Chem. B*, **110** (15), 7783–7792.

42. Garrett, T.M., Koert, U., and Lehn, J.-M. (1992) Binding cooperativity in the self-assembly of double stranded silver(I) trihelicate. *J. Phys. Org. Chem.*, **5**, 529–532.

43. Pfeil, A. and Lehn, J.-M. (1992) Helicate self-organisation: Positive cooperativity in the self assembly of double-helical metal complexes. *J. Chem. Soc., Chem. Commun.*, **1992**, 838–840.

44. Floquet, S., Ouali, N., Bocquet, B. *et al.* (2003) The first self-assembled trimetallic lanthanide helicates driven by positive cooperativity. *Chem. Eur. J.*, **9** (8), 1860–1875.

45. Hamacek, J., Borkovec, M., and Piguet, C. (2006) Simple thermodynamics for unravelling sophisticated self-assembly processes. *Dalton Trans.*, **2006**, 1473–1490.

46. Koper, G. and Borkovec, M. (2001) Binding of metal ions to polyelectrolytes and their oligomeric counterparts: an application of a generalized Potts model. *J. Phys. Chem. B*, **105**, 6666–6674.

47. Dalla Favera, N., Guenee, L., Bernardinelli, G., and Piguet, C. (2009) In search for tuneable intramolecular intermetallic interactions in polynuclear lanthanide complexes. *Dalton Trans.*, **2009**, 7625–7638.

48. Zeckert, K., Hamacek, J., Rivera, J.-P. *et al.* (2004) A simple thermodynamic model for rationalizing the formation of self-assembled multimetallic edifices: Application to triple-stranded helicates. *J. Am. Chem. Soc.*, **126**, 11589–11601.

49. Riis-Johannessen, T., Dalla Favera, N., Todorova, T.K. *et al.* (2009) Understanding, controlling and programming cooperativity in self-assembled polynuclear complexes in solution. *Chem. Eur. J.*, **15**, 12702–12718.

50. Zeckert, K., Hamacek, J., Senegas, J.-M. *et al.* (2005) Predictions, synthetic strategy and isolation of the first linear tetrametallic triple-stranded lanthanide helicate. *Angew. Chem. Int. Ed.*, **44**, 7954–7958.

51. Fyles, T.M. and Tong, C.C. (2007) Predicting speciation in the multi-component equilibrium self-assembly of a metallosupramolecular complex. *New J. Chem.*, **31**, 296–304.

52. Canard, G. and Piguet, C. (2007) The origin of the surprising stabilities of highly charged self-assembled polymetallic complexes in solution. *Inorg. Chem.*, **46**, 3511–3522.

53. Hasenknopf, B., Lehn, J.-M., Kneisel, B.O. *et al.* (1996) Self-assembly of a circular double helicate. *Angew. Chem. Int. Ed.*, **35**, 1838–1840.

54. Scherer, M., Caulder, D.L., Johnson, D.W., and Raymond, K.N. (1999) Triple helicate–tetrahedral cluster interconversion controlled by host–guest interactions. *Angew. Chem. Int. Ed.*, **38**, 1588–1592.

55. Albrecht, M., Osetska, O., Fröhlich, R. *et al.* (2007) Highly efficient near-IR emitting Yb/Yb and Yb/Al helicates. *J. Am. Chem. Soc.*, **129**, 14178–14179.

56. Terazzi, E., Guénée, L., Varin, J. *et al.* (2011) Silver baits for the "miraculous draught" of amphiphilic lanthanide helicates. *Chem. Eur. J.*, **17**, 184–195.

# 4

# Structural Aspects of Helicates

*Martin Berg and Arne Lützen*
*Kekulé Institute of Organic Chemistry and Biochemistry, University of Bonn, Germany*

## 4.1 Introduction

Spirals and helices are virtually ubiquitous, they can be found in giant objects such as spiral galaxies, but also in the shells, horns, teeth, and claws of animals, or in plant structures, as well as in manmade objects like helical springs, wires, and so on. Besides the interesting mechanical properties and the mathematical models that can be used to describe them, the attractiveness of their simply appealing structure has fascinated humans for a long time and caused use of this motif in numerous examples of architecture and art. Helical structures, however, are not only found in macroscopic objects but are also well known structural motifs in molecular sciences, as for example, double-stranded DNA or the α-helical secondary structures of peptides and proteins. Actually, there is even one naturally occurring dinuclear metal complex known that has such a structure [1,2]. In this complex that a red yeast uses as a growth factor, three molecules of rhodo-torulic acid (**1**), a diketopiperazine dihydroxamic acid, act as bridging ligands that wind themselves around two iron(III) centres in a helical fashion (Figure 4.1).

Interestingly, even before the elucidation of the structure of this natural compound a first example of an artificial dinuclear helical zinc(II) complex {[Zn$_2$(**2**)$_2$], Figure 4.2} was reported by J.-H. Fuhrhop and colleagues in 1976 that could be characterized by X-ray analysis [3]. Within the next decade a few further ligands (**3–5**, Figure 4.2) that form oligonuclear helical complexes were reported by G. van Koten [4], K.N. Raymond [5], and J.-M. Lehn [6] who finally coined the term *"helicate"* for coordination compounds

*Metallofoldamers: Supramolecular Architectures from Helicates to Biomimetics*, First Edition.
Edited by Galia Maayan and Markus Albrecht.
© 2013 John Wiley & Sons, Ltd. Published 2013 by John Wiley & Sons, Ltd.

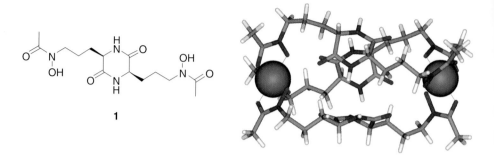

**1**

*Figure 4.1* *Rhodotorulic acid (**1**) and its helical dinuclear complex (Δ,Δ)-[Fe₂(**1**)₃] (PM3-minimized structure based on the proposed structure of Ref. [2b]).*

where at last one ligand strand is wound around two or more metal centres in a helical fashion.

Progress in analytical techniques, synthetic methods, and the Nobel price for supramolecular chemistry and molecular recognition awarded to C.J. Pedersen, D.J. Cram, and J.-M. Lehn in 1987 paved the way for the enormous development of this field which continues today. This also holds true for metallosupramolecular aggregates like helicates that have gained increasingly more attention since the late 1980s, as can be seen by the impressive list of review articles on these interesting compounds [7]. Although helicates are often referred to as model systems that mimic DNA, this is not really true since there are hardly any systems that carry sequential information and there are almost no heterostranded helicates, that is, helicates that are formed from complementary but not self-complementary strands. Nevertheless, they are aesthetically very appealing and they can be used as almost ideal models to study all kind of self-processes, such as self-assembly and self-sorting in terms of self-recognition, or self-discrimination. They also allow for the investigation of secondary thermodynamic factors associated with self-assembly processes like allostery or cooperativity. Furthermore, they can also be employed to study the kinetic effects of the formation of self-assembled structures or regarding their dynamic behaviour during interconversion, or to learn more about template effects. Another important aspect is of course chirality: although by far not as well developed as the synthesis of covalently assembled molecules, stereoselective molecular recognition or the stereocontrolled formation of non-covalently assembled aggregates has also evolved into a major issue in supramolecular chemistry [8]. In this respect, helicates are very important since helical structures are per se chiral objects and the mechanical coupling of two or more metal binding sites in a ligand strand offers an outstanding opportunity to control the relative stereochemistry of stereogenic metal centres which is difficult to achieve otherwise [9].

This overview will focus on the structural aspects involved in the self-assembly of helicates. This comprises structural dynamics, template effects, sequence selectivity, and self-sorting effects with regard to ligand size, number of binding sites, and especially stereoselectivity.

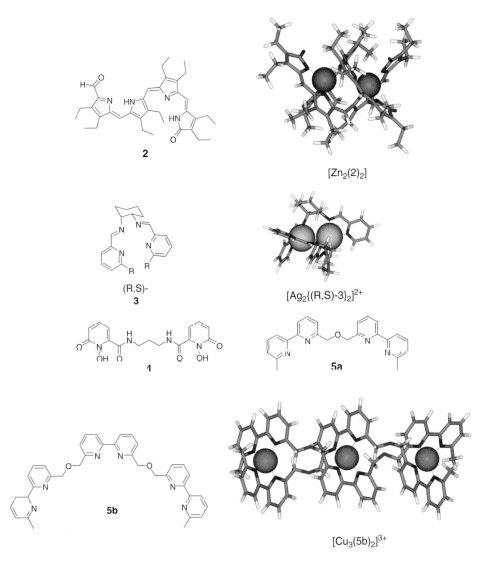

*Figure 4.2*  *First examples for artificial helicates (PM3 minimized structures).*

## 4.2  Structural Dynamics

Usually, helicates are formed in strict self-assembly processes, meaning that the process leads to the thermodynamic minimum structure. Therefore, reversibility of the individual binding events between ligand units and metal centres is necessary in order not to be kinetically trapped before reaching this minimum.

One of the first studies concerning this phenomenon was published by K.N. Raymond's group [10]. They prepared a series of bis(catechol) ligands that form racemic mixtures of

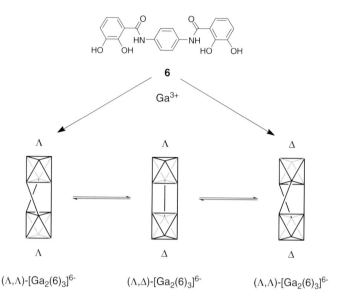

$(\Lambda,\Lambda)$-$[Ga_2(6)_3]^{6-}$      $(\Lambda,\Delta)$-$[Ga_2(6)_3]^{6-}$      $(\Lambda,\Lambda)$-$[Ga_2(6)_3]^{6-}$

**Figure 4.3** *Interconversion of stereoisomeric helicates from ligand* **6**.

$\Delta,\Delta$- and $\Lambda,\Lambda$-configured dinuclear triple-stranded gallium(III) and iron(III) helicates and investigated the interconversion of the two enantiomeric gallium complexes by NMR techniques (Figure 4.3).

Interestingly they found that once dinuclear helicates are formed epimerization of the individual stereogenic metal centres happens rather through a Bailar twist mechanism than through a dissociation–reassociation process because the mechanical coupling between the two metal ion sites is only weak. Hence, the transition state effectively involves the twisting of only one metal centre at a time, leading to the heterochiral $\Delta,\Lambda$-intermediate. However, the lifetime of this intermediate is short enough and its energy high enough compared with its homochiral diastereomers to prevent its detection.

Another result of these studies was that changing the surroundings by for example lowering the pH value has of course a major influence on the outcome of the equilibrium-driven processes and may also result in other reaction pathways, such as a proton-assisted mechanism.

Another interesting study concerning the formation of two very different types of metallosupramolecular aggregates from a single ligand **7** was reported by J.M. Lehn and colleagues (Figure 4.4) [11]. They found that mixing ligand **7** with an appropriate amount of silver(I) ions results in the simultaneous formation and interconversion of a quadruple-stranded helicate $[Ag_{10}(7)_4](OTf)_{10}$ and a [4 × 5]-Ag(I) grid structure $[Ag_{20}(7)_9](OTf)_{20}$ which could both be characterized by X-ray crystallography.

Similarly, M.J. Hannon reported on bis(pyridylimine) ligands that were found to form a very complicated and rapidly interconverting mixture of aggregates upon coordination to silver(I) ions from which two – a triple-stranded helicate and a planar dimer – could be characterized by X-ray crystallography [12]. These two studies might act as examples to show that it becomes more and more difficult to design and predict the exclusive

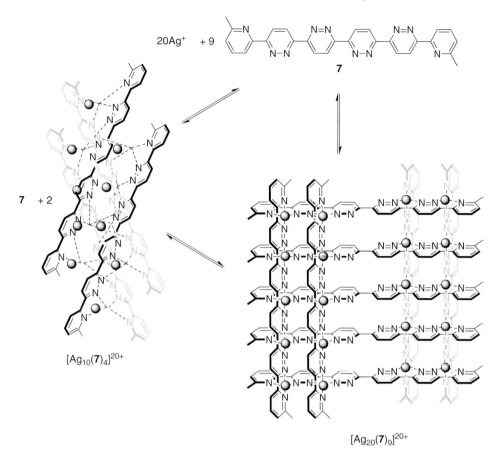

**Figure 4.4**  *Simultaneous formation and interconversion of helical and grid-like metallosupramolecular structures from ligand 7.*

formation of a single metallosupramolecular architecture with either an increasing number of components or an increasing number of possible different kinds of energetically similar non-covalent interactions because the differences in terms of thermodynamic stability become less pronounced.

## 4.3  Template Effects

As we have seen above, changing the equilibrium conditions by adding for example protons can have a marked effect on the outcome of self-assembly processes and their structural dynamics. The addition of components can result in templating effects. One of the first examples for this phenomenon was reported once again by J.-M. Lehn and colleagues when they found that chloride ions can act as templates for the formation of circular double-stranded helicates whereas other anions like tetrafluoroborate do not (Figure 4.5) [13].

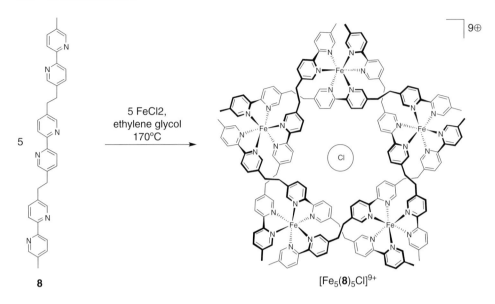

**Figure 4.5** *Counterion template effects I. Chloride ions act as templates in the formation of circular helicates.*

In the same year M. Albrecht found the first indications that alkali metal ions can also have a pronounced templating effect in the formation of dinuclear complexes of bis(catecholate) ligands, as only the use of lithium ions resulted in the formation of well defined dinuclear complexes from ligand **9** (in this case *meso*-helicates which will be discussed in more detail below), as shown by NMR spectroscopy and X-ray crystallography, whereas sodium and potassium ions gave complicated mixtures of aggregates with this ligand alone [14a]. In a series of following studies he was able to find more evidence that this effect was quite general for this kind of (formal) anionic helicates [14]. Interestingly, a template effect was also observed when a mixture of ligands **9** and **10** was investigated which usually gave rise to either *meso*-helicates $[M_2(9)_3]^{n-}$ or helicates $[M_2(10)_3]^{n-}$, respectively (Figure 4.6) [14c].

Later, K.N. Raymond found that this counterion effect can also be extended to a complete structural interconversion between triple-stranded helicates and capsular tetrahedral cluster aggregates by host–guest interactions with alkylammonium ions (Figure 4.7) [15].

## 4.4 Sequence Selectivity

So far, almost all ligands that were used to make helicates were symmetrical. Only one of K.N. Raymonds ligands used in his structural dynamics studies [10] was non-symmetrical because of two different terminal substituents at the catechol units. This difference, however, did not result in a sequence selective formation of a single helicate but rather resulted in a complex mixture of all possible isomers. In order to achieve sequence selectivity it was found to be a better approach to use heterotopic ligand strands, that is ligands

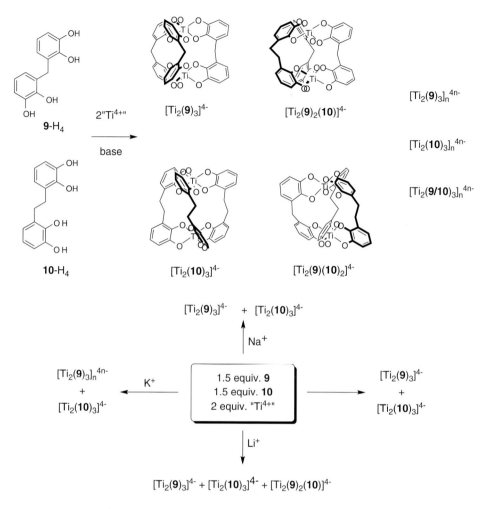

**Figure 4.6** *Counterion template effects II. Alkali metal ions act as templates in the formation of helicates from bis(catecholate) ligands. Top: possible homoleptic {[Ti$_2$(9)$_3$]$^{4-}$ and [Ti$_2$(10)$_3$]$^{4-}$} and heteroleptic {[Ti$_2$(9)$_2$(10)]$^{4-}$ and [Ti$_2$(9)(10)$_2$]$^{4-}$} dinuclear and oligomeric complexes obtained from a 1 : 1 mixture of ligands 9-H$_4$ and 10-H$_4$ and titanium(IV) ions in the presence of alkali metal carbonates (2 eq.) as base. Bottom: detected species obtained with different cations and mixtures of cations.*

with at least two different metal binding sites. In doing so, one can make use of the preference of different metal ions for different donor atoms, different coordination numbers, or charges in order to achieve selectivity.

The first reports on this came from the laboratory of C. Piguet in 1995 [16] who could nicely demonstrate that heteronuclear complexes containing d- and f-block metal ions can be assembled in a selective fashion by using benzimidazole ligands like **12** and **13** depicted in Figure 4.8. Since then the group of C. Piguet has characterized quite a number

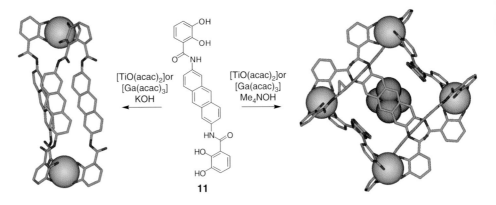

**Figure 4.7** *Structural interconversion of [M₂(**11**)₃] helicates and supramolecular [M₄(**11**)₆] capsules by host–guest interactions (X-ray crystal structure analyses, H atoms, solvent molecules, and counter ions omitted).*

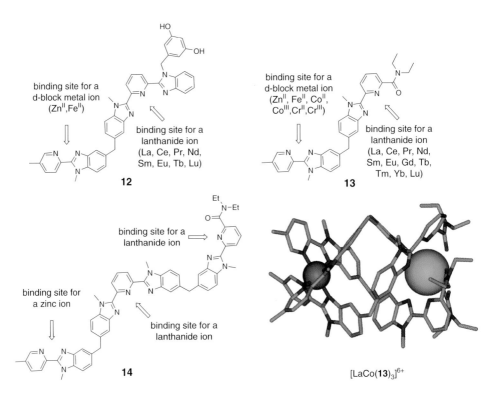

**Figure 4.8** *Sequence selective formation of di- and trinuclear helicates containing d- and f-block metal ions from ligands **12**–**14** carrying different metal binding sites. The structure of [LaCo(**13**)₃](ClO₄)₅.₅(OH)₀.₅ is given as an example (X-ray crystal structure analysis: H atoms, solvent molecules, and counter ions omitted).*

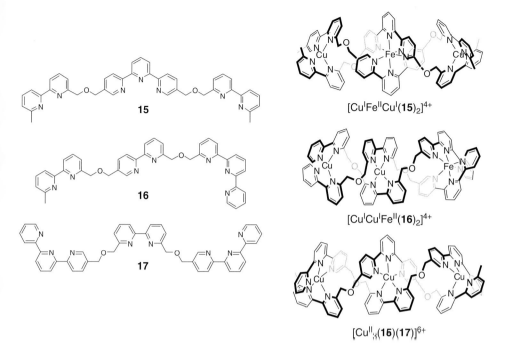

**Figure 4.9** *Sequence selective formation of helicates and self-assembly of heterostranded helicates from ligands carrying different metal binding sites.*

of these complexes very carefully by different spectroscopic (NMR, UV, fluorescence), mass spectrometric, and X-ray diffraction techniques [17]. In 2006 they were able to extend this approach to a triple-stranded trinuclear complex containing two zinc and one europium ion [18]. and quite recently C. Piguet was able to reveal in collaboration with T. Riis-Johannessen that it is possible to access triple-stranded trinuclear helicates with a sequence of 3d-4f-4f-metals in a selective manner using properly designed nitrogen donor ligands like **14** [19].

For oligopyridine ligands this was first demonstrated by J.-M. Lehn in 1996 [20]. Using a series of three different ligands **15–17** he was able to achieve the selective formation of sequence selective heterometallic trinuclear complexes (Figure 4.9), as evidenced by mass spectrometry and X-ray crystal structure analysis. This approach allows also for the formation of the first heterostranded helicate which is still one of the very rare examples for this kind of assembly [20,21].

M. Albrecht was then able to show similar sequence selectivity in bimetallic helicate formation from ligand systems with different oxygen donor ligands (Figure 4.10) like combinations of catechols and amino phenols [22], catechols and carboxylic acid hydrazones [23], 8-hydroxyquinolines and 2-carbamido-8-hydroxyquinolines [24], or catechols and sulfonic acid hydrazones [25].

Other examples for sequence selectivity appeared from the group of K.N. Raymond who used ligands like **18**-$H_4$ containing N-methyl-2,3-dihydroxypyridinone and 2,3-

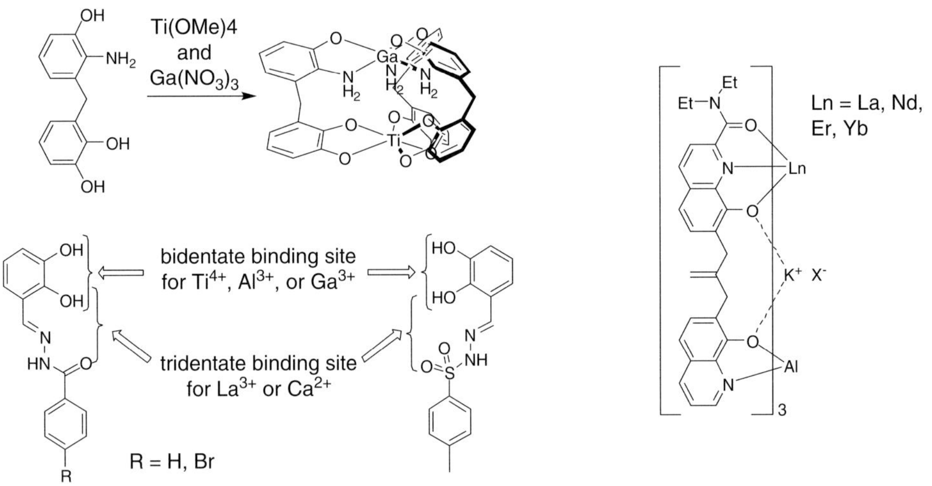

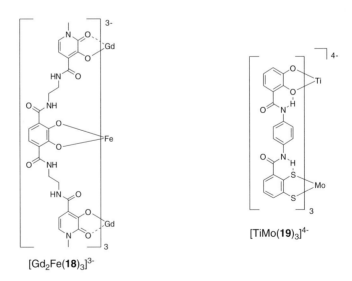

**Figure 4.10** *O-Donor ligands for sequence selective formation of helicates.*

dihydroxy-terephthalamide groups to selectively coordinate gadolinium(III) and iron(III) ions, respectively [26]. Quite recently F.E. Hahn demonstrated that sequence selectivity could also be achieved using ligand **19**-H$_4$ with a catechol and its dithiol derivative as metal binding sites and a mixture of an oxophilic with a thiophilic metal ion [27] (Figure 4.11).

[Gd$_2$Fe(**18**)$_3$]$^{3-}$

[TiMo(**19**)$_3$]$^{4-}$

**Figure 4.11** *Ligands for sequence selective formation of heterometallic triple-stranded helicates: trinuclear gadolinium iron complex [Gd$_2$Fe(**18**)$_3$]$^{3-}$ by K.N. Raymond and binuclear titanium molybdenum complex [TiMo(**19**)$_3$]$^{4-}$ by F.E. Hahn.*

## 4.5 Self-Sorting Effects in Helicate Formation

As mentioned in the introduction helicates are very interesting supramolecular aggregates to study the principle effects associated with self-assembly processes. One example would be self-sorting processes in mixtures of different types of ligands. In principle, one can expect three possible scenarios in such an experiment: (a) there is no selectivity at all so that one would obtain a statistical mixture of all possible assemblies, (b) only homo-stranded helicates are formed that contain only one sort of ligand, or (c) only hetero-stranded helicates are formed that contain different sorts of ligands. The latter two scenarios would both be results of self-sorting effects – (b) would be called self-recognition and (c) self-discrimination. So far these self-sorting phenomena have been investigated by different approaches and we will concentrate on studies with achiral ligands first.

Actually, it should be noticed that J.-M. Lehn's heterostranded helicate $[Cu^{II}_3(\mathbf{15})$ $(\mathbf{17})]^{6+}$ shown in Figure 4.9 is one of the very rare examples for self-discrimination in helicate chemistry [20,21]. The other extreme – self-recognition – is however obviously far more common in helicate chemistry. Already in 1993 J.M. Lehn could demonstrate that only homostranded helicates are formed from a mixture of the four different oligobi-pyridine ligands **20–23** when mixed with the appropriate amount of copper(I) ions according to two factors: (a) the maximum occupancy rule that favours those aggregates where all coordination sites of metal ions and ligands are used and none of them stay unsaturated, and (b) the entropically favourable formation of discrete aggregates compared to polymeric ones at concentrations below the effective molarity (i.e. the concentration where the formation of oligomers and polymers starts to compete energetically with the formation of the discrete assemblies of low nuclearity; Figure 4.12) [28].

In the same study J.-M. Lehn and co-workers also demonstrated that using different linker groups in ligand strands that lead to a different distance between the individual bipyridine metal binding sites also results in self-recognition. A similar observation was also made by K.N. Raymond's group when studying mixtures of bis(catecholate) ligands (**24**-H$_4$)-(**26**-H$_4$) with rigid linkers of different length (Figure 4.13) [29].

Finally, a third approach was reported by Y. Cohen in which ligand strands comprising bipyridine or bithiophene ligand units also showed self-recognition behaviour [30].

Within chiral ligands another assumedly subtle difference between two ligands would be the inversion of stereogenic elements (i.e. the use of stereoisomeric ligands). So far, a number of racemic ligands have been studied in this context. An example published by the group of P.K. Mascharak is ligand **27** (Figure 4.14). This ligand was found to undergo selective self-recognition to give a racemic mixture of helical complexes with quadruple planar coordinated copper(II) ions [31].

The situation gets more complicated if helicates are formed that bear stereogenic metal centres (either in a tetrahedral or an octahedral coordination geometry). In such cases even a completely selective self-sorting process can lead to different diastereoisomeric helicates due to the possibility of the metal ions to adopt a Λ- or Δ-configuration. Such a system was first reported by T.D.P. Stack at the end of the 1990s. When he investigated the chiral, racemic stereoisomers (*R,R*)-**3** and (*S,S*)-**3** of G. van Kotens *meso*-configured bis(pyridylimine) ligand (*R,S*)-**3** he found that the self-assembly proceeds exclusively via a self-recognition process, yielding a racemic pair of homochiral double-stranded

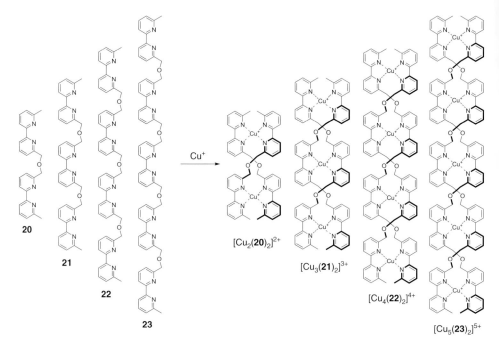

**Figure 4.12** *Self-recognition in helicate assembly from achiral ligands* **20–23** *according to the maximum occupancy rule.*

helicates when complexed with copper(I) ions (Figure 4.15). Interestingly, this self-assembly process is not only selective in terms of self-recognition of the ligand strands, it is also diastereoselective in terms of the fact that the stereochemistry of the ligand strand completely controls the configuration of the newly formed stereogenic metal centres [32].

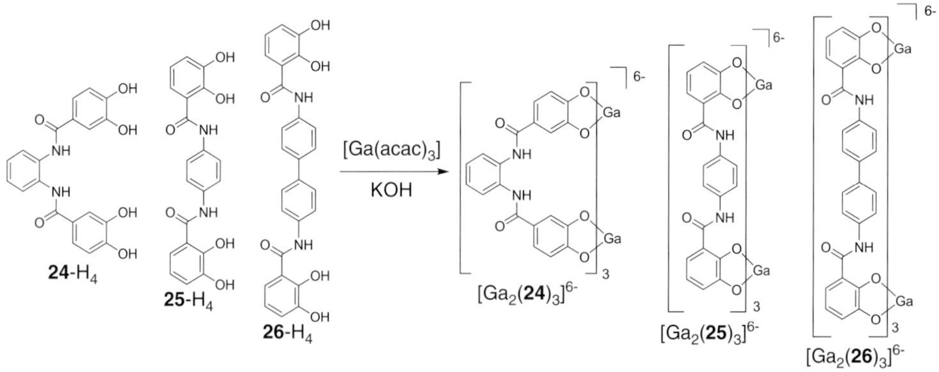

**Figure 4.13** *Self-recognition in helicate assembly from achiral ligands according to avoid unfavourable steric stress due to linkers of different lengths.*

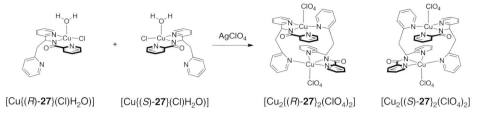

[Cu{(R)-27}(Cl)H₂O)]    [Cu{(S)-27}(Cl)H₂O)]    [Cu₂{(R)-27}₂(ClO₄)₂]    [Cu₂{(S)-27}₂(ClO₄)₂]

**Figure 4.14**  *Self-recognition process of racemic ligand 27 giving rise to a racemic mixture of homochiral double-stranded helicates with non-stereogenic metal centres.*

This behaviour is quite general if the ligand is properly designed and the formation of the helicate does not impose too much sterical stress on the assembly, which finally overcomes the energetic differences between the possible stereoisomers. Rigid bis(bipyridyl)ligands **28–33** based on 2,8-difunctionalized Tröger's base scaffolds, for example, self-assemble in a completely diastereoselective self-recognition manner into dinuclear double-stranded helicates upon coordination to copper(I) or silver(I) ions (Figure 4.16) [33].

In a later study we were able to reveal the almost perfect preorganization of these ligands for the formation of double-stranded helicates [34], but when we investigated the formation of the corresponding dinuclear triple-stranded zinc(II) helicates or dinuclear triple-stranded titanium(IV) complexes with corresponding di(catechol) ligands **34** and **35** we observed almost no selectivity because the rigid conformation of the ligand is just not well preorganized to form these assemblies which ask for a greater opening angle of the V-shaped ligand structure [35].

However, more flexible ligands **36–39** with a binaphthyl core unit that can adjust better to adopt the conformations necessary for the formation of double-stranded *and* triple-stranded helicates were found to undergo a diastereoselective self-assembly of both (Figure 4.17) [36].

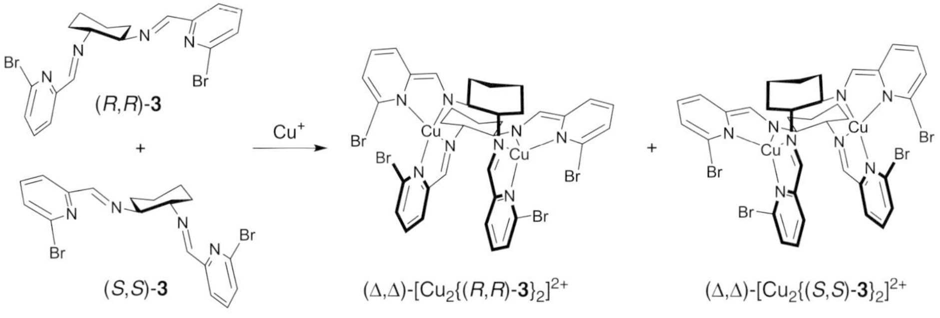

(R,R)-3

(S,S)-3

Cu⁺

(Δ,Δ)-[Cu₂{(R,R)-3}₂]²⁺

(Δ,Δ)-[Cu₂{(S,S)-3}₂]²⁺

**Figure 4.15**  *Diastereoselective self-recognition process of racemic ligands (R,R)-3 and (S,S)-3 giving rise to a racemic mixture of homochiral double-stranded helicates with stereogenic metal centres.*

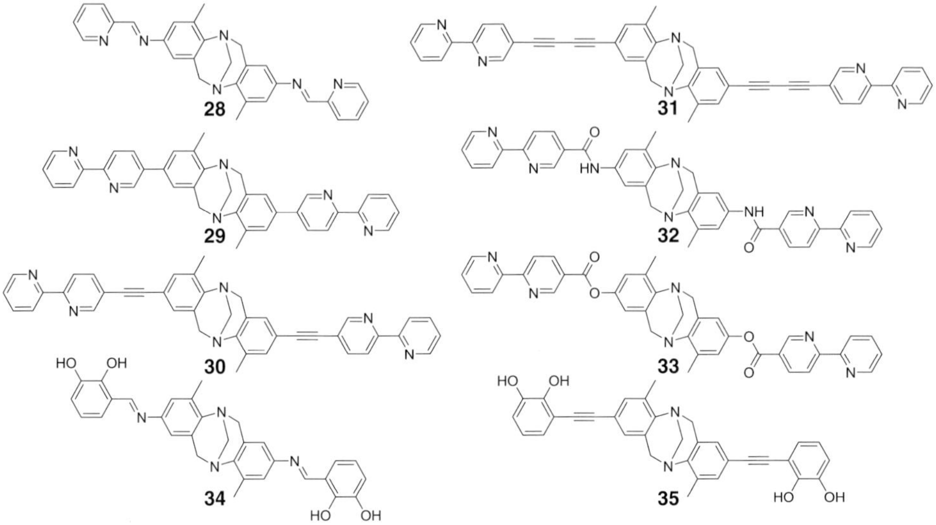

**Figure 4.16** *Racemic ligands based on the Tröger's base scaffold that give rise to racemic mixtures of homochiral double-stranded helicates via diastereoselective self-recognition processes.*

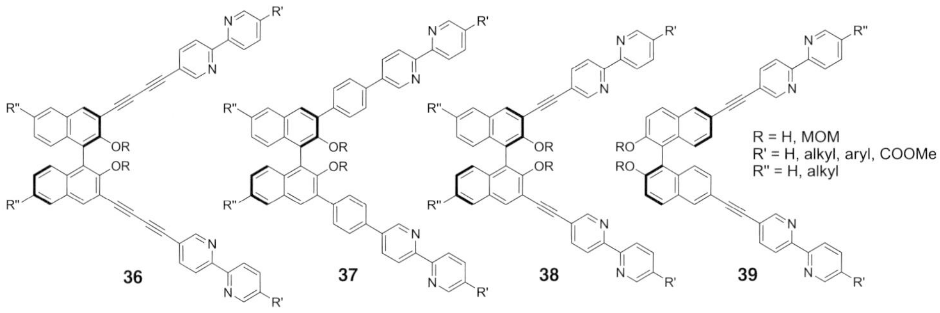

**Figure 4.17** *Racemic ligands 36–39 based on binaphthyl scaffolds that give rise to racemic mixtures of homochiral double- and triple-stranded helicates via diastereoselective self-recognition process.*

## 4.6 Diastereoselectivity I – "*Meso*"-Helicate versus Helicate Formation

As seen in the last chapter the octahedrally or tetrahedrally coordinated metal centres are stereochemical elements. Thus, in a dinuclear assembly they can adopt either a Λ,Λ-, a Δ,Δ- or a Δ,Λ-configuration. Whereas the first two would be a racemic mixture of helicates if no further stereogenic element is present in the ligand structure, the last one is not a real helicate corresponding to the definition given in the introduction but is rather a box-like side by side complex which is called a "*meso*"-helicate [37]. Thus, *meso*-helicates

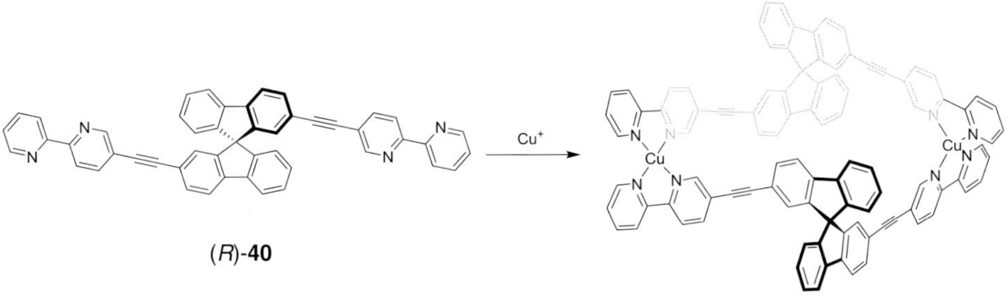

**Figure 4.18** *Diastereoselective self-assembly of an enantiomerically pure "meso"-helicate (Δ,Λ)–[Cu₂{(R)–**40**}₂]²⁺ from 9,9′-spirobifluorene-bridged bis(bipyridine) ligand (R)-**40**.*

and helicates are diastereomers and in fact there have been a number of studies to prepare these *meso*-forms in a selective manner. In 1995 two articles were published within a month about the selective formation of these *meso*-isomers. The first was from the group of M. Albrecht who later found an interesting odd/even effect of the linker length between two catechol units: odd numbers of methylene groups gave *meso*-helicates whereas even numbers gave helicates (see also Figure 4.6) [7e,14a,b,38].

The second report came from the laboratory of M.M. Harding and she reported on a number of naphthyl-bridged bis(bipyridine) ligands that rather gave rise to non-helical metallomacrocycles rather than helicates upon coordination to nickel(II) or zinc(II) ions [39]. Later, further studies were made by the groups of K.N. Raymond [40], K. Gloe [41], and ourselves (Figure 4.18) [42] that further confirmed that *meso*-helicates can be accessed in a selective manner.

A prediction, however, why the formation of *meso*-helicates becomes more favourable than that of the corresponding helical diastereomers is still very difficult, except for the systematically studied systems of M. Albrecht.

## 4.7 Diastereoselectivity II – Enantiomerically Pure Helicates from Chiral Ligands

The previous example described an enantiomerically pure supramolecular aggregate. This section will focus on the preparation and isolation of enantiomerically pure helicates. In fact, there are two strategies to obtain helicates in optically pure form, either by resolution of a racemic mixture or stereoselective self-assembly using enantiomerically pure components.

The first is rather rare because the kinetic lability which is needed for the thermo-dynamically controlled self-assembly to occur usually prevents the separation of enantiomers. Thus, the resolution of racemic mixtures can only be successful with kinetically quite stable assemblies that do not racemize under given conditions due to mechanisms described in Section 4.2. One way to achieve this is by modifying the lability of a metal complex by an electron transfer reaction, such as the oxidation of labile cobalt(II) ions to

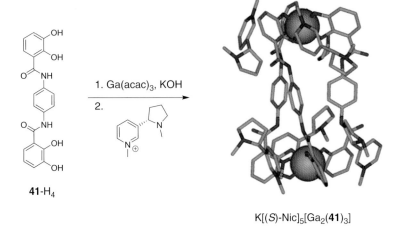

**41**-H$_4$

K[(S)-Nic]$_5$[Ga$_2$(**41**)$_3$]

**Figure 4.19** *Enantiomerically pure helicate K[(S)-Nic]$_5$[Ga$_2$(**41**)$_3$] [(S)-Nic = (S)-N-methylni-cotinium] via diastereoselective self-assembly by ion pair formation (X-ray crystal structure analysis H atoms, solvent molecules, and potassium ion omitted).*

inert cobalt(III). Following this approach or using more inert complexes from the beginning, there are a few reports where helicates could indeed be resolved either by spontaneous resolution by crystallization as a conglomerate [43], or by chromatographic means using enantiomerically pure stationary phases [44].

Much more common, however, are strategies where enantiomerically pure components, especially ligand strands, are used to assemble a certain diastereomer in a stereoselective fashion. In these cases the fact that any epimerization of metal centres results in the formation of energetically different diastereomers rather than enantiomers (as in the example in Section 4.2) prevents a loss of stereochemical integrity since the structural dynamics of the self-assembly process will always lead back to the thermodynamically most favourable diastereomer.

Before we concentrate on the use of enantiomerically pure ligand strands, the only example for the use of a chiral counterion should be mentioned. This was reported in 2001 by K.N. Raymond's group when they used enantiomerically pure (S)-N-methylnicotinium to induce the overall diastereoselective formation of a single optically pure anionic dinuclear triple-stranded helicate from an achiral bis(catecholate) ligand (Figure 4.19) [45].

In principle, there are two design strategies for introducing a chiral element into a ligand strand: either in the outer periphery, or in between two metal binding sites. In fact, both strategies have been successfully realized in a number of examples using different metal binding sites. In the following we will divide the individual approaches by sorting them according to the metal (chelating) binding motifs.

### 4.7.1 2,2′-Bipyridine Ligands

The first example dates back to the year 1991 and was reported by the group of J.-M. Lehn. By introducing methyl groups into the ether linkages of his well known

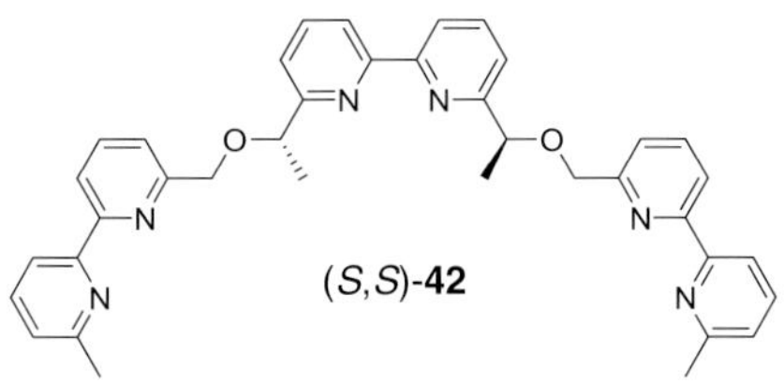

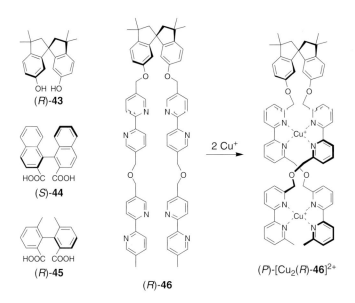

**Figure 4.20** *Enantiomerically pure ligand (S,S)-42 that gave rise to optically pure helicate {Cu₃[(S,S)-42]₂}³⁺ via diastereoselective self-assembly.*

oligo(bipyridine) he was able to prepare an enantiomerically pure ligand (*S,S*)-**42** (Figure 4.20) which gave rise to enantiomerically pure double-stranded helicates upon coordination to silver(I) or copper(I) ions, as revealed by NMR and CD spectroscopy although the absolute configuration of the metal centres could not be assigned unambiguously based on these experimental data [46].

A few years later J.S. Siegel's group came up with systems where two ligand strands were actually covalently coupled to $C_2$-symmetric building blocks. In this way these central building blocks (*R*)-**43**, (*S*)-**44**, or (*R*)-**45** acted as templates to ensure diastereoselective formation of the helicate structures {e.g. (*P*)-[Cu₂{(*R*)-**46**}]²⁺} by translating their stereogenic information into the helix as proven by NMR and CD spectroscopic means (Figure 4.21) [47].

**Figure 4.21** *Enantiomerically pure building blocks (R)-43, (S)-44, and (R)-45 that were used to translate their stereogenic information into optically pure helicate structures like, for example, (P)-[Cu₂(R)-46]²⁺ via diastereoselective self-assembly.*

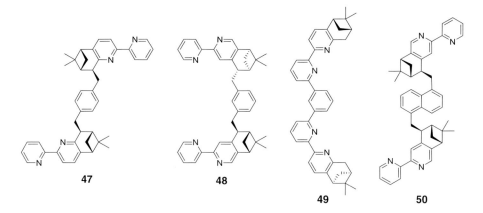

**Figure 4.22** *Enantiomerically pure bis(bipyridyl) ligands **47**, **48**, **49**, and **50** explored by the von Zelewski group to form cyclic (**47**), non-cyclic (**48**, **49**), and polymeric (**50**) helicates via diastereoselective self-assembly.*

A. von Zelewsky then introduced "oligomers" and constitutional isomers of his famous chiragen ligands (chiral 2,2′-bipyridines derived from terpenes like α-pinene) [48] to prepare enantiomerically pure helicates. In a whole series of publications he could demonstrate that using these ligands **47–50** (Figure 4.22) with the bicyclic terpene skeleton annealed to one of the 2,2′-bipyridine rings is a very reliable and robust motif that ensures diastereoselective self-assembly of circular (**47**) [49], non-circular (**48**, **49**) [50], and even polymeric helicates (**50**) [51]. Especially, the last one is exceptional because additional intermolecular interaction in the solid state packing causes the formation of a polymeric aggregate to be more favourable than the formation of discrete dinuclear coordination compounds.

A very similar approach to that of A. von Zelewsky was propagated by the group of E.C. Constable. They prepared chiral pinene-derived bis(bipyridine) ligands connected by flexible alkyl bridges like **51** (Figure 4.23) and could show that these self-assembled completely diastereoselectively to double-stranded helicates upon coordination to silver(I) ions despite the flexible linkers. However, when they investigated the coordination behaviour of these ligands towards copper(I) ions, they found a much more

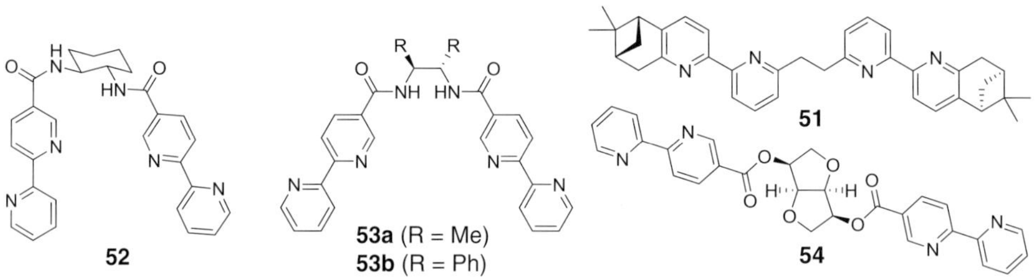

**Figure 4.23** *Enantiomerically pure bis(bipyridyl) ligands **51–54** explored by the groups of E.C. Constable, N.C. Fletcher, and ourselves that were demonstrated to undergo diastereoselective self-assembly to optically pure helicates upon coordination to suitable transition metal ions.*

complex mixture of linear and cyclic aggregates that were formed without any diastereoselectivity [52].

Other approaches focussed on the synthesis and investigation of bis(bipyridine) ligands using chiral central building blocks like *trans*-1,2-diaminocyclohexane (**52**) [53], 1,2-diphenyl- or 1,2-dimethylethylenediamine (**53**) [54], 2,2'-dihydroxy-1,1'-binaphthyl derivatives (**36–39**) [36,55], Tröger's base derivatives (**29–33**) [33–35], or isomannite (**54**) [56], which turned all out to be successful in terms of achieving at least very high degrees of diastereoselectivity (Figures 4.16, 4.17, 4.23). Hence, this approach can be regarded as a very general solution to the problem of controlling the stereochemistry of oligonuclear coordination compounds as long as the ligand strands provide an appropriate degree of rigidity and at the same time ensure a properly adjusted relative orientation of the bipyridine units to allow them to be engaged in the formation of a tetrahedral and/or octahedral coordination sphere around the individual metal centres.

### 4.7.2   2,2':6',2''-Terpyridine and 2,2':6',2'':6'',2-Quaterpyridine Ligands

These classes of ligands were very successfully investigated by the group of E.C. Constable again preparing a number of terpyridine [57] and quaterpyridine ligands [58] that carry terpene moieties in their periphery, which ensure diastereoselective helicate formation (Figure 4.24).

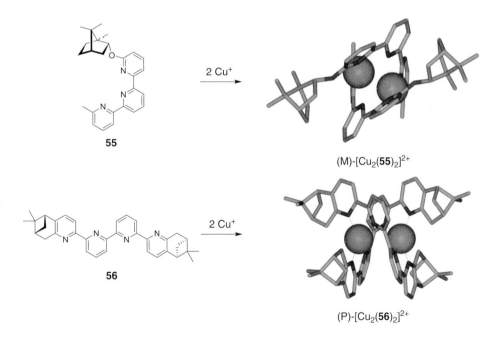

**55**

2 Cu⁺

(M)-[Cu₂(**55**)₂]²⁺

**56**

2 Cu⁺

(P)-[Cu₂(**56**)₂]²⁺

***Figure 4.24*** *Enantiomerically pure terpyridine ligand **55** and quaterpyridine ligand **56** explored by the group of E.C. Constable that were demonstrated to undergo diastereoselective self-assembly to optically pure double-stranded helicates (M)-[Cu₂(**55**)₂]²⁺ and (P)-[Cu₂(**56**)₂]²⁺ upon coordination to copper(I) or silver(I) ions, respectively (X-ray crystal structure analyses, H atoms, solvent molecules, and counterions omitted).*

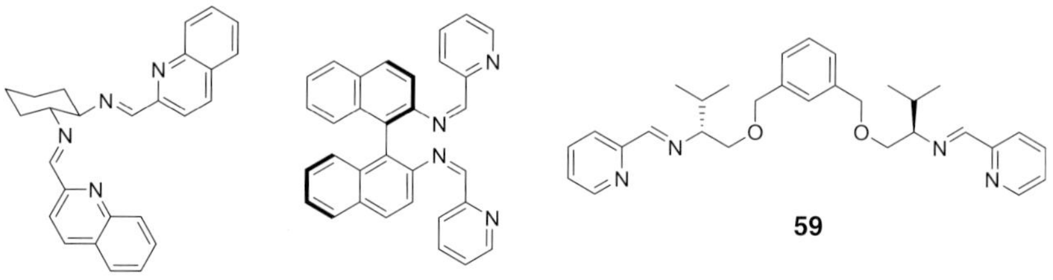

(R,R)-57          (R)-58          59

**Figure 4.25**   *Chiral 2-pyridylimine ligands prepared by L. Fabrizzi, M.J. Hannon, and S.G. Telfer and R. Kuroda that undergo diastereoselective self-assembly to optically pure double-stranded helicates upon coordination to copper(I) or silver(I) ions, respectively.*

### 4.7.3   2-Pyridylimine Ligands

The use of this ligand motif in helicate chemistry dates back to the very first examples of G. van Koten [(R,S)-**3**] in the early 1980s (see Figure 4.3) [4].

Later, chiral versions [(R,R)-**3** and (S,S)-**3**] of the initially prepared *meso*-ligand [(R,S)-**3**] were prepared by T.D.P. Stack to demonstrate the phenomenon of self-recognition (see Figure 4.15) [32]. Varieties of this theme were then explored by the groups of L. Fabrizzi (**57**) [59], M.J. Hannon (**58**) [60], S.G. Telfer and R. Kuroda (**59**) [61], and ourselves (see ligand **28** in Figure 4.16) [33]. An overview of these ligands is shown in Figure 4.25. Again, either a rigid central chiral building block was introduced into the ligand structure to ensure diastereoselectivity, or several stereogenic centres were placed close to the chelating unit in systems with larger, more flexible linkers, similar to J.M. Lehn's approach.

### 4.7.4   Further Hexadentate N-Donor Ligands

In addition to the well known iminopyridines, bi-, ter-, and quaterpyridines, there were also two other reports on the use of enantiomerically pure hexadentate *N*-donor ligands for the diastereoselective self-assembly of helicates. Whereas the pinene-derived ligand **60** of A. von Zelewsky was demonstrated to form an optically pure tetranuclear circular helicate upon coordination to zinc(II) ions [62], the cyclohexanediamine-derived ligand **61** of E.C. Constable was found to form dinuclear helicates [63]. Interestingly, the formation of $[M_2\mathbf{61}_2]^{n+}$ complexes was observed not only with iron(II) and zinc(II) ions as expected, but also with silver(I) ions, which were found to be coordinated by a rather unusually high number of donor atoms, maybe also including an attractive Ag–Ag interaction (Figure 4.26).

### 4.7.5   Oxazoline Ligands

Oxazolines are also powerful ligands that can easily be prepared in optically pure forms by the condensation of appropriate aldehydes and chiral aminoalcohols derived from, for example, α-aminoacids. Using such ligands **62** and **63** derived from phenylalanine and phenylglycine, A.F. Williams was the first to show that they give rise to linear or

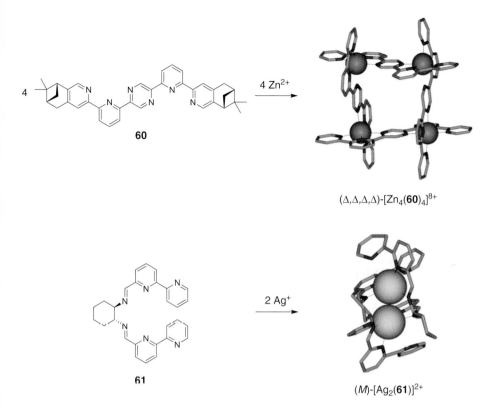

$(\Delta,\Delta,\Delta,\Delta)$-[Zn$_4$(**60**)$_4$]$^{8+}$

$(M)$-[Ag$_2$(**61**)]$^{2+}$

**Figure 4.26** *Chiral hexadentate N-donor ligands **60** and **61** prepared by A. von Zelewski and E.C. Constable, respectively, and their diastereoselective self-assembly to optically pure circular tetranuclear zinc(II) helicate $(\Delta,\Delta,\Delta,\Delta)$-[Zn$_4$(**60**)$_4$]$^{8+}$ and dinuclear silver(I) helicate $(M)$-[Ag$_2$(**61**)]$^{2+}$ (X-ray crystal structure analyses, H atoms, counter ions, and solvent molecules omitted).*

circular helicates [Ag$_2$**62**$_2$]$^{2+}$ and [Ag$_3$**63**$_3$]$^{3+}$ upon coordination to silver(I) ions [64]. Later, F.G. Gelalcha was able to demonstrate that further ligands of this kind, like valine-derived **64**, can also be used to form dinuclear helicates with copper(I) ions in which the copper ions are non-equivalent (Figure 4.27) [65].

### 4.7.6 P-Donor Ligands

In the mid1980s F.A. Cotton and R.D. Peacock were already exploring the formation of chiral dinuclear molybdenum complexes from enantiomerically pure diphosphane ligands [66]. Although the two bridging ligand strands were actually found to be helically wound around the two molybdenum centres in the solid state, and hence can be regarded as helicates, they will not be further discussed here because the metal centres themselves are not stereogenic centres whose sense of chirality is controlled in a diastereoselective manner by the chiral information embedded in the ligand structure. This, however, has also been

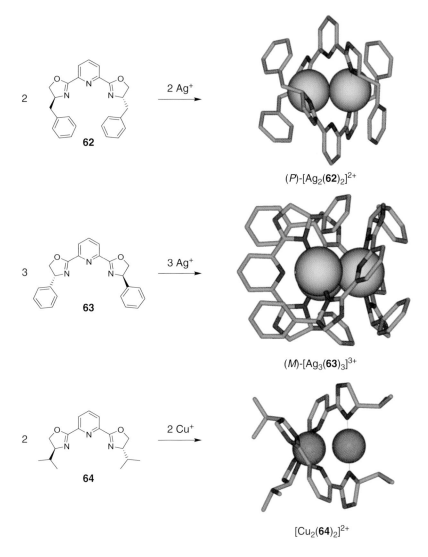

**Figure 4.27** *Amino acid-derived bis(oxazoline) ligands 59–61 and their diastereoselective self-assembly to optically pure dinuclear double-stranded silver(I) and copper(I) helicates (P)-[Ag$_2$(62)$_2$]$^{2+}$ and [Cu$_2$(64)$_2$]$^{2+}$ and circular trinuclear silver(I) helicate (M)-[Ag$_3$(63)$_3$]$^{3+}$ (X-ray crystal structure analyses, H atoms, solvent molecule, and counter ions omitted).*

achieved with chiral *P*-donor ligands and, in fact, in the examples published by the group of S.B. Wild the chiral information is located directly at configurationally stable tertiary phosphine atoms, as depicted in Figure 4.28 [67]. However, besides double-stranded helical complexes with copper(I), silver(I), and gold(I) ions, non-helical side by side stereo-isomers were also formed in most cases.

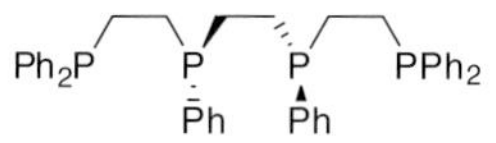

Ph₂P ... P ... P ... PPh₂
Ph    Ph

**65**

*Figure 4.28*    *Chiral P-donor ligand 65 with stereogenic P atoms.*

### 4.7.7  Hydroxamic Acid Ligands

As shown by the naturally occurring complex formed upon coordination of rhodotorulic acid to iron(III) ions, hydroxamic acids are suitable ligands for the formation of helicates.

In an effort to access artificial iron ion binders, A. Shanzer reported on a tertiary amine **66** carrying three identical chains containing a stereogenic centre and two hydroxamic acid functions [68]. Upon coordination to iron(III) ions, this ligand was found to form an optically pure dinuclear complex in a diastereoselective fashion (Figure 4.29).

### 4.7.8  β-Diketonate Ligands

Another class of chelating ligands with a single negative charge are β-diketonates. M. Albrecht prepared an example of these from tartaric acid (**67**) and demonstrated that it can be used to achieve the diastereoselective formation of enantiomerically pure neutral triple-stranded dinuclear iron(III) and gallium(III) helicates (Figure 4.30) [69]. Interestingly, this helicate was found to act as a non-covalently assembled cryptand for lithium ions.

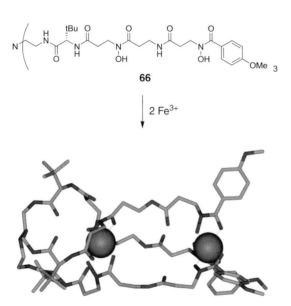

**66**

$\downarrow$ 2 Fe$^{3+}$

*Figure 4.29*    *Triamine 66 with chiral chains containing hydroxamic acid functionalities for the complexation of iron(III) ions in a diastereoselective fashion (PM3-minimized structure, H atoms omitted).*

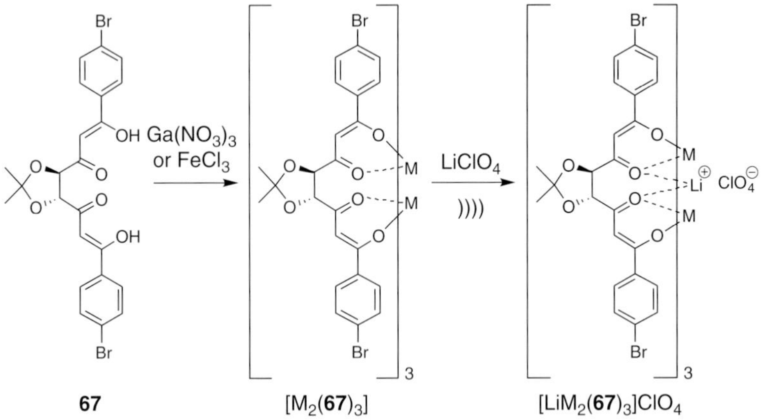

**Figure 4.30** *Tartaric acid-derived bis(β-diketonate) ligand **67**, its triple-stranded dinuclear gallium(III) or iron(III)helicates, and their use as metalla-cryptand for the complexation of lithium ions.*

### 4.7.9 Catecholate Ligands and Other Dianionic Ligand Units

Besides 2,2′-bipyridines, catechol ligands are probably the most often used metal binding motifs in helicate chemistry. Therefore, it does not come as a surprise that enantiomerically pure bis(catechol) ligands have also been prepared and tested with regard to their ability to undergo diastereoselective self-assembly to triple-stranded helicates upon coordination to suitable metal ions, like gallium(III) or titanium(IV) ions. Again, both strategies – either using a chiral building block as the centre between the two catechol units, or putting it in the outer periphery – have been successfully applied. T. D.P. Stack, for instance, prepared chiral ligands in which he used chiral diamines in order to link two catechol groups via amide bonds [70]. Interestingly, he found that ligands with sterically constrained $C_2$-linkers between the amide groups gave rise to tetranuclear clusters rather than dinuclear helicates, which were only obtained by less constrained $C_2$- or $C_3$-linkers (Figure 4.31) [70b].

Complete diastereoselectivity was also observed by M. Albrecht when studying a very simple to prepare dicatechol imine ligand **72** (Figure 4.32), again introducing the chiral information in the centre of the ligand [71]. Interestingly, this ligand has to undergo an inversion of the cyclohexyl moiety to form the less favoured conformer with 1,2-diaxial configured imine groups in order to be able to undergo diastereoselective self-assembly.

In a different approach he was able to derive dicatechol ligand **73** (Figure 4.33) with a rather flexible ethylene linker with chiral 1-phenylethylamine moieties at the two ends. Again this approach proved to be successful since only enantiomerically pure dinuclear titanium helicates [Ti$_2$(**73**)$_3$] were obtained. Unfortunately, he was not able to assign an absolute or relative stereochemistry of the metal centres in this case [72].

A completely different tetraanionic ligand was prepared by E.J. Corey in 1994. Starting from mannitol he was able to prepare an enantiomerically pure tetrol ligand **74** that formed a trinuclear titanium complex in a diastereoselective fashion, as revealed by X-ray analysis (Figure 4.33) [73].

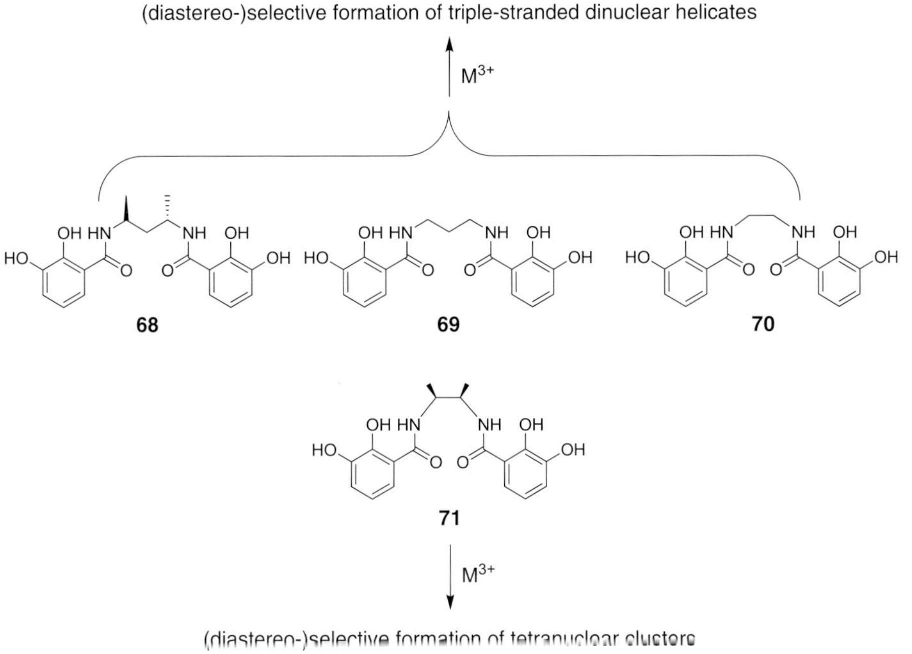

(diastereo-)selective formation of triple-stranded dinuclear helicates

**Figure 4.31** *Bis(catecholate) ligands 68–71 and their self-assembly behaviour upon coordination towards trivalent metal ions like gallium(III).*

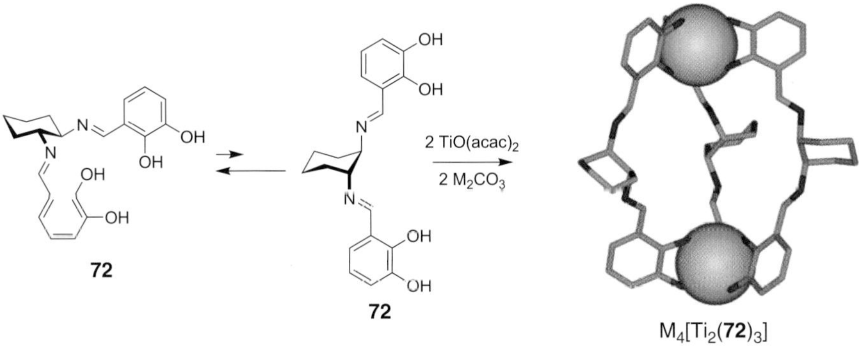

**Figure 4.32** *Bis(catecholate) ligand 72 and its diastereoselective self-assembly behaviour upon coordination towards titanium(IV) ions (X-ray crystal structure analysis, H-atoms, solvent molecules, and counter ions omitted).*

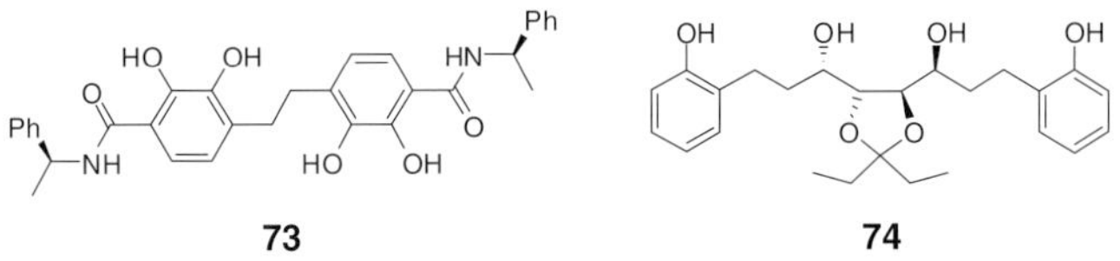

**Figure 4.33** *Bis(catecholate) ligand 73 and tetrol ligand 74.*

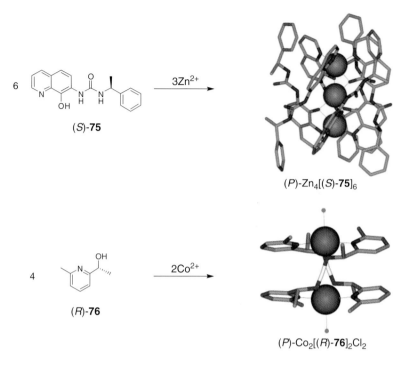

**Figure 4.34** *Hierarchical self-assembly of enantiomerically pure non-covalently assembled helicates (P)-[Zn₃(75)₆] and (L,L)-[Co₂{(R)-76}₄] (X-ray crystal structure analyses, H atoms and solvent molecules omitted).*

### 4.7.10 Non-Covalently Assembled Ligand Strands

The last examples to be reported in this section are actually very interesting because these cases use chiral ligand subunits **75** and **76** that have only one chelating metal binding site (Figure 4.34).

Since these contain additional functional groups, however, that are capable of being engaged in additional non-covalent interactions, like hydrogen bonds, they can organize further to hierarchically assembled helicates with non-covalently assembled ligand strands in a well defined matter. Examples for such systems have been reported by M. Albrecht [74] and by S.G. Telfer and R. Kuroda [75].

### 4.8 Summary and Outlook

Over the last 30 years helicates have certainly been the most studied types of metallosupramolecular aggregates. The wealth of information gathered in all these studies now provides us with a pretty good idea how to control certain structural aspects of these kinds of supramolecular assemblies. We have learned for instance that self-sorting is a very common feature in helicate chemistry – no matter whether based on different numbers of metal binding sites, different types of binding sites, or different distances between the metal binding sites. Furthermore, helicates usually show a high degree of self-sorting when chiral racemic ligands are used in which the ligand strands are designed properly, meaning that the formation of the metal complexes does not induce too much steric stress in the assembly.

Using these principles new ligands can be designed that allow the preparation of oligonuclear complexes of defined length, shape, and stereoselectivity which can be decorated with additional functional groups in defined positions. This, however, is the prerequisite to access functional helicates and actually some functions have already been realized: they have been demonstrated to act as templates in the synthesis of molecular knots [76], they can be used as a new building and ordering principle for mesogenic materials [77], they have been shown to interact with DNA in a specific manner [78], they have been used as receptors [79] (although not yet making use of their chirality), the principle of diastereoselective formation of helicates has been employed for the resolution of racemic ligands using an immobilized enantiomerically pure template [80], and they have been demonstrated to be efficient promoters of chemical reactions [81]. However, one can definitely envision more to come in the bright future of helicate chemistry.

# References

1. Atkin, C.L. and Neilands, J.B. (1968) Rhodotorulic acid, a diketopiperazine dihydroxamic acid with growth-factor activity. I. Isolation and characterization. *Biochemistry*, **7**, 3734–3739.
2. (a) Carrano, C.J. and Raymond, K.N. (1978) Synthesis and characterization of iron complexes of rhodotorulic acid: a novel dihydroxamate siderophore and potential chelating drug. *J. Chem. Soc. Chem. Commun.*, **1978**, 501–502; (b) Carrano, C.J. and Raymond, K.N. (1978) Coordination chemistry of microbial ion transport compounds. 10. Characterization of the complex of rhodotorulic acid, a dihydroxamate siderophore. *J. Am. Chem. Soc.*, **100**, 5371–5374; (c) Carrano, C.J., Cooper, S.R., and Raymond, K.N. (1979) Chemistry of microbial ion transport compounds. 11. Solution equilibria and electrochemistry of ferric rhodotorulate complexes. *J. Am. Chem. Soc.*, **101**, 599–604.
3. Struckmeier, G., Thewalt, U., and Fuhrhop, J. H. (1976) Structure of zinc octaethyl formylbiliverdinate hydrate and ist dehydrated bis-helical dimer. *J. Am. Chem. Soc.*, **98**, 278–279.
4. (a) van Stein, G.C., van der Poel, H., van Koten, G. *et al.* (1980) Quadridentate nitrogen donor ligands acting as bridging Di-bidentates: X-ray crystal and molecular structure of [Ag{μ-(R) (S)-1,2-(2-$C_5H_4N$)-C(H)=N]$_2$-cyclohexane}$_2$]($O_3SCF_3$)$_2$ and the observation of $^3J(^{107,109}Ag-^1H)$ in the $^1H$ N.M.R. spectrum of the dinuclear [$Ag_2L_2$]$^{2+}$ Cation. *J. Chem. Soc. Chem. Commun.*, **1980**, 1016–1018; (b) van Stein, G.C., van Koten, G., Vrieze, K. *et al.* (1984) Strucutral investigations of silver(I) and copper(I) complexes with neutral $N_4$ donor ligands: X-ray crystal and molecular structure of the dimer [Ag{μ-(*R,S*)-1,2-(py-2-CH=N)$_2$Cy}$_2$] ($O_3SCF_3$)$_2$ and $^1H$, $^{13}C$, and INEPT $^{109}Ag$ and $^{15}N$ NMR solution studies. *J. Am. Chem. Soc.*, **106**, 4486–4492.
5. Scarrow, R.C., White, D.L., and Raymond, K.N. (1985) Ferric ion sequestering aents. 14. 1-hydroxy-2(1*H*)-pyridinone complexes: properties and structure of a novel Fe-Fe dimer. *J. Am. Chem. Soc.*, **107**, 6540–6546.
6. Lehn, J.-M., Rigault, A., Siegel, J. *et al.* (1987) Spontaneous assembly of double-stranded helicates from oligobipyridine ligands and copper(I) cations: Structure of an inorganic double helix. *Proc. Natl Acad. Sci. USA*, **84**, 2565–2569.

7. (a) Williams, A. (1997) Helical complexes and beyond. *Chem. Eur. J.*, **3**, 15–19; (b) Piguet, C., Bernardinelli, G., and Hopfgartner, G. (1997) Helicates as versatile supramolecular complexes. *Chem. Rev.*, **97**, 2005–2062. (c) Albrecht, M. (1998) Dicatechol ligands: novel building-blocks for metallo-supramolecular chemistry. *Chem. Soc. Rev.*, **27**, 281–287. (d) Caulder, D.L. and Raymond, K.N. (1999) Supermolecules by Design. *Acc. Chem. Res.*, **32**, 975–982. (e) Albrecht, M. (2000) How do they know? Influencing the relative stereochemistry of the complex units of dinuclear triple-stranded helicate-type complexes. *Chem. Eur. J.*, **6**, 3485–3489. (f) Albrecht, M. (2000) From molecular diversity to template-directed self-assembly – new trends in metallo-supramolecular chemistry. *J. Inclus. Phenom. Macrocycl. Chem.*, **36**, 127–151. (g) Albrecht, M. (2001) "Let's twist again" – double-stranded, triple-stranded, and circular helicates. *Chem. Rev.*, **101**, 3457–3497. (h) Bünzli, J.-C.G. and Piguet, C. (2002) Lanthanide-containing molecular and supramolecular polymetallic functional assemblies. *Chem. Rev.*, **102**, 1897–1928. (i) Mamula, O. and von Zelewski, A. (2003) Supramolecuklar coordination compounds with chiral pyridine and polypyridine ligands derived from terpenes. *Coord. Chem. Rev.*, **242**, 87–95. (j) Hannon, M.J. and Childs, L.J. (2004) Helices and helicates: beautiful supramolcular motifs with emerging applications. *Supramol. Chem.*, **16**, 7–22. (k) Albrecht, M. (2004) Supramolecular templating in the formation of helicates. *Top. Curr. Chem.*, **248**, 105–139. (l) Piguet, C., Borkovec, M., Hamkacek, J., and Zeckert, K. (2005) Strict self-assembly of polymetallic helicates: the concept behind the semantics. *Coord. Chem. Rev.*, **249**, 705–726. (m) He, C., Zhao, Y., Guo, D. *et al.* (2007) Chirality transfer through helical motifs in coordination compounds. *Eur. J. Inorg. Chem.*, **2007**, 3451–3463. (n) Albrecht, M. and Fröhlich, R. (2007) Symmetry driven self-assembly of metallo-supramolecular architectures. *Bull. Chem. Soc. Jpn*, **80**, 797–808.

8. *For an overview on different aspects of supramolecular chirality see*: Crego-Calama, M. and Reinhoudt, D.N. (2006) *Top. Curr. Chem.*, **265**.

9. *Reviews and books on the stereoselective synthesis of coordination compounds:* (a) von Zelewsky, A. (1996) *Stereoschemistry of Coordination Compounds*, John Wiley & Sons, Inc., New York, (b) von Zelewsky, A. (1999) Stereoselective synthesis of coordination compounds. *Coord. Chem. Rev.*, **190–192**, 811–825; (c) Knof, U. and von Zelewsky, A. (1999) Predetermined chirality at metal centers. *Angew. Chem. Int. Ed.*, **38**, 302–322; (d) von Zelewsky, A. and Mamula, O. (2000) The bright future of stereoselective synthesis of co-ordination compounds. *J. Chem. Soc. Dalton Trans.*, **2000**, 219–231; (e) Amouri, H. and Gruselle, M. (2008) *Chirality in Transition Metal Chemistry*, John Wiley & Sons, Inc., New York.

10. (a) Kersting, B., Meyer, M., Powers, R.E., and Raymond, K.N. (1996) Dinuclear catecholate helicates: their inversion mechanism. *J. Am. Chem. Soc.*, **118**, 7221–7222; (b) Meyer, M., Kersting, B., Powers, R.E., and Raymond, K.N. (1997) Rearrangement reactions in dinuclear triple helicates. *Inorg. Chem.*, **36**, 5179–5191.

11. Baxter, P.N.W., Lehn, J.-M., Baum, G., and Fenske, D. (2000) Self-assembly and structure of interconverting multinuclear inorganic arrays: A [4×5]-$Ag_{20}^I$ grid and an $Ag_{10}^I$ quadruple helicate. *Chem. Eur. J.*, **6**, 4510–4517.

12. Hamblin, J., Jackson, A., Alcock, N.W., and Hannon, M.J. (2002) Triple helicates and planar dimers arising from silver(I) coordination to directly linked bis-pyrdiylimine ligands. *J. Chem. Soc. Dalton Trans.*, **2002**, 1635–1641.

13. Hasenknopf, B., Lehn, J.-M., Kneisel, B.O. *et al.* (1996) Self-assembly of a circular double helicate. *Angew. Chem. Int. Ed.*, **35**, 1838–1840.

14. (a) Albrecht, M. and Kotila, S. (1996) Counter-ion induced self-assembly of a meso-helicate type molecular box. *Chem. Commun.*, **1996**, 2309–2310; (b) Albrecht, M. (1997) Self-assembly of dinuclear $CH_2$-bridged titanium(IV)/catecholate complexes: influence of the counterions and of methyl substitutents in the ligands. *Chem. Eur. J.*, **3**, 1466–1471; (c) Albrecht, M., Schneider, M., and Röttele, H. (1999) Template-directed self-recognition

of alkyl-bridged bis(catechol) ligands in the formation of helicate-type complexes. *Angew. Chem. Int. Ed.*, **38**, 557–559.

15. Scherer, M., Caulder, D.L., Johnson, D.W., and Raymond, K.N. (1999) Triple helicate – tetrahedral cluster interconversion controlled by host-guest interactions. *Angew. Chem. Int. Ed.*, **38**, 1588–1592.

16. (a) Piguet, C., Rivara-Minten, E., Hopfgartner, G., and Bünzli, J.-C.G. (1995) Structural and photophysical properties of pseudo-tricapped trigonal prismatic lanthanide building blocks controlled by zinc(II) in heterodinuclear d-f complexes. *Helv. Chim. Acta*, **78**, 1541–1566; (b) Piguet, C., Rivara-Minten, E., Hopfgartner, G., and Bünzli, J.-C.G. *et al.* (1995) Molecular magnetism and iron(II) spin-state equilibrium as structural probes in heterodinuclaer d-f complexes. *Helv. Chim. Acta*, **78**, 1651–1672.

17. (a) Piguet, C., Bünzli, J.-C.G., Bernardinelli, G. *et al.* (1996) Lanthanide podates with predetermined structural and photophysical properties: strongly luminescent self-assembled heterodinuclear d-f complexes with a segmental ligand containing heterocyclic imines and carboxamide binding units. *J. Am. Chem. Soc.*, **118**, 6681–6697; (b) Piguet, C., Rivara-Minten, E., Bernardinelli, G. *et al.* (1997) Non-covalent lanthanide podates with predetermined physicochemical properties: iron(II) spin-state equilibria in self-assembled heterodinuclear d-f supramolecular complexes. *J. Chem. Soc. Dalton Trans.*, **1997**, 421–433; (c) Rigault, S., Piguet, C., Bernardinelli, G., and Hopfgartner, G. (1998) Lanthanide-assisted self-assembly of an inert, metal-containing nonadentate tripodal receptor. *Angew. Chem. Int. Ed.*, **37**, 169–172; (d) Cantuel, M., Bernardinelli, G., Imbert, D. *et al.* (2002) A kinetically inert and optically active $Cr^{III}$ partner in thermodynamically self-assembled heterodimetallic non-covalent d-f podates. *J. Chem. Soc. Dalton Trans.*, **2002**, 1929–1940.

18. Cantuel, M., Gumy, F., Bünzli, J.-C.G., and Piguet, C. (2006) Encapsulation of labile trivalent lanthanides into a homobimetallic chromium(III)-containing triple-stranded helicate. Synthesis, characterization, and divergent intramolecular energy transfers. *J. Chem. Soc. Dalton Trans.*, **2006**, 2647–2660.

19. Riis-Johannessen, T., Bernardinelli, G., Filinchuk, Y. *et al.* (2009) Self-assembly of the first discrete 3d-4f-4f triple-stranded helicate. *Inorg. Chem.*, **48**, 5512–5525.

20. Smith, V.C.M. and Lehn, J.-M. (1996) Helicate self-assembly from heterotopic ligand strands of specific binding site sequence. *Chem. Commun.*, **1996**, 2733–2734.

21. Hasenknopf, B., Lehn, J.-M., Baum, G., and Fenske, D. (1996) Self-assembly of a heteroduplex helicate from two different ligand strands and Cu(II) cations. *Proc. Natl Acad. Sci. USA*, **93**, 1397–1400.

22. Albrecht, M. and Fröhlich, R. (1997) Controlling the orientation of sequential ligands in the self-assembly of binuclear coordination compounds. *J. Am. Chem. Soc.*, **119**, 1656–1661.

23. Albrecht, M., Liu, Y., Zhu, S.S. *et al.* (2009) Self-assembly of heterodinuclear triple-stranded helicates: control by coordination number and charge. *Chem. Commun.*, **2009**, 1195–1197.

24. Albrecht, M., Osetska, O., Bünzli, J.-C.G. *et al.* (2009) Homo- and heterodinuclear helicates of lanthanide(III), zinc(II) and aluminium(III) based 8-hydroxyquinoline ligands. *Chem. Eur. J.*, **15**, 8791–8799.

25. Albrecht, M., Latorre, I., Liu, Y., and Fröhlich, R. (2010) Changing the overall shape of heterodinuclear helicates *via* substitution of acylhydrazones by tosylhydrazones. *Z. Naturforsch.*, **65b**, 311–316.

26. Pierre, V.C., Botta, M., Aime, S., and Raymond, K.N. (2006) Fe(II)-templated Gd(III) self-assemblies – a new route toward macromolecular MRI contrast agents. *J. Am. Chem. Soc.*, **128**, 9272–9273.

27. Hahn, F.E., Offermann, M., Schulze Isfort, C. *et al.* (2008) Heterobimetallic triple-stranded helicates with directional benzene-o-dithiol/catechol ligands. *Angew. Chem. Int. Ed.*, **47**, 6794–6797.

28. Krämer, R., Lehn, J.-M., and Marquis-Rigault, A. (1993) Self-recognition in helicate self-assembly: spontaneous formation of helical metal complexes from mixtures of ligands and metal ions. *Proc. Natl Acad. Sci. USA*, **90**, 5394–5398.

29. Caulder, D.L. and Raymond, K.N. (1997) Supramolecular self-recognition and self-assembly in Gallium(III) catecholamide triple helices. *Angew. Chem. Int. Ed.*, **36**, 1440–1442.

30. (a) Shaul, M. and Cohen, Y. (1999) Novel phenanthroline-containing trinuclear double-stranded helicates: self-recognition between helicates with phenanthroline and bipyridine binding sites. *J. Org. Chem.*, **64**, 9358–9364; (b) Greenwald, M., Wessely, D., Katz, E. *et al.* (2000) From homoleptic to heteroleptic double stranded copper(I) helicates: the role of self-recognition in self-assembly processes. *J. Org. Chem.*, **65**, 1050–1058.

31. Rowland, J.M., Olmstead, M.M., and Mascharak, P.K. (2002) Chiral monomeric and homochiral dimeric copper(II) complexes of a new chiral ligand, N-(1,2-Bis(2-pyridyl)ethyl)pyridine-2-carboxamide: an example of molecular self-recognition. *Inorg. Chem.*, **41**, 1545–1549.

32. Masood, M.A., Enemark, E.E., and Stack, T.D.P. (1998) Ligand self-recognition in the self-assembly of a $[\{Cu(L)\}_2]^{2+}$ complex: the role of chirality. *Angew. Chem. Int. Ed.*, **37**, 928–932.

33. Kiehne, U., Weilandt, T., and Lützen, A. (2007) Diastereoselective self-assembly of double-stranded helicates from Tröger's base derivatives. *Org. Lett.*, **9**, 1283–1286.

34. Dalla Favera, N., Kiehne, U., Bunzen, J. *et al.* (2010) Intermetallic interactions within solvated polynuclear complexes: a misunderstood concept. *Angew. Chem. Int. Ed.*, **49**, 125–128.

35. (a) Kiehne, U., Weilandt, T., and Lützen, A. (2008) Self-assembly of dinuclear double- and triple-stranded helicates from bis(bipyridine) ligands derived from Tröger's base analogues. *Eur. J. Org. Chem.*, **2008**, 2056–2064; (b) Kiehne, U. and Lützen, A. (2007) Synthesis of bis (catechol)-ligands of Tröger's base derivatives and their dinuclear triple-stranded complexes with titanium(IV)-ions. *Eur. J. Org. Chem.*, **2007**, 5703–5711.

36. (a) Lützen, A., Hapke, M., Griep-Raming, J. *et al.* (2002) Synthesis and stereoselective self-assembly of double and triple-stranded helicates. *Angew. Chem. Int. Ed.*, **41**, 2086–2089; (b) Bunzen, J., Bruhn, T., Bringmann, G., and Lützen, A. (2009) Synthesis and helicate formation of a new family of BINOL-based bis(bipyrdine) ligands. *J. Am. Chem. Soc.*, **131**, 3621–3630; (c) Bunzen, J., Hovorka, R., and Lützen, A. (2009) Surprising substituent effects on the self-assembly of helicates from bis(bipyridine)-BINOL ligands. *J. Org. Chem.*, **74**, 5228–5236.

37. Actually, the term "*meso*" is kind of misleading because it sometimes describes even less symmetrical assemblies compared to diasteromeric helicate structures.

38. Albrecht, M. and Kotila, S. (1995) Formation of "*meso*-helicate" by self-assembly of three bis(catecholate) ligands and two titanium(IV) ions. *Angew. Chem. Int. Ed.*, **34**, 2134–2137.

39. Bilyk, A., Harding, M.M., Turner, P., and Hambley, T.W. (1995) Octahedral non-helical bis(bipyridyl) metallomacrocycles. *J. Chem. Soc. Dalton Trans.*, **1995**, 2549–2553.

40. Xu, J., Parac, T.N., and Raymond, K.N. (1999) *meso* Myths: what drives assembly of helical versus *meso*-[M$_2$L$_3$] clusters? *Angew. Chem. Int. Ed.*, **38**, 2878–2882.

41. Jeazet, H.B.T., Gloe, K., Doert, T. *et al.* (2010) Self-assembly of neutral hexanuclear circular copper(II) *meso*-helicates: topological control by sulfate ions. *Chem. Commun.*, **46**, 2373–2375.

42. Piehler, T. and Lützen, A. (2010) Diastereoselective self-assembly of enantiomerically pure C$_2$-symmetric dinuclear silver(I) and copper(I) complexes from a bis(2,2′-bipyridine) substituted 9,9′-spirobifluorene ligand. *Z. Naturforsch. B.*, **65b**, 329–336.

43. Krämer, R., Lehn, J.-M., De Cian, A., and Fischer, J. (1993) Self-assembly, structure, and spontaneous resolution of a trinuclear triple helix from an oligopyridine ligand and Ni$^{II}$ ions. *Angew. Chem. Int. Ed.*, **32**, 703–706.

44. (a) Charboniere, L.J., Bernardinelli, G., Piguet, C. *et al.* (1994) Synthesis, structure, and resolution of a dinuclear Co$^{III}$ triple helix. *J. Chem. Soc., Chem. Commun.*, **1994**, 1419–1420; (b) Hasenknopf, B. and Lehn, J.-M. (1996) Trinuclear double helicates of iron(II) and nickel(II): self-assembly and resolution into helical enantiomers. *Helv. Chim. Acta*, **79**, 1643–1650; (c) Rutherford, T.J., Pellegrini, P.A., Aldrich-Wright, J. *et al.* (1998) Isolation of enantiomers of a range of tris(bidentate)ruthenium(II)species using chromatographic resolution and stereoretentive synthetic methods. *Eur. J. Inorg. Chem.*, **1998**, 1677–1688; (d) Hannon, M. J., Meistermann, I., Isaac, C.J., *et al.* (2001) Paper: a cheap yet efficient chiral stationary phase for chromatographic resolution of metallo-supramolecular helicates. *Chem. Commun.*, **2001**, 1078–1079.

45. Yeh, R.M., Ziegler, M., Johnson, D.W. *et al.* (2001) Imposition of chirality in a dinuclear triple-stranded helicate by ion pair formation. *Inorg. Chem.*, **40**, 2216–2217.

46. Zarges, W., Hall, J., Lehn, J.-M., and Bolm, C. (1991) Helicity induction in helicate self-organisation from chiral tris(bipyridine) ligand strands. *Helv. Chim. Acta*, **74**, 1843–1852.

47. (a) Woods, C.R., Benaglia, M., Cozzi, F., and Siegel, J.S. (1996) Enantioselective synthesis of copper(I) bipyridine based helicates by chiral templating of secondary structure: transmission of stereochemistry on the nenometer scale. *Angew. Chem. Int. Ed.*, **35**, 1830–1833; (b) Annunziata, R., Benaglia, M., Cinquini, M. *et al.* (2001) Long-distance propagation of stereochemical information by stereoselective synthesis of copper(I) bipyridine helicates. *Eur. J. Org. Chem.*, **2001**, 173–180.

48. (a) Hayoz, P. and von Zelewsky, A. (1992) New versatile optically active bipyridines as building blocks for helicating and caging ligands. *Tetrahedron Lett.*, **33**, 5165–5168; (b) Hayoz, P., von Zelewsky, A., and Stoeckll-Evans, H. (1993) Stereoselective synthesis of octahedral complexes with predetermined helical chirality. *J. Am. Chem. Soc.*, **115**, 5111–5114; (c) Mürner, H., von Zelewsky, A., and Stoeckli Evans, H. (1996) Octahedral complexes with predetermined helical chirality: xylene-bridged bis([4,5]-pineno-2,2′-bipyridine) ligands (chiragen[*o*-, *m*-, *p*-xyl]= with ruthenium(II). *Inorg. Chem.*, **35**, 3931–3935; (d) Mürner, H., Belser, P., and von Zelewsky, A. (1996) New configurationally stable chiral building blocks for polynuclear coordination compounds: Ru (chiragen[X])Cl$_2$. *J. Am. Chem. Soc.*, **118**, 7989–7994.

49. (a) Mamula, O., von Zelewsky, A., and Bernardinelli, G. (1998) Completely stereospecific self-assembly of a circular helicate. *Angew. Chem. Int. Ed.*, **37**, 290–293; (b) Mamula, O., Monlien, F.J., Porquet, A. *et al.* (2001) Self-assembly of multinuclear coordination species with chiral bipyridine ligands: complexes of 5,6-CHIRAGEN (*o,m,p*-xylidene) ligands and equilibrium behaviour in solution. *Chem. Eur. J.*, **7**, 533–539; (c) Mamula, O., von Zelewsky, A., Brodard, P. *et al.* (2005) Helicates of chiragen-type ligands and their aptitude for chiral self-recognition. *Chem. Eur. J.*, **11**, 3049–3057.

50. (a) Mürner, H., von Zelewski, A., and Hopfgartner, G. (1998) Dinuclear metal complexes of Cd (II), Zn(II) and Fe(II) with triple-helical structure and predetermined chirality. *Inorg. Chim. Acta*, **271**, 36–39; (b) Perret-Aebi, L.-E., von Zelewsky, A., and Neels, A. (2009) Diastereoselective preparation of Cu(I) and Ag(I) double helices by the use of chiral bis-bipyridine ligands. *New J. Chem.*, **33**, 462–465.

51. Mamula, O., von Zelewsky, A., Bark, T., and Bernardinelli, G. (1999) Stereoselective synthesis of coordination compounds: self-assembly of a polymeric double helix with controlled chirality. *Angew. Chem. Int. Ed.*, **38**, 2945–2948.

52. Baum, G., Constable, E.C., Fenske, D. *et al.* (1999) Chiral 1,2-ethanediyl-spaced quaterpyridines give a library of cyclic and double helices with copper(I). *Chem. Commun.*, **1999**, 195–196.

53. Prabaharan, R., Fletcher, N.C., and Nieuwenhuyzen, M. (2002) Self-assembled triple helices with preferential helicity. *J. Chem. Soc. Dalton Trans.*, **2002**, 602–608.

54. Prabaharan, R. and Fletcher, N.C. (2003) The stereoselective coordination chemistry of the helicating ligand *N,N'*-bis(-2,2'-dipyridyl-5-yl)carbonyl-(S/R,S/R)-1,2-diphenylethylenediamine. *Inorg. Chim. Acta*, **355**, 449–453.

55. He, Y., Bian, Z., Kang, C., and Gao, L. (2010) Stereoselective and hierarchical self-assembly from nanotubular homochiral helical coordination polymers to supramolecular gels. *Chem. Commun.*, **46**, 5695–5697.

56. Kiehne, U. and Lützen, A. (2007) Diastereoselective self-assembly of dinuclear double- and triple-stranded helicates from a D–isomannide derivative. *Org. Lett.*, **9**, 5333–5336.

57. (a) Constable, E.C., Kulke, T., Neuburger, M., and Zehnder, M. (1997) Diastereoselective formation of P and M dicopper(I) double helicates with chiral 2,2':6',2''-terpyridines. *Chem. Commun.*, **1997**, 489–490; (b) Constable, E.C., Kulke, T., Neuburger, M., and Zehnder, M. (1997) Chiral 2,2':6',2''-terpyridine ligands for metallosupramolecular chemistry. Part 1. Synthesis of ligands and structural characterization of 4'-([(1S)-endo]-bornyloxy)-2,2':6',2''-terpyridine, 4'-([(1R)-endo]-bornyloxy)-2,2':6',2''-terpyridine, 4'-quininyl-epi-quininyl-2,2':6',2''-terpyridine, and 4'-(2,2':6',2''-terpyridinyl)-(1S)-10-camphorsulfonate. *New J. Chem.*, **21**, 633–646; (c) Constable, E.C., Kulke, T., Neuburger, M., and Zehnder, M. (1997) Chiral 2,2':6',2''-terpyridine ligandsfor metallosupramolecular chemistry. Part 2. Metal complexes of 4'-([(1S)-endo]-bornyloxy)-2,2':6',2''-terpyridine, 4'-([(1R)-endo]-bornyloxy)-2,2':6',2''-terpyridine, 4'-quininyl-epi-quininyl-2,2':6',2''-terpyridine, and 4'-(2,2':6',2''-terpyridinyl)-(1S)-10-camphorsulfonate. *New J. Chem.*, **21**, 1091–1102; (d) Baum, G., Constable, E.C., Fenske, D. *et al.* (1998) Solvent control in the formation of mononuclear and dinuclear double-helical silver(I)-2,2':6',2''-terpyridine complexes. *Chem. Commun.*, **1998**, 2659–2660; (e) Baum, G., Constable, E.C., Fenske, D. *et al.* (2000) Regio- and diastereo-selective formation of dicopper(I) and disilver(I) double helicates with chiral 6-substituted 2,2':6',2''-terpyridines. *J. Chem. Soc. Dalton Trans.*, **2000**, 945–959.

58. (a) Baum, G., Constable, E.C., Fenske, D., and Kulke, T. (1997) Diastereoselective formation of disilver(I) double helicates with chiral 2,2':6',2''':6'',2''-quaterpyridines. *Chem. Commun.*, **1997**, 2043–2044; (b) Constable, E.C., Kulke, T., Baum, G., and Fenske, D. (1998) Diastereoselective formation of chirale helicates. *Inorg. Chem. Commun.*, **1**, 80–82; (c) Baum, G., Constable, E.C., Fenske, D. *et al.* (1999) Stereoselective double-helicate assembly from chiral 2,2':6',2'',6'',2''-quaterpyridines and tetrahedral metal centres. *Chem. Eur. J.*, **5**, 1862–1873.

59. Amendola, V., Fabrizzi, L., Mangano, C. *et al.* (2000) M and P double helical complexes of copper(I) with bis-imino bis-quinoline enantiomerically pure chiral ligands. *Inorg. Chem.*, **39**, 5803–5806.

60. Hamblin, J., Childs, L.J., Alcock, N.W., and Hannon, M.J. (2002) Directed one-pot synthesis of enantiopure dinuclear silver(I) and copper(I) metallo-supramolecular double helicates. *J. Chem. Soc. Dalton Trans.*, **2002**, 164–169.

61. (a) Telfer, S.G., Kuroda, R., and Sato, T. (2003) Stereoselective formation of dinuclear complexes with anomalous CD spectra. *Chem. Commun.*, **2003**, 1064–1065; (b) Telfer, S.G., Tajima, N., and Kuroda, R. (2004) CD spectra of polynuclear complexes of diimine ligands: theoretical and experimental evidence for the importance of internuclear exciton coupling. *J. Am. Chem. Soc.*, **126**, 1408–1418.

62. Bark, T., Düggeli, M., Stoeckli-Evans, H., and von Zelewsky, A. (2001) Designed molecules for self-assembly: the controlled formation of two chiral self-assembled polynuclear species with predetermined configuration. *Angew. Chem. Int. Ed.*, **40**, 2848–2851.

63. Constable, E.C., Zhang, G., Housecroft, C.E. *et al.* (2010) Diastereoselective assembly of helicates incorporating a hexadentate chiral scaffold. *Eur. J. Inorg. Chem.*, **2010**, 2000–2011.

64. (a) Provent, C., Hewage, S., Brand, G. *et al.* (1997) Enantioselective formation of double and triple helicates of silver(I): the role of stacking interactions. *Angew. Chem. Int. Ed.*, **36**, 1287–1289; (b) Provent, C., Rivara-Minten, E., Hewage, S., *et al.* (1999) Solution equilibria of

enantiopure helicates: the role of concentration, solvent and stacking interactions in self-assembly. *Chem. Eur. J.*, **5**, 3487–3494.

65. Gelalcha, F.G., Schulz, M., Kluge, R., and Sieler, J. (2002) Molecular self-assembly: diastereoselective synthesis and structural characterisation of a novel binuclear copper(I) double helicate. *J. Chem. Soc. Dalton Trans.*, **2002**, 2517–2521.

66. (a) Agaskar, P.A., Cotton, F.A., Fraser, I.F., and Peacock, R.D. (1984) Configuration chirality of metal-metal multiple bonds: preparation and circular dichroism spectrum of tetrachlorobis-[(*S*, *S*)-2,3-(diphenylphosphino)butane]dimolybdenum ($Mo_2Cl_4(S,S\text{-dppb})_2$). *J. Am. Chem. Soc.*, **106**, 1851–1853; (b) Agaskar, P.A., Cotton, F.A., Fraser, I.F. *et al.* (1986) Synthesis, structures, and circular dichroism spectra of β-$Mo_2X_4(S,S\text{-dppb})_2$ (X=Cl, Br; *S,S*-dppb=(2*S*,3*S*)-bis(diphenylphosphino)butane). *Inorg. Chem.*, **25**, 2511–2519; (c) Christie, S., Fraser, I.F., McVitie, A., and Peacock, R.D. (1986) Isomerization reactions of bridged and chelated dimolybdenum complexes. *Polyhedron*, **5**, 35–37; (d) Fraser, I.F., McVitie, A., and Peacock, R.D. (1986) Circular dichroism of configurationally chiral molybdenum complexes. *Polyhedron*, **5**, 39–45.

67. (a) Airey, A.L., Swiegers, G.F., Willis, A.C., and Wild, S.B. (1995) Self-assembly of homochiral double helix and side-by-side helix conformers of a double-stranded disilver(I)-tetra(tertiary phosphine) complex. *J. Chem. Soc. Chem. Commun.*, **1995**, 695–696; (b) Cook, V.C., Willis, A.C., Zank, J., and Wild, S.B. (2002) Synthesis and resolution of (*R*\*,*R*\*)-(±)1,1,4,7,10, 10-hexaphenyl-1,10-diarsa-4,7-diphosphadecane: new ligand for the stereoselective self-assembly of dicopper(I), disilver(I), and digold(I) helicates. *Inorg. Chem.*, **41**, 1897–1906; (c) Bowyer, P. K., Cook, V.C., Gharib-Naseri, N. *et al.* (2002) Configurationally homogeneous diastereomers of a linear hexa(tertiary phosphine): enantioselective self-assembly of a double-stranded parallel helicate of the typ (*P*)-[$Cu_3(hexaphos)_2$]($PF_6)_3$. *Proc. Natl Acad. Sci. USA*, **99**, 4877–4822; (d) Airey, A.L., Swiegers, G.F., Willis, A.C., and Wild, S.B. (1997) Self-assembly of homochiral double helix and side-by-side helix conformers of double-stranded disilver(I)- and digold(I)-tetra (tertiary phosphine) helicates. *Inorg. Chem.*, **36**, 1588–1597.

68. Libman, J., Tor, Y., and Shanzer, A. (1987) Helical ferric ion binders. *J. Am. Chem. Soc.*, **109**, 5880–5881.

69. Albrecht, M., Schmid, S., deGroot, M. *et al.* (2003) Self-assembly of an unpolar enantiomerically pure helicate-type metalla-cryptand. *Chem. Commun.*, **2003**, 2526–2527.

70. (a) Enemark, E.J. and Stack, T.D.P. (1995) Synthesis and structural characterization of a stereospecific dinuclear gallium triple helix: use of the trans-influence in metal-assisted self-assembly. *Angew. Chem. Int. Ed.*, **34**, 996–998; (b) Enemark, E.J. and Stack, T.D.P. (1998) Stereospecificity and self-selectivity in the generation of a chiral molecular tetrahedron by metal-assisted self-assembly. *Angew. Chem. Int. Ed.*, **37**, 932–935.

71. Albrecht, M., Janser, I., Fleischhauer, J. *et al.* (2004) An enantiomerically pure dinuclear triple-stranded helicate: X-ray structure, CD-spectroscopy and DFT calculations. *Mendeleev Commun.*, **14**, 250–253.

72. Albrecht, M. (1996) Synthesis of a chiral alkyl-bridged bis(catecholamide) ligand for the self-assembly of enantiomerically pure helicates. *Synlett*, **1996**, 565–567.

73. Corey, E.J., Cywin, C.L., and Noe, M.C. (1994) Synthesis and X-ray structure of a novel chiral trinuclear titanium-tetrol complex. *Tetrahedron Lett.*, **35**, 69–72.

74. Albrecht, M., Witt, K., Röttele, H., and Fröhlich, R. (2001) Stereoselective formation of a trinuclear hexa-stranded helicate-type zinc(II) complex. *Chem. Commun.*, **2001**, 1330–1331.

75. (a) Telfer, S.G., Sato, T., and Kuroda, R. (2004) Noncovalent ligand strands for transition-metal helicates: the straightforward and stereoselective self-assembly of dinuclear double-stranded helicates using hydrogen bonding. *Angew. Chem. Int. Ed.*, **43**, 581–584; (b) Telfer, S.G. and Kuroda, R. (2005) The versatile, efficient, and stereoselective self-assembly of transition-metal helicates by using hydrogen-bonds. *Chem. Eur. J.*, **11**, 57–68.

76. (a) Dietrich-Buchecker, C.O. and Sauvage, J.-P. (1987) Interlocking of molecular threads: from the statistical approach to the templated synthesis of catenands. *Chem. Rev.*, **87**, 795–810; (b) Sauvage, J.-P. (1990) Interlacing molecular threads on transition metals: catenands, catenates, and knots. *Acc. Chem. Res.*, **23**, 319–327.

77. (a) Tschierske, C. (2000) Oligopyridine liquid crystals – novel building blocks for supramolecular architectures based on metal coordination and hydrogen bonding. *Angew. Chem. Int. Ed.*, **39**, 2454–2458; (b) Ziessel, R. (2001) Schiff-based bipyridine ligands. Unusual coordination features and mesomorphic behaviour. *Coord. Chem. Res.*, **216/217**, 195–223.

78. (a) Hannon, M.J., Moreno, V., Prieto, M.J. *et al.* (2001) Intramolecular DNA coiling mediated by a metallo-supramolecular cylinder. *Angew. Chem. Int. Ed.*, **40**, 880–884; (b) Meistermann, I., Moreno, V., Prieto, M.J. *et al.* (2002) Intramolecular DNA coiling mediated by metallosupramolecular cylinders: differential binding of P and M helical enantiomers. *Proc. Natl Acad. Sci. USA*, **99**, 5069–5074.

79. (a) Beer, P.D., Wheeler, J.W., and Moore, C.P. (1992) Copper(I) and silver(I) homometallic complexes of new bis(2,2′bipyridine) ligands. *J. Chem. Soc. Dalton Trans.*, **1992**, 2667–2673; (b) Bilyk, A. and Harding, M.M. (1994) Assembly of [2+2] bimetallic macrocyclic complexes from bis(bipyridyl) ligands and metal ions. *J. Chem. Soc. Dalton Trans.*, **1994**, 77–82; (c) Bilyk, A. and Harding, M.M. (1995) Guest-induced assembly of a chiral [2+2] metallomacrocycle. *J. Chem. Soc. Chem. Commun.*, **1995**, 1697–1698; (d) Albrecht, M., Röttele, H., and Burger, P. (1996) Alkali-metal cation binding by self-assembled cryptand-type supermolecules. *Chem. Eur. J.*, **2**, 1264–1268; (e) Houghton, M.A., Bilyk, A., Harding, M.M. *et al.* (1997) Effect of guest molecules, metal ions and linker length on the assembly of chiral [2+2] metallomacrocycles: solution studies and crystal structures. *J. Chem. Soc. Dalton Trans.*, **1997**, 2725–2733; (f) McMorran, D.A. and Steel, P.J. (1998) The first coordinatively saturated, quadruply stranded helicate and its encapsulation of a hexafluorophosphate anion. *Angew. Chem. Int. Ed.*, **37**, 3295–3297; (g) Zhu, X., He, C., Dong, D. *et al.* (2010) Cerium-based triple-stranded helicates as luminescent chemosensors for the selective sensing of magnesium ions. *J. Chem. Soc. Dalton Trans.*, **39**, 10051–10055.

80. Bunzen, J., Kiehne, U., Benkhäuser-Schunk, C., and Lützen, A. (2009) Immobilization of bis(bipyridine) BINOL ligands and their use in chiral resolution. *Org. Lett.*, **11**, 4786–4789.

81. (a) Kwong, H.-L., Yeung, H.-L., Lee, W.-S., and Wong, W.-T. (2006) Stereoselective formation of a single-stranded helicate: Structure of a bis(palladium-allyl)quaterpyridine complex and its use in catalytic enantioselective allylic substitution. *Chem. Commun.*, **2006**, 4841–4843; (b) Sham, K.-C., Yeung, H.-L., Yiu, S.-M. *et al.* (2010) New binuclear double-stranded manganese helicates as catalysts for alkene epoxidation. *J. Chem. Soc. Dalton Trans.*, **39**, 9469–9471; (c) Martínez-Calvo, M., Vázquez López, M., Pedrido, R. *et al.* (2010) Endogenous arene hydroxylation promoted by copper(I) cluster helicates. *Chem. Eur. J.*, **16**, 14175–14180.

# 5

# Helical Structures Featuring Thiolato Donors

*F. Ekkehardt Hahn and Dennis Lewing*
*Institut für Anorganische und Analytische Chemie, Westfälische*
*Wilhelms-Universität Münster, Germany*

## 5.1 Introduction

Metallosupramolecular structures became a field of intensive research after Lehn *et al.* demonstrated the spontaneous formation of dinuclar helicates from bipyridine and Cu$^I$ [1]. In general, metallosupramolecular chemistry deals with the construction of highly complex functional chemical systems which are held together by noncovalent forces [2]. Spontaneous self-assembly reactions leading to metallosupramolecular assemblies have attracted much interest over the last decades [3] and different structural motifs have been obtained by transition metal-directed self-assembly reactions [4].

Metallohelicates belong to the first metallosupramolecular assemblies studied in detail [1]. Most of these helicates contain ligands with N- or O-donors like oligopyrimidines or catecholates [5]. The coordination chemistry of related polydentate ligands featuring sulfur donors has been less studied. We became interested in the coordination chemistry of bis- and tris(benzene-*o*-dithiolato) and mixed benzene-*o*-dithiolato/catecholato ligands. Such ligands can exhibit an interesting and diverse coordination chemistry.

In hexacoordinated complexes, octahedral (*OC*) coordination is by far the most common coordination geometry encountered. This is mainly due to the minimization of steric interactions between the ligands, and *OC* geometry often leads to a maximization of the ligand field stabilization energy. The second, less common structural type associated with the coordination number 6 is trigonal-prismatic (*TP*) coordination geometry.

*Metallofoldamers: Supramolecular Architectures from Helicates to Biomimetics*, First Edition.
Edited by Galia Maayan and Markus Albrecht.
© 2013 John Wiley & Sons, Ltd. Published 2013 by John Wiley & Sons, Ltd.

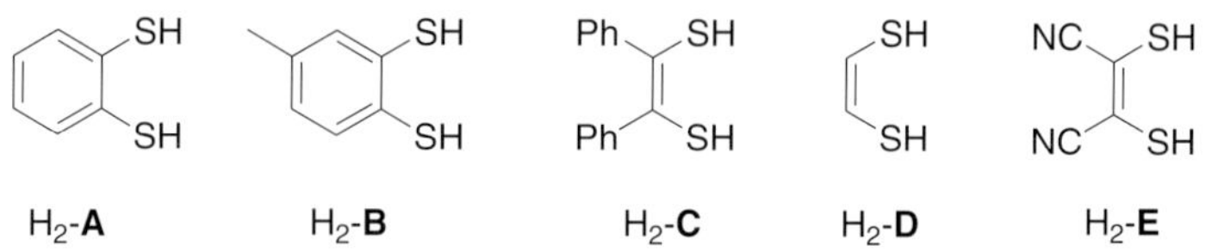
SH
SH
H2-A
SH
SH
H2-B
Ph
SH
Ph
SH
H2-C
SH
SH
H2-D
NC
SH
NC
SH
H2-E

**Figure 5.1** *1,2-Dithiol ligands used for the preparation of trigonal-prismatic complexes.*

*TP* coordination geometry was initially only known in the solid state for compounds such as $MoS_2$ [6], $WS_2$ [7] and NiAs [8]. Later, reports appeared on molecular compounds like $[Re(S_2C_2Ph_2)_3]$ [9], $[Mo(S_2C_2H_2)_3]$ [10] and $[V(S_2C_2Ph_2)3]$ [11], featuring a metal with *TP* coordination geometry. In addition, *TP* coordination geometry has been observed for a few permethylated complexes such as $[W(CH_3)_6]$ [12,13], $[Re(CH_3)_6]$ [13], $[Mo(CH_3)_6]$ [14], $[Li(tmed)]_2[Zr(CH_3)_6]$ [15] and $[Li(OEt_2)_3][Ta(CH_3)_6]$ [13] and for some complexes with macrobicyclic tris(catecholylamide) ligands [16]. These observations led to an ongoing debate regarding the reasons responsible for the preference of *TP* over *OC* coordination geometry. Most *TP* complexes, however, have been observed with unsaturated *o*-dithiolato ligands obtained after deprotonation of *o*-dithiols like $H_2$-**A**-$H_2$-**E** (Figure 5.1).

Molybdenum or tungsten complexes with three benzene-*o*-dithiolato ligands $\mathbf{A}^{2-}$ have been shown to change their coordination geometry from *TP* to *OC* and vice versa depending on the oxidation state of the metal center. A series of mononuclear complexes of type $[M(bdt)_3]^{n-}$ (M = Mo [17], W [18], bdt = benzene-*o*-dithiolate, $n = 0, 1, 2$) have been prepared and characterized by X-ray diffraction. Figure 5.2 shows the molecular structures of $[W(bdt)_3]^{n-}$ ($n = 2, 1, 0$). Complex $[W^{VI}(bdt)_3]$ shows a crystallographically imposed almost perfect *TP* geometry around the metal center (Figure 5.2, right). The ligand bending at the sulfur atoms is thought to be a consequence of a second-order Jahn–Teller distortion [19]. The complex anion $[W^{V}(bdt)_3]^{-}$, obtained after a one-electron reduction of $[W^{VI}(bdt)_3]$, exhibits a distorted octahedral geometry (Figure 5.2, middle). The dianionic complex $[W^{IV}(bdt)_3]^{2-}$ exhibits a more or less distorted *TP* coordination geometry. This facile change of the coordination geometry indicates that the energy required to modify the twist angle φ is small and factors different from the d-electron count at the metal center may influence the coordination geometry. A study on the influence of the counter cations on the coordination geometry of 3,6-dichlorobenzene-1,2-dithiolato complexes of tungsten and molybdenum corroborated this assumption [20].

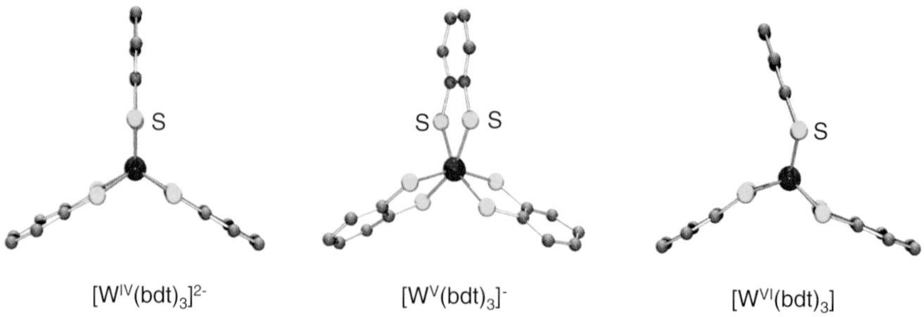
S
S S
S
[W^IV(bdt)3]2-
[W^V(bdt)3]-
[W^VI(bdt)3]

**Figure 5.2** *Molecular structures of the complexes* $[W(bdt)_3]^{n-}$ *(n = 2, 1, 0).*

The situation becomes further complicated by the non-innocent nature of the bdt$^{2-}$ ligand. Wieghardt *et al.* showed that an intramolecular redox reaction can convert the coordinated bdt$^{2-}$ ligand into a coordinated *o*-dithio-benzosemiquinonate(1−) radical anion. Under such conditions the observable *spectroscopic oxidation state* of a metal center may differ from the *formal oxidation state*. The geometrical changes within the bdt$^{n-}$ ligand associated with intramolecular redox reactions have been outlined for some hexacoordinated [17c] and tetracoordinated square-planar complexes [21]. Likewise, manganese complexes containing two toluene-3,4-dithiolato ligands **B**$^{2-}$ (Figure 5.1) show a change in their coordination geometry during the course of a redox reaction. The tetrahedral complex [Mn$^{II}$(**B**)$_2$]$^{2-}$ converts into the square-planar complex [Mn$^{III}$(**B**)$_2$]$^{-}$ by a formal one-electron oxidation [22]. In summary and regardless of the ultimate reasons, changes in coordination geometry have been observed in the series of homologous molybdenum and tungsten tris(benzene-*o*-dithiolate) complexes, depending on the *formal* oxidation state of the metal center.

We intended to study this type of geometry change in dinuclear (helical) complexes generated from bis(benzene-*o*-dithiolato) ligands. Related dicatecholato ligands are known and have been used for the synthesis of several structural motifs. Particularly important among these are the dinuclear triple-stranded helicates [5] of type [M$_2$L$_3$]$^{n-}$ and tetranuclear tetrahedral clusters [23] [M$_4$L$_6$]$^{m-}$ (L = dicatecholate ligand).

The first poly(benzene-*o*-dithiol) ligands were described in 1995 [24]. The synthetic methodology was subsequently improved [25] and a general synthetic strategy for the preparation of bis(benzene-*o*-dithiols ligands with different spacers is presented in Scheme 5.1.

A number of bis(benzene-*o*-dithiol) ligands featuring different aromatic or aliphatic spacers (Figure 5.3) have been prepared using the methodology depicted in Scheme 5.1. The majority of these have been obtained from 2,3-dimercaptobenzoic acid and a suitable diamine via the formation of two amide bonds [25]. The introduction of either an aromatic or a purely aliphatic spacer like in H$_2$-**1** (Figure 5.3) is equally possible [25]. The use of 2,3-dimercaptobenzoic acid and different tripodal amines leads, after removal of the *S*-protection groups, to the tripodal hexadentate ligands H$_6$-**7** [26], H$_6$-**8** [27] and H$_6$-**9** [28] (Figure 5.3) which, in spite of their topological similarity, form different complexes with transition metals.

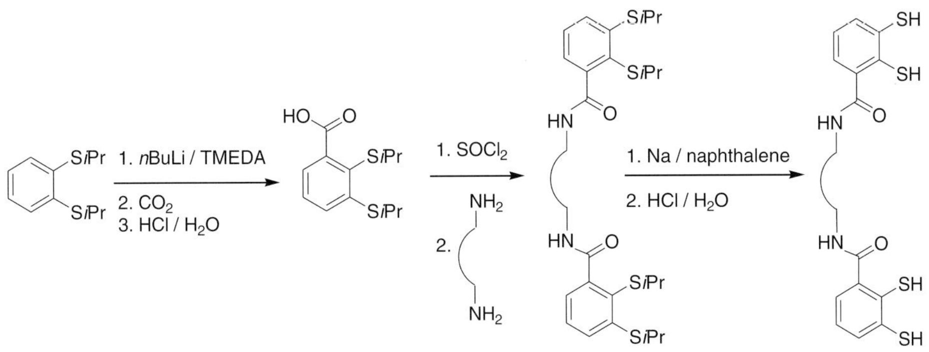

**Scheme 5.1** *General procedure for the synthesis of bis(benzene-o-dithiol) ligands.*

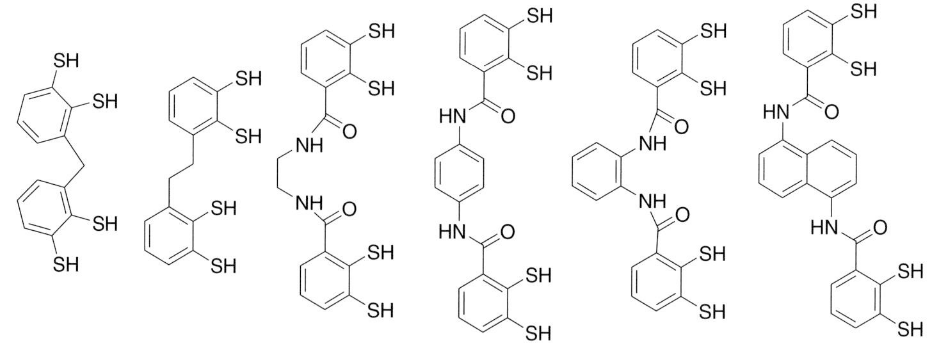

$H_4$-**1** $H_4$-**2** $H_4$-**3** $H_4$-**4** $H_4$-**5** $H_4$-**6**

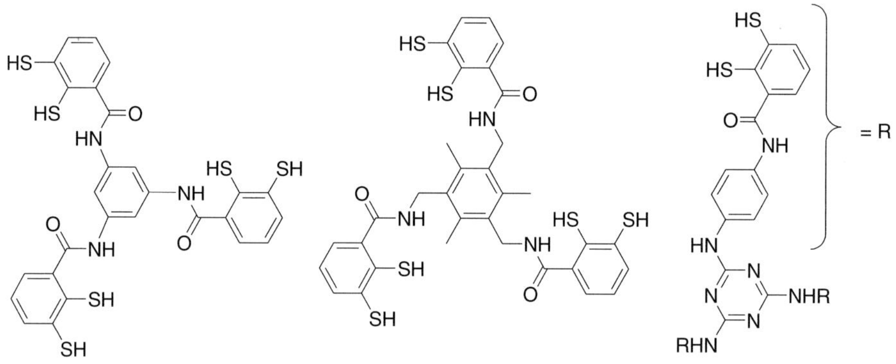

$H_6$-**7** $H_6$-**8** $H_6$-**9**

***Figure 5.3*** *Bis- and tris(benzene-o-dithiol) ligands.*

Apart from ligands containing exclusively benzene-*o*-dithiol groups, mixed benzene-*o*-dithiol/catechol ligands like $H_4$-**10**–$H_4$-**13** (Figure 5.4) have been prepared from 2,3-dimercaptobenzoic acid and functionalized catechol using standard protection group chemistry [29].

## 5.2 Coordination Chemistry of Bis- and Tris(Benzene-o-Dithiolato) Ligands

### 5.2.1 Mononuclear Chelate Complexes

Depending on the topology of the ligand and the preferred coordination geometry of a given metal ion, polydentate ligands can form several different coordination compounds. Ligands with long and flexible backbones, such as $H_4$-**14** and $H_4$-**15**, react with $Co^{III}$ ions under the formation of mononuclear chelate complexes. Both complexes $(NEt_4)[Co(\mathbf{14})]$ and $(NEt_4)[Co(\mathbf{15})]$ have been obtained in a thermodynamically controlled ligand transfer reaction utilizing $(NEt_4)_2[CoCl_4]$ and the halophilic di(titanocene) complexes of $\mathbf{14}^{4-}$ and $\mathbf{15}^{4-}$ (Scheme 5.2) [30].

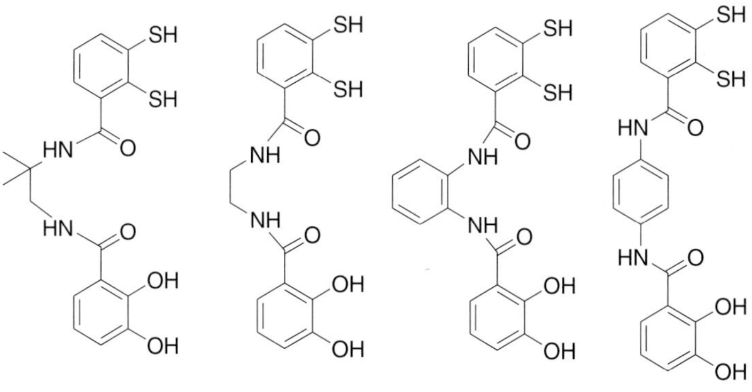


$H_4$-**10** $H_4$-**11** $H_4$-**12** $H_4$-**13**

***Figure 5.4*** *Polydentate benzene-o-dithiol/catechol ligands.*

The rigid ligand backbone of complex $(NEt_4)$[Co(**14**)] probably causes the limited stability of this complex since it does not permit a strain-free square-planar coordination geometry at the metal center. In addition, the halophilicity of the titanium atoms also determine the stability of $(NEt_4)$[Co(**14**)]. The reaction of $[(Cp_2Ti)_2$(**14**)] with a source of chloride ion like $NMe_4Cl$ yielded complex $(NMe_4)$[Ti(Cp)(**14**)] (Figure 5.5) [31]. The reaction is most likely initiated by the substitution of at least one benzene-*o*-dithiolato group from one titanium center by chloride ions. Several reaction pathways are conceivable for the following substitution of a cyclopentadienyl group at the other titanium center by the liberated benzene-*o*-dithiolato donor [32] and the formal elimination of unstable $[Ti(Cp)_3Cl]$. In contrast to the situation in the strained square-planar anion [Co(**14**)]$^-$, the two benzene-*o*-dithiolato donors in [Ti(Cp)(**14**)]$^-$ occupy the basal plane of a square-pyramidal complex anion and are bent into an *exo*/*endo* conformation relative to the cyclopentadienyl ligand. Probably due to the steric requirements of the phenylene

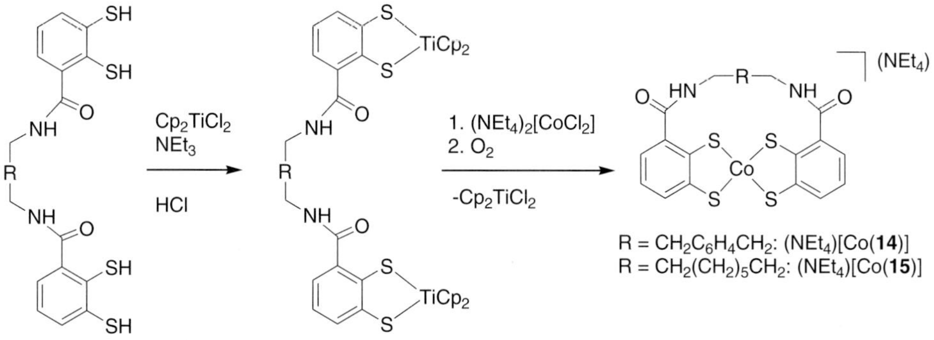


R = $CH_2C_6H_4CH_2$: $H_4$-**14**
R = $CH_2(CH_2)_5CH_2$: $H_4$-**15**

R = $CH_2C_6H_4CH_2$: $[(Cp_2Ti)_2$(**14**)]
R = $CH_2(CH_2)_5CH_2$: $[(Cp_2Ti)_2$(**15**)]

***Scheme 5.2*** *Synthesis of di(titanocene) complexes of ligands* **14**$^{4-}$ *and* **15**$^{4-}$ *and ligand transfer to give the* $Co^{III}$ *complexes.*

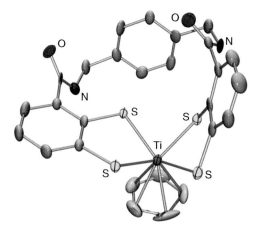

**Figure 5.5**    *Molecular structure of the anion [TiCp(**14**)]⁻.*

backbone, the *endo* bend is larger than the *exo* bend. An inverse situation has been observed in the related anion [Ti(Cp)(bdt)$_2$]⁻ with two unbridged benzene-*o*-dithiolato ligands [33].

Mononuclear chelate complexes have also been obtained from the tris(benzene-*o*-dithiola) ligand H$_6$-**8** and metal ions that are capable of coordinating three benzene-*o*-dithiolato units [27]. Ligand H$_6$-**7** with a shorter linker between the donor groups and the central phenylene moiety, however, does not yield a mononuclear complex but reacts with octahedrally coordinated metal centers under the formation of a tetranuclear cluster of type [M$^{IV}_4$(**7**)$_4$]$^{8-}$ (see below). Complex (Ph$_4$As)$_2$[Ti(**8**)] was obtained from the reaction of equimolar amounts of H$_6$-**8** and [Ti(OPr)$_4$] in the presence of an excess of Li$_2$CO$_3$ followed by cation exchange with Ph$_4$AsCl [27]. Figure 5.6 shows the molecular structure of the complex anion [Ti(**8**)]$^{2-}$. The metal center is sourrounded by the six sulfur donors in a distorted octahedral fashion similar to the geometry observed for the metal complexes of related tri(catecholato) ligands [34,35]. The orientation of the amide groups indicated the presence of intramolecular N—H···S hydrogen bonds which probably have a

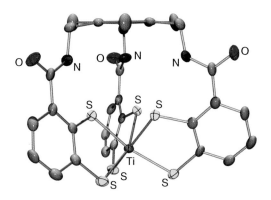

**Figure 5.6**    *Molecular structure of the complex anion [Ti(**8**)]$^{2-}$.*

structure-determining effect similarly to the situation observed for complexes of related tricatecholamide ligands [36].

### 5.2.2 Dinuclear Double-Stranded Complexes

Bis(benzene-*o*-dithiol) ligands with an inflexible or short spacer unit are not capable of forming mononuclear chelate complexes like the previously discussed derivatives $H_4$-**14** and $H_4$-**15** or tris(benzene-*o*-dithiol) $H_6$-**8**. Benzene-*o*-dithiolato ligands mostly yield square-planar complexes with metal ions from the first transition period [37]. The formation of square-planar complexes is also expected for bis(benzene-*o*-dithiolato) ligands even if the ligand topology prevents the formation of a chelate complex. In fact, the bis (benzene-*o*-dithiol) ligands $H_4$-**2** and $H_4$-**3** (Figure 5.3) react with $Ni^{II}$ to give dinuclear double-stranded complexes containing two square-planar $Ni^{II}$ or $Co^{II}$ ions (Figure 5.7) [25]. Both $Ni^{II}$ complexes undergo aerial oxidation to give the $Ni^{III}$ derivatives. Molecular structure determinations (Figure 5.7) established the expected square-planar coordination geometry for the $Ni^{II}$ and $Ni^{III}$ complexes. The $Ni^{II}$ and $Ni^{III}$ complex anions with ligand $3^{4-}$ possess very similar molecular structures. The major difference is found in the orientation of the amide N—H groups which are pointing towards the sulfur atoms in the $Ni^{II}$ complex. This was taken as an indication for weak intramolecular N—H···S hydrogen

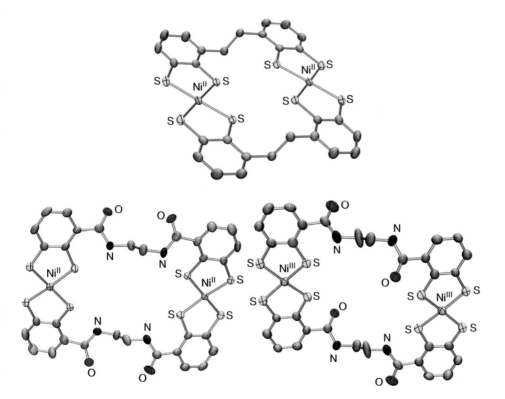

**Figure 5.7** *Molecular structures of the complex anions $[Ni^{II}_2(2)_2]^{4-}$ (top), $[Ni^{II}_2(3)_2]^{4-}$ (bottom left) and $[Ni^{III}_2(3)_2]^{2-}$ (bottom right).*

bonding, which apparently does not exist in the electron-poorer Ni$^{III}$ complex. Cyclic voltammetry studies show quasi-reversible two electron transfer waves for the reduction Ni$^{III}$ → Ni$^{II}$ with a peak potential recorded at −1160 mV for (NEt$_4$)$_2$[Ni$_2$(**2**)$_2$] and −785 mV for Na$_4$[Ni$_2$(**3**)$_2$] (vs Fc/Fc$^+$). The reduction of complex anion [Ni$_2$(**3**)$_2$]$^{2-}$ is facilitated by electron withdrawing diamide bridges while the complex anion with alkyl spacers is more difficult to reduce.

Reduction of the length of the spacer in H$_4$-**2** from an ethylene group to a methylene group leads to the bis(benzene-*o*-dithiol) ligand H$_4$-**1** [38]. The analogous dicatechol ligand has been described to react with [TiO(acac)$_2$] with formation of a triple stranded *meso*-helicate [5d]. Reaction of H$_4$-**1** with CoCl$_2$·6H$_2$O in the presence of Li$_2$CO$_3$ as a base followed by aerial oxidation and cation exchange with PNPCl did not yield the expected helicate but instead the salt (PNP)$_4$[Co$_4$(**1**)$_4$] featuring a tetranuclear complex anion. The complex anion (Figure 5.8) is composed of four Co$^{III}$ ions each coordinated by two benzene-*o*-dithiolato groups from two different **1**$^{4-}$ ligands, leading to a metal-lamacrocycle. Two different types of cobalt centers are observed in the complex anion with the alkyl spacers arranged in an *anti* conformation (Co1, Co3) or *syn* conformation (Co2, Co4). While the analogous dicatechol ligand H$_4$- **1** forms triple-standed *meso*-helicates with Ti$^{IV}$ [5d], helicate formation is not possible with **1**$^{4-}$ since the longer MS bonds and the short bridging unit between the bdt$^{2-}$ donor groups would lead to short intrastrand S···S contacts. Rotation of one bdt$^{2-}$ group about the Ar—CH$_2$ bond minimizes these interactions, leading to the metallamacrocycle.

In the search for dinuclear triple-stranded helicates with bis(benzene-*o*-dithiolato) ligands, compound H$_4$-**5** (Figure 5.3), featuring a larger spacer between the donors than H$_4$-**1**, was deprotonated with Li$_2$CO$_3$ and reacted with [Ti(OEt)$_4$] followed by cation exchange with (Ph$_4$As)Cl. Again, the reaction did not yield a triple-standed dinuclear complex but instead gave the dinuclear double-stranded complex anion [Ti$_2$(**5**)$_2$(μ-OCH$_3$)$_2$]$^{2-}$ (Figure 5.9) [39]. The two metal centers are connected by two ligand strands **5**$^{4-}$ in addition to two bridging methoxo ligands which are part of a central four-membered Ti$_2$(μ-OCH$_3$)$_2$ ring. The coordination geometry at the metal centers is best described as strongly distorted octahedral. The two metal centers assume different configurations (Λ and Δ) which leads to a *meso*-complex anion. As can be seen from Figure 5.9, the amide groups are not oriented in a coplanar fashion with the benzene-*o*-dithiolate groups

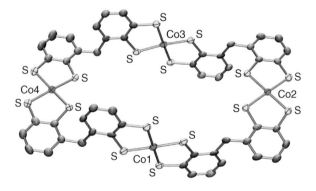

**Figure 5.8**   *Molecular structure of the complex anion [Co$^{III}$$_4$(**1**)$_4$]$^{4-}$.*

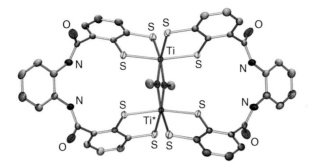

**Figure 5.9**   *Molecular structure of the dinuclear complex anion [Ti$_2$(5)$_2$($\mu$-OCH3)$_2$]$^{2-}$.*

they are connected to, which excludes the presence of intramolecular N—H$\cdots$S hydrogen bonds. A similar bonding situation as in [Ti$_2$(**5**)$_2$($\mu$-OCH3)$_2$]$^{2-}$ has been found in the Ti$^{IV}$ complex bearing the analogous dicatecholato ligand where, however, strong intramolecular N—H$\cdots$O hydrogen bonds have been observed [39].

### 5.2.3   Dinuclear Triple-Stranded Complexes

Metallohelicates with N- and O-donors belong to the most thoroughly studied compounds in metallosupramolecular chemistry [5,40]. While the first metallohelicates derived from dicatechol ligands were reported almost 20 years ago [5], the first dinuclear triple-stranded complexes bearing bis(benzene-*o*-dithiolato) ligands were only described in 2006. Reaction of compounds H$_4$-**3** [39,41] or H$_4$-**4** [39] with [Ti(OC$_2$H$_5$)$_4$] in the presence of Li$_2$CO$_3$ yielded the salts Li$_4$[Ti$_2$(**3**)$_3$] and Li$_4$[Ti$_2$(**4**)$_3$]. Cation exchange with (PNP)Cl gave the salts Li(PNP)$_3$[Ti$_2$(**3**)$_3$] and (PNP)$_4$[Ti$_2$(**4**)$_3$] which could be crystallized from DMF as DMF solvates. In the case of Li$_4$[Ti$_2$(**3**)$_3$] only three of the four lithium cations could be exchanged, even if a large excess of (PNP)Cl was used in the salt metathesis reaction [39,41].

The $^1$H NMR spectra for both compounds show only one set of signals for the protons of the three ligand strands, indicating C3-symmetry for the complex anions in solution. The resonance signal for the NH protons in the complex anions is shifted downfield relative to the equivalent signal for the free ligands, indicating the presence of N—H$\cdots$S hydrogen bonds in a six-membered ring formed between the amide protons and the *o*-thiolato sulfur atoms. Both salts crystallize in centrosymmetric space groups. Figure 5.10 shows the molecular structures of the complex anions [Ti$_2$(**3**)$_3$]$^{4-}$ ($\Delta,\Delta$ isomer) and [Ti$_2$(**4**)$_3$]$^{4-}$ ($\Lambda,\Lambda$ isomer). The molecular structure of the anion [Ti$_2$(**4**)$_3$]$^{4-}$ resembles that reported for the dinuclear triple-stranded helicate with the corresponding dicatecholato ligand [5h]. In both complex anions [Ti$_2$(**3**)$_3$]$^{4-}$ and [Ti$_2$(**4**)$_3$]$^{4-}$ short nonbonding intraligand N$\cdots$S distances are found between the amide nitrogen atoms and the *ortho*-sulfur atoms of the adjacent benzene-*o*-dithiolato group. Adjacent benzene-*o*-dithiolato and amide subunits are arranged in a nearly coplanar fashion. The presence of intrastrand N—H$\cdots$S hydrogen bonds was already indicated by $^1$H NMR spectroscopy, and the observation of a small dihedral angle between the above-mentioned groups supports their existence. As a consequence, the Ti—S bond length to the sulfur atoms in

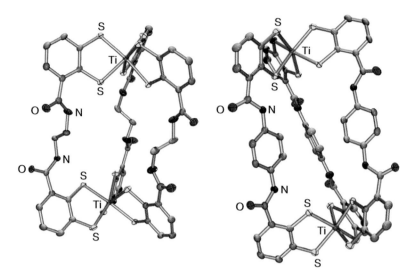

**Figure 5.10** *Molecular structure of the helical complex anions Δ,Δ-[Ti₂(3)₃]⁴⁻ (left) and Λ,Λ-[Ti₂(4)₃]⁴⁻ (right).*

the *ortho*-position to the amide functions is elongated, while the Ti–S bond length to the sulfur atoms in the *meta*-position to the amide group is shorter [39]. The helical twist angles differ significantly in $[Ti_2(3)_3]^{4-}$ (12°) and $[Ti_2(4)_3]^{4-}$ (60°) with the small value for $[Ti_2(3)_3]^{4-}$ possibly resulting from a tendency to reduce the electrostatic repulsion between the double-negatively charged $TiS_2^{2-}$ moieties by a maximization of the Ti⋯Ti distance. The remaining lithium cation in compound $Li(PNP)_3[Ti_2(3)_3]\cdot 3DMF\cdot H_2O$ is coordinated by two carbonyl groups of two different anions $[Ti_2(3)_3]^{4-}$ plus a water and a DMF molecule. This leads to indefinite chains $Li\cdots[Ti_2(3)_3]^{4-}\cdots Li\cdots[Ti_2(3)_3]^{4-}$ in the crystal lattice. No interactions between the cations and the anion have been observed in compound $(PNP)_3[Ti_2(4)_3]\cdot 3DMF$.

Besides dinuclear triple-standed helicates, tetrahedral tetranuclear clusters derived from dicatecholato ligands have been reported [23]. Such clusters have found much interest owing to their ability to serve as container molecules for selected catalytic transformations [23e,f]. The factors governing the preferred formation of a tetrahedral cluster of type $[M_4(F)_6]^{m-}$ versus a triple helicate $[M_2(F)_3]^{n-}$ ($H_4$-**F** = 1,5-naphthalenediamido-linked dicatechol ligand, Scheme 5.3) starting from an identical stoichiometry M:$H_4$-**F** = 2:3 have been discussed [42]. Molecular modeling and experimental studies demonstrated that the presence of the naphthalene spacer in $H_4$-**F** causes the catechol binding units to be offset from one another when the ligand adopts the conformation required for helicate formation, thereby disfavoring the formation of a helicate. We have prepared the bis(benzene-*o*-dithiol) ligand $H_4$-**6** which also contains the 1,5-naphthalenediamido spacer linking two benzene-*o*-dithiol groups. In spite of the identical spacers present in $H_4$-**F** and $H_4$-**6**, the latter was surprisingly found to yield dinuclear complexes $[Ti_2(6)_3]^{4-}$ when reacted with $[Ti(OPr)_4]$ in methanol (Scheme 5.3) [43].

Three equivalents of $H_4$-**6** reacted with two equivalents of $[Ti(OPr)_4]$ in the presence of $Na_2CO_3$ in methanol with the formation of a deep red solution, typical for the

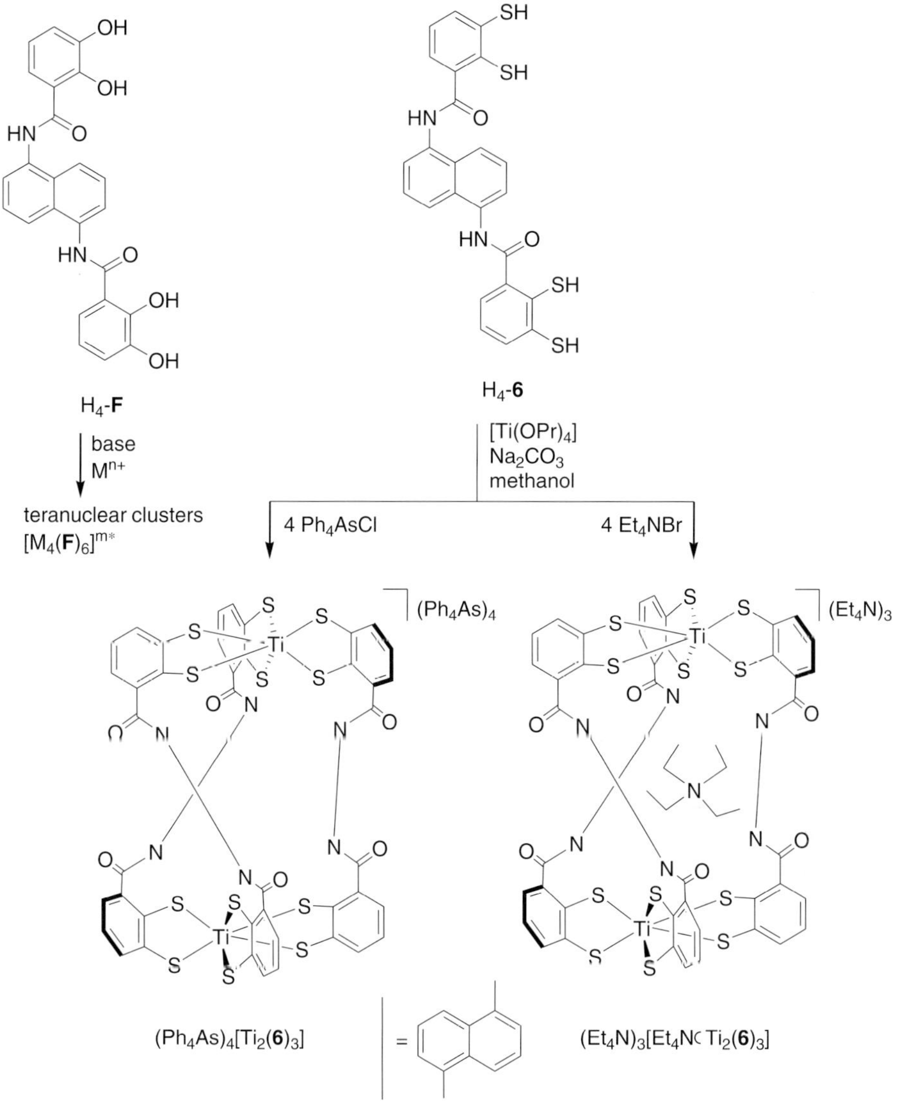

**Scheme 5.3** *Reactions of dicatechol and bis(benzene-o-dithiol) ligands containing the 1,5-naphthalenediamido spacer.*

$\{Ti(bdt)_3\}^{2-}$ chromophore. The addition of four equivalents of $Ph_4AsCl$ to the red solution yielded a deep red precipitate which was identified as the dinuclear complex $(Ph_4As)_4[Ti_2(\mathbf{6})_3]$. The formation of the tetrahedral cluster $(Ph_4As)_8[Ti_4(\mathbf{6})_6]$ which had been obtained under similar reaction conditions with the dicatechol ligand $H_4$-**F** [23b] was not observed. The molecular structure of the tetraanion $[Ti_2(\mathbf{6})_3]^{2-}$ is shown in Figure 5.11 (left). The complex anion resides on a crystallographic inversion center located at the midpoint of the central naphthalene group. Each titanium atom is coordinated by six

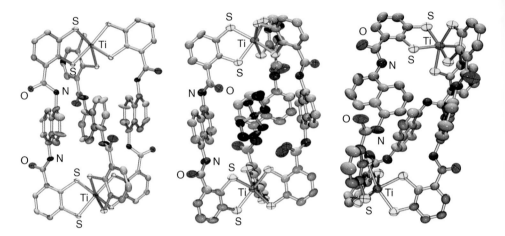

**Figure 5.11** *Molecular structures of the dinuclear complex anions meso-[Ti$_2$(**6**)$_3$]$^{4-}$ (left), meso-[Et$_4$N⊂Ti$_2$(**6**)$_3$]$^{3-}$ (middle, the encapsulated NEt$_4^+$ cation is drawn in black) and Λ,Λ-[Ti$_2$(**6**)$_3$]$^{4-}$.*

sulfur atoms in a strongly distorted octahedral fashion. The titanium atoms exhibit opposite absolute configurations (Λ and Δ) due to the inversion center present. Consequently [Ti$_2$(**6**)$_3$]$^{4-}$ is an achiral dinuclear triple-stranded *meso*-complex.

The nonbonding intrastrand N···S separations in the range 3.026–3.043 Å are indicative of the presence of only weak or no N—H···S hydrogen bonds. This assumption is confirmed by the observation of Ti–S$_{ortho}$ distances which are not longer than the Ti–S$_{meta}$ distances, as would be expected in the case of formation of N—H···S$_{ortho}$ hydrogen bonds [39]. The NMR spectra of (Ph$_4$As)$_4$[Ti$_2$(**6**)$_3$] are rather simple, exhibiting only resonances for one ligand strand, which was taken as an indication that the complex anion adopts the C$_3$-symmetry in solution.

Raymond and coworkers described the influence of ammonium templates on the coordination chemistry of an anthracene bridged dicatechol ligand. Molecular modeling studies showed the capability of the ligand to form both helicates of type [M$_2$L$_3$]$^{m-}$ and tetrahedral clusters of type [M$_4$L$_6$]$^{n-}$. Reaction of the ligand with Ti$^{4+}$ or Ga$^{3+}$ gave dinuclear triple-stranded helicates, whereas the presence of a tetraalkylammonium salt caused the formation of a tetrahedral cluster with one template ion in the interior of its cavity [44]. Based on these results we investigated the influence of the tetraethylammonium template on the reaction of H$_4$-**6** with Ti$^{4+}$. It was hoped that the ammonium template might induce the formation of the tetranuclear [Ti$_4$(**6**)$_6$]$^{8-}$ cluster anion instead of the dinuclear complex anion [Ti$_2$(**6**)$_3$]$^{4-}$. The reaction of two equivalents of [Ti(OPr)$_4$] with three equivalents of H$_4$-**6** in the presence of four equivalents of Et$_4$NBr (Scheme 5.3) in methanol yielded a dark red precipitate which was characterized by X-ray diffraction as the dinuclear complex (Et$_4$N)$_3$[Et$_4$N⊂Ti$_2$(**6**)$_3$].

Complex (Et$_4$N)$_3$[Et$_4$N⊂Ti$_2$(**6**)$_3$] crystallized in a centrosymmetric space group. Figure 5.11 (middle) shows the molecular structure of the anion [Et$_4$N⊂Ti$_2$(**6**)$_3$]$^{3-}$ that contains one molecule of the tetraethylammonium cation located between the two metal atoms. The two titanium atoms in the complex anion are each coordinated by three benzene-*o*-dithiolato moieties differing slightly in their coordination geometry and exhibiting opposite chiralities. This leads again to the formation of a *meso*-complex. Additional bond parameters in the [Et$_4$N⊂Ti$_2$(**6**)$_3$]$^{3-}$ anion are similar to those found for the [Ti$_2$(**6**)$_3$]$^{4-}$ anion. Long and variable N···S separations (range 2.981–4.132 Å) exclude the presence of N—H···S hydrogen bonds. The encapsulated tetraethylammonium cation does not lead to a second set of resonances for the NEt$_4$$^+$ cations, which would indicate that it is rapidly exchanged in solution. This assumption is corroborated by the observation of only one set of resonances for all four tetraethylammonium cations and one set of resonances for the three ligand strands in the NMR spectra.

Finally, the complex anion [Ti$_2$(**6**)$_3$]$^{4-}$ was prepared in the presence of Bu$_4$N$^+$ cations, leading to the compound (Bu$_4$N)$_4$[Ti$_2$(**6**)$_3$] [43b]. The molecular structure of the complex anion is shown in Figure 5.11 (right). Again, each titanium atom is surrounded by six sulfur atoms in a strongly distorted octahedral fashion. Surprisingly in this case, both titanium atoms possess the same absolute configuration Λ, and therefore this tetraanion [Ti$_2$(**6**)$_3$]$^{4-}$ is a dinuclear, triple-stranded helicate with a calculated helix angle of 58.2°. Since (Bu$_4$N)$_4$[Ti$_2$(**6**)$_3$] crystallized in the centrosymmetric space group *C*2/*c*, both the Λ,Λ and Δ,Δ enantiomers are present in the same crystal. A comparison of the molecular structure of the helical (Bu$_4$N)$_4$[Ti$_2$(**6**)$_3$] to both *meso*-complexes (Ph$_4$As)$_4$[Ti$_2$(**6**)$_3$] and (Et$_4$N)$_3$[Et$_4$N⊂Ti$_2$(**6**)$_3$] revealed similar Ti–S distances and S—Ti—S bite angles. Since only the size of the counter ions varies in the three compounds, they are most likely responsible for the structural differences with the encapsulated tetraethyl ammonium cation, accounting for the structural differences between the two *meso*-complexes (Ph$_4$As)$_4$[Ti$_2$(**6**)$_3$] and (Et$_4$N)$_3$[Et$_4$N⊂Ti$_2$(**6**)$_3$].

Ligand H$_4$-**6** is the first bis(benzene-*o*-dithiolato) derivative which has been shown to be capable of forming isomeric complexes (*meso*-complex and helicate) in metal-directed self-assembly reactions with Ti$^{4+}$. The formation of isomeric complex anions [Ti$_2$(**6**)$_3$]$^{4-}$ is most likely caused by the different cations present during their synthesis. The influence of the counterions on the coordination geometry of mononuclear tungsten and molybdenum complexes of type [M(bdt)$_3$]$^{n-}$ has been previously studied [20].

A detailed analysis using density functional theory showed different mechanisms for the Λ → Δ interconversion of tris(catecholato) and tris(benzene-*o*-dithiolato) titanium complexes [43b]. The Λ → Δ interconversion of the mononuclear [Ti(bdt)$_3$]$^{2-}$ complex proceeds via a C$_{3h}$-symmetric transition structure while a D$_{3h}$ transition state was previously found for the [Ti(cat)$_3$]$^{4-}$ complex anion. The DFT calculations also reveal a small activation energy for the Λ → Δ interconversion which transforms a *meso*-complex anion of type [Ti$_2$(**3**)]$^{4-}$ into the helicate.

Initially it was surprising that the dicatechol ligand H$_4$-**F** reacts with metal ions under the formation of the tetranuclear cluster anion [Et$_4$N⊂M$_4$(**F**)$_6$]$^{11-}$ (M = Ga$^{III}$, Fe$^{III}$) containing an encapsulated tetraethylammonium cation [23] while the analogous bis(benzene-*o*-dithiol) ligand H$_4$-**6**, both in the absence and in the presence of tetraalkylammonium cations, gave only the dinuclear complexes [Ti$_2$(**6**)$_3$]$^{4-}$ and [Et$_4$N⊂Ti$_2$(**6**)$_3$]$^{3-}$, respectively. A close inspection of the coordination mode of dicatecholato and bis(benzene-*o*-dithiolato) ligands reveals significant differences which might well explain the different coordination chemistry. The MO$_2$C$_6$H$_4$ fragment in dinuclear or tetranuclear tetrahedral complexes with dicatecholato ligands is almost planar, while the TiS$_2$C$_6$H$_4$ fragments in [Ti$_2$(**6**)$_3$]$^{4-}$, [Et$_4$N⊂Ti$_2$(**6**)$_3$]$^{3-}$ and other [TiL(bdt)$_x$] complexes [33] is severely bent about the S···S vector (see Figures 5.2 and 5.5). This bend changes the approach angle as defined by Raymond *et al.* [42] and removes the geometric factors favoring $M_4L_6$ cluster formation over the entropically favored formation of dinuclear complexes of type $M_2L_3$. In addition, the nature of the donor atoms and particularly the longer Ti—S distances compared to the Ti—O distances and the higher flexibility in the sulfur–sulfur bite angle can aid in the formation of the $M_2L_3$ helicates versus $M_4L_6$ clusters.

### 5.2.4 Coordination Chemistry of Tripodal Tris(Benzene-o-Dithiolato) Ligands

While 1,5-naphthalenediamido-linked dicatecholato ligands form tetrahedral $M_4L_6$ clusters, planar $C_3$-symmetric tricatecholato ligands have been shown by Raymond [23b,42] and Albrecht [45] to yield tetranuclear tetrahedral clustes of type $M_4L_4$ with selected metal ions. This ligand building principle was transferred to $C_3$-symmetric tris(benzene-*o*-dithiolato) ligands [26]. While ligand $H_6$-**8** (Figure 5.3) with a flexible backbone reacts with $Ti^{IV}$ ions with formation of a mononuclear chelate complex (Figure 5.6) [27], the more rigid essentially planar tris(benzene-*o*-dithiol) ligand $H_6$-**7** (Figure 5.3) is incapable of forming mononuclear chelate complexes.

Reaction of $H_6$-**7** with $[Ti(OPr)_4]$ in methanol, in the presence of $Li_2CO_3/K_2CO_3$, led to the formation of a dark red solution. The complex $Li_xK_{8-x}[Ti_4(\mathbf{7})_4]$ was not isolated. Instead the alkali metal cations were exchanged for tetraethylammonium cations. The use of the organic cations led to a red precipitate which was isolated. Since a ligand:metal ratio of 1 : 1 was used and ligand $H_6$-**7** is not capable of forming mononuclear complexes, formation of the tetranuclear complex $(Et_4N)_8[Ti_4(\mathbf{7})_4]$ was assumed [26]. The small number of resonances in the ¹H NMR spectrum of $(Et_4N)_8[Ti_4(\mathbf{7})_4]$ (in DMF-$d_7$/CD₃CN) indicated the presence of only one highly symmetric species in solution. In the case of encapsulation of some tetraethylammonium cations within the cavity of the complex anion $[Ti_4(\mathbf{7})_4]^{8-}$, two sets of proton resonances would be expected, with a highfield shift for those of the encapsulated cations [23a]. However, only one set of sharp resonances was observed for all tetraethylammonium cations at ambient temperature.

In a second attempt, ligand $H_6$-**7** was reacted with $[Ti(OPr)_4]$ in the presence of $Li_2CO_3/K_2CO_3$ followed by the addition of $Me_4NCl$. This reaction yielded a dark red solid. The ESI (negative ions) mass spectrum showed peaks for the anions $\{(Me_4N)_3[Ti_4(\mathbf{7})_4]\}^{5-}$ and $\{(Me_4N)_5[Ti_4(\mathbf{7})_4]\}^{3-}$ with the correct isotope distribution, respectively. Recrystallization of the solid from DMF/CH₃CN yielded red crystals of LiK $(Me_4N)_6[Ti_4(\mathbf{7})_4]\cdot6DMF$ (Scheme 5.4) [26].

The complex $LiK(Me_4N)_6[Ti_4(\mathbf{7})_4]\cdot6DMF$ crystallized in the monoclinic space group $C2/c$ with the octaanion residing on a crystallographic twofold axis (Figure 5.12, top). The asymmetric unit contains half of the octaanion, three $Me_4N^+$ cations and three DMF molecules. The potassium cation resides on the twofold axis passing through the center of the octaanion. One half of a positive charge per asymmetric unit could not be located and it was assumed either that each asymmetric unit contains half of a disordered lithium cation or that a lithium cation is also located on the twofold axis.

Each titanium atom in the anion $[Ti_4(\mathbf{7})_4]^{8-}$ is coordinated by six sulfur atoms in a strongly distorted octahedral fashion. Two titanium atoms adopt the $\Delta$ configuration (Ti1, Ti1*) and the other two the $\Lambda$ configuration (Ti2, Ti2*), which is a rare feature for supramolecular tetrahedral clusters of type $[M_4(L)_4]^{n-}$ normally possessing four homochiral

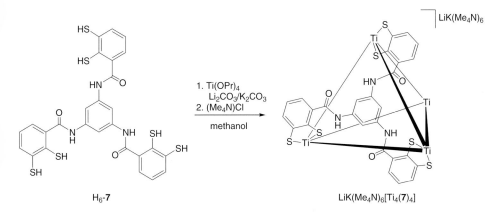

**Scheme 5.4** *Synthesis of the tetranuclear complex LiK(Me₄N)₆[Ti₄(7)₄].*

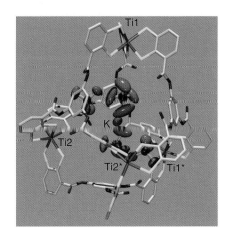

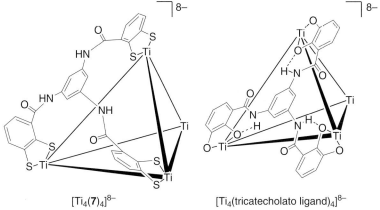

$[Ti_4(7)_4]^{8-}$          $[Ti_4(tricatecholato\ ligand)_4]^{8-}$

**Figure 5.12** *Molecular structure of the $[Ti_4(7)_4]^{8-}$ anion with the encapsulated cations (top, Me₄N⁺ cations in black) and schematic representation of the $C_3$-symmetric ligands in the anions $[Ti_4(7)_4]^{8-}$ and the analogous octaanion with the tri(catecholato) ligand (bottom).*

metal centers. Long and variable N···S separations in the range 3.309–4.127 Å confirm the absence of intramolecular N—H···S hydrogen bonds in the solid state, as was already concluded from the $^1$H NMR spectrum for the situation in solution. The $C_2S_2Ti$ heterocycles are bent along the S—S vector (range of dihedral angles between the $C_6H_3S_2$ and $S_2Ti$ planes 2.7–15.2°), as was observed earlier for related complexes of $bdt^{2-}$ [39,41,43,46].

The anion $[Ti_4(\mathbf{7})_4]^{8-}$ encapsulates not only a potassium cation but also four additional $Me_4N^+$ cations (Figure 5.12, top) and is therefore best described as $[K(Me_4N)_4 \subset Ti_4(\mathbf{7})_4]^{3-}$. This behavior is in remarkable contrast to the situation described for a complex anion with the analogous tricatecholato ligand [42a] where no encapsulation of any cations has been observed due to the limited space within the cluster anion.

The tricatecholato ligand in $[Ti_4(tricatecholato\ ligand)_4]^{8-}$ is essentially planar with strong intramolecular N—H···O hydrogen bonds which cause coplanar orientation of the phenylene backbone with the catecholato donor groups (Figure 5.12, bottom right) [42a]. Ligand $[\mathbf{7}]^{6-}$ in $[Ti_4(\mathbf{7})_4]^{8-}$ is not planar and no N—H···S hydrogen bonds have been observed. The benzene-$o$-dithiolato donor groups are oriented essentially perpendicular to the central phenylene group (range of dihedral angles between the $C_6H_3$ and $C_6H_3S_2$ planes 73.5–89.9°; see ligand bridging Ti1, Ti1$^*$ and Ti2$^*$ in Figure 5.12, top). This ligand conformation (Figure 5.12, bottom left) together with the nonplanar $C_2S_2Ti$ heterocycles generates a much larger and much more open cavity in $[Ti_4(\mathbf{7})_4]^{8-}$ than was found for $[Ti_4(tricatecholato\ ligand)_4]^{8-}$, which in turn allows the encapsulation of the five cations (one $K^+$ and four $NMe_4^+$).

Encouraged by the encapsulation of tetramethylammonium cations observed with $[Ti_4(\mathbf{7})_4]^{8-}$, encapsulation of the larger benzimidazolium cation was attempted. Reaction of ligand $H_6$-$\mathbf{7}$ with $[Ti(OPr)_4]$ and $Li_2CO_3/K_2CO_3$ in methanol followed by the addition of four equivalents of $N,N'$-dimethylbenzimidazolium bromide gave a red precipitate. The $^1$H NMR spectrum of this solid in DMF-$d_7$/CD$_2$Cl$_2$ showed the signals for the $[Ti_4(\mathbf{7})_4]^{8-}$ complex anion and four broad resonances for the benzimidazolium cations. Compared to the $^1$H NMR spectrum of benzimidazolium bromide, measured in the absence of complex anion $[Ti_4(\mathbf{7})_4]^{8-}$, all resonances for the benzimidazolium cations are shifted highfield in the presence of $[Ti_4(\mathbf{7})_4]^{8-}$. Both the highfield shift of the resonances of the benzimidazolium cation and the observed line broadening were taken as indications for a fast exchange of benzimidazolium cations between the inside and the outside of the octanuclear $[Ti_4(\mathbf{7})_4]^{8-}$ octaanion [26].

Surprising results were obtained with the tripodal ligand $H_6$-$\mathbf{9}$ (Figure 5.3) [28]. Equimolar amounts of $H_6$-$\mathbf{9}$ and $Ti(OPr)_4$ react in the in the presence of $Na_2CO_3$ in methanol to give a deep red solution ($\lambda_{max} = 540$ nm), typical for the $[Ti(bdt)_3]^{2-}$ chromophore [46]. The addition of $(Me_3PhN)Cl$ to the methanolic solution yielded a deep red precipitate which was shown to contain a dinuclear complex anion in compound $(Me_3PhN)_4[Ti_2(\mathbf{9})_2]$ (Scheme 5.5) instead of the expected $M_4L_4$ complex [28].

Figure 5.13 (left) shows the molecular structure of the trianion $[Me_3PhN \subset Ti_2(\mathbf{9})_2]^{3-}$ with one $Me_3PhN^+$ cation encapsulated within the interior of the cavity. To the best of our knowledge, the complex anion $[Ti_2(\mathbf{9})_2]^{4-}$ constitutes the first example of two $C_3$-symmetric ligands forming a dinuclear complex, with two ligand arms of each ligand coordinating to one metal center and the remaining one coordinating to the second metal center (and vice versa).

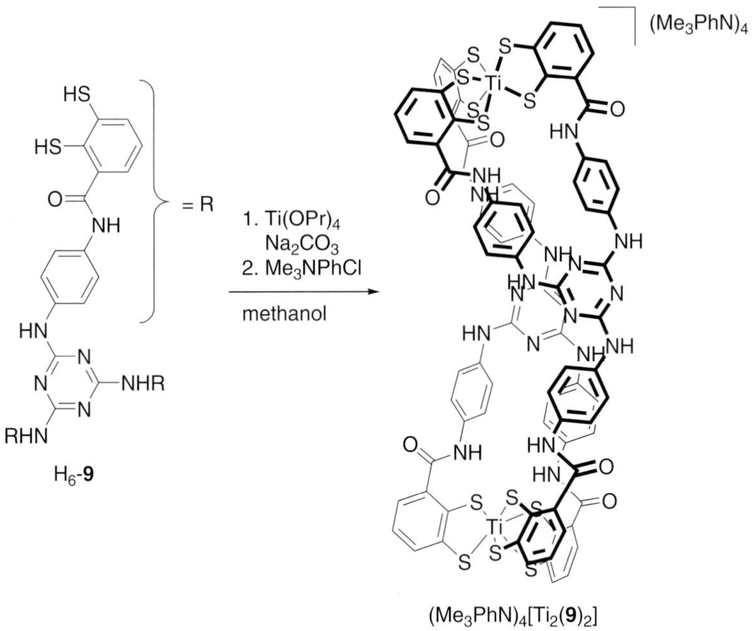

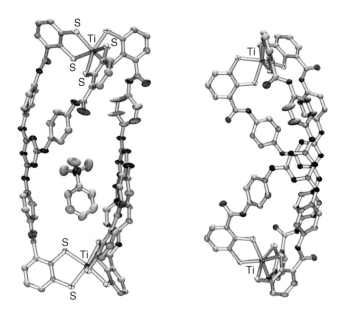

**Scheme 5.5**  *Synthesis of the dinuclear complex (Me₃PhN)₄[Ti₂(9)₂].*

**Figure 5.13**  *Molecular structure of the complex anion [Me₃PhN⊂Ti₂(9)₂]³⁻ (left) and side view of the anion showing its banana shape (right).*

Each titanium atom in $[Ti_2(\mathbf{9})_2]^{4-}$ is coordinated by six sulfur atoms in a strongly distorted octahedral fashion. Both titanium atoms exhibit the same absolute configuration, namely $\Delta$ in the depicted complex anion. Each $\{TiS_6\}$ moiety in complex anion $[Ti_2(\mathbf{9})_2]^{4-}$ shows two long $N \cdots S$ separations ($\sim$3.5 Å) and one short $N \cdots S$ separation ($\sim$2.9 Å), indicative for the formation of just one $N-H \cdots S$ hydrogen bond per $\{TiS_6\}$ unit.

A side view of the complex anion (Figure 5.13, right) shows the uncommon banana-like shape of the anion. Remarkable is the very long $Ti \cdots Ti$ distance of about 19 Å. Since the intramolecular separation of the triazine moieties measures about 9 Å, the two ligands span a large groove and the complex anion can serve as a molecular host. The molecular structure determination shows one $Me_3PhN^+$ cation located inside this cavity (Figure 5.13, left), while the other three cations remain outside maintaining no remarkable interactions with the complex anion.

Encapsulation and release from the cavity in solution is too fast to be detected by NMR spectroscopy. Even at low temperatures neither a broadening of the proton resonances nor a chemical shift to a higher field was observed for the $Me_3PhN^+$ cation, as would be expected for an encapsulated cation.

## 5.3 Coordination Chemistry of Mixed Bis(Benzene-*o*-Dithiol)/Catechol Ligands

Tetradentate ligands with different donor groups (directional ligands) [47,48] like benzene-*o*-dithiol/catechol derivatives are of special interest, since they offer the opportunity to prepare heterodinuclear complexes (different preferences of the donor groups for different metal ions) as well as complexes with a different orientation of the ligand strands (parallel or antiparallel). The versatile coordination chemistry of bis(benzene-*o*-dithiol) ligands described in Section 5.2 prompted attempts to substitute one of the benzene-*o*-dithiol groups for a catechol group, leading to the ligands $H_4$-**10**–$H_4$-**13** (Figure 5.4).

### 5.3.1 Dinuclear Double-Stranded Complexes

Initial experiments regarding the coordination chemisty of directional benzene-*o*-dithiol/catechol ligands were undertaken with the ligand $H_4$-**11** which was reacted with $[TiO(acac)_2]$ and $Na_2CO_3$. Since the analogous bis(benzene-*o*-dithiol) ligand $H_4$-**3** [41] (Figure 5.3) and its catechol anologue [40a] with the same topology as $H_4$-**11** both form dinuclear triple-stranded helicates $[Ti_2L_3]^{4-}$, the formation of a triple-stranded helicate $Na_4[Ti_2(\mathbf{11})_3]$ was expected. Surprisingly, all attempts to prepare the triple-standed complex anion $[Ti_2(\mathbf{11})_3]^{4-}$ failed and the double-stranded complex $Na_2[Ti_2(\mathbf{11})_2(\mu\text{-}OCH_3)_2]$ was exclusively obtained [49].

The directionality of the ligand $\mathbf{11}^{4-}$ enables the formation of two regioisomeric complex anions $[Ti_2(\mathbf{11})_2(\mu\text{-}OCH_3)_2]^{2-}$, one with a parallel orientation of the ligand strands and one with an antiparallel orientation. In addition, the chirality at the metal centers allows the generation of up to seven stereoisomers (Figure 5.14). Since the ligand $H_4$-**11** is achiral, six of the seven stereoisomers form three pairs of enantiomers.

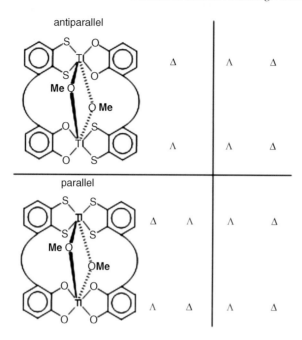

**Figure 5.14** *Schematic representation of the possible isomers of the anion* [Ti$_2$(**11**)$_2$($\mu$-OCH$_3$)$_2$]$^{2-}$.

The observation of six NH resonances, two for each isomer/pair of enantiomers, in the $^1$H NMR spectrum of (Ph$_4$As)$_2$[Ti$_2$(**11**)$_2$($\mu$-OCH$_3$)$_2$] demonstrated that three of the four possible isomers/pairs of enantiomers are formed in solution, which means that at least one isomer with a parallel and one isomer with an antiparallel orientation of the ligand strands are present in solution.

The X-ray diffraction analysis of compound (AsPh$_4$)$_2$[Ti$_2$(**11**)($\mu$-OCH$_3$)$_2$] did not provide conclusive results [49]. The compound crystallized in the centrosymmetric space group $P\bar{1}$ with $Z = 1$. Consequently, the complex dianion must reside on a crystallographic inversion center. Provided that there is no crystallographic disorder, the complex dianion with the parallel orientation of the ligand strands, which was detected in solution by $^1$H NMR spectroscopy, cannot lie on an inversion center. The detection of [Ti$_2$(**11**)($\mu$-OCH$_3$)$_2$]$^{2-}$ on a crystallographic inversion center could therefore indicate that only complex anions with an antiparallel orientation of the ligand stands are present in the crystals of (AsPh$_4$)$_2$[Ti$_2$(**11**)($\mu$-OCH$_3$)$_2$]. Alternatively, all three isomers present in solution could co-crystallize in a disordered lattice. Unfortunately, the latter situation was found. The [Ti$_2$(**11**)($\mu$-OCH$_3$)$_2$]$^{2-}$ dianion is severely disordered in the solid state. All donor atoms (S and O) coordinated to the titanium atoms are disordered. As a consequence each benzene-*o*-dithiolato group can be replaced with a catecholato group and vice versa. Therefore, no unambigous assignment can be made regarding the donor atoms for each titanium atom and the orientation of the ligand strands cannot be determined. The crystal structure determination only confirms the chemical composition of the dianion to be [Ti$_2$(**11**)$_2$($\mu$-OCH$_3$)$_2$]$^{2-}$.

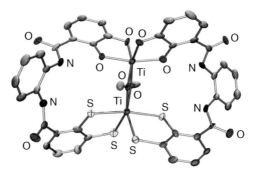

**Figure 5.15** *Molecular structure of the anion* $[Ti_2(12)_2(\mu\text{-}OCH_3)_2]^{2-}$.

The double-stranded complex anion $[Ti_2(12)_2(\mu\text{-}OCH_3)_2]^{2-}$ was obtained from [TiO (acac)$_2$] and the *o*-phenylenediamine bridged benzene-*o*-dithiol/catechol ligand H$_4$-**12** in methanol at room temperature [49]. Contrary to the observation for the anion $[Ti_2(11)_2(\mu\text{-}OCH_3)_2]^{2-}$, $^1$H NMR spectroscopy indicated the formation of only one pair of enantiomers in solution. Only the amide protons linked to the catechoylamido groups form strong hydrogen bonds with the catecholato oxygen atoms, which leads to a downfield shift of the resonance for these protons in the complex anion relative to the resonance of the free ligand H$_4$-**12**. The formation of N—H···S hydrogen bonds was not observed for the complex $[Ti_2(12)_2(\mu\text{-}OCH_3)_2]^{2-}$ in solution. An X-ray diffraction analysis with single crystals of (PNP)$_2$[Ti$_2$(**12**)$_2$($\mu$-OCH$_3$)$_2$] revealed a parallel orientation of the ligand strands (Figure 5.15). The reasons for the parallel orientation of the ligand strands is still subject to speculation. It is reasonable to assume that the formation of strong N—H···O hydrogen bonds, as observed in the solid state, has an influence on the orientation of the ligand strands.

Preliminary experiments with an excess of the ligand H$_4$-**12** and [TiO(acac)$_2$]/Na$_2$CO$_3$ at elevated temperature indicated, via $^1$H NMR spectroscopy, the formation of a dinuclear triple-stranded complex. We assigned this behavior to the elevated reaction temperature. $^1$H NMR investigations showed that the isomers with a parallel orientation and with an antiparallel orientation of ligand strands coexist in solution [49].

### 5.3.2 Dinuclear Triple-Stranded Complexes

Ligands H$_4$-**10** and H$_4$-**13** react with [TiO(acac)$_2$] and Na$_2$CO$_3$ in methanol to give exclusively the triple-stranded helicates Na$_4$[Ti$_2$(**10**)$_3$] [29,49] and Na$_4$[Ti$_2$(**13**)$_3$] [49]. Attempts to prepare a dinuclear double-stranded complex Na$_2$[Ti$_2$(**10**)$_2$($\mu$-OCH$_3$)$_2$] related to Na$_2$[Ti$_2$(**11**)$_2$($\mu$-OCH$_3$)$_2$] failed, although H$_4$-**10** and H$_4$-**11** have the same ligand backbone, except for the methyl groups of the bridging unit in H$_4$-**10**. Attempts to exchange all four sodium cations by PNP$^+$ cations failed in both cases, even after addition of up to 10 equivalents of (PNP)Cl to methanolic solutions of Na$_4$[Ti$_2$(**10**)$_3$] and Na$_4$[Ti$_2$(**13**)$_3$]. Only three sodium cations could be exchanged, resulting in the formation of the complexes Na(PNP)$_3$[Ti$_2$(**10**)$_3$] [29,49] and Na(PNP)$_3$[Ti$_2$(**13**)$_3$] [49].

$^1$H NMR spectra of Na(PNP)$_3$[Ti$_2$(**10**)$_3$] [29,49] showed only one set of signals for the ligand strands, which demonstrated the exclusive formation of the geometrical isomer

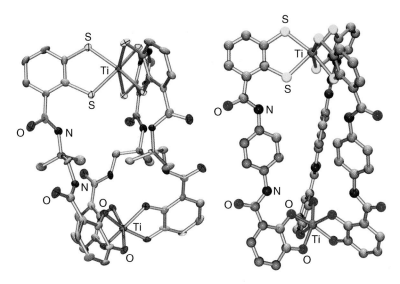

**Figure 5.16** *Molecular structures of the anions $[Ti_2(10)_3]^{4-}$ (left) and $[Ti_2(13)_3]^{4-}$ (right).*

with a parallel orientation of the ligand strands. A strong downfield shift of the catecholyamide NH proton resonance in the anion $[Ti_2(10)_3]^{4-}$ relative to the free ligand $H_4$- **10** indicates the formation of strong N—H···O hydrogen bonds, whereas no N—H···S hydrogen bonds were detected in the $[Ti_2(10)_3]^{4-}$ anion. While NMR spectroscopy showed that only one pair of stereoisomers ($\Lambda,\Delta/\Delta,\Lambda$ or $\Lambda,\Lambda/\Delta,\Delta$) is present in solution, an X-ray diffraction study was carried out to show that both metals in the anion $[Ti_2(10)_3]^{4-}$ adopt the same absolute configuration (Figure 5.16, left). Upon crystallization, a spontaneous resolution into crystals containing exclusively the $\Lambda,\Lambda$ or the $\Delta,\Delta$ isomer (space group *P*1) occurred. The unit cell of Na(PNP)$_3$[Ti$_2$(**10**)$_3$] contains one tetraanion $[Ti_2(10)_3]^{4-}$, three PNP$^+$ cations, one sodium cation and additional solvent molecules. The sodium cation acts as a bridge between two $[Ti_2(10)_3]^{4-}$ tetraanions by coordination of the amide carbonyl functions of two different tetraanions and this leads to the indefinite polymeric chains Na–$[Ti_2(10)_3]^{4-}$···Na···$[Ti_2(10)_3]^{4-}$ in the solid state.

The reaction of three equivalents of $H_4$-**13** with two equivalents of $[TiO(acac)_2]$ and Na$_2$CO$_3$ in methanol at room temperature under an argon atmosphere gave a deep red solution of Na$_4$[Ti$_2$(**13**)$_3$] [49]. As was observed with Na$_4$[Ti$_2$(**10**)$_3$] [29], only three of the four sodium cations can be substituted with PNP$^+$ cations. The ligand strands in Na(PNP)$_3$[Ti$_2$(**13**)$_3$] are also oriented in a parallel fashion (Figure 5.16, right) and strong N—H···O hydrogen bonds were detected by $^1$H NMR spectroscopy in solution and by X-ray crystalography in the solid state. In contrast to the situation in Na(PNP)$_3$[Ti$_2$(**10**)$_3$], two tetraanions $[Ti_2(13)_3]^{4-}$ are connected via a sodium cation that is coordinated by the three *meta* oxygen atoms of the catecholato groups of each complex tetraanion, thereby forming a central NaO$_6$ octahedron in an anionic pentanuclear complex {$[Ti_2(13)_3]$···Na···$[Ti_2(13)_3]$} [49].

The different binding preferences of the donor groups in the benzene-*o*-dithiol/catechol ligand $H_4$-**13** for the selective assembly of a heterobimetallic helical complexes have been

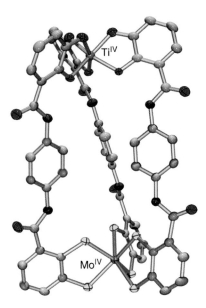

***Figure 5.17*** *Molecular structure of anion $\Lambda,\Lambda$-[TiMo($\mathbf{13}$)$_3$]$^{4-}$.*

studied. The reaction of three equivalents of H$_4$-$\mathbf{13}$ with one equivalent of both [TiO (acac)$_2$] and [MoCl$_4$(CH$_3$CN)$_2$] in the presence of Li$_2$CO$_3$ and Na$_2$CO$_3$ gives a dark green solution. It was assumed that the presence of two different donor groups in the ligand aids in the selective formation of the heterobimetallic complex [TiMo($\mathbf{13}$)$_3$]$^{4-}$.

The addition of four equivalents of (PNP)Cl to the methanolic solution yields, after 12 h, a dark green precipitate of (PNP)$_4$[TiMo($\mathbf{13}$)$_3$] which is only soluble in DMF. Crystals of Li$_{0.5}$(PNP)$_{3.5}$[TiMo($\mathbf{13}$)$_3$]·4.5DMF·1.5EtOH·H$_2$O have been obtained by vapor diffusion of diethyl ether into a solution of (PNP)$_4$[TiMo($\mathbf{13}$)$_3$] in DMF [50]. The (PNP)$_4$[TiMo($\mathbf{13}$)$_3$] used for the crystallization experiments was found to contain some traces of lithium cations, which were not detected by NMR spectroscopy. Structure analysis (Figure 5.17) revealed that the different binding preferences of the donor groups in $\mathbf{13}^{4-}$ for the two different metal ions indeed leads to a triple-standed heterodinuclear complex with parallel orientation of the ligand strands. The complex tetraanion [TiMo($\mathbf{13}$)$_3$]$^{4-}$ contains distorted *OC* {Ti(cat)$_3$} ($\phi = 40.4°$) and distorted *TP* {Mo(bdt)$_3$} polyhedra ($\phi = 23.8°$). The helical twist between the two $\Lambda,\Lambda$-configurated metal centers [Mo$\cdots$Ti 12.423(15) Å] in [TiMo ($\mathbf{13}$)$_3$]$^{4-}$ measures 42.6°. Since the compound crystallized in a centrosymmetric space group, the $\Delta,\Delta$-enantiomer is also present in the crystal lattice.

In the solid state, two enantiomeric helicates, $\Lambda,\Lambda$-[TiMo($\mathbf{13}$)$_3$]$^{4-}$ and $\Delta,\Delta$-[TiMo ($\mathbf{13}$)$_3$]$^{4-}$, are connected via a PNP$^+$ cation which resides on a crystallographic inversion center. This leads to a tetranuclear unit {[TiMo($\mathbf{13}$)$_3$]$\cdots$(PNP)$\cdots$[TiMo($\mathbf{13}$)$_3$]}$^{7-}$ (Figure 5.18, left). The complete formula for this assembly is {(PNP)$_3$[TiMo($\mathbf{13}$)$_3$]$\cdots$(PNP)$\cdots$ [TiMo($\mathbf{13}$)$_3$](PNP)$_3$}$^-$. The lithium cation missing for charge neutrality could not be located by X-ray crystallography.

The importance of alkali metal cations for the stereoselective self-organization of dinuclear triple-stranded helicates has been described several times [5c,d,f,h,40a]. The

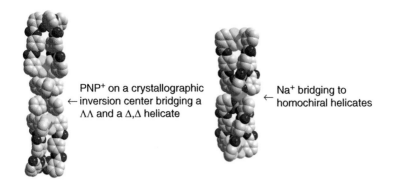

PNP⁺ on a crystallographic
← inversion center bridging a
ΛΛ and a Δ,Δ helicate

Na⁺ bridging to
← homochiral helicates

**Figure 5.18** *Molecular structures of the supramolecular aggregates {Λ,Λ-[TiMo(13)₃]···(PNP)···Δ,Δ-[TiMo(13)₃]}⁷⁻ (left) and {Δ,Δ-[TiMo(13)₃]···Na···Δ,Δ-[TiMo(13)₃]}⁷⁻ (right).*

aggregation of helicates of type $[TiMo(\mathbf{13})_3]^{4-}$ is also controlled by the nature and number of available alkali metal cations. The addition of only two equivalents of (PNP)Cl to the methanolic solution containing the components Ti$^{IV}$:Mo$^{IV}$:$\mathbf{13}^{4-}$ in the stochiometric ratio 1 : 1 : 3, followed by 12 h of stirring yielded the compound Li$_{1.5}$Na$_{0.5}$(PNP)$_2$[TiMo (**13**)₃] (Scheme 5.6). Crystals of the composition Li$_{1.5}$Na$_{0.5}$(PNP)$_2$[TiMo(**13**)₃]·6H$_2$O were analyzed by X ray diffraction. This analysis revealed that the compound crystallized in the acentric monoclinic space group, $C2$. The geometric parameters of the anions $[TiMo(\mathbf{13})_3]^{4-}$ are essentially unchanged when compared to (PNP)$_4$[TiMo(**13**)₃]. Aggregation of the helical $[TiMo(\mathbf{3})_3]^{4-}$ tetraanions in the solid state is, however, determined by the cations present. As described above, in (PNP)$_4$[TiMo(**13**)₃], a PNP cation residing on a crystallographic inversion center bridges two enantiomeric helicates via {Mo(bdt)₃}/PNP interactions, leading to $\{\Lambda,\Lambda\text{-}[TiMo(\mathbf{13})_3]\cdots(PNP)\cdots\Delta,\Delta\text{-}[TiMo(\mathbf{13})_3]\}^{7-}$ (Figure 5.18, left). In Li$_{1.5}$Na$_{0.5}$(PNP)$_2$[TiMo[(**13**)₃], however, two {Ti(cat)₃} polyhedra interact with a sodium cation, leading to dimers of the type $\{\Delta,\Delta\text{-}[TiMo(\mathbf{13})_3]\cdots Na\cdots\Delta,\Delta\text{-}[TiMo(\mathbf{13})_3]\}^{7-}$, where two homochiral helicates are connected via sodium···O$_{catecholate}$ interactions (Figure 5.18, right).

## 5.4 Subcomponent Self-Assembly Reactions

Previous sections discussed "classical" self-assembly reactions of pre-designed ligands with suitable metal ions. In spite of its successful application, this strategy is limited by the often time-consuming synthesis of suitable ligands and their intrinsic stability. As an alternative the "subcomponent self-assembly" methodology was recently developed. It is based on the reversible condensation reaction of suitable amines with metal-coordinated aldehydes under the formation of the thermodynamically most stable product [51]. While this strategy led to impressive results such as the assembly of an unlockable-relockable molecular cage [52a] and a self-assembled cage to encapsulate white phosphorus [52b], it is still limited by the need for stable precursor complexes that allow a reversible imine formation within the metal coordination sphere [52,53].

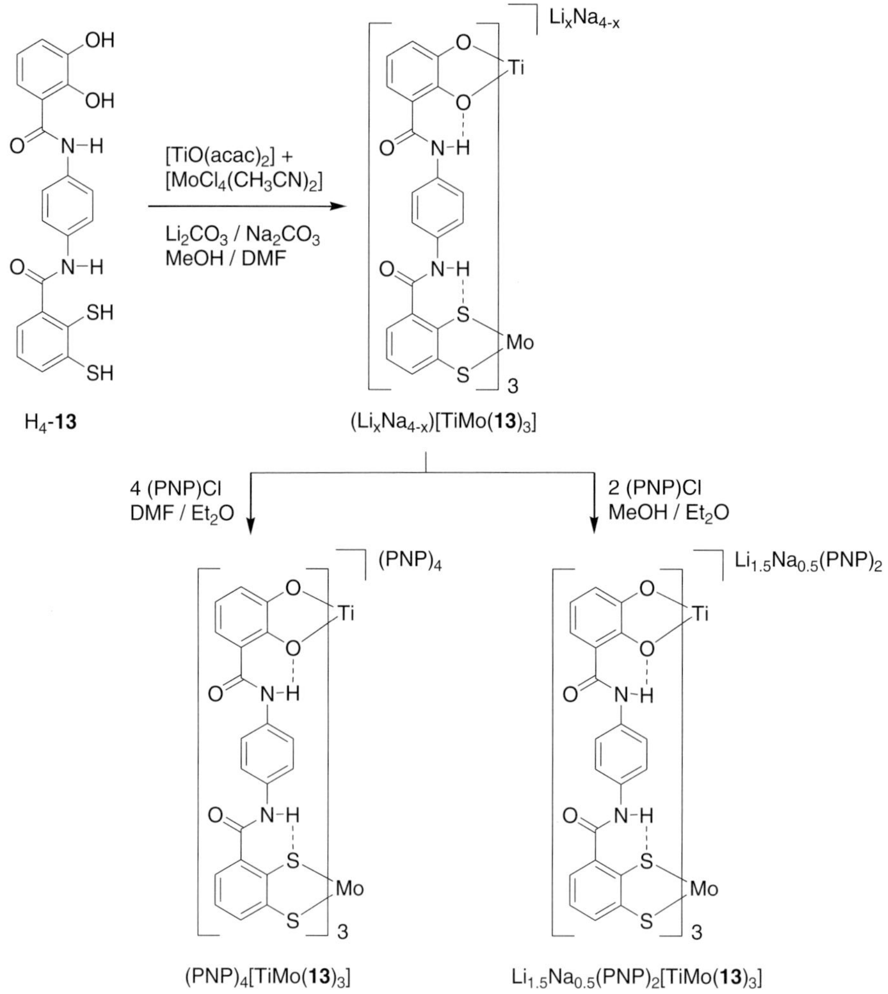

**Scheme 5.6** *Synthesis of the heterobimetallic helicates (PNP)₄[TiMo(**13**)₃] and Li₁.₅Na₀.₅(PNP)₂[TiMo(**13**)₃].*

Transmetallation of supramolecular structures obtained by classical or subcomponent assembly constitutes a promising strategy to overcome these shortcomings. For example, supramolecular structures containing ligands not accessible by conventional organic synthesis could be obtained by subcomponent self-assembly, and transmetallation would subsequently allow the use of these ligands in classical supramolecular chemistry.

We studied the incorporation of the sulfur donor function into Schiff base ligands. In contrast to most Schiff bases, the thiosalicylaldimine (*o*-mercaptobenzaldimine, NS) subunit is not accessible by direct condensation of an amine with the corresponding *o*-mercaptobenzaldehyde, which instead leads to 1,5-dithiocins [53,54]. However, treatment of a preformed complex bearing a 2-thiolatobenzaldehyde ligand with an appropriate primary amines leads in a template-controlled reaction to the desired complexes with NS donor functions [55].

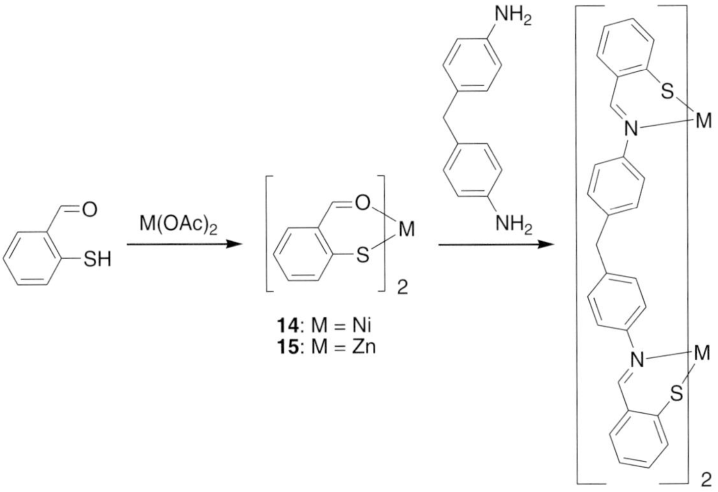

**Scheme 5.7** *Synthesis of complexes **14** and **15** followed by subcomponent self-assembly to the dinuclear complexes **16** and **17**.*

We found that nickel(II) and zinc(II) are excellent templates for this type of chemistry, as indicated by the facile synthesis and stability of the 2-thiolatobenzaldehyde complexes **14** and **15** (Scheme 5.7) [56]. The molecular structure of the square-planar nickel complex **14** has been determined, showing a *cis*-configuration of the oxygen donors. The zinc complex **15** most likely possesses a tetrahedral coordination geometry. These features make complexes **14** and **15** ideal building blocks for a subsequent subcomponent self-assembly reaction via a double Schiff base condensation, leading to the dinuclear complexes of type [M$_2$L$_2$] (**16**: M = Ni; **17**: M = Zn) where L is a bis(bidentate) N,S^N,S ligand.

In fact, reaction of the nickel complex **14** with 4,4′-diaminodiphenyl-methane afforded [Ni$_2$(N,S^N,S)$_2$] **16** as a brown solid. The NMR spectra of complex **16** showed only signals for half of each ligand strand, indicating the formation of a highly symmetric compound. The molecular structure of **16** was unequivocally established by an X-ray crystal structure determination (Figure 5.19, left) which revealed the formation of a dinuclear double-stranded nickel helicate. The two nickel centers in **16** are each coordinated by two thiolato-imine units in a nearly perfect square-planar geometry. Both six-membered C$_3$NSNi rings are significantly bent along the N···S vector, placing the parent phenyl groups on opposing sides of the plane defined by the four donor atoms (see the coordination environment for the bottom nickel atom in Figure 5.19, left). The ligand strands are wrapped around the Ni···Ni axis in a helical manner with a twist angle of 159.5°.

Interestingly, the subcomponent self-assembly of zinc precursor **15** with 4,4′-diaminodiphenylmethane proceeds more readily, enabling an efficient one-pot synthesis of **17** at ambient temperature (Scheme 5.7). *In situ* preparation of zinc complex **15** from *o*-mercaptobenzaldehyde and zinc acetate was followed by the addition of 4,4′-diaminodiphenylmethane to generate [Zn$_2$(N,S^N,S)$_2$] **17** as a yellow solid. In addition to the expected signals for a dinuclear double-stranded zinc helicate, a second set of signals with lower

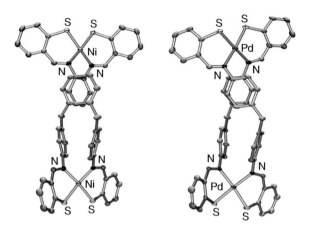

**Figure 5.19** *Molecular structure of dinuclear complexes $[M_2(N,S{\wedge}N,S)_2]$ viewed along the $M{\cdots}M$ axis ($M = Ni^{II}$: **16**, left, $M = Pd^{II}$: **18**, right).*

intensity (ca. 10%) was observed in the $^1$H NMR spectrum of **17**, which can be assigned to the corresponding trinuclear $[Zn_3(N,S{\wedge}N,S)_3]$ species, as indicated by MALDI MS data. 2D EXSY NMR experiments also confirmed the presence of an equilibrium between the dinuclear and trinuclear zinc species. A similar equilibrium between dimeric and trimeric copper complexes bearing related Schiff base ligands has been reported [57].

The subcomponent self-assembly protocol is not applicable for the synthesis of the dipalladium helicate **18**, most likely due to the reduced reactivity of the mononuclear palladium bis(2-thiolatobenzaldehyde) complex towards diamines. Fortunately, both dinuclear helicates **16** and **17** undergo complete metal exchange with palladium(II) via the simple addition of two equivalents of palladium acetate to solutions of the complexes at ambient temperature, affording $[Pd_2(N,S{\wedge}N,S)_2]$ **18** (Scheme 5.8). These transmetallations can be followed visually by the fast color change from either dark brown (**16**, absorption of the $\{NiN_2S_2\}$ chromophore at $\lambda_{max} = 501$ nm) or bright yellow (**17**, $\lambda_{max} = 426$ nm) to deep orange for dipalladium species **18** ($\lambda_{max} = 463$ nm).

A single-crystal X-ray diffraction analysis of **18** (Figure 5.19, right) confirmed the formation of a dinuclear double-stranded palladium complex, which shows structural features comparable to the nickel analogue **16**. Both palladium atoms are coordinated in a slightly distorted square-planar fashion and the helical twist angle measures 152.1°. The chelate rings are not planar, but bent along the $N{\cdots}S$ vector. Although both complexes **16** and **18** are stabilized by intra- and intermolecular interactions, the latter play a more important role for the dipalladium complex **18**. The metallosupramolecular transmetallation protocol was completed by the reaction of the dizinc complex **17** with nickel acetate, leading to the dinickel complex **16**.

The thermodynamic driving force for these metal exchange reactions can be attributed to the preferred coordination of the soft (N,S) binding groups to the softer metal center available [58], as shown earlier for transmetallations involving mononuclear zinc, nickel [59] or palladium [60] complexes with sulphur–nitrogen ligation. In order to gain more insight into the principles of the transmetallation, exchange reactions involving helicates **16**, **17** and **18** were investigated.

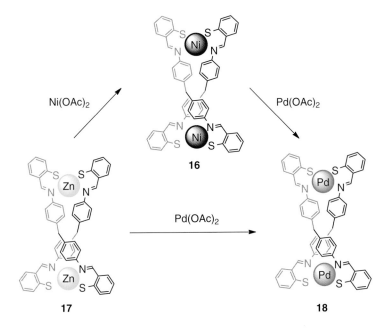

**16**

**17**                        **18**

***Scheme 5.8***    *Transmetallations involving dinuclear helicates **16–18**.*

Figure 5.20 shows the characteristic CH$_2$ region of the $^1$H NMR spectra of a 1 : 1 mixture of the dinickel helicate **16** and the dizinc helicate **17** (including traces of the trinuclear species) 30 min (Figure 5.20a) and 18 h (Figure 5.20b) after dissolving the two complexes in DMF-$d_7$. The first spectrum (Figure 5.20a) is dominated by the resonances

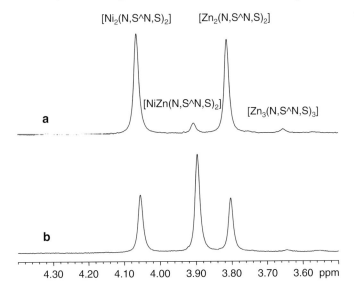

***Figure 5.20***    *The CH$_2$ resonance signals in the $^1$H NMR spectra of a 1 : 1 mixture of **16** and **17** after (a) 30 min and (b) 18 h in DMF-d$_7$.*

for the homodinuclear complexes with a weak resonance at $\delta = 3.90$ ppm appearing for the heterodinuclear [NiZn(N,S^N,S)$_2$] complex. The intensity of this resonance increases with time and becomes the dominant resonance in the spectrum after 18 h (Figure 5.20b). Formation of the heterodinuclear complex is consistent with the MALDI MS data recorded for this reaction and the analogous ones involving the metal exchanges of **16** or **17** with **18**.

The formation of heterodinuclear species from the reaction of two homodinuclear helicates verifies that the helicates interact and exchange metals while in equilibrium and, thereby, confirms the reversibility of the metal exchange. This reversibility represents the essential criterion for self-assembly processes and enables the formation of the thermodynamically most stable product from a number of possible aggregates.

## 5.5   Summary and Outlook

Multidentate ligands featuring benzene-*o*-dithiolato donor groups have been shown to be suitable building blocks for the generation of metallosupramolecular assemblies. Both symmetric bis(benzene-*o*-ditholato) ligands and unsymmetric benzene-*o*-dithiolato/catecholato ligands form dinuclear double- or triple-stranded complexes with suitable metal ions. Some analogies to homologous dicatechol ligands were observed. Heterobimetallic complexes can be obtained from the unsymmetrical benzene-*o*-dithiolato/catecholato ligand by utilizing the different binding preferences of the benzene-*o*-dithiolato and catecholato donor groups. The $C_3$-symmetrical tris(benzene-*o*-dithiolato) ligands yield dinuclear, double-stranded complexes or tetranuclear, tetrahedral clusters depending on the spacer between the donor groups.

The coordination chemistry of the benzene-*o*-dithiolato ligand differs significantly from that of the analogous catecholato donor. Besides the differences in the metric parameters (C—O vs C—S and M—O vs M—S bonds), benzene-*o*-dithiolato complexes often feature a non-planar S—C—C—S—M chelate ring, and the coordination geometry around the metal center in tris(benzene-*o*-dithiolato) complexes can fluctuate between *OC* and *TP* depending on the number of d-electrons at the metal center. It can be expected that these features will lead to new and interesting coordination compounds with benzene-*o*-dithiolato donor groups in the future.

## References

1. (a) Lehn, J.-M., Rigault, A., Siegel, J. *et al.* (1987) Spontaneous assembly of double-stranded helicates from oligobipyridine ligands and copper(I) cations: Structure of an inorganic double helix. *Proc. Natl Acad. Sci. USA*, **84**, 2565–2569; (b) Krämer, R., Lehn, J.-M., and Marquis-Rigault, A. (1993) Self-recognition in helicate self-assembly: Spontaneous formation of helical metal complexes from mixtures of ligands and metal ions. *Proc. Natl Acad. Sci. USA*, **90**, 5394–5398; (c) Hasenknopf, B., Lehn, J.-M., Baum, G., and Fenske, D. (1996) Self-assembly of a heteroduplex helicate from two different ligand strands and Cu(II) cations. *Proc. Natl Acad. Sci. USA*, **93**, 1397–1400.
2. Lehn, J.-M. (1995) *Supramolecular Chemistry: Concepts and Perspectives*, VCH, Weinheim.

3. Lehn, J.-M. (2007) From supramolecular chemistry towards constitutional dynamic chemistry and adaptive chemistry. *Chem. Soc. Rev.*, **36**, 151–160.

4. (a) Jones, C.J. (1998) Transition metals as structural components in the construction of molecular containers. *Chem. Soc. Rev.*, **27**, 289–299; (b) Leininger, S., Olenyuk, B., and Stang, P.J. (2000) Self-assembly of discrete cyclic nanostructures mediated by transition metals. *Chem. Rev.*, **100**, 853–908; (c) Swiegers, G.F. and Malefetse, T.J. (2000) New self-assembled structural motifs in coordination chemistry. *Chem. Rev.*, **100**, 3483–3537; (d) Würthner, F., You, C.-C., and Saha-Möller, C.R. (2004) Metallosupramolecular sqares: from structure to function. *Chem. Soc. Rev.*, **33**, 133–146; (e) Fujita, M., Tominaga, M., Hori, A., and Therrien, B. (2005) Coordination assemblies from a Pd(II)-cornered square complex. *Acc. Chem. Res.*, **38**, 371–380; (f) Saalfrank, R.W., Maid, H., and Scheurer, A. (2008) Supramolecular coodination chemistry: the synergistic effect of serendipity and rational design. *Angew. Chem. Int. Ed.*, **47**, 8794–8824; (g) Yoshizawa, M., Klosterman, J.K., and Fujita, M. (2009) Functional molecular flasks: new properties and reactions with discrete, self-assembled hosts. *Angew. Chem. Int. Ed.*, **48**, 3418–3438.

5. (a) Albrecht, M. and Kotila, S. (1995) Formation of a "*meso*-Helicate" by self-assembly of three bis(catecholate) ligands and two titanium(IV) Ions. *Angew. Chem. Int. Ed. Engl.*, **34**, 2134–2137; (b) Enemark, E.J., and Stack, D.T.P. (1995) Synthesis and structural characterization of a stereospecific dinuclear gallium triple helix: use of the *trans*-influence in metal-assisted self-assembly. *Angew. Chem. Int. Ed. Engl.*, **34**, 996–998; (c) Albrecht, M. and Kotila, S. (1996) Stabilization of an unusual coordination geometry at $Li^+$ in the interior of a cryptand-type helicate. *Angew. Chem. Int. Ed.*, **35**, 1208–1210; (d) Albrecht, M. and Kotila, S. (1996) Counter-ion induced self-assembly of a *meso*-helicate type molecular box. *Chem. Commun.*, **1996**, 2309–2310; (e) Enemark, E.J. and Stack, T.D.P. (1996) Spectral and structural characterization of two ferric coordination modes of a simple bis(catecholamide) ligand: metal-assisted self-assembly in a siderophore analog. *Inorg. Chem.*, **35**, 2719–2720; (f) Kersting, B., Meyer, M., Powers, R.E., and Raymond, K.N. (1996) Dinuclear catecholate helicates: their inversion mechanism. *J. Am. Chem. Soc.*, **118**, 7221–7222; (g) Caulder, D.L. and Raymond, K.N. (1997) Superamolecular self-recognition and self-assembly in gallium(III) catecholamide triple helices. *Angew. Chem. Int. Ed.*, **36**, 1440–1442; (h) Meyer, M., Kersting, B., Powers, R.E., and Raymond, K.N. (1997) Rearrangement reactions in dinuclear triple helicates. *Inorg. Chem.*, **36**, 5179–5191; (i) Albrecht, M., Schneider, M., and Fröhlich, R. (1998) Self-assembly of a triple-stranded helicate from a rigid di(catechol) ligand and formation of its dimer in the solid state. *New J. Chem.*, **22**, 753–754; (j) Albrecht, M., Schneider, M., and Röttele, H. (1999) Template-directed self-recognition of alkyl-bridged bis(catechol) ligands in the formation of helicate-type complexes. *Angew. Chem. Int. Ed.*, **38**, 557–559; (k) Albrecht, M. and Schneider, M. (2002) Dinuclear triple-stranded helicates from rigid oligo-*p*-phenylene ligands: self-assembly and ligand self-recognition. *Eur. J. Inorg. Chem.*, 1301–1306; (l) Albrecht, M., Janser, I., Houjou, H., and Fröhlich, R. (2004) Long-range stereocontrol in the self-assembly of two-nanometer-dimensioned triple-stranded dinuclear helicates. *Chem. Eur. J.*, **10**, 2839–2850; (m) For reviews on this subject see: Piguet, C., Bernardinelli, G., and Hopfgartner, G. (1997) Helicates as versatile supramolecular complexes. *Chem. Rev.*, **97**, 2005–2062; (n) Albrecht, M. (2000) How do they know? Influencing the relative stereochemistry of the complex units of dinuclear triple-stranded helicate-type complexes. *Chem. Eur. J.*, **6**, 3485–3489; (o) Albrecht, M. (2001) "Let's twist again"—double-stranded, triple-stranded, and circular helicates. *Chem. Rev.*, **101**, 3457–3497; (p) Bünzli, J.-C.G. and Piguet, C. (2002) Lanthanide-containing molecular and supramolecular polymetallic functional assemblies. *Chem. Rev.*, **102**, 1897–1928.

6. Dickinson, R.G. and Pauling, L. (1923) The crystal structure of molybdenite. *J. Am. Chem. Soc.*, **45**, 1466–1471.

7. van Arkel, A.E. (1926) Über die Kristalstruktur der Verbindungen Manganfluorid, Bleijodid und Wolframsulfid. *Recl. Trav. Chim. Pays-Bas.*, **45**, 437–444.

8. Aminoff, G. (1923) Untersuchungen über die Kristallstrukturen von Wurtzit und Rotnickelkies. *Z. Kristallogr.*, **58**, 203–219.

9. Eisenberg, R. and Ibers, J.A. (1965) Trigonal prismatic coordination. The molecular structure of tris(*cis*-1,2-diphenylethene-1,2-dithiolato)rhenium. *J. Am. Chem. Soc.*, **87**, 3776–3778.

10. Smith, A.E., Schrauzer, G.N., Mayweg, V.P., and Heinrich, W. (1965) The crystal and molecular structure of $MoS_6C_6H_6$. *J. Am. Chem. Soc.*, **87**, 5798–5799.

11. (a) Eisenberg, R., Stiefel, E.I., Rosenberg, R.C., and Gray, H.B. (1966) Six-coordinate trigonal-prismatic complexes of first-row transition metals. *J. Am. Chem. Soc.*, **88**, 2874–2876; (b) Eisenberg, R. and Gray, H.B. (1967) Trigonal-prismatic coordination. The crystal and molecular structure of tris(*cis*-1,2-diphenylethylene-1,2-dithiolato)vanadium. *Inorg. Chem.*, **6**, 1844–1849.

12. Pfennig, V. and Seppelt, K. (1996) Crystal and molecular structures of hexamethyltungsten and hexamethylrhenium. *Science*, **271**, 626–628.

13. Kleinhenz, S., Pfennig, V., and Seppelt, K. (1998) Preparation and structures of [$W(CH_3)_6$], [$Re$-$(CH_3)_6$], [$Nb(CH_3)_6$]$^-$, and [$Ta(CH_3)_6$]$^-$. *Chem. Eur. J.*, **4**, 1687–1691.

14. Roessler, B. and Seppelt, K. (2000) [$Mo(CH_3)_6$] and [$Mo(CH_3)_7$]$^-$. *Angew. Chem. Int. Ed.*, **39**, 1259–1261.

15. Morse, P.M. and Girolami, G.S. (1989) Are $d^0$ $ML_6$ complexes always octahedral? The X-ray structure of trigonal-prismatic [$Li(tmed)$]$_2$[$ZrMe_6$]. *J. Am. Chem. Soc.*, **111**, 4114–4116.

16. (a) Garrett, T.M., McMurry, T.J., Hosseini, M.W. *et al.* (1991) Synthesis and characterization of macrobicyclic iron(III) sequestering agents. *J. Am. Chem. Soc.*, **113**, 2965–2977; (b) Karpishin, T.B., Stack, T.D.P., and Raymond, K.N. (1993) Octahedral *vs.* trigonal prismatic geometry in a series of catechol macrobicyclic ligand–metal complexes. *J. Am. Chem. Soc.*, **115**, 182–192.

17. (a) n = 0: Cowie, M. and Bennett, M.J. (1976) Trigonal-prismatic vs. octahedral coordination in a series of tris(benzene-1,2-dithiolato) complexes. 1. Crystal and molecular structure of tris (benzene-1,2-dithiolato)molybdenum(VI), $Mo(S_2C_6H_4)_3$. *Inorg. Chem.*, **15**, 1584–1589; (b) n = 1: Cervilla, A., Llopis, E., Marco, D., and Pérez, F. (2001) X-ray structure of ($Bu_4^nN$)[$Mo$ (1,2-benzenedithiolate)$_3$]. trigonal-prismatic versus octahedral coordination in tris(1,2-benzenedithiolate) complexes. *Inorg. Chem.*, **40**, 6525–6528; (c) n = 2: Schulze Isfort, C., Pape, T., and Hahn, F.E. (2005) Synthesis and X-ray molecular structure of ($PNP$)$_2$[$Mo^{IV}(C_6H_4S_2$-1,2)$_3$] completing the structural characterization of the series [$Mo(C_6H_4S_2$-1,2)$_3$]$^{n-}$ (n = 0, 1, 2). *Eur. J. Inorg. Chem.*, 2607–2611.

18. (a) n = 0: Huynh, H.V., Lügger, T., and Hahn, F.E. (2002) Synthesis and X-ray molecular structure of [$W^{VI}(C_6H_4S_2$-1,2)$_3$] completing the structural characterization of the series [$W(C_6H_4S_2$-1,2)$_3$]$^{n-}$ (n = 0, 1, 2): trigonal-prismatic versus octahedral coordination in tris(benzene-1,2-dithiolato) complexes. *Eur. J. Inorg. Chem.*, 3007–3009; (b) Burrow, T.E., Morris, R.H., Hills, A., Hughes, D.L., and Richards, R.L. (1993) Structure of dimethyl(phenyl)phosphonium tris (1,2-benzenedithiolato)tungsten(V). *Acta Crystallogr. Sect. C*, **49**, 1591–1594; (c) Lorber, C., Donahue, J.P., Goddard, C.A. *et al.* (1998) Synthesis, structures, and oxo transfer reactivity of bis(dithiolene)tungsten(IV,VI) complexes related to the active sites of tungstoenzymes. *J. Am. Chem. Soc.*, **120**, 8102–8112.

19. Campbell, S. and Harris, S. (1996) New explanation for ligand bending in transition metal tris (dithiolate) complexes. *Inorg. Chem.*, **35**, 3285–3288.

20. Sugimoto, H., Furukawa, Y., Tarumizu, M. *et al.* (2005) Synthesis and crystal structures of [$W(3,6$-dichloro-1,2-benzenedithiolate)$_3$]$^{n-}$ (n = 1, 2) and [$Mo(3,6$-dichloro-1,2-benzenedithiolate)$_3$]$^{2-}$: Dependence of the coordination geometry on the oxidation number and counter-cation in trigonal-prismatic and octahedral structures. *Eur. J. Inorg. Chem.*, 3088–3092.

21. (a) Gosh, P., Begum, A., Herebian, D. *et al.* (2003) Coordinated *o*-dithio- and *o*-iminothiobenzosemiquinonate(1−) π radicals in [$M^{II}$(bpy)(L·)]($PF_6$) complexes. *Angew. Chem. Int. Ed.*, **42**, 563–567; (b) Ray, K., Weyhermüller, T., Goossens, A. *et al.* (2003) Do S,S′-coordinated *o*-dithiobenzosemiquinonate(1−) radicals exist in coordination compounds? The [$Au^{III}$(1,2-$C_6H_4S_2$)$_2$]$^{1−/0}$ couple. *Inorg. Chem.*, **42**, 4082–4087; (c) Bachler, V., Olbrich, G., Neese, F., and Wieghardt, K. (2002) Theoretical evidence for the singlet diradical character of square planar nickel complexes containing two *o*-semiquinonato type ligands. *Inorg. Chem.*, **41**, 4179–4193.

22. (a) Henkel, G., Greiwe, K., and Krebs, B. (1985) [$Mn(S_2C_6H_3Me)_2$]$^{n−}$: Mononuclear manganese complexes with square-planar ($n = 1$) and distorted tetrahedral (n=2) sulfur coordination. *Angew. Chem. Int. Ed. Engl.*, **24**, 117–118; (b) Greiwe, K., Krebs, B., and Henkel, G. (1989) Preparation, structure, and properties of manganese toluene-3,4-dithiolate complexes in different oxidation states. *Inorg. Chem.*, **28**, 3713–3720.

23. (a) Caulder, D.L., Powers, R.E., Parac, T.N., and Raymond, K.N. (1998) The self-assembly of a predesigned tetrahedral $M_4L_6$ supramolecular cluster. *Angew. Chem. Int. Ed.*, **37**, 1840–1843; (b) Caulder, D.L., Brückner, C., Powers, R.E. *et al.* (2001) Design, formation and properties of tetrahedral $M_4L_4$ and $M_4L_6$ supramolecular clusters. *J. Am. Chem. Soc.*, **123**, 8923–8938; (c) Leung, D.H., Fiedler, D., Bergman, R.G., and Raymond, K.N. (2004) Selective C—H bond activation by a supramolecular host–guest assembly. *Angew. Chem. Int. Ed.*, **43**, 963–966; (d) Fiedler, D., Pagliero, D., Brumaghim, J.L. *et al.* (2004) *Inorg. Chem.*, **43**, 846–848; (e) Fiedler, D., Leung, D.H., Bergman, R.G. *et al.* (2005) *Acc. Chem. Res.*, **38**, 351–360; (f) Pluth, M.D. and Raymond, K.N. (2007) Reversible guest exchange mechanisms in supramolecular host–guest assemblies. *Chem. Soc. Rev.*, **36**, 161–171.

24. (a) Hahn, F.E. and Seidel, W.W. (1995) Synthesis and coordination chemistry of *ortho*-functionalized dimercaptobenzene: Building blocks for tripodal hexathiol ligands. *Angew. Chem. Int. Ed. Engl.*, **34**, 2700–2703; (b) Huynh, H.V., Seidel, W.W., Lügger, T. *et al.* (2002) *ortho*-Lithiation of benzene-1,2-dithiol: A methodology for *ortho*-functionalization of benzene-1,2-dithiol. *Z. Naturforsch. B*, **57b**, 1401–1408.

25. Huynh, H.V., Schulze Isfort, C., Seidel, W.W. *et al.* (2002) Dinuclear complexes with bis(benzenedithiolate) ligands. *Chem. Eur. J.*, **8**, 1327–1335.

26. Birkmann, B., Fröhlich, R., and Hahn, F.E. (2009) Assembly of a tetranuclear host with a tris (benzene-*o*-dithiolato) ligand. *Chem. Eur. J.*, **15**, 9325–9329.

27. Birkmann, B., Seidel, W.W., Pape, T. *et al.* (2009) Coordination chemistry of the sulfur analog of tricatechol siderophores. *Dalton Trans.*, 7350–7352.

28. Hupka, F. and Hahn, F.E. (2010) A banana-shaped dinuclear complex with a tris(benzene-*o*-dithiolato) ligand. *Chem. Commun.*, **46**, 3744–3746.

29. Hahn, F.E., Schulze Isfort, C., and Pape, T. (2004) Λ dinuclear, triple-stranded helicate with a diamide-bridged catechol/benzenedithiol ligand. *Angew. Chem. Int. Ed.*, **43**, 4807–4810.

30. Seidel, W.W. and Hahn, F.E. (1999) Chelate complexes of cobalt(III) with bis(dithiolate) ligands: backbone influence on the electronic properties and the reactivity of the metal center. *J. Chem. Soc. Dalton Trans.*, 2237–2241.

31. Seidel, W.W., Hahn, F.E., and Lügger, T. (1998) Coordination chemistry of *N*-Alkylbenzamide-2,3-dithiolates as an approach to poly(dithiolate) ligands: 1,B-bis[(2,3-dimercaptobenzamido) methyl]benzene and its chelate complex with the ($C_5H_5$)Ti fragment. *Inorg. Chem.*, **37**, 6587–6596.

32. James, T.A. and McCleverty, J.A. (1970) Transition-metal dithiolene complexes. Part XVII. Cleavage of the ring from π-cyclopentadienyl–metal complexes by 1,2-dithiolato- and dithiocarbamato-ligands. *J. Chem. Soc. A.*, 3318–3321.

33. Köpf, H., Lange, K., and Pickardt, J. (1991) [$Ph_4P$][CpTi(1,2-$S_2C_6H_4$)$_2$]: Synthese und Molekülstruktur eines anionischen, tetragonal-pyramidal koordinierten Benzol-1,2-dithiolato-Komplexes des Titans. *J. Organomet. Chem.*, **420**, 345–352.

34. (a) Karpishin, T.B. and Raymond, K.N. (1992) The First Structural Characterization of a Metal–Enterobactin Complex: [V(enterobactin)]$^{2-}$. *Angew. Chem. Int. Ed. Engl.*, **31**, 466–468; (b) Karpishin, T.B., Dewey, T.M., and Raymond, K.N. (1993) The vanadium(IV) enterobactin complex: Structural, spectroscopic, and electrochemical characterization. *J. Am. Chem. Soc.*, **115**, 1842–1851.

35. Hahn, F.E., Rupprecht, S., and Moock, K.H. (1991) The titanium(IV) complex of a hexadentate tricatechol ligand: Synthesis, crystal structure and electrochemistry. *J. Chem. Soc. Chem. Commun.*, 224–225.

36. (a) Bulls, A.R., Pippin, C.G., Hahn, F.E., and Raymond, K.N. (1990) Synthesis and characterization of a series of vanadium–tunichrome B 1 analogues. Crystal structure of a tris (catecholamide) complex of vanadium. *J. Am. Chem. Soc.*, **112**, 2627–2632; (b) Karpishin, T.B., Stack, T.D.P., and Raymond, K.N. (1993) Stereoselectivity in chiral Fe$^{III}$ and Ga$^{III}$ tris(catecholate) complexes effected by nonbonded, weakly polar interactions. *J. Am. Chem. Soc.*, **115**, 6115–6125.

37. Alves, H., Simão, D., Santos, I.C. *et al.* (2004) A series of transition metal bis(dicyanobenzene-dithiolate) complexes [M(dcbdt)$_2$] (M = Fe, Co, Ni, Pd, Pt, Cu, Au and Zn). *Eur. J. Inorg. Chem.*, 1318–1329.

38. Hahn, F.E., Kreickmann, T., and Pape, T. (2006) Self-assembly of a tetranuclear Co$^{III}$-metallacycle from the reaction of a bis(benzene-*o*-dithiolato) ligand with Co$^{II}$ and subsequent aerial oxidation. *Eur. J. Inorg. Chem.*, 535–539.

39. Kreickmann, T., Diedrich, C., Pape, T. *et al.* (2006) Metallosupramolecular chemistry with bis-(benzene-*o*-dithiolato) ligands. *J. Am. Chem. Soc.*, **128**, 11808–11819.

40. (a) Albrecht, M. (1998) Dicatechol ligands: novel building-blocks for metallo-supramolecular chemistry. *Chem. Soc. Rev.*, **27**, 281–288; (b) Albrecht, M., Janser, I., and Fröhlich, R. (2005) Catechol imine ligands: from helicates to supramolecular tetrahedra. *Chem. Commun.*, **2005**, 157–165.

41. Hahn, F.E., Kreickmann, T., and Pape, T. (2006) A dinuclear triple-stranded helicate with a bis (benzene-*o*-dithiolato) ligand. *Dalton Trans.*, 769–771.

42. (a) Brückner, C., Powers, R.E., and Raymond, K.N. (1998) Symmetry-driven rational design of a tetrahedral supramolecular Ti$_4$L$_4$ cluster. *Angew. Chem. Int. Ed.*, **37**, 1837–1839; (b) Caulder, D.L., and Raymond, K.N. (1999) The rational design of high symmetry coordination clusters. *J. Chem. Soc. Dalton Trans.*, 1185–1200; (c) Caulder, D.L., and Raymond, K.N. (1999) Supermolecules by design. *Acc. Chem. Res.*, **32**, 975–982.

43. (a) Hahn, F.E., Birkmann, B., and Pape, T. (2008) Self-assembly reactions with a bis(benzene-*o*-dithiolato) ligand. *Dalton Trans.*, 2100–2102; (b) Birkmann, B., Ehlers, A.W., Fröhlich, R. *et al.* (2009) Metallosupramolecular complexes derived from bis(benzene-*o*-dithiol) ligands. *Chem. Eur. J.*, **15**, 4301–4311.

44. Scherer, M., Caulder, D.L., Johnson, D.W., and Raymond, K.N. (1999) Triple helicate – tetrahedral cluster interconversion controlled by host–guest interactions. *Angew. Chem. Int. Ed.*, **38**, 1588–1592.

45. (a) Albrecht, M., Janser, I., Meyer, S. *et al.* (2003) A metallosupramolecular tetrahedron with a huge internal cavity. *Chem. Commun.*, 2854–2855; (b) Albrecht, M., Janser, I., Runsink, J. *et al.* (2004) Selecting different complexes from a dynamic combinatorial library of coordination compounds. *Angew. Chem. Int. Ed.*, **43**, 6662–6666; (c) Albrecht, M., Janser, I., and Fröhlich, R. (2005) Catechol imine ligands: from helicates to supramolecular tetrahedra. *Chem. Commun.*, 157–165; (d) Albrecht, M., Janser, I., Burk, S., and Weis, P. (2006) Self-assembly and host–guest chemistry of big metallosupramolecular M$_4$L$_4$ tetrahedra. *Dalton Trans.*, 2875–2880; (e) Albrecht, M., and Fröhlich, R. (2007) Symmetry driven self-assembly of metallo-supramolecular architectures. *Bull. Chem. Soc. Jpn*, **80**, 797–808; (f) Albrecht, M., Burk, S., Weis, P., Schalley, C.A., and Kogej, M. (2007) The Wittig reaction as a key step in the

preparation of triangular ligands for the self-assembly of molecular $M_4L_4$ tetrahedra. *Synthesis*, 3736–3741.

46. Könemeann, M., Stüer, W., Kirschbaum, K., and Giolando, D.M. (1994) Synthesis, crystal structure and electrochemistry of bis(*N,N*-dimethylammonium) tris(1,2-benzenedithiolato)titanate(IV). *Polyhedron*, **13**, 1415–1425.

47. (a) Directional ligands with different donor groups: Piguet, C., Hopfgartner, G., Williams, A.F., and Bünzli, J.-C.G. (1995) Self-assembly of the first heterodinuclear d–f triple helix in solution. *J. Chem. Soc. Chem. Commun.*, 491–493; (b) Albrecht, M., and Fröhlich, R. (1997) Controlling the orientation of sequential ligands in the self-assembly of binuclear coordination compounds. *J. Am. Chem. Soc.*, **119**, 1656–1661.

48. (a) Directional ligands with an unsymmetric bridge between the donor groups: Albrecht, M., Napp, M., Schneider, M. *et al.* (2001) Dinuclear titanium(IV) complexes from amino acid bridged dicatechol ligands: formation, structure, and conformational analysis. *Chem. Eur. J.*, **7**, 3966–3975; (b) Hannon, M.J., Bunce, S., Clarke, A.J., and Alcock, N.W. (1999) Spacer control of directionality in supramolecular helicates using an inexpensive approach. *Angew. Chem. Int. Ed.*, **38**, 1277–1278.

49. Schulze Isfort, C., Kreickmann, T., Pape, T. *et al.* (2007) Helical complexes containing diamide-bridged benzene-*o*-dithiolato/catecholato ligands. *Chem. Eur. J.*, **13**, 2344–2357.

50. Hahn, F.E., Offermann, M., Schulze Isfort, C. *et al.* (2008) Heterobimetallic triple-stranded helicates with directional benzene-*o*-dithiol/catechol ligands. *Angew. Chem. Int. Ed.*, **47**, 6794–6797.

51. (a) Nitschke, J.R., Schulz, D., Bernardinelli, G., and Gérard, D. (2004) Selection rules for helicate ligand component self-assembly: Steric, pH, charge, and solvent effects. *J. Am. Chem. Soc.*, **126**, 16538–16543; (b) Schulz, D., and Nitschke, J.R. (2006) Designing multistep transformations using the Hammett equation: Imine exchange on a copper(I) template. *J. Am. Chem. Soc.*, **128**, 9887–9892; (c) Sarma, R.J., and Nitschke, J.R. (2008) Self-assembly in systems of subcomponents: Simple rules, subtle consequences. *Angew. Chem. Int. Ed.*, **47**, 377–380; (d) Nitschke, J.R. (2007) Construction, substitution, and sorting of metallo-organic structures via subcomponent self-assembly. *Acc. Chem. Res.*, **40**, 103–112.

52. (a) Mal, P., Schulz, D., Beyeh, K. *et al.* (2008) An unlockable–relockable iron cage by subcomponent self-assembly. *Angew. Chem. Int. Ed.*, **47**, 8297–8301; (b) Mal, P., Breiner, B., Rissanen, K., and Nitschke, J.R. (2009) White phosphorus is air-stable within a self-assembled tetrahedral capsule. *Science*, **324**, 1697–1699.

53. Krinsky, J.L., Arnold, J., and Bergman, R.G. (2007) Platinum group thiophenoxyimine complexes: Syntheses and crystallographical/computational studies. *Organometallics*, **26**, 897–909.

54. (a) Corrigan, M.F., Rae, I.D., and West, B.O. (1978) The constitution of *N*-substituted 2-(iminomethyl)benzenethiols (*o*-mercaptobenzaldimines). *Aust. J. Chem.*, **31**, 587–594; (b) Still, I.W. J., Natividad-Preyra, R., and Toste, D.F. (1999) A versatile synthetic route to 1,5-dithiocins from *o*-mercapto aromatic aldehydes. *Can. J. Chem.*, **77**, 113–121.

55. Marini, P.J., Murray, K.S., and West, B.O. (1983) Iron complexes of *N*-substituted thiosalicylindeneimines. Part 1. Synthesis and reactions with oxygen and carbon monoxide. *J. Chem. Soc. Dalton Trans.*, 143–151.

56. Dömer, J., Slootweg, J.C., Hupka, F. *et al.* (2010) Subcomponent assembly and transmetalation of dinuclear helicates. *Angew. Chem. Int. Ed.*, **49**, 6430–6433.

57. Childs, L.J., Alcock, N.W., and Hannon, M.J. (2002) Assembly of a nanoscale chiral ball through supramolecular aggregation of bowl-shaped triangular helicates. *Angew. Chem. Int. Ed.*, **41**, 4244–4247.

58. Pearson, R.G. (1963) Hard and soft acids and bases. *J. Am. Chem. Soc.*, **85**, 3533–3539.

59. Stenson, P.A., Board, A., Marin-Becerra, A. *et al.* (2008) Molecular and electronic structures of one-electron oxidized $Ni^{II}$–(dithiosalicylidenediamine) complexes: $Ni^{III}$–thiolate versus $Ni^{II}$–thiyl radical states. *Chem. Eur. J.*, **14**, 2564–2576.

60. Reichert, R. and Schläpfer, C.W. (1977) Metallaustausch in quadratisch planaren Nickel- und Palladium-Mercaptoäthylamin-Komplexen. *Helv. Chim. Acta.*, **60**, 722–729.

# 6

# Photophysical Properties and Applications of Lanthanoid Helicates

*Jean-Claude G. Bünzli*

*Institute of Chemical Sciences and Engineering, École Polytechnique Fédérale de Lausanne, Switzerland*
*Photovoltaic Materials, Department of Materials Chemistry, Korea University, Sejong Campus*

## List of Acronyms and Abbreviations

| | |
|---|---|
| 2PA | two-photon absorption |
| 3PA | three-photon absorption |
| Ab | antibody |
| AO | acridine orange |
| CD | circular dichroism |
| CPL | circularly polarized luminescence |
| DNA | deoxyribonucleic acid |
| dpa | 2,6-pyridine dicarboxylate |
| dtpa | diethylenetriaminepentaacetate |
| EB | ethidium bromide |
| ED | electric dipole |
| edta | ethylenediamine tetraacetate |
| EQ | electric quadrupole |
| ER | oestrogen receptor |
| ES-MS | electrospray mass spectrometry |

*Metallofoldamers: Supramolecular Architectures from Helicates to Biomimetics*, First Edition.
Edited by Galia Maayan and Markus Albrecht.
© 2013 John Wiley & Sons, Ltd. Published 2013 by John Wiley & Sons, Ltd.

ESA        excited state absorption
ETU        energy transfer upconversion
GM        Göppert–Mayer unit: $10^{-50}\,cm^4\,s\,photon^{-1}$ for 2PA and $10^{-83}\,cm^6$ $s^2photon^{-2}$ for 3PA
Her2/*neu*        human epidermal growth factor receptor
HHH        head to head to head arrangement of ligand strands
IgG        immunoglobulin G
JO        Judd–Ofelt
LIS        lanthanoid-induced shift
LMCT        ligand-to-metal charge transfer
LOD        limit of detection
mAb        monoclonal antibody
MALDI-TOF        matrix-assisted laser desorption – ionization time of flight mass spectrometry
MD        magnetic dipole
MLCT        metal-to-ligand charge transfer
NIR        near infrared
NMR        nuclear magnetic resonance
NP        nanoparticle
pet        photoinduced electron transfer
PCR        polymerase chain reaction
TL        total luminescence
tpy        terpyridine
TRD        time-resolved detection
TRLM        time-resolved luminescence microscopy
WST-1        water-soluble tetrazolium salts
YAG        yttrium aluminum garnet

## 6.1 Introduction

A helicate is a discrete coordination compound constituted by one or more organic covalent strand(s) wrapped helically around at least two metal centres defining the helical axis [1]. Additionally, weak interactions programmed between the ligand strands lead to supramolecular recognition of the metal ions. The first helicates synthesized by J.-M. Lehn in 1987 contained d-transition metal ions. The fact that ions with spherical electronic density such as $Cu^I$, $Zn^{II}$ or $Ga^{III}$ easily self-assemble with polytopic ligands bearing adequate coordination groups to yield double- or triple-stranded polynuclear helicates opened perspectives for tailoring helical edifices in which the metal ions are trivalent lanthanoid ions ($Ln^{III}$). These ions have a $[Xe]4f^n$ electronic configuration, so that 4f electrons are shielded from external interactions by the filled $5s^25p^6$ electronic subshells. Consequently, they are large, spherical and labile so that the polytopic ligands encapsulating them have to be carefully designed. Interest for lanthanoid helical complexes started in 1992 when the group of E. Constable showed the helical wrapping of sexipyridine around $Eu^{III}$ in $[Eu(NO_3)_2(L1)](PF_6)$ [2]. The same year, Xing Ya Cheng *et al.* and C. Piguet *et al.* reported the first tripled-stranded dinuclear lanthanoid helicates. The Chinese

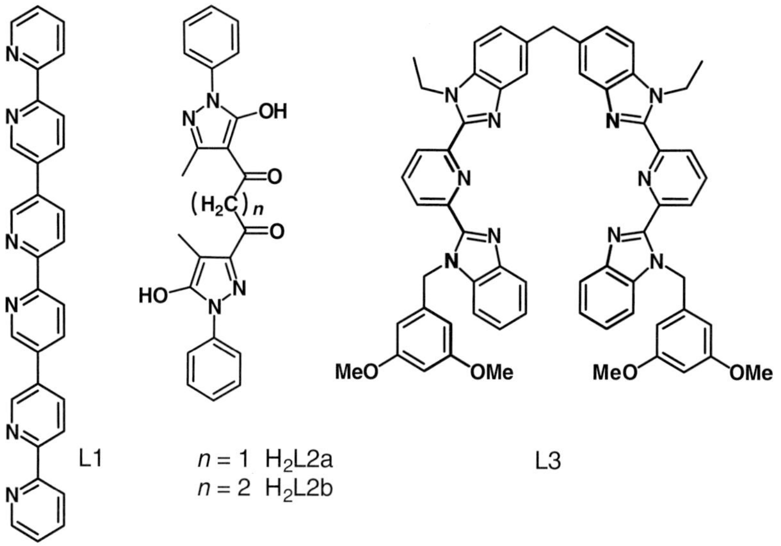

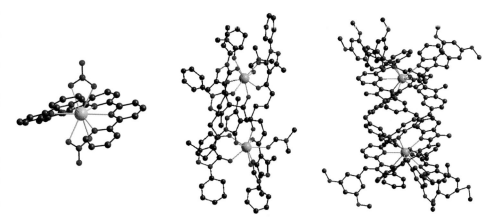

**Scheme 6.1** *Ligands used for isolating the first lanthanoid helical complexes and helicates.*

authors isolated [Ln$_2$(L2a)$_3$(DMF)$_2$] complexes with a dianionic pyrazolonate ligand [3], the structure of the Sm$^{III}$ complex being subsequently published in 1995 [4], while the second team self-assembled a Eu$^{III}$ helicate from a neutral bis(benzimidazolepyridine) ligand [5] (Figure 6.1); the photophysical properties of the highly charged species [Eu$_2$(L3a)$_3$]$^{6+}$ were deciphered the next year [6]. Overall, the bis(benzimidazolepyridine) framework proved to be quite versatile and its subsequent modifications yielded a wealth of new exciting molecular objects in which various interactions and properties, magnetic or spectroscopic, could be tailored. The thermodynamically controlled self-assembly of

**Figure 6.1** *X-Ray molecular structures of the lanthanoid helical complex [Eu(NO$_3$)$_2$(L1)]$^-$ [2] and of the first lanthanoid helicates, [Sm$_2$(L2a)$_3$(DMF)$_2$] [4] and [Eu$_2$(L3)$_3$]$^{6+}$ [5]; see Scheme 6.1 for ligand formulae. Redrawn from crystal structure data.*

lanthanoid helicates with this class of ligands is dealt with in Chapter 3 of this book. Moreover, programmed subtle differences between the two coordination sites of ditopic ligands L3 result in the recognition of two unlike lanthanoid ions [7], while reducing the denticity of one coordination site to two leads to the assembly of heterometallic d-f helicates [8]. Finally the ligand backbone can be expanded to accommodate three or even four coordination units, allowing the isolation of homometallic [9] or heterometallic [10] 4f-4f-4f helicates, or 4f-3d-4f [11] or 3d-4f-4f [12] edifices, or even a rare homometallic tetranuclear helicate [13].

With respect to photophysical properties, the introduction of lanthanoid ions into helicates has two major advantages. The first one is the highly protective coordination environment generated by the wrapping of the ligand strands while the second lies in the intermetallic communication that can be programmed along the helical axis. Energy transfer between metal ions can be rationally programmed and leads to the control of photophysical properties of one ion by the other. For instance, d-transition metal ions with long excited-state lifetimes such as $Cr^{III}$ can populate the emissive states of near infrared (NIR) emitting lanthanoid ions resulting in a shift of their apparent lifetimes from the micro- to the millisecond range [14]. Another example stems from a pair of $Cr^{III}$ ions sequentially transferring energy onto a single $Er^{III}$ ion to generate molecular upconversion of NIR to visible light [15].

In this chapter, we describe basic photophysical properties of lanthanoid helicates, along with photonic applications in biosciences and luminescent materials. Literature is covered until August 2011 but only compounds for which quantitative luminescence data (quantum yield, sensitization efficiency, lifetime) are available are taken into consideration. A summary of lanthanoid photophysics can be found in reference [16]. The reader interested in full details is referred to descriptions in earlier works by G. S. Ofelt [17], B. R. Judd [18], G. H. Dieke [19], W. T. Carnall [20], C. A. Morrison [21] or S. Hüfner [22]. In addition, the reviews by C. Görller-Walrand and K. Binnemans on the rationalization of ligand–field parameterization [23] and spectral intensities of f-f transitions [24] are very useful and easy to read, with careful and precise definitions as well as very helpful dimensional analysis. More recently, the book edited by G. K. Liu and B. Jacquier sheds light on lanthanoid-containing optical materials [25] while Judd–Ofelt theory has been summarized in an elegant way by B. M. Walsh [26]. Applications of lanthanoid luminescence in biosciences are summarized in a multiauthor book edited by P. Hänninen and H. Härmä, published in 2011 [27].

Other relevant references deal with spin-orbit constants [28], energy level diagrams [29], symmetry-related selection rules derived from group-theoretical considerations [30], listings of f-f spectra [20,31], determination of radiative lifetimes [32], Judd–Ofelt parameterization of f-f transition intensities [31,33], determination of radiative lifetimes [34], mechanisms of energy transfer [35], influence of high-energy vibrations [36] or interconfigurational $4f^n \leftrightarrow 4f^{n-1}5d^1$ transitions [37].

In view of the small absorption coefficients of f-f transitions, a sensitization process has to be used, which is sketched in Figure 6.2. $Q_{Ln}^L$ is the overall quantum yield, obtained upon indirect excitation into the ligand levels, while $Q_{Ln}^{Ln}$ is the intrinsic quantum yield measured upon indirect excitation into the excited f levels. They are related by the sensitization efficiency $\eta_{sens}$. When $Q_{Ln}^{Ln}$ cannot be determined, it may be estimated with the observed and radiative lifetimes (see Equation 6.1); in turn, the radiative lifetime can be

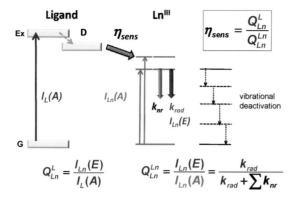

**Figure 6.2** *Simplified scheme demonstrating sensitization of Ln^III luminescence by a ligand. Key: A = absorption, E = emission, G = ground state, Ex = excited state, D = donor state; other parameters are defined in the text.*

extracted from absorption spectra (Equation 6.2) or, in the case of Eu^III, from emission spectra (Equation 6.3).

$$Q_{Ln}^{Ln} = \frac{k_{rad}}{k_{obs}} = \frac{\tau_{obs}}{\tau_{rad}} \tag{6.1}$$

$$\frac{1}{\tau_{rad}} = 2303 \times \frac{8\pi c n^2 \tilde{v}^2 (2J+1)}{N_A (2J'+1)} \int \varepsilon(\tilde{v}) d\tilde{v} \tag{6.2}$$

$$\frac{1}{\tau_{rad}} = A_{MD,0} \cdot n^3 \left(\frac{I_{tot}}{I_{MD}}\right) \tag{6.3}$$

with $A_{MD,0}$ being equal to $14.65\,s^{-1}$; ($I_{tot}/I_{MD}$) is the total integrated emission from the Eu($^5D_0$) level to the $^7F_J$ manifold ($J = 0$–6) divided by the integrated intensity of the $^5D_0 \rightarrow {}^7F_1$ transition.

## 6.2 Homometallic Lanthanoid Helicates

Tridentate coordinating units of the $C_2$-symmetrical ligands described in Scheme 6.2 are connected by saturated methylene or cyclohexane units which are poor electronic relays. It is therefore anticipated that the two ends of the helical edifices [Ln$_2$(L)$_3$]$^{n+}$ ($n = 0$ or 6) will behave independently and that trends in their photophysical properties will parallel those found in mononuclear complexes in which the metal ion has a similar environment [38]. Ligands sketched in Schemes 6.1 and 6.2 have been tailored to self-assemble with Ln^III ions either in aprotic (L3, L4) or protic solvents leading to nine-coordinate chemical environments with idealized trigonal symmetry. Single-crystal X-ray structure determinations of dinuclear helicates with ligands L3 [6], L4 [39] and H$_2$L5 [40] confirm coordination geometries derived from tricapped trigonal prismatic polyhedra. Furthermore, NMR spectra and solution structure determination based on lanthanoid-induced shifts and relaxation times [41] fully agree with this picture [40,42]. The only exception is ligand H$_2$L16

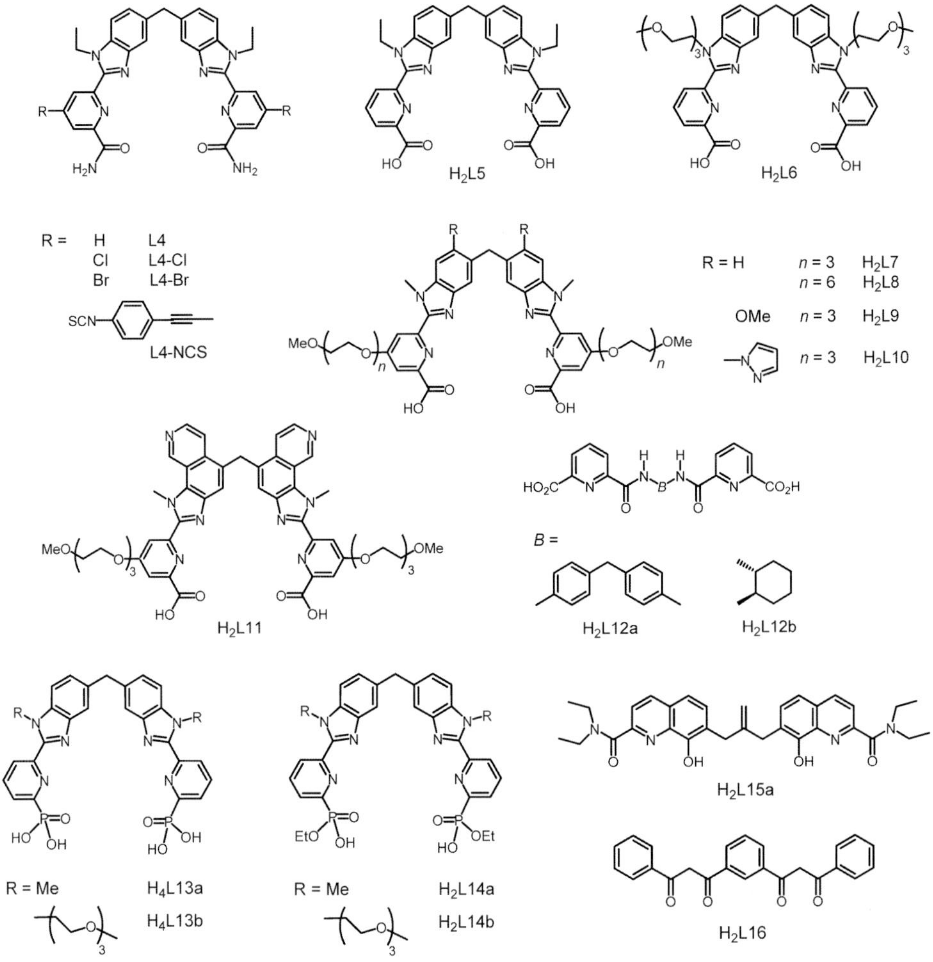

R =  H    L4
     Cl   L4-Cl
     Br   L4-Br

SCN— ≡—
L4-NCS

H₂L5

H₂L6

$R = H$   $n = 3$   H₂L7
          $n = 6$   H₂L8

OMe   $n = 3$   H₂L9

$n = 3$   H₂L10

H₂L11

$B =$

H₂L12a

H₂L12b

$R = Me$   H₄L13a
           H₄L13b

$R = Me$   H₂L14a
           H₂L14b

H₂L15a

H₂L16

**Scheme 6.2**  *Ditopic bis(tridentate) ligands for homometallic dinuclear lanthanoid helicates.*

which provides a hexacoordinate environment to the Ln$^{III}$ ions in [Ln₂(L16)₃] so that the coordination sphere is completed either with bound water molecules or with an additional ligand strand leading to an anionic quadruple-stranded [Eu(L16)₄]⁻ helicate [43]. The coordination environments provided by the ditopic ligands are either N₉ (L3) or N₃O₆ (all the other ligands, except H₂L16), which allows a systematic discussion of essential photophysical parameters. Sections 6.2.1–6.2.5 focus on triple-stranded dinuclear helicates for which extensive photophysical data are at hand. Other helicates are dealt with in Section 6.2.6.

### 6.2.1  Influence of the Triplet-State Energy on Quantum Yields

Typical photophysical parameters for the [Eu₂(L)₃]$^{n+}$ helicates are reported in Table 6.1. They include the energy of 0-phonon component of the triplet state of the bonded ligands, $E_{0-0}(T^*)$, which often plays a decisive role in the energy transfer process, the lifetime of

***Table 6.1*** *Experimental photophysical parameters at 295 K (unless otherwise stated) for [$Eu_2(L)_3$]$^{n+}$ dinuclear helicates. Whenever available, standard deviations ($2\sigma$) are given within parentheses; estimated errors on quantum yields are $\pm 10$–$15\%$. See Schemes 6.1 and 6.2 for ligand formulae.*

| Ligand | Solvent | $E_{0\text{-}0}(T^*)$ (cm$^{-1}$ at 77 K) | $\tau_{obs}$ (ms) | $Q_{Eu}^{L}$ (%) | $E(^5D_0 \rightarrow {}^7F_0)$ (cm$^{-1}$) | Refs. |
|---|---|---|---|---|---|---|
| L3 | MeCN | 19 880[a] | 0.24(1)[a], 2.04(5) (77 K) | 0.16[b] | 17 236[a,d] | [6,39] |
| L4 | MeCN | 21 280[a] | 2.11(5)[a] | 8.6[b] | 17 232[a,c,d] | [39] |
| L4-Cl | MeCN | 21 200 | 1.62(1) | 10.8[b] | 17 226 | [42] |
| L4-NCS | MeCN | 20 090 | 1.62(1) | 26.7[b] | 17 222[a] | [44] |
| (L5)$^{2-}$ | H$_2$O, pH 7.0 | 20 660[e] | 2.54 | 24(3) | 17 232 | [40] |
| (L6)$^{2-}$ | H$_2$O, pH 7.4 | 20 920 | 2.2(1) | 11(1) | 17 235 | [45] |
| (L7)$^{2-}$ | H$_2$O, pH 7.4 | 21 900 | 2.43(9) | 21(2) | 17 234 | [46] |
| | [a] | n.a. | 2.36(1) | 24(1) | 17 233 | [47] |
| | [f] | 21 800 | 2.50(3) | 23(1) | n.a. | [48] |
| | [g] | 21 800 | 2.40(2) | 28(1) | n.a. | [48] |
| (L8)$^{2-}$ | H$_2$O, pH 7.4 | 21 800 | 2.43(3) | 19(2) | 17 233 | [49] |
| (L9)$^{2-}$ | H$_2$O, pH 7.4 | 19 950 | 0.54(2) | 0.35 (5) | 17 237 | [50] |
| (L10)$^{2-}$ | H$_2$O, pH 7.4 | 21 150 | 2.52(2) | 15(2) | 17 242 | [50] |
| (L11)$^{2-}$ | H$_2$O, pH 7.4 | 20 800 | 2.30(2) | 9.0(9) | 17 242 | [50] |
| (L12a)$^{2-}$ | H$_2$O | n.a. | 1.48 | n.a. | 17 226 | [51] |
| (L12b)$^{2-}$ | H$_2$O | n.a. | 1.56 | n.a. | 17 221 | [51] |
| (L13a)$^{4-}$ | H$_2$O, pH 7.4 | 21 300 | 2.54(2) | 4.6(7) | n.a. | [52] |
| (L13b)$^{4-}$ | H$_2$O, pH 7.4 | 22 200 | 1.9(1) | 6(2) | n.a. | [52] |
| (L14a)$^{2-}$ | H$_2$O, pH 7.4 | 21 400 | 3.23(1) | 25(2) | 17 241 | [52] |
| (L14b)$^{2-}$ | H$_2$O, pH 7.4 | 21 250 | 2.1(2) | 2.5(3) | n.a. | [52] |
| (L16)$^{2-}$ | DMF | 20 400 | 0.22 | 5 | n.a. | [43] |

[a]Solid state sample.

[b]Some originally published quantum yield data were in error because the value used for the reference [Eu(tpy)$_3$]$^{3+}$ was largely underestimated due to an instrumental problem; the correct value is 32(1)%. Consequently, quantum yield data have been recalculated.

[c]Recalculated from 77 K data using a 1 cm$^{-1}$/24 K temperature dependence.

[d]Average value of three different sites.

[e]At 10 K.

[f]Monodisperse [Eu$_2$(L7)$_3$]@SiO$_2$ nanoparticles (55 $\pm$ 5 nm).

[g]Monodisperse [Eu$_2$(L7)$_3$]@SiO$_2$-NH$_2$ nanoparticles (90 $\pm$ 10 nm).

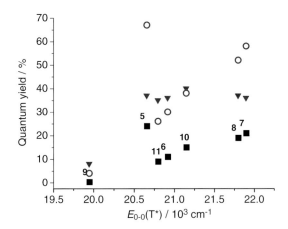

**Figure 6.3**   *Quantum yields for aqueous solutions of the dicarboxylate [Eu$_2$L$_3$] helicates versus $E_{0-0}(T^*)$ at 77 K. Ligand numbering according to Table 6.1. Black squares: overall quantum yields. Red triangles: intrinsic quantum yields. Blue circles, sensitization efficiencies.*

the $^5D_0$ excited state, $\tau_{obs}$, the overall quantum yield $Q_{Eu}^L$ and the energy of the $^5D_0 \rightarrow {}^7F_0$ transition which reflects the nephelauxetic effect created by the metal–ion surroundings.

Referring to Figure 6.2, the main donor state in the case of Eu$^{III}$ and Tb$^{III}$ is often a ligand triplet state so that attempts have been made to relate the energy of this state with the value of the overall quantum yield. To put it simple, if there is no other energy migration paths operating, $E_{0-0}(T^*)$ should lie at least 1500 cm$^{-1}$ above the emitting level to avoid too much back transfer and the ideal energy gap between donor and receiving levels lies between 2000 and 4000 cm$^{-1}$. This however has to be taken as a guideline only as illustrated by the graph in Figure 6.3 in which quantum yields are reported versus triplet state energies. The correlation is good for the helicates with the dicarboxylate ligands and generally speaking, the quantum yield increases monotonically up to a triplet state energy of 22 000 cm$^{-1}$, which is about the energy of the Eu($^5D_2$) level. If the main energy transfer mechanism is a double electron exchange (Dexter mechanism), which is usually the case in coordination compounds, the corresponding selection rule on $J$ quantum number is $\Delta J = 0, \pm 1$ with $J = J' = 0$ excluded. This means that energy will be mainly funnelled onto the $^5D_1$ level located at 19 000 cm$^{-1}$, either from $^7F_0$ or from $^7F_1$ (which has about 30% population at 295 K). Effectively, $Q_{Eu}^L$ values start to be sizeable when $E_{0-0}(T^*) > 20\,800$ cm$^{-1}$, corresponding to an energy difference of about 1800 cm$^{-1}$ with respect to the $^5D_0$ level. The quantum yield, however, continues to increase with increasing energy gaps larger than 1800 cm$^{-1}$, which means that transfer onto higher excited state(s) also occurs, involving for instance $^5D_2$.

There is, however, a serious exception for [Eu$_2$(L5)$_3$], the reason for which is difficult to explain. The triplet state energy for [Eu$_2$(L5)$_3$] has been determined at 10 K while the data for the other helicates are at 77 K, but it is doubtful that there is a very large energy shift between these two temperatures. Further, one may note that, since quantum yields are determined at room temperature, triplet state energies should also be measured at this temperature, which is often not possible since emission spectra of the helicates rarely display triplet-state phosphorescence at room temperature, a reason why 77 K data are used. The dependence of $E_{0\text{-}0}(T^*)$ upon temperature is not precisely known (typically 1000–3000 cm$^{-1}$ between 77 and 295 K) but one may hypothesize that in the case of [Eu$_2$(L5)$_3$] it shifts the triplet state to an energy more favourable for energy transfer. Another explanation could be found in the fact that the overall quantum yield depends on two main factors: the energy transfer efficiency from the ligand and the ability of the edifice to prevent nonradiative deactivation. The latter is not taken into consideration in the correlation. However, looking into the lifetime data one sees that they are very similar (2.2–2.5 ms) for all helicates with the exception of [Eu$_2$(L9)$_3$], meaning that radiationless processes are very comparable so that this factor does not seem to play a role for [Eu$_2$(L5)$_3$] having a large quantum yield. But the explanation for the short lifetime of [Eu$_2$(L9)$_3$] lies in the resonance between the triplet state energy and the Eu($^5$D$_1$) level, leading to energy back transfer, short lifetime and low quantum yield.

That the simple relationship between $E_{0\text{-}0}(T^*)$ and $Q_{Eu}^L$ cannot rationalize all quantum yield values is illustrated by data for the other helicates reported in Table 6.1 which would not fit well into the monotonous correlation. Solvent effects have also to be taken into consideration. Finally, deviations from the monotonous correlation are found within the phosphonate helicate series. For instance, [Eu$_2$(L14a)$_3$] is an exception similar to [Eu$_2$(L5)$_3$], while the quantum yield of the other chelates is very low with respect to their triplet state energy.

Among the cationic helicates with neutral ligands L3–L4 measured in acetonitrile, [Eu$_2$(L4-NCS)$_3$]$^{6+}$ has much larger quantum yield than predicted by the correlation, while [Eu$_2$(L4)$_3$]$^{6+}$ and [Eu$_2$(L4-Cl)$_3$]$^{6+}$ have slightly smaller quantum yields. The case of [Eu$_2$(L3)$_3$]$^{6+}$ is similar to [Eu$_2$(L9)$_3$], with a triplet state energy close to the energy of the Eu($^5$D$_1$) level. In addition, it has been shown for a mononuclear analogue that the EuN$_9$ coordination favours a low-lying LMCT state [53] which strongly deactivates the ligand singlet state [54].

A more detailed way of looking into the data reported in Table 6.1 is to evaluate the radiative lifetimes and intrinsic quantum yields, and, subsequently, the sensitization efficiencies. These parameters are reported in Table 6.2 for helicates with carboxylate and phosphonate ligands. Interestingly, the radiative lifetimes for the seven solutions of dicarboxylate helicates lie in a narrow 6.2–6.9 ms range, with an average of 6.6 ms. This is by no means unexpected since they all feature the same N$_6$O$_3$ chemical environment with very similar Eu–ligand distances. In turn, the intrinsic quantum yields $Q_{Eu}^{Eu}$ are very similar, 35–40% (Figure 6.3, see triangles), except for [Eu$_2$(L9)$_3$] for which $Q_{Eu}^{Eu}$ is much smaller; this is due to the short observed lifetime discussed above. For the other dicarboxylate helicates, differences in overall quantum yields have therefore to be traced back to differences in energy transfer efficiencies ($\eta_{sens}$; Figure 6.3, circles) which vary depending on subtle electronic differences in the ligands. In effect, $\eta_{sens}$ and $Q_{Eu}^L$ value dependences are perfectly parallel and the large quantum efficiency displayed by [Eu$_2$(L5)$_3$] with respect to $E_{0\text{-}0}(T^*)$ can be entirely ascribed to a more efficient energy transfer. However, the very poor quantum yield of [Eu$_2$(L9)$_3$] results both from an inefficient energy transfer and from large nonradiative deactivation rate constants, as reflected by the very small value of $Q_{Eu}^{Eu}$.

Although radiative lifetimes for helicates with phosphonate ligands are only available for two chelates, an interesting observation can be made: they are much shorter

***Table 6.2*** *Derived photophysical parameters for triple-stranded $[Eu_2(L)_3]^{n+}$ dinuclear helicates in $H_2O$ at pH 7.0–7.4 and 295 K unless otherwise stated.*

| Ligand | $\tau_{rad}$ (ms)[a] | $Q_{Eu}^{Eu}$ (%)[a] | $\eta_{sens}$ (%)[a] | Ref. |
|---|---|---|---|---|
| $(L5)^{2-}$ | 6.8 | 37 | 67 | [40] |
| $(L6)^{2-}$ | 6.2 | 36 | 30 | [45] |
| $(L7)^{2-}$ | 6.9 | 36 | 58 | [46] |
| $(L7)^{2-b}$ | 4.9 | 48 | 50 | [47] |
| $(L7)^{2-c}$ | $3.9^d$ | 60 | 38 | [48] |
| $(L7)^{2-e}$ | $3.6^d$ | 63 | 44 | [48] |
| $(L8)^{2-}$ | 6.6 | 37 | 52 | [49] |
| $(L9)^{2-}$ | 6.8 | 8 | 4 | [50] |
| $(L10)^{2-}$ | 6.4 | 40 | 38 | [50] |
| $(L11)^{2-}$ | 6.7 | 35 | 26 | [50] |
| $(H_2L13b)^{2-}$ | 4.2 | 45 | 13 | [52] |
| $(L14a)^{2-}$ | 4.8 | 67 | 37 | [52] |

[a]Estimated errors: $\tau_{rad}$ ±12%, $Q_{Eu}^{Eu}$ ±12%, $\eta_{sens}$ ±16%.
[b]Solid state sample.
[c]Monodisperse $[Eu_2(L7)_3]@SiO_2$ nanoparticles (55 ± 5 nm).
[d]Recalculated, see Section 6.2.2.
[e]Monodisperse $[Eu_2(L7)_3]@SiO_2$-$NH_2$ nanoparticles (90 ± 10 nm).

than for the dicarboxylate complexes (about $-30\%$) despite a seemingly similar coordination environment ($N_3O_6$). But phosphonate groups are more strongly coordinating than carboxylates, which explains this shortening. In addition, the observed lifetimes are somewhat longer, probably due to the larger hydrophobicity of phosphonate versus carboxylate anions, resulting in substantially larger $Q_{Eu}^{Eu}$ values [52]. Further, the sensitization efficiency has a tendency to be rather low, so that overall quantum yields are smaller than those of the carboxylate helicates or at most comparable, in the case of $[Eu_2(L14a)_3]$.

Photophysical parameters for the $[Ln_2(L)_3]^{n+}$ edifices with lanthanoid ions other than $Eu^{III}$ are listed in Table 6.3. The more abundant data are found for $Tb^{III}$ and the $Q_{Ln}^L$ versus $E_{0-0}(T^*)$ correlation is depicted in Figure 6.4 for helicates encompassing this ion and dissolved in water. The first excited level of $Tb^{III}$, $^5D_4$, is located at an energy close to $20\,500\,cm^{-1}$, therefore helicates with ligands having $E_{0-0}(T^*) < 21\,500\,cm^{-1}$ are likely to be poorly luminescent, which is indeed the case for ligands $H_2Li$ with $i = 5, 9, 10, 11, 14a$ and 14b. Energy back-transfer occurs, as demonstrated by the $Tb(^5D_4)$ lifetime, which is short at room temperature ($10–120\,\mu s$ for the dicarboxylate helicates) but considerably longer at 77 K with values between 1.9 and 2.6 ms, in line with a temperature-dependent nonradiative deactivation process. When the energy gap is around $1500\,cm^{-1}$, for $[Tb_2(L7)_3]$ and $[Tb_2(L8)_3]$, the edifices are consequently more luminescent, although lifetimes are also quite temperature-dependent, but their room temperature values are around $650–660\,\mu s$. Moreover $[Tb_2(Li)_3]$ ($i = 6, 13a, 13b$) do not fit at all into the correlation. For $[Tb_2(L6)_3]$, lifetime dependence is a bit comparable to those with ligands $H_2Li$ ($i = 7, 8$), $\tau_{obs}$ increasing from 0.39 ms at room temperature to 1.85 ms at 77 K, so that one may additionally invoke a less good energy transfer to explain the very low quantum yield value of only 0.34%. The cases of the helicates with $H_4L13a,b$ are equally difficult to

**Table 6.3** *Photophysical parameters at 295 K (unless otherwise stated) for triple-stranded* $[Ln_2(L)_3]$ *(Ln ≠ Eu) dinuclear helicates upon ligand excitation.*[a]

| Ligand | Ln | Solvent | $E_{0-0}(T^*)$ (cm$^{-1}$ at 77 K) | $\tau_{obs}$ (μs) | $Q^L_{Ln}$ (%) | Ref. |
|---|---|---|---|---|---|---|
| L4 | Tb | MeCN | 20 930 | 61(3) | 9.8[b] | [39] |
| (L5)$^{2-}$ | Sm | H$_2$O, pH 7.5 | 20 660 | 42.0(4) | 0.14 | [55] |
| | Tb | H$_2$O | 20 660 | 50 | 1.2[b] | [40] |
| | Yb | D$_2$O, pD 7.5 | 20 660 | 40(2) | 1.8 | [55] |
| (L6)$^{2-}$ | Tb | H$_2$O, pH 7.4 | 22 100 | 390(40) | 0.34(4) | [45] |
| (L7)$^{2-}$ | Nd | H$_2$O, pH 7.4 | 22 050$^c$ | 0.21(2) | 0.031(6) | [46] |
| | Sm | H$_2$O, pH 7.4 | 22 050$^c$ | 30.4(4) | 0.38(6) | [46] |
| | Tb | H$_2$O, pH 7.4 | 22 050 | 650(20) | 11(2) | [46] |
| | Yb | H$_2$O, pH 7.4 | 22 050$^c$ | 4.40(7) | 0.15(3) | [46] |
| (L8)$^{2-}$ | Tb | H$_2$O, pH 7.4 | 21 900 | 660(20) | 10(1) | [49] |
| (L9)$^{2-}$ | Tb | H$_2$O, pH 7.4 | 19 550 | 10.5(8) | $^d$ | [50] |
| | Yb | H$_2$O, pH 7.4 | 19 550 | 4.28(2) | 0.15(2) | [50] |
| (L10)$^{2-}$ | Tb | H$_2$O, pH 7.4 | 21 150 | 120(10) | 2.5(3) | [50] |
| (L11)$^{2-}$ | Tb | H$_2$O, pH 7.4 | 20 800 | 40(3) | 0.31(5) | [50] |
| | Yb | H$_2$O, pH 7.4 | 20 800 | 4.33(3) | 0.16(2) | [50] |
| (H$_2$L13a)$^{2-}$ | Tb | H$_2$O, pH 7.4 | 22 570$^e$ | 220(50) | 2.7(4) | [52] |
| (H$_2$L13b)$^{2-}$ | Tb | H$_2$O, pH 7.4 | 22 285 | n.a. | 0.20(7) | [52] |
| (L14a)$^{2-}$ | Tb | H$_2$O, pH 7.4 | 20 400$^e$ | 130(10) | 2.5(4) | [52] |
| (L14b)$^{2-}$ | Tb | H$_2$O, pH 7.4 | 21 050 | 460(10) | 1.6(6) | [52] |

[a]Whenever available, standard deviations (2σ) are given within parentheses for lifetime data while uncertainties on quantum yields are in the range ±10–15%.
[b]Some originally published quantum yield data were in error because the value used for the reference $[Tb(tpy)_3]^{3+}$ was largely underestimated due to an instrumental problem; the correct value is 35(1)%. Consequently, quantum yield data have been recalculated.
[c]Estimated from Tb value.
[d]Too small to be measured.
[e]Estimated from Lu value.

rationalize as far as absolute values of $Q^L_{Tb}$ are concerned, but from lifetime dependence, it is also clear that back transfer is operating.

The three quantum yields available for Yb$^{III}$ helicates in water are the same (0.15–0.16%), consistent with the large energy difference between the Yb($^2F_{5/2_{5/2}}$) level and $E_{0-0}(T^*)$, 10–12 000 cm$^{-1}$, so that small variations in $E_{0-0}(T^*)$ have little influence. Quantum yield values are essentially determined by nonradiative deactivations which are quite efficient in view of the small energy gap between Yb($^2F_{5/2_{5/2}}$) and Yb($^2F_{7/2_{7/2}}$), approximately 10 000 cm$^{-1}$. Similar considerations apply to Sm$^{III}$ (gap 7500 cm$^{-1}$) and Nd$^{III}$ (5400 cm$^{-1}$).

### 6.2.2 Radiative Lifetime and Nephelauxetic Effect

It is clear from Equation 6.1 that, if one wishes to increase the intrinsic quantum yield, one has to maximize $\tau_{obs}$ by minimizing nonradiative deactivation and/or to decrease $\tau_{rad}$. The first action is rather well understood and mastered, at least for visible-emitting ions, that is for those ions which have a large energy gap between the emitting level and the highest sublevel of the ground (or receiving) state. High energy vibrations are relatively

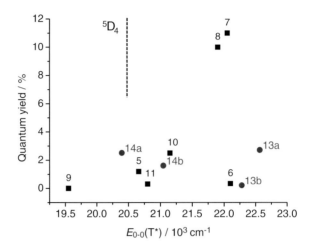

**Figure 6.4** *Relationship between the experimental overall quantum yield of [Tb$_2$L$_3$] helicates in water and the energy of the ligand triplet state. Squares: dicarboxylates. Circles: phosphonates (ligand numbering according to Table 6.3).*

easy to avoid and a protective and rigid environment will prevent solvent molecules to quench the excited state by collisional deactivations. However, this becomes problematic for NIR-emitting ions, particularly Nd$^{III}$ (gap 5400 cm$^{-1}$) and Er$^{III}$ (6600 cm$^{-1}$) so that deuteration or halogenation of ligands is required for coordination compounds to display reasonably large quantum yields [35,36,56,57]. Alternatively, inorganic matrices are better suited since their phonon energies can be quite low (<600 cm$^{-1}$).

Equations 6.2 and 6.3 give a clue on one way of decreasing $\tau_{rad}$ by increasing the refractive index. This is perfectly well illustrated with helicate [Eu$_2$(L7)$_3$] the radiative lifetime of which amounts to 6.9 ms in water ($n = 1.33$) and 4.9 ms in the solid state (Table 6.2) [47]. Taking the commonly accepted value $n = 1.5$ for solid state samples of complexes with organic ligands, one notes, referring to Equation 6.3, that the ratio $4.9/6.9 = 0.71$ is almost identical to $(1.33/1.5)^3 = 0.70$. The lifetimes of the aqueous solution, $\tau_{obs} = 2.43(9)$ ms, and the solid state sample, 2.36(1) ms, are the same within experimental errors so that the decrease in $\tau_{rad}$ leads to a one-third increase in $Q_{Eu}^{Eu}$, from 36 to 48%. Unfortunately, the sensitization efficiency decreases, from 58 to 50% so that *in fine*, the solid state sample presents only a 15% increase in overall quantum yield compared with the solution. When the same helicate is doped into silica nanoparticles [48], analysis of the photophysical parameters becomes more intricate because the $I_{tot}/I_{MD}$ ratio changes, which was not the case between solution (4.01) and solid state (4.14) samples. Here this ratio amounts to 5.38 in pure silica nanoparticles and to 5.90 in silica nanoparticles derivatized with amine groups for bioanalytical purposes. Therefore the refractive index correction ($n = 1.475$ for pure silica nanoparticles;[58] in the original paper the authors used 1.45[48]; data are recalculated), contributes to increase $\tau_{rad}$ by only 5% with respect to the solid state sample while the larger $I_{tot}/I_{MD}$ ratios decrease it by 25 and 32%, respectively. The larger $I_{tot}/I_{MD}$ ratios stem essentially from more intense hypersensitive transitions, possibly reflecting a lowering of symmetry of the metal ion sites due to interaction of the ligand strands with silica. Indeed, a lower symmetry would favour more orbital mixing since wavefunctions must have the same symmetry in order to combine, which is more easily achieved in low-symmetry edifices.

Another handle for decreasing $\tau_{rad}$ is to induce more orbital mixing by increasing the polarizability of the Ln–ligand bonds, which leads to a larger nephelauxetic effect. This concept has been proposed by C. K. Jørgensen [59] which rationalized several experimental findings showing that interelectronic repulsion is smaller for transition metal complexes than for gaseous ions. This corresponds to an expansion of the metal orbitals the electrons of which delocalize on the ligands and the word "nephelauxetic" was logically coined because in Greek it means "cloud expansion." The so-called nephelauxetic parameter is simply the ratio of the electronic repulsion Racah parameter B in the complex and in gaseous phase. It is linked to the covalency of the metal–ligand bonds and can be estimated from absorption spectra. For Eu$^{III}$ complexes, there is a simpler way of accessing the magnitude of this effect by determination of the energy of the $^5D_0$ level through measuring the highly forbidden $^5D_0 \rightarrow {}^7F_0$ transition. Since this transition is often very faint, its energy is commonly determined with the help of laser-excited excitation spectra. A phenomenological relationship has been put forward which relates $E(^5D_0 \rightarrow {}^7F_0) = E_0$ (expressed in cm$^{-1}$) to the sum of the contributions of each coordinated group to the overall nephelauxetic effect [60] (Equation 6.4):

$$E_0 = 17\,374 + C \sum_i n_i \delta_i \tag{6.4}$$

where 17 374 cm$^{-1}$ is the value for gaseous Eu$^{III}$ calculated by taking electrostatic and spin-orbit interactions into account, $C$ is a factor depending on the coordination number, $\delta_i$ is the nephelauxetic effect generated by each coordinated group and $n_i$ is the number of such groups in the first coordination sphere. Some relevant parameters are gathered in Table 6.4. The effects generated by individual groups is small and depends on the Ln-X bond length, which may vary if coordinating groups are involved in hydrogen bonding (e.g. water molecules). As a result, the relationship does not always give very precise estimates. It is, however, a useful guideline.

For estimating the nephelauxetic effects in the helicates listed, one needs to know the increment generated by aromatic heterocyclic imine groups (pyridine, imidazole), which has not been documented in the initial publications [60]. An average value of $-15.3$ cm$^{-1}$ has therefore been proposed upon analysing the photophysical properties of a Zn-Ln helicate with ligands bearing benzimidazolepyridine moieties [61] and this value exactly

**Table 6.4** *Nephelauxetic parameters* $C \times \delta_i$ *(cm$^{-1}$) for some coordinating groups and coordination numbers (CN) [60].*

| CN | O(CO$_2^-$) | O(ether) | O(amide) | O(H$_2$O) | O(diket)[a] | O(NO$_3^-$) | O(OH) | N(am)[b] | N(ar)[c] | Cl$^-$ |
|---|---|---|---|---|---|---|---|---|---|---|
| 8 | −18.2 | −11.8 | −16.6 | −11.0 | −14.7 | −14.1 | −12.3 | −12.8 | −16.2 | −22.5 |
| 9 | −17.2 | −11–1 | −15.7 | −10.4 | −13.9 | −13–3 | −11.6 | −12.1 | −15.3 | −21.2 |
| 10 | −16.3 | −10.5 | −14.9 | −9.9 | −13.2 | −12.6 | −11.0 | −11.5 | −14.5 | −20.1 |

[a]$\beta$-Diketonate.
[b]Amine.
[c]Aromatic heterocyclic imine, from ref. [61], see text.

reproduces $E_0$ for $[Eu_2(L3)_3]^{6+}$, $17\,236\,cm^{-1}$. For the amide ligand L4, inducing a $N_6O_3$ coordination environment, one calculates $E_0 = 17\,235\,cm^{-1}$, in line with the observed value of $17\,232\,cm^{-1}$. Substitution of pyridine by Cl and NCS apparently affects the nephelauxetic contribution of this moiety leading to overall effects larger by 9 and $13\,cm^{-1}$, respectively. When the amide group is substituted with a carboxylate function (ligands $H_2Li$, $i = 5$–11) one calculates $E_0 = 17\,231\,cm^{-1}$, in excellent agreement with the range of values observed for $i = 5$–8 ($17\,232$–$17\,235\,cm^{-1}$). When the fused benzene ring of benz-imidazole groups is substituted by aromatic rings, the nephelauxetic effect decreases leading to $E_0$ energies which are 6–$11\,cm^{-1}$ smaller than predicted. In view of the limited number of substitutions, this trend is difficult to quantify. Finally, the $N_3O_6$ environment provided by ligands $H_2L12a,b$ results in a calculated $E_0 = 17\,229\,cm^{-1}$, in reasonable agreement with the experimental values of $17\,226$ and $17\,221\,cm^{-1}$, respectively. From all these data, one sees that the accuracy of the relationship is around 5% (i.e., discrepancies range between 1 and $11\,cm^{-1}$, representing 1–8% of the total nephelauxetic effect) which, given the average increments taken into consideration, is satisfactory. The small span of nephelauxetic effects displayed by the helicates for which relevant experimental data are available is unfortunately insufficient to check if the trend in these effects is indeed reflected in $\tau_{rad}$ values, especially taking into account the relatively large experimental uncertainties on the latter parameter. More data will be needed in this respect.

### 6.2.3 Site-Symmetry Analysis

As briefly outlined in Section 6.2.1, analysis of the fine structure of the emission bands taking into account group-theory selection rules gives access to the symmetry of the $Eu^{III}$ site. Such analysis has been carried out for helicates $[Eu_2(Li)_3]$ ($i = 5$–9, 11) in frozen aqueous solutions at 10 K because emission lines are narrower at low temperature. Paramagnetic induced NMR shifts and relaxation times point to $[Eu_2(L5)_3]$ having an averaged structure with $D_3$ symmetry at room temperature, so that emission spectra have been analysed with respect to this point group of symmetry. A typical spectrum is shown on Figure 6.5 and we describe its partial analysis only. In $D_3$ symmetry, the

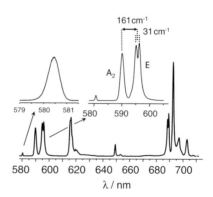

**Figure 6.5** *High-resolution emission spectrum of a frozen solution of $[Eu_2(L7)_3]$ in water/ glycerol 90 : 10 (v/v) at 10 K.* Reproduced with permission from ref. [63]; © 2008 New York Academy of Sciences.

**Table 6.5** *Parameters (in cm$^{-1}$) for the symmetry analysis of the metal ion sites in [Eu$_2$(Li)$_3$] (i = 5–8, 11). Frozen solutions in water/glycerol 90:10 (v/v) at 10 K, unless otherwise stated.*

| Helicate | $^5D_0 \rightarrow {}^7F_0$ fwhh[a] | $^5D_0 \rightarrow {}^7F_1$ $\Delta E(A_2$-E) | $^5D_0 \rightarrow {}^7F_1$ $\Delta E$(E-E) | Ref. |
|---|---|---|---|---|
| [Eu$_2$(L5)$_3$] | 15[b] | 156 | 28 | [40] |
| [Eu$_2$(L6)$_3$] | 19[b] | 162 | 38 | [45] |
| [Eu$_2$(L7)$_3$] | 17[b] | 161 | 31 | [46] |
| [Eu$_2$(L8)$_3$] | 15 | 161 | 31 | [49] |
| [Eu$_2$(L9)$_3$] | 17 | 160 | 31 | [50] |
| [Eu$_2$(L10)$_3$] | 20 | 181 | 26 | [50] |
| [Eu$_2$(L11)$_3$] | 20 | 177 | 34 | [50] |
| [LaEu(L17)$_3$]$^{6+c}$ | 7 | 131 | 16 | [64] |
| [Eu$_2$(L17)$_3$]$^{6+d}$ | 10 | 125 | 33 | [64] |

[a]Full width at half height.
[b]At 295 K.
[c]Solid state sample.
[d]Single crystal.

$^5D_0 \rightarrow {}^7F_0$ transition is forbidden, so that it is observed with a very faint intensity. It features only one relatively symmetrical component, indicative of the two metal ion sites having the same chemical environment. The $^7F_1$ level is split into two components, A$_2$ and E (degenerate twice) and the magnitude of this splitting is directly proportional to the strength of the crystal field (in fact it is proportional to the $B_0^2$ ligand–field parameter [62]). The additional splitting seen in the *E* component is indicative of a slight distortion from the idealized $D_3$ symmetry. The corresponding parameters are reported in Table 6.5. The A$_2$-E separation varies negligibly along the series with ligands $i = 5$–9, with an average value of 160(2) cm$^{-1}$. It increases significantly (10–13%) in [Eu$_2$(L10)$_3$], [Eu$_2$(L11)$_3$] in which the benzimidazole moieties bear aromatic substituents. As far as distortion from the idealized symmetry is concerned, the splittings lie in a narrow 26–31 cm$^{-1}$ range for all helicates with the exception of [Eu$_2$(L6)$_3$] (38 cm$^{-1}$) and [Eu$_2$(L11)$_3$] (34 cm$^{-1}$), two compounds the ligands of which bear relatively large substituents on the benzimidazole units. Altogether, these splittings point to relatively small deviations from the ternary symmetry, which may well be induced by the freezing of the solutions.

Heterometallic 4f-4f′ helicates have been obtained with the unsymmetrical ditopic ligands sketched on Scheme 6.3. In solution, mixtures of several species are obtained but heterometallic [LnLn′(L)$_3$]$^{6+}$ are always favoured [7,65,66]. A luminescence study is only available for helicates with ligand L17, with which crystal structures for the pairs of lanthanoids LaEu, LaTb, PrEr, PrLu and EuEu could be solved. These structural data show that: (i) the ligands wrap around the metal centres in a head to head to head (*HHH*) arrangement, (ii) a pseudo axial $C_3$ symmetry is maintained in the edifices, (iii) the Ln-X distances for a given Ln$^{III}$ ion in a given coordination site are the same as in reference homometallic edifices and (iv) the smaller ion is coordinated in the N$_6$O$_3$ cavity [64]. The energy of the $^5D_0 \rightarrow {}^7F_0$ transition for [LaEu(L17)$_3$]$^{6+}$ is within 1 cm$^{-1}$ of the one reported for [Eu$_2$(L4)$_3$]$^{6+}$ (Table 6.1): 17 230 cm$^{-1}$ and in agreement with the calculated value of 17 235 cm$^{-1}$. However, both $\Delta E$(A$_2$-E) and $\Delta E$(E-E) are substantially smaller. In the case of a single crystal of [Eu$_2$(L17)$_3$]$^{6+}$ only one $^5D_0 \rightarrow {}^7F_0$ transition is observed at 17 232 cm$^{-1}$ (compare 17 236 and 17 232 cm$^{-1}$ for helicates with L3 and L4,

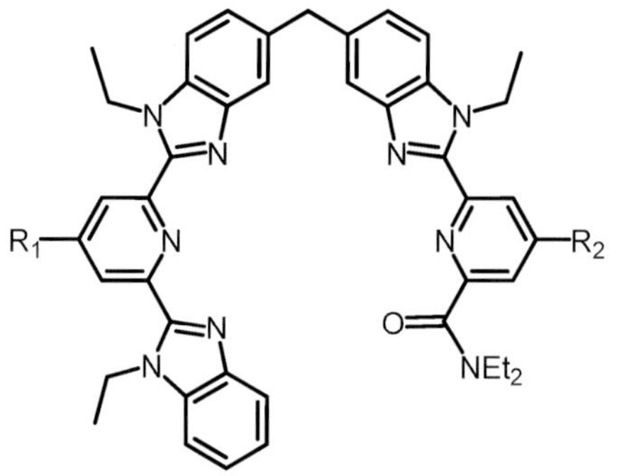

| R$_1$ | R$_2$ | |
|---|---|---|
| H | H | L17 |
| H | NEt$_2$ | L18a |
| H | Cl | L18b |
| NEt$_2$ | H | L19a |
| Cl | H | L19b |

**Scheme 6.3** *Unsymmetrical ligands for heteronuclear 4f-4f helicates.*

respectively). This is due to the very low quantum yield of Eu$^{III}$ in a N$_9$ coordination site (see above). The $\Delta E$(A$_2$-E) splitting is comparable to the value found for LaEu while $\Delta E$ (E-E) is larger and comparable to the splitting found for the helicates with symmetrical ligands (Table 6.5). The observed lifetimes for these two helicates are the same, $2.20 \pm 0.01$ (LaEu) and $2.26 \pm 0.05$ ms (EuEu) with a small temperature dependence, again pointing to the observed emission spectra arising mainly from the Eu(N$_6$O$_3$) sites.

### 6.2.4 Energy Transfer between Lanthanoid Ions

Bimetallic 4f-4f' helicates are well suited for studying the energy transfer between two different lanthanoid ions. Since the intermetallic distance is in the order of 90 pm, the transfer is likely to be governed by the dipole–dipole (Förster) mechanism. In the absence of a transfer, both ions are excited by a transfer from ligand donor states and emit light with their characteristic lifetime. When one ion transfers energy onto the acceptor, its lifetime is shortened, while the lifetime of the acceptor remains the same, provided that the rate constant of the transfer process is larger than the rate constant of the deactivation of the acceptor (Figure 6.6). Within this hypothesis, the following simplified equations hold to estimate the efficiency of transfer between the donor ion D and the acceptor ionA (Equations (6.5,6.6)):

$$k_{\text{et}} = k_{\text{obs}}^{\text{D}} - k_0^{\text{D}} = \frac{1}{\tau_{\text{obs}}^{\text{D}}} - \frac{1}{\tau_0^{\text{D}}} \tag{6.5}$$

$$\eta_{\text{et}} = 1 - \frac{k_0^{\text{D}}}{k_{\text{obs}}^{\text{D}}} = 1 - \frac{\tau_{\text{obs}}^{\text{D}}}{\tau_0^{\text{D}}} = \frac{1}{1 + (R_{\text{DA}}/R_0)^6} \tag{6.6}$$

in which $\tau_{\text{obs}}^{\text{D}}$ and $\tau_0^{\text{D}}$ are the lifetimes of the donor in the presence and in the absence of the acceptor, respectively, $R_{\text{DA}}$ is the distance between the donor and the acceptor and $R_0$ is the critical distance for 50% transfer, which depends on: (i) an orientation factor $\kappa$ having an isotropic limit of 2/3, (ii) the intrinsic quantum yield $Q_{\text{D}}$ of the donor in absence of the acceptor, which is sometimes estimated as being equal to $\tau_0^{\text{D}}(\text{H}_2\text{O})/\tau_0^{\text{D}}(\text{D}_2\text{O})$, (iii) the refractive index $n$ of the medium and (iv) the overlap integral $J_{ov}$ between

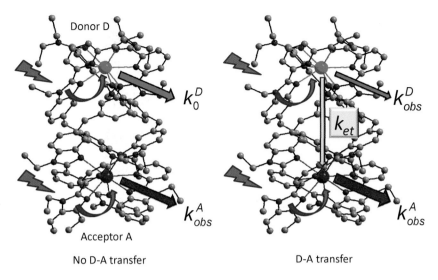

**Figure 6.6** *$Ln^{III}$ to $Ln^{III}$ energy transfer in an heterobimetallic helicate* (the crystal structure is drawn after data reported for $[LaTb(L17)_3]^{6+}$ in Ref. [64]).

the emission spectrum $E(\tilde{\nu})$ of the donor and the absorption spectrum $\varepsilon(\tilde{\nu})$ of the acceptor (Equations (6.7,6.8)):

$$R_0^6 = 8.75 \times 10^{-25} (\kappa^2 \cdot Q_D \cdot n^{-4} \cdot J_{ov}) \tag{6.7}$$

$$J_{ov} = \frac{\int \varepsilon(\tilde{\nu}) \cdot E(\tilde{\nu}) \cdot (\tilde{\nu})^{-4} d\tilde{\nu}}{\int E(\tilde{\nu}) d\tilde{\nu}} \tag{6.8}$$

In Equation 6.6, the ratio $\dfrac{\tau_{obs}^D}{\tau_0^D}$ can be substituted with the intensity ratio $\dfrac{I_{obs}^D}{I_0^D}$. An estimate of $R_0$ is therefore accessible from the experimental optical and structural properties of the system. If a crystal structure is at hand, the problem simplifies in that $R_{DA}$ is known and, if the lifetimes of Equation 6.6 can be measured, then calculation of $R_0$ is straightforward. These equations can also be used when the donor and/or the acceptor are other entities, such as organic chromophores. The methodology allows one to determine distances between chromophores and metal ion sites in biological molecules [67], particularly in metalloproteins, for instance when $Ca^{II}$ or $Zn^{II}$ are replaced with $Ln^{III}$ ions [68].

$Tb^{III}$ with an emissive level $^5D_4$ located at $20\,500\,cm^{-1}$ is well suited for transferring energy on a number of other lanthanoid ions, in particular $Eu^{III}$, the energy difference with $Eu(^5D_1)$ and $Eu(^5D_0)$ being 1500 and $3250\,cm^{-1}$, respectively, that is within an optimal range to avoid back energy transfer. An early study involved helicates with ligand L3 (Scheme 6.1) which was reacted with equimolar amounts of $Eu^{III}$ and $Tb^{III}$ in acetonitrile, yielding solutions with a mixture of $[Eu_2(L3)_3]^{6+}$, $[Tb_2(L3)_3]^{6+}$ and $[EuTb(L3)_3]^{6+}$ species in a 1/3, 1/3, 1/3 proportion, as determined by ES-MS [6]. While the luminescence decay for $Eu(^5D_0)$ is monoexponential with an associated lifetime matching the one of

**Table 6.6** Parameters for the calculation of $R_{DA}$ in $[Ln_2(L12a,b)_3]$ helicates by $Eu^{III}$ to $Nd^{III}$ energy transfer in aqueous solutions [51].

| Parameter[a] | $(L12a)^{2-}$ | $(L12b)^{2-}$ | Parameter[a] | $(L12a)^{2-}$ | $(L12b)^{2-}$ |
|---|---|---|---|---|---|
| $\tau_0^{Eu}$ (ms) | 1.48 | 1.56 | $Q_{Eu}$ | 0.40 | 0.80 |
| $\tau_{obs}^{Eu}$ (ms) | 1.26 | 0.25 | $10^{17} \times J$ ($cm^6 \times mol^{-1}$) | 1.22 | 1.22 |
| $\eta_{et}$ | 0.15 | 0.843 | $R_0$ (Å) | 9.90 | 9.85 |
| $k_{et}$ ($s^{-1}$) | 118 | 3359 | $r_{DA}$ (Å) | 13.2 | 7.4 |

[a]Estimated uncertainties: $\pm5$–10% on decay rates, $\pm12\%$ on $Q_{Eu}$ and $J$, $\pm0.8$ Å on $r_{DA}$. Note: a change in $R_0^6$ by a factor of 2 leads to $r_{DA}$ estimate changing by only 12%.

the homometallic helicate, a biexponential decay was found for $Tb^{III}$, corresponding to a long lifetime identical to the one of the homometallic complex and to a shorter one evidencing $Tb^{III}$ to $Eu^{III}$ energy transfer. In view of the peculiar properties of this system, with $E_{0-0}(T^*)$ close to $Tb(^5D_4)$, measurements have been performed on frozen acetonitrile solutions at 77 K using both direct f-f excitation and excitation in the ligand bands (data within parentheses). $Tb^{III}$ lifetimes are 0.66 and 0.16 ms (0.71 and 0.20 ms), which translates into $\eta_{et} = 76\%$ (72%) and given the Tb–Tb distance determined by X-ray crystallography, 8.876 Å, $R_0$ is calculated as being 10.7 Å (10.4 Å). The latter distance is slightly larger compared with observed range of values for $Tb^{III}$–$Eu^{III}$ pairs (6–9 Å) [30]. This stems from the special arrangement of the $Ln^{III}$ ions on the symmetry axis of the molecule, which results in little interference from ligands along the energy-transfer path.

Another study involved $Eu^{III}$ to $Nd^{III}$ energy transfer to determine intermetallic distances in $[EuNd(L12a,b)_3]$ in aqueous solution [51]. The data obtained (Table 6.6) are in reasonable agreement with model calculations based on molecular mechanics assuming ionic bonding between the metal ion and the ligands: 9.9 and 8.0 Å for helicates with $(L12a)^{2-}$ and $(L12b)^{2-}$, respectively.

More energy transfer phenomena will be discussed in Sections 6.3.2 and 6.3.3.

### 6.2.5 Lanthanoid Luminescent Bioprobes

In addition to displaying sharp, therefore easily recognizable, emission lines, analytical probes containing trivalent lanthanoid ions have the advantage of suffering little or no photobleaching and presenting long lifetimes of the metal-centred luminescence, in the milli- to microsecond range. This decisive advantage translates into the possibility of using cheap equipment for time-resolved detection of the luminescence. Both bioanalyses such as immunoassays [56,69–71] and bioimaging [56,71–75] benefit from this feature.

Since dinuclear helicates contain two luminescent ions and since they are easy to assemble in aqueous solutions under physiological conditions, it was logical to test if they would be good bioprobes. To this end, their solubility has been improved by the addition of short polyoxyethylene fragments on the benzimidazole moiety ($H_2L6$; refer to Scheme 6.2) or a pyridine group ($H_2Li$, $i = 7$–11). Several properties of the resulting helicates have to be checked before using them as bioprobes. The first is their thermodynamic stability, which has been assessed by the determination of conditional stability constants at pH 7.4 in 0.1 M Tris-HCl. In most cases, the triple-stranded helicate is the main species, with a speciation larger than 92%, except for $[Eu_2(L6)_3]$ and $[Eu_2(L9)_3]$. In view of the

**Table 6.7**  *Percentages of [Eu$_2$L$_3$] helicate in aqueous solution at pH 7.4 (Tris-HCl 0.1 M) containing a total ligand concentration of 4.5 × 10$^{-4}$ M.*

| Ligand | (L6)$^{2-}$ | (L7)$^{2-}$ | (L8)$^{2-}$ | (L9)$^{2-}$ | (L10)$^{2-}$ | (L11)$^{2-}$ |
|---|---|---|---|---|---|---|
| % Eu$_2$L$_3$ | 82.0$^a$ | 97.1 | 92.0$^b$ | 88.7 | 92.5 | 94.6 |
| Ref. | [45] | [46] | [49] | [50] | [50] | [50] |

$^a$For total ligand concentration 1 × 10$^{-4}$ M.
$^b$For total ligand concentration 1 × 10$^{-3}$ M.

data reported in Table 6.7, the development of lanthanoid bioprobes has focused on helicates with ligands (L7)$^{2-}$ and, to a lesser extent, (L11)$^{2-}$. The thermodynamic stability and kinetic inertness of the [Eu$_2$(L7)$_3$] helicate has further been assessed as a function of pH and by competitive reactions with various chelating agents (edta, dtpa), anions (citrate, L-ascorbate) and cations (Ca$^{II}$, Zn$^{II}$). It has proved to be satisfying: addition of 100 equivalent of exogenous reactants resulted in no, or at most 10%, loss in luminescence properties after 24–96 h. The time for a bioanalysis being usually <1 h, the slight dissociation observed after 24 h in some cases should not be a problem. In fact the stability of [Eu$_2$(L7)$_3$] is intermediate between that of the polyaminocarboxylate with dtpa and the well known macrocyclic [Eu(dota)]$^-$ complex [46]. The helicate is however sensitive to the addition of one equivalent of 3d metal ions, such as Mn$^{II}$, Co$^{II}$ or Cu$^{II}$, but not Fe$^{II/III}$ [76].

A second important point is the cytotoxicity of the helicates, which has been tested by incubating several cell lines in the presence of various concentrations of helicates and measuring the cell viability by WST-1 assays. These assays encompassed 5D10 mouse hybridoma, Jurkat human T leukemia, HeLa cervical adenocarcinoma, MCF-7 human breast carcinoma and nonmalignant epithelial HaCat (human keratinocytes) cells. They were conducted for [Eu$_2$(L6)$_3$] [45], [Eu$_2$(L7)$_3$] [46], [Eu$_2$(L8)$_3$] (HeLa only) [49] and [Eu$_2$(L11)$_3$] (HeLa only)[50] and evidenced no cytotoxicity up to a concentration of 500 μM. Due to its smaller solubility, [Eu$_2$(L5)$_3$] was tested only up to 0.125 μM but was also found to have no influence on cell growth and morphology [63]. With these data at hand, combined with the interesting photophysical properties described above, the helicates have been tested on three aspects: luminescent cell staining, specific targeting of cancerous cells and tissues and DNA analysis.

### 6.2.5.1  Cell Penetration and Staining

Internalization experiments have involved helicates with (L5)$^{2-}$ [63] and (L6)$^{2-}$ (Ln = Eu) [45], (L7)$^{2-}$ (Ln = Sm, Eu, Tb, Yb) [46,77] and (L11)$^{2-}$ (Ln = Eu) [50]. Only a few relevant features are described here. All tested helicates penetrate into the various cell lines mentioned above without altering their morphology, as shown in Figure 6.7 (top panel, top row): although time-resolved luminescence microscopy (TRLM) reveals that the cells contain a sizeable amount of the Eu$^{III}$ chelate (bottom panel), no swollen nuclei or granules are visible, contrary to what happens when the cells are incubated in presence of acridine orange (top panel, bottom row). From the TRLM images (bottom panel), it is obvious that the helicates are not uptaken in the nucleus of the cells. The first bright images can be seen for helicate loading concentrations >5 μM and already after 10–15 min incubation. The chelates are first uptaken in isolated vesicles which diffuse into the cytoplasm and

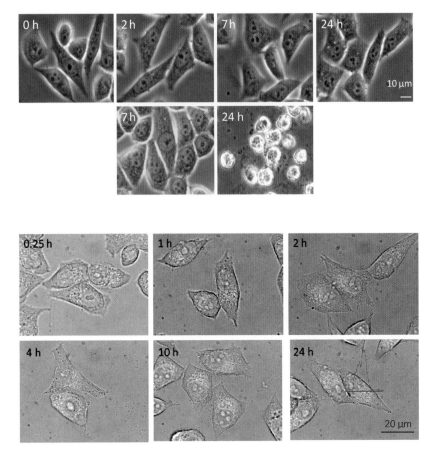

**Figure 6.7** *Top panel: phase contrast images of HeLa cells exposed to 100 μM of [Eu₂(L7)₃] (top row) or to 3.2 μM acridine orange (bottom row) for 0–24 h at 37°C. Bottom panel: time-resolved luminescence microscopy images of live HeLa cells incubated with 100 μM of [Eu₂(L7)₃] for 0.25–24 h at 37°C.* Reproduced from ref. [77] with permission, © 2008 Royal Society of Chemistry.

eventually gather around the nucleus, on one side of the cells. The vesicles have sizes ranging from 0.5 to 2.0 μm. The helicates remain in the vesicles, that is they do not diffuse into the cytoplasm or into the nucleus. Co-localization experiments with several organic dyes led to the conclusion that the majority of [Eu₂(L7)₃] stained vesicles are localized within the endoplasmic reticulum and not in the Golgi apparatus [77].

The mechanism of uptake has been identified as an endocytotic process by superposing images with the helicate and with a fluorescent organic marker for secondary endosomes and lysosomes. Further evidence was gathered by incubating the cells at low temperature, at which endocytosis does not take place, and by adding endocytosis inhibitors. The concentration of [Eu₂(L7)₃] in cells was estimated from time-resolved luminescence assays similar to luminescence immunoassays and was found to be on the order of $8 \times 10^{-16}$ mol cell$^{-1}$. The leakage of helicates out of the cells was determined with time-lapse experiments and found to be around 30% after 24 h. Furthermore, helicates with $Sm^{III}$, $Tb^{III}$ and $Yb^{III}$ behave in the same way as $[Eu_2(L7)_3]$. Since the luminescence lifetimes are different, some time-based discrimination is feasible, in addition to wavelength discrimination [77].

The helicate stains have to be excited at around 320 nm, which is a wavelength at which cell damage may be induced. Additionally, confocal microscopes usually have excitation wavelengths $\geq 405$ nm. For this reason, $H_2L11$ (Scheme 6.2) was developed, with the hope of shifting the excitation wavelength more towards the visible range. Indeed, with respect to $(L7)^{2-}$, the absorption maximum of $(L11)^{2-}$ is red-shifted by 28 nm, from 322 to 350 nm; and another, slightly weaker absorption band occurs at 370 nm (Figure 6.8, left). This results in brighter confocal luminescence microscopy images, as shown on the right side of Figure 6.8, while maintaining the same biochemical behaviour (cytotoxicity, internalization mechanism, localization, egress) as the helicate with $(L7)^{2-}$.

Another and potentially better way of shifting the excitation wavelength is to excite the lanthanoid bioprobes via two- or three-photon processes. The excitation wavelength can

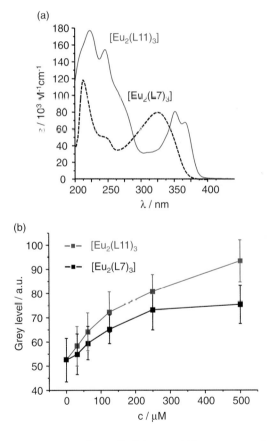

**Figure 6.8** (a) Absorption spectra of [Eu₂(Li)₃] (i = 7, 11) in aqueous solution at pH 7.4 (Tris-HCl 0.1 M). (b) Integrated luminescence intensity from confocal microscopy images of HeLa cells incubated at 37°C with various concentrations of [Eu₂(L7)₃] and [Eu₂(L11)₃]. Reprinted with permission from [50]. Copyright 2009 WILEY-VCH Verlag GmbH & Co. KGaA, Weinheim.

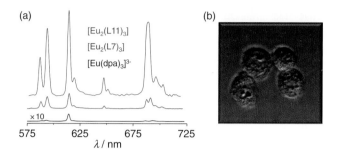

**Figure 6.9** *(a) 2PA-([Eu₂(L11)₃]) and 3PA-([Eu₂(L7)₃], [Eu(dpa)₃]³⁻)] excited emission spectra (λ_exc = 800 nm) of the helicates tested for cell staining and, as a comparison, the tris(dipicolinate) mononuclear complex. (b) Superposition of bright field and 2PA luminescence confocal microscopy images of HeLa cells incubated with [Eu₂(L11)₃] 200 μM and excited at 750 nm. Redrawn with permission from [78]. Copyright 2010 American Chemical Society.*

therefore be as long as 700–800 nm, that is in the transparency window of biological tissues. Helicates [Ln₂(L7)₃] and [Ln₂(L11)₃] (Ln = Eu, Tb) can indeed be excited in this way, the former by a three-photon process (3PA) and the latter by two-photon absorption (2PA) [78]. Luminescence spectra for the Eu$^{III}$ helicates obtained upon excitation at 800 nm are shown in Figure 6.9 (left). The multiphoton cross-sections remain modest (Table 6.8), but they are much larger than for the reference mononuclear tris(dipicolinate) complexes. Note that units of 2PA and 3PA cross-sections are not the same and therefore the corresponding data cannot be directly compared, so that there is no contradiction in the fact that the emission spectrum of [Eu₂(L11)₃] shown in Figure 6.9 has a much larger intensity compared to emission from [Eu₂(L7)₃] despite a seemingly threefold increase only in multiphoton cross-section! It is not known why [Eu₂(L11)₃] does not present 3PA absorption, which would correspond to a spectral range (267 nm) where its molar absorption coefficient is larger than for [Eu₂(L7)₃]. Further, instrumentation limitations prevented the determination of $\sigma^{(2)}$ at a wavelength of 760 nm corresponding to the

**Table 6.8** *Multiphoton absorption cross sections for mono- and di-nuclear complexes in aqueous solutions at pH 7.4 (Tris-HCl 0.1 M) determined upon irradiation at 800 nm[78].*

| Compound | Cs₃[Eu(dpa)₃] | Cs₃[Tb(dpa)₃] | [Eu₂(L7)₃] | [Tb₂(L7)₃] | [Eu₂(L11)₃] |
|---|---|---|---|---|---|
| $\lambda_{abs}^{max}$ (nm) | 271 | 271 | 322 | 322 | 350 |
| $\varepsilon_{abs}^{max}$ (M⁻¹ cm⁻¹) | 13 200 | 13 200 | 77 300 | 77 300 | 80 300 |
| $Q_{TPA}{}^{a}$ | 0.29 | 0.21 | 0.21 | 0.11 | 0.09 |
| $\lambda_{abs}^{eff}$ (nm)$^{b}$ | 267 | 267 | 267 | 267 | 400 |
| $\varepsilon_{abs}^{eff}$ (M⁻¹ cm⁻¹) | 9500 | 9500 | 42 700 | 42 700 | 920 |
| Process | 3PA | 3PA | 3PA | 3PA | 2PA |
| $\sigma^{(n)}$ (GM) | 0.011 | 0.014 | 0.26 | 0.28 | 0.75 |

$^{a}$Quantum yield upon one-photon excitation.
$^{b}$(n = 2 or 3).

maximum of the one-photon excitation spectrum (375 nm); extrapolation based on molar absorption coefficients yields $\sigma^{(2)}$ (760 nm) = 9 GM, which is sizeable, but still much smaller than the cross-sections reported for mononuclear complexes with ligands optimized for multiphoton absorption [79]. Despite this low performance, HeLa cells have been incubated in the presence of 200 μM [Eu$_2$(L11)$_3$] and their image recorded on a luminescence multiphoton confocal microscope upon excitation at 750 nm; internalized helicates are quite visible, which opens interesting perspectives (Figure 6.9, right) [78].

All these experiments show that dinuclear lanthanoid helicates qualify as efficient and versatile luminescent stains for imaging live cells by time-resolved luminescence microscopy, a technique which produces highly contrasted images in view of the elimination of the fluorescence background by time discrimination. However, the staining is nonspecific, the helicates being uptaken by several cell lines, malignant or not. A more targeted approach is needed for the specific detection of cancerous cells, which is described in the next section.

### 6.2.5.2 Specific Bioimaging of Breast Cancer Cells and Tissues

Attaining specificity in the detection of a given cell line by simple chemical probes is difficult, even almost impossible. A much easier way is to make use of biochemical reactants which target specific markers expressed by the cells and to attach a luminescent tag onto these reactants, usually monoclonal antibodies. To this end, helicates have to be modified for coupling with the biochemical molecules [75]. The coupling is achieved either directly, that is the lanthanoid probe is linked to a monoclonal antibody mAb, or indirectly with the chelate being covalently bound first to avidin (or biotin); the resulting duplex is then fixed onto a biotinylated (or avidin derivatized) monoclonal antibody, B-mAb, via the strong avidin–biotin interaction (log $K \approx 10^{15}$) [80], as described in Figure 6.10. Alternatively, avidin may be substituted by streptavidin [81] or bovine serum albumin (BSA) [82].

A relatively easy way to achieve bioconjugation is to graft a carboxylic acid group onto the ligand, for example H$_4$L20 (Scheme 6.4), which will then react with amine groups of the biochemical molecule. This has been done, yielding [Ln(H$_2$L20)$_3$] avidin conjugates (LnL20-A). MALDI-TOF spectra and a quantitative UV-vis analysis point to the binding being covalent, with an average of 3.2 helicates bound per avidin molecule. The properties of the Eu$^{III}$ helicate and bioconjugate have been checked. The addition of carboxylic acid groups leads to a decrease in quantum yield to 15 ± 2% for [Eu(H$_2$L20)$_3$] and 9.3 ± 0.9% for EuL20-A, as compared with 21 ± 2% for [Eu$_2$(L7)$_3$]. But all other photophysical properties are maintained, in particular the Eu($^5$D$_0$) lifetime (2.45 ± 0.04 and 2.17 ± 0.01 ms, respectively) and the overall shape of the emission spectra. Most important too, the biochemical activity of avidin is unchanged. Finally, bioaffinity assays demonstrate that LnL20-A linked to a specific monoclonal antibody can target the 5D10 antigen expressed by MCF-7 human breast cancer cells and that it performs as well as a commercially available lanthanoid tag and much better than an all-organic conjugate [83].

The method is particularly effective in the detection of cultured cells grown in microchannels, a way of achieving screening tests more rapidly (cells grow more rapidly in fibronectin-coated microchannels than in a bulk culture medium) and in a more cost-

(a)

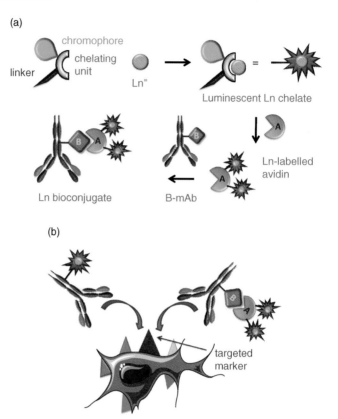

**Figure 6.10** *(a) Synthesis of a monoclonal antibody decorated with luminescent lanthanoid chelates. (b) Two ways of specifically detecting a biomarker expressed by a cell.* Redrawn with permission from [83]. Copyright 2010 American Chemical Society.

effective way, since only very small volumes of reactants and expensive lanthanoid-decorated bioconjugates are needed. An interesting development of this technology has been the simultaneous immunohistochemical detection of two biomarkers expressed by cells in human breast cancer tissue sections. These markers are oestrogen receptors (ER)

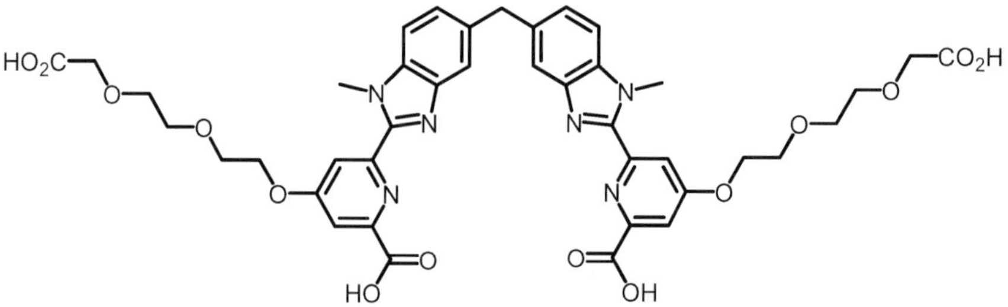

H₄L20

**Scheme 6.4** *Ligand used for conjugating helicates to avidin or monoclonal antibodies.*

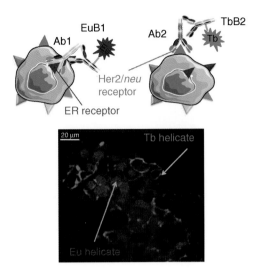

**Figure 6.11** *Top: principle of the dual immunohistochemical detection of biomarkers expressed by human breast cancer cells. Ab1 = anti-human ER mouse mAb. Ab2 = polyclonal rabbit anti-human c-erbB-2 oncoprotein Ab. EuB1 and TbB2: Bottom: Luminescent assays of ER and Her2/neu receptors expressed in a cancerous breast tissue section.* Redrawn with permission from [83]. Copyright 2010 The Royal Society of Chemistry.

expressed by the nucleus and human epidermal growth factor receptors (Her2/*neu*) expressed by the cell membrane. The relative concentrations of the two receptors are relevant for deciding which therapy has to be applied to the patient. For the analysis, 4-μm sections of formalin-fixed breast carcinomas are pressed against the microchannel and the two markers are visualized by a dual indirect immunohistochemical luminescent assay under time-resolved conditions, as illustrated in Figure 6.11. The two bioconjugates used are a goat anti-mouse IgG antibody labelled with [Eu$_2$(H$_2$L20)$_3$] (EuB1) and a goat anti-rabbit IgG antibody labelled with [Tb$_2$(H$_2$L20)$_3$] (TbB2). The first and simplest microfluidic devices designed allowed the test to be completed in 20 min versus >2 h in the hospital and using only one-fifth of the expensive reactants [83,84].

Another approach to selective immunohistochemical assays is to incorporate the luminescent helicates into silica nanoparticles (NP) decorated with linking groups. The photophysical properties of monodispersed [Eu$_2$(L7)$_3$]@SiO$_2$ and [Eu$_2$(L7)$_3$]@SiO$_2$/NH$_2$ NPs are reported in Table 6.1. Nanoparticles with NH$_2$-decorated silica contain an average of 2380 molecules of helicates per NP and can be conjugated to avidin or to goat anti-mouse IgG polyclonal antibody. The latter bioconjugate detects biotinylated 5D10 monoclonal antibodies expressed by MCF-7 breast cancer cells and bright images of these cells are obtained by time-resolved luminescence microscopy [48].

*6.2.5.3 DNA Analysis*

The emission intensity and lifetime of the [Eu$_2$(L5)$_3$] helicate are severely quenched by molecules such as ethidium bromide (EB) and acridine orange (AO), which are selective fluorescent stains for nucleic acids. When nucleic acids such as DNA are added to a quenched solution of the $Eu^{III}$ helicate containing EB or AO, the organic stain intercalates between the DNA strands and the initial $Eu^{III}$ luminescence is restored.

The quenching of a luminophore is collisional and/or static in nature and is modelled by Stern–Volmer theory. For collisional quenching resulting from particle diffusion in solution (Equation 6.9):

$$\frac{I_0}{I} = 1 + k_q \tau_0 [Q] = 1 + K_D [Q] \tag{6.9}$$

in which $I_0$ and $I$ are the emission intensities in absence and in presence of the quencher $Q$, respectively, $k_q$ is the bimolecular rate constant, $\tau_0$ the observed lifetime in absence of quencher and $K_D$ the dynamic quenching constant. Intuitively, $1/K_D$ is the quencher concentration for which 50% of the luminescence intensity is lost. When collisional quenching occurs, the lifetime decreases in parallel to the luminescence intensity (Equation 6.10):

$$\frac{I_0}{I} = \frac{\tau_0}{\tau} \tag{6.10}$$

Static quenching results when the luminophore and the quencher form a nonluminescent ground-state complex with association constant $K_S$ (Equation 6.11):

$$\frac{I_0}{I} = 1 + K_S [Q] \tag{6.11}$$

In this case, the lifetime of the emitting species remains unchanged since the effect of the process is simply to remove a fraction of luminescent molecules from observation. Therefore $\tau_0/\tau = 1$. When both dynamic and static quenching occurs, the equations combine into Equation 6.12:

$$\frac{I_0}{I} - 1 = (K_D + K_S)[Q] + K_D \cdot K_S [Q]^2 \tag{6.12}$$

In this case, lifetime dependence follows Equation 6.9.

Two situations are illustrated in Figure 6.12. On the left side, both intensities and lifetimes vary in the same way with increasing concentration of AO, pointing to a pure dynamic quenching, with $K_D = 6.7(1) \times 10^5 \, M^{-1}$ and $k_q = 2.7(1) \times 10^8 \, M^{-1} s^{-1}$. On the right side, $I_0/I$ clearly deviates from linearity with an upward curvature while the $\tau_0/\tau$ plot is linear; this is diagnostic for EB inducing simultaneous dynamic and static quenching, with associated constants $K_D = 3.03(5) \times 10^4 \, M^{-1}$, $k_q = 1.23(4) \times 10^7 \, M^{-1} s^{-1}$ and $K_S = 2.0(1) \times 10^3 \, M^{-1}$.

The analytical method has been tested for five different types of DNA and limits of detection are reported in Table 6.9. The method is rugged in that it is insensitive to pH in the range 3–10 and results are affected by less than $\pm 5\%$ upon the addition of a 1000-fold excess of potentially interfering substances such as glucose, sodium hydrogenophosphate, edta, $Ca^{II}$, $Mg^{II}$, $Zn^{II}$, $Fe^{II}$, $Fe^{III}$, citrate and urate. In contrast, the addition of bovine serum

**Table 6.9** *Limits of detection (LOD) for quantifying various types of DNA by the AO/[Eu$_2$(L5)$_3$] method [76].*

| Sample | Actin sense-2 DNA | Plasmid DNA | λDNA/HindIII | Sheared salmon sperm DNA |
|---|---|---|---|---|
| LOD (ng μl$^{-1}$) | 0.32 | 0.18 | 0.57 | 0.66 |

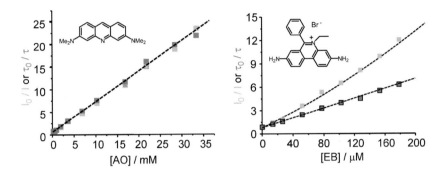

**Figure 6.12** *Quenching of [Eu$_2$(L5)$_3$] 5 μM in water (pH 7.4, Tris-HCl) by acridine orange (left; AO) and ethidium bromide (right; EB).* Reproduced with permission from [76]. Copyright 2008 The Royal Society of Chemistry.

albumin, sodium dodecylsulfate, Mn$^{II}$ and ascorbate alters the results by ±5–10% while Cu$^{II}$ and Co$^{II}$ have a much larger influence. But the concentrations of exogenous substances used (0.1 mM) are far larger than those found in biological systems. As a matter of fact, quantification of DNA in cell lysate results in an excellent linearity of the emission intensity versus the number of cells in the range 500–10 000 cells. Finally, the method is also applicable to PCR products with small numbers of base pairs: comparisons with conventional methods based on UV-vis absorption or fluorescence of PicoGreen© dye give the same results [76].

### 6.2.6 Other Investigated Helicates

#### 6.2.6.1 Dinuclear Triple-Stranded Helicates

The pyridine-2-6-dicarboxamide based ligand L21 (Scheme 6.5) reacts with Ln(CF$_3$SO$_3$)$_3$ (Ln = Sm, Eu, Tb, Lu) in acetonitrile to give rise to triple-stranded helicates [Ln$_2$(L21)$_3$] with a speciation of about 84% at stoichiometric ratio and total ligand concentration of $10^{-5}$ M, the other major species being a 2 : 2 species. Spectrophotometric absorption and luminescence data allowed the determination of stability constants for both species, which are large, in the range 19–22 for log $\beta_{22}$ and 26–29 for log $\beta_{23}$. The Eu($^5$D$_0$) decay is monoexponential when the concentration of the 2 : 2 species is low and the corresponding lifetime amounts to 1.57 ms. Monitoring Eu luminescence also led to the conclusion that the formation of the helicate is fast, since the emission intensity does not change after only a few seconds [85].

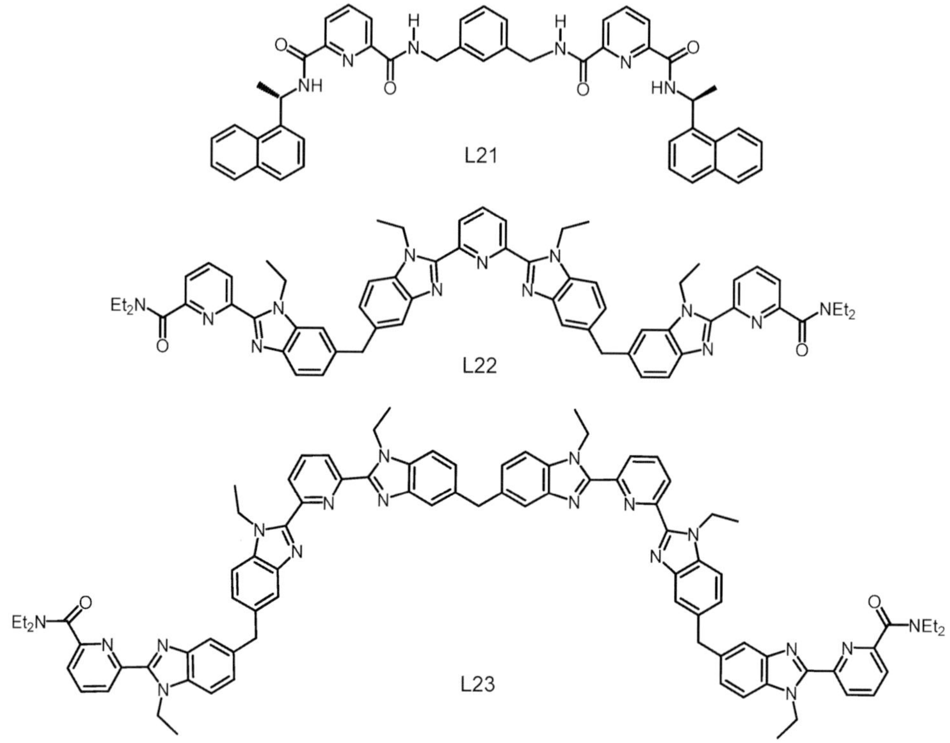

L21

L22

L23

**Scheme 6.5** *Ligands for polynuclear helicates described in Section 6.2.6.*

### 6.2.6.2 Tri- and Tetranuclear Triple-Stranded Helicates

The tritopic ligand L22 self-assembles with lanthanoid triflates in acetonitrile to produce triple-stranded helicates $[Ln_2(L22)_3]^{6+}$ and $[Ln_3(L22)_3]^{9+}$ (Ln = La, Nd, Eu, Tb, Ho, Tm, Lu) with stability constants which do not vary along the series, $\log \beta_{23} \approx 26$ and $\log \beta_{33} \approx 35$ [86]. In the trinuclear helicates, the metal ions are embedded into two different coordination sites, the two end $N_6O_3$ cavities and the central $N_9$ cavity. This is perfectly reflected in the $^5D_0 \leftarrow {}^7F_0$ excitation spectra reported in Figure 6.13. They have been measured under selective analysis of the high-energy components of the $^5D_0 \rightarrow {}^7F_1$ transitions which differ sufficiently in energy, as shown by the corresponding emission spectra recorded under selective excitation of the $^5D_0 \rightarrow {}^7F_0$ transitions. To avoid as much as possible energy transfer between the sites, the excitation wavelength has not been set exactly on the maxima of the two components. At room temperature the energies of the two $^5D_0 \rightarrow {}^7F_0$ transitions amount to 17 239 and 17 225 cm$^{-1}$, in reasonable agreement with the observed values for $[Eu_2(L3)_3]^{6+}$, 17 236 cm$^{-1}$ and for $[Eu_2(L4)_3]^{6+}$, 17 232 cm$^{-1}$ (cf. Table 6.1). That is, the higher energy value corresponds to the $N_9$ chemical environment, a fact substantiated by the much weaker emission from this site due to LMCT quenching. Analysis of the ligand–field splitting of the $^7F_1$ level in terms of pseudo-$D_3$ symmetry at room temperature and 10 K (data in parentheses) gives $\Delta E(A_2 - E) = 100$ (75) and 123 (134) cm$^{-1}$ as well as $\Delta E(E - E) = 31$ (42) and 38 (38) cm$^{-1}$ for the $N_6O_3$ and $N_9$ sites,

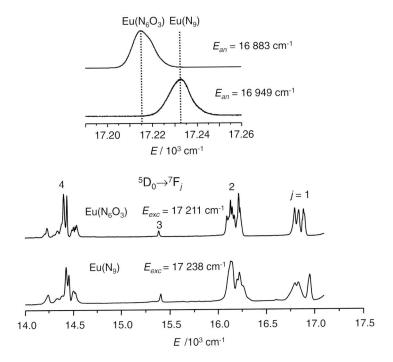

**Figure 6.13** *Top: excitation profiles of the $^5D_0 \leftarrow {}^7F_0$ transition of $[Eu_3(L22)_3]^{9+}$ obtained by selectively monitoring components of the $^5D_0 \rightarrow {}^7F_1$ transition. Bottom: corresponding emission spectra under selective excitation of the $^5D_0 \leftarrow {}^7F_0$ transitions. Solid state sample at 10 K.* Redrawn with permission from [86]. Copyright 2003 WILEY-VCH Verlag GmbH & Co. KGaA, Weinheim.

respectively, evidencing a somewhat larger distortion from the idealized $D_3$ symmetry for the latter site.

The overall quantum yield in acetonitrile amounts to only 0.3%, that is about 30 times less than the one of the reference helicate $[Eu_2(L4)_3]^{6+}$ with $N_6O_3$ coordination sites (emission from the central site contributes little to the overall luminescence). The $^5D_0(N_6O_3)$ and $^5D_0(N_9)$ lifetimes are the same at 10 K, 2.3(1) and 2.2(1) ms and decrease to 1.7 ms at room temperature. Interestingly the temperature dependence of $\tau(N_9)$ is less pronounced than for $[Eu_2(L3)_3]^{6+}$, suggesting that the slide of the ligand strands around the $N_9$ site shifts the LMCT at slightly higher energy. However, the much lower than expected quantum yield probably reflects an intramolecular quenching of the $Eu(N_6O_3)$ luminescence from this state located on the neighbouring $EuN_9$ moiety [86].

Heterometallic 4f-4f′-4f helicates have also been obtained in solution and the affinity of the two sites for a range of $Ln^{III}$ ions (Ln = La, Nd, Sm, Eu, Yb, Lu, Y) has been determined. In particular a La-Eu compound could be isolated and crystallized (Figure 6.14). One important question was to determine if the crystals were homogeneous or contained different species. Chemical elemental analysis corresponds to $[La_{0.93}Eu_{2.07}(L22)_3]$ $(CF_3SO_3)_9$, while the formula that can be deduced from single-crystal X-ray crystallography is $[La_{0.96}Eu_{2.04}(L22)_3](CF_3SO_3)_9$. Refinement of the diffraction data leads to the

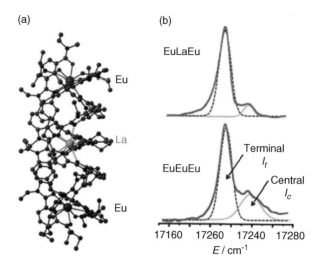

**Figure 6.14** (a) Molecular structure of the [EuLaEu(L22)₃]⁹⁺ helicate; (b) Eu(⁵D₀ ← ⁷F₀) excitation spectra at 15 K. Redrawn with permission from [10]. Copyright 2004 WILEY-VCH Verlag GmbH & Co. KGaA, Weinheim.

conclusion that scrambling among the sites occurs and that the populations are 0.74 : 0.26 (La:Eu) for the central sites and about 0.11 : 0.89 for the end sites. In solution, NMR spectroscopy points to the presence of three main species, EuLaEu (48%), EuEuEu (23%) and LaLaEu (19%), which explains Eu occupancy of the central site and La occupancy of about 10% in each of the terminal sites in the crystals.

To get more insight into this problem, the Eu(⁵D₀ → ⁷F₀) transition has been analysed in detail for both the homometallic EuEuEu and the heterometallic EuLaEu helicates (Figure 6.14). Gaussian decomposition allowed the authors to determine the relative emission efficiency ρ of the two sites in EuEuEu (Equation 6.13):

$$\frac{I_c}{I_c + I_t} = 0.27 = \frac{\rho_c}{\rho_c + 2\rho_t} \quad \text{therefore} \quad \frac{\rho_c}{\rho_t} = 0.74 \tag{6.13}$$

For the heterometallic helicate, $I_c/(I_c + I_t) = 0.094$. Taking the La population equal to 10% in each terminal site and setting α as being equal to the population of EuEuEu, one gets Equation 6.14:

$$\frac{I_c}{I_c + I_t} = \frac{0.8\alpha\rho_c}{0.8\alpha\rho_c + 1.8\rho_t} \tag{6.14}$$

which solves for $\alpha = 0.32$. From this, a population of $0.8\alpha = 0.26$ is calculated for the Eu occupancy of the central site corresponding to [La₀.₉₄Eu₂.₀₆(L22)₃]⁹⁺, a formula compatible with elemental analysis and X-ray diffraction data. This means that crystals are a mixture of 54% EuLaEu, 20% EuEuEu and 26% LaLaEu, matching very well the solution speciation [10].

The bis(benzimidazole) framework can be expanded to feature four tridentate coordination sites as in L23 (Scheme 6.5). The ligand self-assembles in acetonitrile with lanthanoid triflates to give triple-stranded tetranuclear helicates $[Ln_4(L23)_3](CF_3SO_3)_{12}$. Similarly to the trinuclear helicate described above, microscopic stability constants have been determined for $Eu^{III}$, leading to a statistical analysis of the stepwise formation of the helical structure and to the determination of the affinity of the sites for the metal ions. The overall structure can be looked at as a pseudo-$D_3$ edifice, an hypothesis which is substantiated by the $Eu(^5D_0)$ emission spectrum: from the fine structure of the $^5D_0 \rightarrow {}^7F_1$ transition, one infers that the $^7F_1$ level splits into two main components, labelled $A_2$ and E in $D_3$ symmetry, the latter further displaying two subcomponents. At 10 K, $\Delta E(A_2 - E) = 75$ cm$^{-1}$ and $\Delta E(E - E) = 42$ cm$^{-1}$ for the $N_6O_3$ terminal sites; these data are identical to the ones for the trinuclear helicate at 10 K. High-resolution excitation spectra of the $^5D_0 \leftarrow {}^7F_0$ transition reveal a main component at 17 226 cm$^{-1}$ ($N_6O_3$) at room temperature, with an energy virtually equal to the one of the corresponding site in the trinuclear compound $[Eu_3(L22)_3]^{9+}$ (17 225 cm$^{-1}$). The component from the $N_9$ central sites is more difficult to detect at room temperature; it occurs at 17 235 cm$^{-1}$ at 10 K, which compares well with 17 238 cm$^{-1}$ for $[Eu_3(L22)_3]^{9+}$ at this temperature. Estimation of this energy at room temperature and solving equations for the two sites leads to the conclusion that the nephelauxetic effect generated by the pyridine groups is much larger than the one stemming from the benzimidazole moieties [87].

## 6.3 Heterometallic d-f Helicates

The work performed on helicates containing f- and d-transition metal ions has been initiated for several reasons: (i) facilitating the self-assembly process by prearranging the ligand with the help of the sterically-demanding d-transition metal ion ($Cr^{II}$, $Fe^{II}$, $Co^{II}$, $Zn^{II}$, $Ru^{II}$, $Os^{II}$; $Cr^{III}$) so that the *HHH* species is privileged, (ii) isolating enantiomerically pure helicates with $Cr^{III}$, (iii) controlling the $Fe^{II}$ spin-crossover parameters and (iv) tuning the photophysical properties of the $Ln^{III}$ ions by the d-transition metal partner. To these ends, pentadentate ditopic ligands were prepared (Scheme 6.6). The self-assembly of the *HHH*-[MLnL$_3$] helicates proceeds smoothly and with a very high thermodynamic control. Structural aspects in solid state (X-ray crystallography) and solution (LIS NMR), as well as paramagnetic properties have been thoroughly investigated [38]. Here, we focus on photophysical properties only.

### 6.3.1 Basic Photophysical Properties

The ligand-centred photophysical properties of the helicates with a spectroscopically silent nd-transition partner are comparable to those of the previously discussed homobimetallic helicates. When a nd-transition ion amenable to form MLCT states with the ligand strands is introduced into the helical edifices, specific absorption and, possibly, emission bands from these states are also observed, as well as d-d transitions. Selected data are collected in Table 6.10. Lanthanoid complexation has small influence on the ligand levels when M = Zn but a more pronounced one for the other transition metal ions.

In order to confirm the structural data obtained in solid state and/or solution, high-resolution excitation spectra of the $Eu(^5D_0 \leftarrow {}^7F_0)$ transition and emission spectra of the Eu $(^5D_0 \rightarrow {}^7F_1)$ transition have been recorded. Relevant data are collected in Table 6.11. The energies of the $^5D_0 \leftarrow {}^7F_0$ transition calculated with Equation 6.4 and the parameters

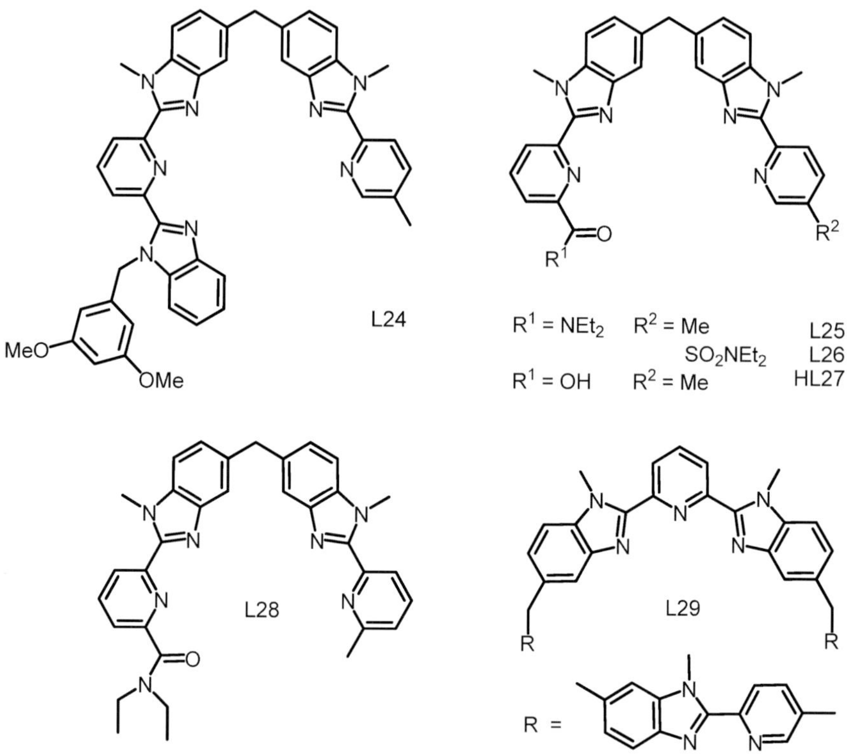

**Scheme 6.6** *Ligand for assembling bimetallic 4f-nd helicates.*

listed in Table 6.4 amount to 17 236, 17 235 and 17 231 cm$^{-1}$ for N$_9$, N$_6$O$_3$(amide) and N$_6$O$_3$(carboxylate) environments, respectively. Experimental data are in good agreement for L24 but somewhat lower than predicted for L25 and L26 (17 224–17 229 cm$^{-1}$), possibly due to the large spin delocalization evidenced by NMR and causing a larger nephelauxetic effect. In contrast, the experimental nephelauxetic effect for HL27 is 5 cm$^{-1}$ smaller than the calculated one.

All $^5D_0 \rightarrow {}^7F_J$ transitions can be analyzed on the basis of a distorted $D_3$ symmetry as seen from the splitting of the $^7F_1$ level into two sublevels labelled A$_2$ and E. There are, however large differences between the various helicates and, also, between solid state and solution samples, particularly in the case of L26. Looking at the $\Delta E(A_2 - E)$ energy difference for the Zn$^{II}$ helicates, which is directly proportional to the $B_0^2$ ligand–field parameter [23], the strength of the ligand field induced by the various ligands at low temperature increases in the series L45 (93 cm$^{-1}$) < L26 (118) < L25 (127) < HL27 (138) < L28 (146). This can be understood when considering the inner sphere composition: a N$_9$ environment generates a weaker field than a N$_6$O$_3$ one. The weaker field induced by L26 with respect to L25 arises from the less distorted coordination polyhedron in [EuZn(L26)$_3$]$^{5+}$ compared with [EuZn(L25)$_3$]$^{5+}$ while the largest field observed for the helicate with L28, with respect to the carboxylic acid HL27, results from the weaker coordination of the 3d transition metal, allowing a tighter wrapping of the ligand strands around the lanthanoid ion.

**Table 6.10** Ligand-centered photophysical properties of selected $[MLn(L25)_3]^{n+}$ helicates.

| Ln | M | State | $T(K)^a$ | $E(^*\pi\leftarrow\pi)$ (cm$^{-1}$) | $E(MLCT) + E(d\text{-}d)$ (cm$^{-1}$) | $E(^1\pi\pi^*)$ (cm$^{-1}$) | $E(^3\pi\pi^*)$ (cm$^{-1}$) | Ref. |
|---|---|---|---|---|---|---|---|---|
| La | Zn$^{II}$ | Solid | 77 | 29 600, 26 200 | | 22 600 | 19 700, 18 900 | [8] |
| Lu | Zn$^{II}$ | solid | 77 | 30 300, 25 650 | | 22 200 | 19 700, 18 900 | |
| La | Fe$^{II}$ | CHCl$_3$ | 293 | 30 550, 26 880 | 18 800 | 23 000$^c$ | 19 000$^b$ | [88] |
| Eu | Fe$^{II}$ | CHCl$_3$ | 293 | 30 270, 27 020 | 18 870 | 23 000$^c$ | 19 000$^b$ | [61] |
| La | Zn$^{II}$ | Solid | 77 | 31 000 | | 22 600 | 19 960, 19 050 | |
| Gd | Zn$^{II}$ | Solid | 77 | 31 250 | | 22 600 | 19 960, 19 050 | |
| La | Fe$^{II}$ | MeCN | 293 | 29 940, 28 570 | 19 050 | 22 220$^c$ | 18 520$^b$ | [89] |
| Lu | Fe$^{II}$ | MeCN | 293 | 29 940, 28 570 | 19 080 | 22 220$^c$ | 18 520$^b$ | |
| La | Cr$^{II}$ | MeCN | 293 | 40 486, 30 121 | 23 640, 16 475, 14 368, 11 710, 9090, 8368 | | | [90] |
| La | Cr$^{III}$ | MeCN | 293 | 40 486, 30 030 | 25 700, 22 800, 21 600, 20 000 | | | |
| Lu | Cr$^{III}$ | MeCN | 293 | 40 486, 29 940 | 25 700, 22 800, 21 600, 20 000 | | | |
| Lu | Ru$^{II}$ | MeCN | 296 | 39 525, 29 940 | 21 100 | $^e$ | 14 730$^c$ | [91] |
| Gd | Ru$^{II}$ | MeCN | 293 | 40 160, 30 120 | 21 185 | 25 800$^d$ | 24 300, 20 700$^d$ | [92] |
| Lu | Os$^{II}$ | MeCN | 293 | 29 850 | 19 920, 14 500 | $^e$ | 12 400$^c$ | [93] |

$^a$Except for $^*\pi\leftarrow\pi$ transitions of solid-state samples, recorded by reflectance spectroscopy at room temperature.
$^b$Very weak signal.
$^c$From the $^3$MLCT state.
$^d$At 77 K.

***Table 6.11*** *High-resolution analysis of the Eu($^5D_0 \leftarrow {}^7F_0$) and Eu($^5D_0 \rightarrow {}^7F_1$) transitions for [MLnL$_3$]$^{n+}$ and [MEuML$_3$]$^{n+}$ helicates.*

| Ligand | M | State | T/K | $^5D_0 \rightarrow {}^7F_0$ E (cm$^{-1}$) | $^5D_0 \rightarrow {}^7F_1$ | | Ref. |
|--------|---|-------|-----|------|------|------|------|
| | | | | | $\Delta E(A_2$-E) (cm$^{-1}$) | $\Delta E$(E-E) (cm$^{-1}$) | |
| L24 | Zn$^{II}$ | MeCN | 10 | 17 224 | 94 | 43 | [8] |
| | Zn$^{II}$ | MeCN | 295 | 12 236 | n.a. | n.a. | |
| L25 | Zn$^{II}$ | Solid | 10 | 17 220 | 127 | 21 | [61] |
| | Zn$^{II}$ | Solid | 295 | 17 229 | 140 | 55 | |
| | Zn$^{IIa}$ | Solid | 295 | 17 226 | 134 | 63 | |
| | Cr$^{III}$ | Solid | 10 | 17 216 | 100 | 48 | [90] |
| HL27 | Zn$^{II}$ | Solid | 10 | 17 224 | 138 | 42 | [94] |
| | Zn$^{II}$ | Solid | 295 | 17 235 | 145 | 35 | |
| | Zn$^{II}$ | MeCN | 295 | 17 237 | 149 | n.a. | |
| L26 | Zn$^{II}$ | Solid | 10 | 17 221 | 118 | 37 | [95] |
| | Zn$^{II}$ | Solid | 295 | 17 225 | 117 | n.a. | |
| | Zn$^{II}$ | MeCN | 295 | 17 224 | 82 | 49 | |
| L28 | Zn$^{II}$ | Solid | 13 | 17 221 | 146 | 35 | [96] |
| | Zn$^{II}$ | MeCN | 295 | 17 232 | 147 | n.a. | |
| | Fe$^{II}$ | Solid | 13 | 17 221 | 144 | 37 | |
| L29 | Zn$^{II}$Zn$^{II}$ | Solid | 10 | 17 221 | 98 | 32 | [11] |
| | Cr$^{III}$Cr$^{III}$ | Solid | 10 | 17 218 | 80 | 34 | |

[a]Eu-doped (2%) [GdZn(L25)$_3$]$^{5+}$.

Quantum yields have only been determined for Eu$^{III}$ helicates (Table 6.12). They span a wide range, from 0.01% in [EuZn(L24)$_3$]$^{5+}$, because the N$_9$ environment made up of benzimidazolepyridine units is known to generate a rather low-lying and quenching LMCT state (see above), to a high 32% for the sparingly soluble carboxylate [EuZn(L27)$_3$]$^{2+}$. The robustness of the triple helical edifices is exemplified by the fact that adding up to 0.93 M water to [EuZn(L25)$_3$]$^{5+}$ does not alter either the $^5D_0$ lifetime or the quantum yield [61]. Similarly, the quantum yield of [EuCr (L25)$_3$]$^{5+}$ in acetonitrile (3.2%) remains unchanged up to 3 M of added water [90]. If the same experiment is conducted on the even more robust [EuZn(L27)$_3$]$^{2+}$ helicate, the quantum yield drops to 87% of its initial value when 2 M of water are added, probably in view of the second sphere effect of fast diffusing O–H vibrators and then further decreases slowly to reach about 80% of its initial value for a water concentration of 10 M. The quantum yield in pure water is about half that in acetonitrile but the lifetime is still long, 2.43 ms. With $\tau$(D$_2$O) = 4.48 ms, one calculates a hydration number $q \approx 0$, using the equation of Supkowski and Horrocks (Equation 6.15) [97], a remarkable result in that, even in pure water, there is no inner sphere interaction with the solvent [94].

$$q = 1.11 \times \left( \frac{1}{\tau(H_2O)} - \frac{1}{\tau(D_2O)} - 0.31 \right) \tag{6.15}$$

## 6.3.2 Eu$^{III}$-to-Cr$^{III}$ Energy Transfer

When one metal ion imbedded in the helicate plays the role of a donor for sensitizing the emission of a second accepting metal ion, the characteristic lifetimes of the excited state is affected by the intermetallic communication process according to Equations (6.5)–(6.8) given in Section 6.2.4. The chromium helicate [EuCr$^{III}$(L25)$_3$]$^{6+}$ displays both intense Cr($^2$E $\rightarrow$ $^4$A$_2$) and Eu($^5$D$_0$ $\rightarrow$ $^7$F$_j$) emission at 10 K. The Eu($^5$D$_0$) lifetime is considerably shorter than in [EuZn(L25)$_3$]$^{5+}$, which is assigned to unidirectional Eu $\rightarrow$ Cr transfer along the $C_3$ axis [90]. Determination of $\eta_{et}$(CrEu) for two types of crystalline samples (anhydrous and hydrated, see Table 6.13) provides similar efficiencies for this transfer, in the range 65–78%. The efficiency for the anhydrous sample is slightly temperature-dependent while that for the hydrated sample is independent of temperature. The energy transfer yield is also insensitive to the nature of the sample, as shown by the same calculation carried out for 10$^{-4}$ M solutions in acetonitrile using either quantum yields (see Table 6.12), $\eta_{et}$(CrEu) = 66%, or lifetimes (65%), a proof of the dipole–dipolar mechanism operating in these fairly rigid triple-helical structures. The calculated $R_0$ distance is $\approx$10.3 Å. In the case of Tb$^{III}$, the transfer is quantitative, no $^5$D$_4$ luminescence being observed even upon direct Tb$^{III}$ excitation, because of the near resonance between the Tb($^5$D$_4$) and Cr($^4$T$_2$) electronic levels. It is also noteworthy that the Cr$^{III}$ luminescence is heavily quenched at room temperature, as shown with the GdCr helicate for which the Cr($^2$E) lifetime dramatically decreases between 10 and 295 K.

Furthermore, divergent Cr $\leftarrow$ Ln $\rightarrow$ Cr intramolecular axial energy transfers could be evidenced in [Cr$_2$Ln(L29)$_3$](CF$_3$SO$_3$)$_9$ with temperature independent global yields of 90% (Eu) and >99.9% (Tb) [11]. These two situations are summarized in Figure 6.15.

**Table 6.12** *Quantum yields and lifetimes of [MLnL$_3$]$^{n+}$ (Ln = Eu, Tb) determined upon ligand excitation.*

| L | Ln | M | $\tau(^5D_j)$ (ms)$^a$ | $c$ (M)$^b$ | $Q_{Ln}^L$ (%)$^c$ | Refs. |
|---|---|---|---|---|---|---|
| L24 | Eu | Zn | 2.30(5) | 10$^{-4}$ | $\approx$0.01 | [8,61] |
| | Tb | Zn | 1.17(4) | | n.a. | [8] |
| L25 | Eu | Zn | 2.56(10) | 10$^{-3}$ | 4.2 | [61] |
| | | | | 10$^{-4}$ | 9.3 | [61] |
| | Tb | Zn | 1.89(6) | | n.a. | [61] |
| | Eu | Cr$^{III}$ | 0.55(4)$^d$ | 10$^{-4}$ | 3.2 | [90] |
| L26 | Eu | Zn | 2.35(2) | 10$^{-3}$ | 8.2 | [95] |
| HL27 | Eu | Zn | 2.99(9) | 10$^{-4}$ | 32 | [94] |
| | | | 2.43(2)$^d$ | 10$^{-4e}$ | 15 | [94] |
| L28 | Eu | Zn | 2.63(1) | 10$^{-3}$ | 7.4 | [96] |
| | Eu | Fe$^{II}$ | 0.28(1) | 10$^{-3}$ | 0.03 | [96] |

$^a$Solid state sample, at 10 or 13 K, ligand excitation.
$^b$In acetonitrile.
$^c$In acetonitrile at room temperature; recalculated by using the most recent values reported for the [Ln(tpy)$_3$]$^{3+}$ internal references, 32 ± 1% (Eu) and 35 ± 1% (Tb) in acetonitrile. Uncertainty: ±10–15%.
$^d$For [EuCr(L25)$_3$](CF$_3$SO$_3$)$_6$·4H$_2$O; 0.75(1) ms for [EuCr(L25)$_3$](CF$_3$SO$_3$)$_6$·4MeCN; 0.87(4) ms for a solution 10$^{-4}$ M in acetonitrile.
$^e$In H$_2$O.

***Table 6.13*** *Lifetimes and energy transfer yields in microcrystalline [MLn(L25)$_3$]$^{n+}$ helicates at 10 K (first number) and 295 K (second number) under ligand excitation [90].*

| LnM[a] | $\tau\,(^5D_0)$ (ms) | $\tau(^2E)$ (ms) | $\eta_{et}$(LnCr) (%) |
|---|---|---|---|
| GdCr (1) | | 3.66(3), 0.29(1) | |
| EuZn (2) | 2.53(1), 1.67(2) | | |
| EuCr (3) | 0.55(4), 0.59(1) | 3.46(1), 0.09(1) | 78(5), 65(3) |
| EuZn (4) | 2.19(1), 1.98(1) | | |
| EuCr (5) | 0.75(1), 0.66(1) | 3.12(1), 0.05(1) | 66(2), 67(2) |
| TbCr (6) | [b] | 3.39(1), 0.17(1) | 100, 100 |

[a](1) = [GdCr(L25)$_3$](CF$_3$SO$_3$)$_6$(H$_2$O)$_6$; (2) = [EuZn(L25)$_3$]ClO$_4$(CF$_3$SO$_3$)$_4$(MeCN)$_4$; (3) = [EuCr(L25)$_3$] (CF$_3$SO$_3$)$_6$(MeCN)$_4$; (4) = [EuZn(L25)$_3$](ClO$_4$)$_5$(H$_2$O)$_2$; (5) = [EuCr(L25)$_3$](CF$_3$SO$_3$)$_6$(H$_2$O)$_4$; (6) = [TbCr (L25)$_3$](CF$_3$SO$_3$)$_6$(H$_2$O)$_3$.
[b]No Tb emission.

### 6.3.3 Control of f-Metal Ion Properties by d-Transition Metal Ions

Heterometallic d-f molecular helicates are appealing because electronic communication between metal ions along the molecular axis can be induced and controlled, which represents an additional tool in the hands of synthetic chemists for tuning optical and/or magnetic properties. With respect to luminescence, motivations for resorting to d-metal containing chromophores are to shift the excitation wavelength from the UV to the visible range, which is particularly appreciated when dealing with biological systems, to lengthen the apparent lifetime of the Ln$^{III}$ emitting ion and to efficiently transfer energy onto the lanthanoid ion.

### 6.3.3.1 Switching Eu$^{III}$ Luminescence On and Off

Control of Eu$^{III}$ luminescence is exemplified in EuFe$^{II}$ helicates by simple modification of the ligand. For instance, if a methyl group is introduced in the 5-position of

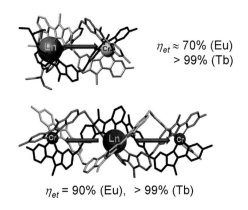

$\eta_{et} \approx 70\%$ (Eu)
$> 99\%$ (Tb)

$\eta_{et} = 90\%$ (Eu), $> 99\%$ (Tb)

***Figure 6.15*** *Di- and trinuclear triple-stranded Cr$^{III}$-Ln$^{III}$ helicates experiencing intramolecular axial energy transfers. Redrawn from crystallographic data published in Refs. [11,90].*

the pyridine in L25 (Scheme 6.6), the resulting $[LnFe(L25)_3]^{5+}$ helicate displays spin-crossover behaviour. In acetonitrile at room temperature, the high-spin fraction of $[EuFe(L25)_3]^{5+}$ amounts to 15% and the transition temperature is estimated to be 359 K by extrapolation of data recorded in the range 243–333 K. The LMCT of the pure low-spin complex $[EuFe(L25)_3]^{5+}$, accounting for its violet colour, extends from 430 to 630 nm, with a molar absorption coefficient reaching 5800 $M^{-1}cm^{-1}$ at 530 nm. As a result, the $^5D_0$ luminescence in $[EuFe(L25)_3]^{5+}$ is totally quenched as a consequence of the overlap between the $Eu^{III}$ emission spectrum and the broad LMCT absorption band [89]. A similar quenching also happens for $[EuFe(L24)_3]^{5+}$ [88]. In contrast, the steric constraint linked to the introduction of the methyl group in the 6-position of the pyridine in L28 leads to the exclusive formation of the high-spin $[EuFe(L28)_3]^{5+}$ helicate, both in solution and in the solid state. This is confirmed by the weak LMCT absorption band ($\varepsilon \approx 500\,M^{-1}cm^{-1}$) around 450 nm, responsible for the yellow colour of the helicate and the faint d-d transitions at 910 (14 $M^{-1}cm^{-1}$) and 1130 nm (11 $M^{-1}cm^{-1}$). Consequently, $[EuFe(L28)_3]^{5+}$ is luminescent, even at room temperature, yet less than the corresponding $Zn^{II}$ complex, as indicated by the $Eu(^5D_0)$ lifetime which drops from 2.63 ms (ligand excitation, solid state, 13 K) in the zinc helicate to 0.28 ms in the iron complex due to partial directional $Eu^{III} \rightarrow Fe^{II}_{HS}$ energy transfer, the latter ion acting as a semi-transparent partner. The situation is summarized in Figure 6.16.

### 6.3.3.2 *Tuning the Lifetime of NIR-Emitting Ln*$^{III}$ *Ions*

The $^2E \rightarrow {}^4A_2$ and $^4T_2 \rightarrow {}^4A_2$ transitions of $Cr^{III}$ around 700 nm are of great importance in technology and are used in ruby lasers and to sensitize the luminescence of $Nd^{III}$ in yttrium aluminium garnet (YAG) lasers since the corresponding emission bands overlap considerably with the absorption spectrum of neodymium. The $Cr(^2E)$ level can be tuned by ligand-field effects and routinely activates the luminescence of NIR-emitting ions ($Nd^{III}$, $Er^{III}$, $Yb^{III}$) [36]. Depending on the energy of the 3d levels, $Cr^{III}$ can also sensitize the luminescence of visible-emitting ions such as $Eu^{III}$ and $Tb^{III}$, an example being heteronuclear complexes $CrLn_2$ with a 2-methyl-8-hydroxyquinoline bridge featuring Cr-Eu distances of 3.3–3.4 Å [98].

Because of its transparency to biological tissue, NIR light is appreciated in bioprobes and many $Ln^{III}$ ions seem to be ideally suited for this purpose. There are however two problems with these probes [36]: (i) due to the small energy gap, the excited states are readily de-activated by all kind of vibrational oscillators, even by those residing in the outer coordination sphere, and (ii) the excited state lifetime is usually short, limiting the ease of application of time-resolved detection. The latter inconvenience may be remedied by populating the excited state of the NIR-emitting $Ln^{III}$ ions by a slow emitting donor. Transition metal ions such as $Cr^{III}$ meet the necessary criteria and the triple helical receptors dealt with in this chapter are adequate hosts for inducing such an energy transfer. The $[LnCr^{III}(L25)_3]^{6+}$ and $[LnRu^{II}(L25)_3]^{5+}$ complexes (Ln = Nd, Gd, Er, Yb), have therefore been proposed for this purpose [14,92]. Relevant data are reported in Table 6.14. The situation can be mathematically described as follows. In the absence of energy transfer, the excited states of the metal ions have deactivation rates $k^{Ln}$ and $k^M$ which are the sum of the radiative and nonradiative rate constants. When the energy transfer rate $\eta_{et}^{M,Ln}$

(a)

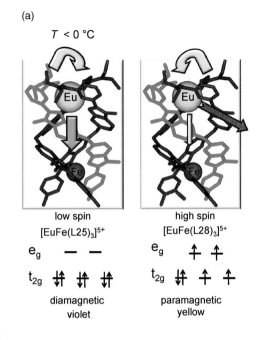

(b)

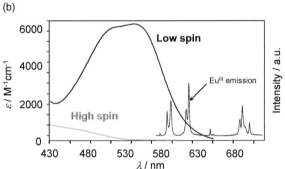

**Figure 6.16** *(a) Energy transfer processes in low-spin [EuFe(L25)₃]⁵⁺ (left) and high-spin [EuFe(L28)₃]⁵⁺ (right); (b) Absorption spectra at 293 K of the helicates in acetonitrile and emission spectrum of the high-spin [EuFe(L28)₃]⁵⁺ complex at 13 K.* Adapted with permission from [96]. Copyright 2001 WILEY-VCH Verlag GmbH & Co. KGaA, Weinheim.

is taken into account, one gets Equations (6.16,6.17):

$$d[M^*]/dt = -\left(k_{et}^{M,Ln} + k^M\right) \cdot [M^*] \tag{6.16}$$

$$d[Ln^*]/dt = k_{et}^{M,Ln} \cdot [M^*] - k^{Ln}[Ln^*] \tag{6.17}$$

After integration, Equation 6.17 transforms into Equation 6.18:

$$[M^*] = [M_0^*] \cdot e^{-\left(k_{et}^{M,Ln} + k^M\right) \cdot t} \tag{6.18}$$

**Table 6.14** Experimental deactivation rates, calculated energy transfer efficiencies and critical distances for 50% transfer ($R_0^{M,Ln}$) in microcrystalline samples of $[LnM(L25)_3]^{n+}$ helicates ($M = Cr^{III}$, $n = 6$; $M = Ru^{II}$, $n = 5$) [92].

| LnM | $T$ (K) | $k_{obs}^M$ (s$^{-1}$) | $k^M$ (s$^{-1}$)[a] | $k_{app}^{Ln}$ (s$^{-1}$) | $k^{Ln}$ (s$^{-1}$)[b] | $k_{et}^{M,Ln}$ (s$^{-1}$) | $\eta_{et}^{M,Ln}$ (%) | $R_0^{M,Ln}$ (Å) |
|---|---|---|---|---|---|---|---|---|
| NdCr | 10 | $2.13 \times 10^3$ | $2.73 \times 10^2$ | $2.13 \times 10^3$ | $6.84 \times 10^5$ | $1.86 \times 10^3$ | 87 | 12.8 |
| YbCr | 10 | $5.10 \times 10^2$ | $2.73 \times 10^2$ | $5.10 \times 10^2$ | $5.00 \times 10^4$ | $2.37 \times 10^2$ | 46 | 9.1 |
| NdRu | 10 | $1.55 \times 10^5$ | $1.04 \times 10^5$ | $1.65 \times 10^5$ | $6.84 \times 10^5$ | $5.10 \times 10^4$ | 33 | 8.1 |
| ErRu | 10 | $1.44 \times 10^5$ | $1.04 \times 10^5$ | $1.44 \times 10^5$ | n.a. | $4.00 \times 10^4$ | 28 | 7.7 |
| YbRu | 10 | $1.27 \times 10^5$ | $1.04 \times 10^5$ | $4.4 \times 10^4$ | $5.00 \times 10^4$ | $2.30 \times 10^4$ | 18 | 7.1 |
| NdCr | 295 | $8.33 \times 10^3$ | $3.45 \times 10^3$ | $8.33 \times 10^3$ | $6.25 \times 10^5$ | $4.88 \times 10^3$ | 59 | 12.8 |
| YbCr | 295 | $4.17 \times 10^3$ | $3.45 \times 10^3$ | $4.17 \times 10^2$ | $4.35 \times 10^4$ | $7.20 \times 10^2$ | 17 | 9.1 |
| NdRu | 295 | $3.45 \times 10^6$ | $1.16 \times 10^6$ | $6.94 \times 10^5$ | n.a. | $2.99 \times 10^6$ | 66 | 8.1 |
| ErRu | 295 | $1.85 \times 10^6$ | $1.16 \times 10^6$ | $1.85 \times 10^6$ | n.a. | $6.90 \times 10^5$ | 37 | 7.7 |
| YbRu | 295 | $1.68 \times 10^6$ | $1.16 \times 10^6$ | $5.71 \times 10^4$ | n.a. | $5.20 \times 10^5$ | 31 | 7.1 |

[a]Measured on GdM samples.
[b]Measured on LnZn samples.

$$[Ln^*] = [M_0^*] \cdot \frac{k_{et}^{M,Ln}}{k^{Ln} - (k_{et}^{M,Ln} + k^M)} \cdot \left(e^{-(k_{et}^{M,Ln}+k^M)\cdot t} - e^{-k^{Ln}\cdot t}\right) \qquad (6.19)$$

As expected, Equation 6.19 reveals that the decay rate of the excited state of the donor $M^*$ increases when energy is transferred onto the acceptor. The experimental decay of the donor metal ion thus corresponds to the sum of the two deactivation rate constants $k_{obs}^M = k^M + k_{et}^{M,Ln}$, which translates into a reduced lifetime $\tau^M$ (Equation 6.20):

$$\tau^M = \left(k_{obs}^M\right)^{-1} = \left(k^M + k_{et}^{M,Ln}\right)^{-1} \qquad (6.20)$$

Interpretation of Equation 6.19 depends on the magnitude of $k_{et}^{M,Ln}$ which controls the population rate of the $Ln^*$ excited state. In fact, the decay profile of $Ln^*$ after initial excitation of the donor depends on the relative magnitudes of the rate constants $k_{obs}^M = k^M + k_{et}^{M,Ln}$ and $k^{Ln}$.

Two limiting cases can be described. The first refers to a situation for which $k_{obs}^M \gg k^{Ln}$, that is the $Ln^*$ level is almost completely populated before any significant lanthanoid-centred deactivation starts. As a consequence, the experimental deactivation rate $k_{app}^{Ln}$ is identical to the one found in the absence of intermetallic communication, $k^{Ln}$. Introducing $k_{obs}^M \gg k^{Ln}$ into Equation 6.19 gives Equation 6.21, in which the time dependence of the luminescence decay corresponds to an apparent rate constant $k_{app}^{Ln} = k^{Ln}$:

$$[Ln^*] = [M_0^*] \cdot \frac{k_{et}^{M,Ln}}{k_{et}^{M,Ln} + k^M} \cdot e^{-k^{Ln}\cdot t} \quad \text{for} \quad k_{obs}^M \gg k^{Ln} \qquad (6.21)$$

This situation occurs for several d-f pairs, because the intrinsic deactivation rates of the d-block donors $k^M$ are often considerably larger than the deactivation of the NIR-emitting Ln-centred excited states. This holds for instance for the $Ru \to Yb$ transfer in $[YbRu(L25)_3]^{5+}$ (Figure 6.17, top). As expected, the experimental decay rate of the donor $k_{obs}^{Ru} = 1.2 \times 10^5 \, s^{-1}$ is larger than $k^{Ru} = 1.0 \times 10^5 \, s^{-1}$ (measured for $[RuGd(L25)_3]^{5+}$), which is diagnostic for the existence of an axial $Ru \to Yb$ energy transfer. Since $k_{obs}^{Ru} = 1.2 \times 10^5 \, s^{-1} > k^{Yb} = 5.0 \times 10^4 \, s^{-1}$ (measured for $[ZnYb(L25)_3]^{5+}$), Equation 6.21 predicts that the experimental Yb-centred decay rate recorded for $[RuYb(L25)_3]^{5+}$ should be equal to $k^{Yb}$, which is the case, within experimental error: $k_{app}^{Yb} = 4.4 \times 10^4 \, s^{-1}$.

The second limiting case arises when $k_{obs}^M \ll k^{Ln}$, that is when the Ln-centred excited state relaxes almost instantaneously after being populated by the slow-decaying d-block chromophore. Therefore, de-excitation of the donor M ion ($k_{obs}^M = k^M + k_{et}^{M,Ln}$) controls the overall deactivation process, and the apparent Ln-centred deactivation rate $k_{app}^{Ln}$ should be equal to $k_{obs}^M$. Introducing the condition $k_{obs}^M \ll k^{Ln}$ into Equation 6.19 provides a simplified Equation 6.22, which points to the time dependence of the $Ln^*$ luminescence decay corresponding to $k_{app}^{Ln} = k_{obs}^M$:

$$[Ln^*] = [M_0^*] \cdot \frac{k_{et}^{M,Ln}}{k^{Ln}} \cdot e^{-(k_{et}^{M,Ln}+k^M)\cdot t} = [M_0^*] \cdot \frac{k_{et}^{M,Ln}}{k^{Ln}} \cdot e^{-k_{obs}^M\cdot t} \quad \text{for} \quad k_{obs}^M \ll k^{Ln} \qquad (6.22)$$

This situation is illustrated when $Ru^{II}$ is replaced with $Cr^{III}$ as the donor in $[CrYb(L25)_3]^{6+}$. The combination of the intrinsic deactivation rate of the Cr-centred donor

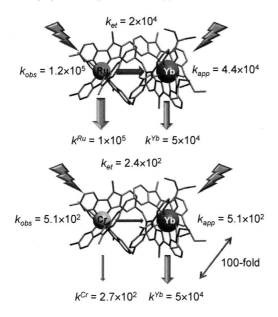

**Figure 6.17** *Energy migration processes in [YbM(L25)$_3$]$^{n+}$ helicates (M = Ru$^{II}$, Cr$^{III}$). Rate constants are given in s$^{-1}$. Redrawn from data published in Ref. [92].*

levels $k^{Cr} = 2.7 \times 10^2\,s^{-1}$ (measured for [CrGd(L25)$_3$]$^{6+}$), with the rate of energy transfer $k_{et}^{Cr,Yb} = 2.4 \times 10^2\,s^{-1}$ gives $k_{obs}^{Cr} = k^{Cr} + k_{et}^{Cr,Yb} = 5.1 \times 10^2\,s^{-1}$. The latter value is small compared with the inherent rate of deactivation of Yb($^2F_{5/2_{5/2}}$), $k^{Yb} = 5.0 \times 10^4\,s^{-1}$ (measured for [ZnYb(L25)$_3$]$^{5+}$); this case is illustrated on the bottom of Figure 6.17. As predicted by Equation 6.22, the experimental decay of the Yb-centred emission amounts to $k_{app}^{Yb} = 5.1 \times 10^2\,s^{-1}$, which exactly matches the slow deactivation rate of the Cr$^{III}$ chromophore. These rate constants can be transformed into characteristic excited lifetimes, thus leading to $\tau_{app}^{Yb} = 23\,\mu s$, when Yb$^{III}$ is sensitized by Ru$^{II}$ in [RuYb(L25)$_3$]$^{5+}$, and $\tau_{app}^{Yb} = 1.96\,ms$, when Yb$^{III}$ is sensitized by Cr$^{III}$ in the isostructural complex [YbCrL25)$_3$]$^{6+}$. Such a lengthening of the lanthanoid-centred NIR luminescence by two orders of magnitude demonstrates the tuning capacity of the dinuclear supramolecular edifices and may be valuable for improving the sensitivity of time-gated bioanalyses, provided that a judicious choice of the donor is made. At low temperature, at which phonon-assisted transfer would be minimized, Cr$^{III}$ is a better donor than Ru$^{II}$ and the efficiency of the process decreases in the order Nd$^{III}$ > Er$^{III}$ > Yb$^{III}$. The latter order is understandable in that the overlap integral between the emission spectrum of the donor and the absorption spectrum of the acceptor is largest for Nd$^{III}$ owing to more lanthanoid-centred excited states in the 13–15 000 cm$^{-1}$ range as compared to Er$^{III}$ and Yb$^{III}$. As a consequence, the largest critical distance $R_0$ for 50% transfer (see Equations (6.6)–(6.8)) occurs for [NdCrL25)$_3$]$^{6+}$. It is also noteworthy that the apparent rate constants for the Ln$^{III}$ ions are smaller by one to two orders of magnitude compared with the situation in which the lanthanoid ion is not populated by energy transfer [92].

### 6.3.3.3 *Molecular Upconversion in a CrErCr Helicate*

Upconversion is a phenomenon in which absorption of two (or more) photons of same energy by a chromophore results in the emission of light with a shorter wavelength. The phenomenon was evidenced by F. Auzel in 1966 and several lanthanoid ions are amenable to this phenomenon, particularly $Er^{III}$ which is extensively provided in telecommunications [99], upconverting nanoparticles for luminescent immunoassays [56,75], and which is being tested in energy conversion of the NIR solar spectrum [100]. Several mechanisms lead to upconversion and two of them, excited state absorption (ESA) and energy transfer upconversion (ETU) are sketched in Figure 6.18.

The tritopic ligand L29 (Scheme 6.6) self-assembles with triads of ions to generate molecular upconversion in the CrErCr helicate, both in solid state and in acetonitrile solution [15]. NIR excitation of this helicate in the $Cr(^2T_1, {}^2E \leftarrow {}^4A_2)$ absorption band around 750 nm with a Ti-sapphire laser indeed results in the green emission of $Er^{III}$ at 540 nm. The mechanism involved is described on the upper right part of Figure 6.18: there is sequential energy transfer, first from the $^2E$ level of an excited $Cr^{III}$ ion, leading to $Er(^4I_{9/2_{9/2}})$, followed by transfer from the second $Cr^{III}$ ion, inducing the $Er^{III}(^2G, {}^4F)_{9/2} \leftarrow {}^4I_{9/2_{9/2}}$ transition (or, alternatively, the $^4F_{5/2_{5/2}} \leftarrow {}^4I_{11/2_{11/2}}$ transition); fast nonradiative deactivation then populates the emissive $^4S_{3/2_{3/2}}$ level. At 77 K, the yield of the Cr to Er energy transfer is 50%. Logarithmic plots of the intensity of the upconverted green emission versus the laser power yield straight lines with slopes 1.8 (solid state) and 2.1 (acetonitrile solution), diagnostic of upconversion. The implied mechanism is most probably ETU. This phenomenon is made possible for the following reasons: (i) the high local concentration of the $Cr^{III}$ sensitizers (equivalent to 0.55 M), (ii) the

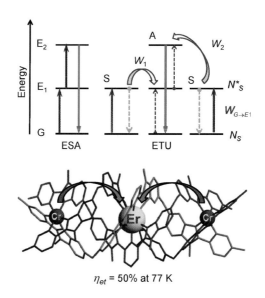

$\eta_{et}$ = 50% at 77 K

**Figure 6.18** *Energy scheme for excited state absorption (ESA) and energy transfer (ETU) mechanisms and molecular up-conversion in [CrErCr(L29)₃] (CF₃SO₃)₉.* Redrawn with permission from [15]. Copyright 2011 WILEY-VCH Verlag GmbH & Co. KGaA, Weinheim.

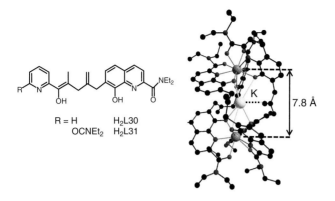

**Figure 6.19**   *Formula of ligands L30 and L31 and structure of [KYb₂(L31)₃]⁺.* Adapted with permission from [108]. Copyright 2007 American Chemical Society.

reasonably large $Cr \rightarrow Er$ transfer efficiency, (iii) the large $Cr(^4T_2)$–$Cr(^2E)$ energy gap $(7600\,cm^{-1})$ so that only the $Cr(^2E)$ level can feed the $Er^{III}$ ion and (iv) the long lifetime of $Cr(^2E)$. This observation is remarkable in that many attempts of designing molecular complexes with organic ligands for $Er^{III}$ upconversion failed, except in two cases, because of the very easy deactivation of $Er^{III}$ by vibrations such as C—H oscillators.

### 6.3.4   Sensitizing NIR-Emitting Lanthanoid Ions

A strongly chelating group which adequately sensitizes the luminescence of NIR-emitting lanthanoid ions is 8-hydroxyquinolinate [101]. This chromophore has been used either directly [102], after suitable derivatization [103], or grafted onto tripodal [104] or tetrapodal [105] ligands or onto benzoxazole [106] or BODIPY [107] frameworks. Finally, two of these units have been fused together into ligands L30, L31 to yield heterometallic trinuclear helicates (Figure 6.19) $[KAlYb(L30)_3]^+$, $[KYb_2(L30)_3]^+$ [108], $[KLn_2(L31)_3]^+$, $[KAlLn(L31)_3]^+$ (Ln — Nd, Er, Yb) and $[M^I Yb_2(L31)_3]^+$ (K = Rb, Cs) [109] which have been fully characterized by NMR, ES-MS and X-ray crystallography. Relevant spectroscopic parameters are reported in Table 6.15. Although modest, quantum yields are sizeable for complexes with organic ligands in view of the problematic presence of C—H vibrators [36,101]. Comparing KAlNd with KLn₂ helicates, one notes that $Al^{III}$ has a small beneficial influence on the quantum yields and lifetimes. In contrast, quantum

**Table 6.15**   *Luminescence lifetimes and quantum yields for microcrystalline samples of heterometallic trinuclear helicates with ligands L30 and L31 [108,109].*

| Ligand | Sample | $\tau_{Ln}$ (μs) | $Q^L_{Ln}$ (%) | Ligand | Sample | $\tau_{Ln}$ (μs) | $Q^L_{Ln}$ (%) |
|---|---|---|---|---|---|---|---|
| L30 | KAlYb | 22.6(2) | 1.17(3) | L31 | KAlNd | 1.29(1) | 0.29(1) |
| L31 | KYb₂ | 18.8(1) | 1.04(7) | | KAlEr | 2.58(3) | 0.028(4) |
| | RbYb₂ | 18.3(2) | 1.96(9) | | KNd₂ | 1.11(1) | 0.24(1) |
| | CsYb₂ | 17.3(1) | 1.74(2) | | KEr₂ | 2.44(6) | 0.026(4) |

yields are largely modulated by the alkaline ion in $KLn_2$ chelates, $RbYb_2$ having the largest value, almost twice as large as $KYb_2$.

## 6.4    Chiral Helicates

In view of their importance in functional biomolecules, chiral structures have always fascinated chemists and biologists. Many chiral polynuclear helical systems and coordination polymers (see next section) have been proposed, particularly with the aim of gaining control on helix inversion [110] but, apart from CD spectra, few quantitative photophysical data are at hand. For instance, trinuclear $Zn_2Ln$ [111] and tetranuclear $Zn_3La$ metallohelicenes [110,112] have been designed for this purpose, but they are not helicates since the metal ions are not located on an internuclear axis. Generally speaking, enantiomerically pure lanthanoid complexes are not easy to isolate in view of the large lability of these trivalent cations. Chiral isomers are characterized by circular dichroism (CD) and circularly polarized luminescence (CPL). CD probes the ground state chirality through measurement of either the coordinated chromophore absorption or the intra-configurational f-f transitions. The extent of the effect is characterized by the absorption dissymmetry factor defined from the difference in molar absorption coefficient $\Delta\varepsilon$ between left (L) and right (R) circularly polarized light [113] (Equation 6.23):

$$g_{abs} = \frac{2\Delta\varepsilon}{\varepsilon} = \frac{2(\varepsilon_L - \varepsilon_R)}{\varepsilon_L + \varepsilon_R} \tag{6.23}$$

CPL is the emissive counterpart of CD and therefore probes the excited state chirality; it also reflects the molecular motions taking place between absorption and emission. In this case, the parameter of interest is the luminescence dissymmetry factor defined as a function of the emission intensities (Equation 6.24):

$$g_{lum} = \frac{2\Delta I}{I} = \frac{2(I_L - I_R)}{I_L + I_R} \tag{6.24}$$

Theoretically, $g_{lum}$ can be related to the electric and magnetic dipole transition moments $\mu^{ge}$ and $m^{ge}$ ($g$ denotes the ground state and $e$ the excited state; Equation 6.25):

$$g_{lum}(\lambda) = 4\frac{f_{CPL}(\lambda)}{f_{TL}(\lambda)} \cdot \frac{\mu^{ge} \cdot m^{ge}}{(\mu^{ge})^2} \tag{6.25}$$

where $f_{CPL}(\lambda)$ and $f_{TL}(\lambda)$ are the line shapes for CPL and total luminescence (TL) signals. Since $\mu^{ge}$ is much larger than $m^{ge}$, CPL usually focuses on Laporte's allowed magnetic dipole transitions. For lanthanoids, the best suited transitions are $Sm(^4G_{5/2} \rightarrow {}^6H_J, J = 7/2, 5/2)$, $Eu(^5D_0 \rightarrow {}^7F_1)$, $Tb(^5D_4 \rightarrow {}^7F_J, J = 3 - 5)$, $Gd(^8S_{7/2} \rightarrow {}^6P_{7/2})$, $Dy(^4F_{9/2} \rightarrow {}^6H_{11/2})$ and $Yb(^2F_{5/2} \rightarrow {}^2F_{7/2})$. One advantage of lanthanoid complexes over organic chiral probes is their often large luminescence dissymmetry factors, which can easily reach $\pm0.5$. A common standard for CPL, tris(3-trifluoroacetyl-$d$-camphorato)europium has $g_{lum} = -0.78$ [114], whereas tetrakis(3-heptafluoro-butylryl-(+)-camphorato)europium displays a record $g_{lum} = +1.35$ [115]. This is to be compared with $\pm 10^{-3}$ to $\pm 10^{-2}$ for organic molecules, including helicenes. One drawback of CPL resides in the weakness of the signals but substantial improvements are being made, particularly with respect to the excitation wavelength used for Eu$^{III}$, and instrumentation can be expanded to use TRD so that CPL may develop as an essential tool for the enantiomeric recognition of biological substrates [114].

There are not many reports on optically active lanthanoid helicates. The first one is concerned with a Nd$^{III}$ helicate with the *R,R* isomer of (L12b)$^{2-}$ (Scheme 6.2), [Nd$_2$(L12c)$_3$]. The corresponding f-f transitions are optically active indicating that the chiral linking group induces chirality at the metal centre. Molecular dynamic calculations suggest a Δ,Δ configuration [51].

Examples of f-f helicates are also documented with ligands H$_2$L32a-c derived from *R, R*-tartaric acid (Scheme 6.7). Both the neutral triple-stranded [Ln$_2$(L32a-c)$_3$] and cationic quadruple-stranded [Eu$_2$(L32c)$_4$]$^{2+}$ helicates display CD spectra in dichloromethane which arise from the ligand. Luminescence from Eu$^{III}$, Tb$^{III}$ and, to a lesser extent, Ho$^{III}$ is sensitized but no CPL spectra are reported [116].

T. Gunnlaugsson and collaborators have succeeded in getting enantiomerically pure f-f homometallic triple-stranded helicates by f-directed synthesis with ligands L33 [117] and L34 [118]. Pairs of *R,R*-[Ln$_2$(L)$_3$]$^{6+}$ and *S,S*-[Ln$_2$(L)$_3$]$^{6+}$ helicates have identical NMR spectra, which proves their appearance as pairs of enantiomers as confirmed by the corresponding CD spectra. Luminescence of Eu$^{III}$ is somewhat sensitized (quantum yields <1%) and the overall emission spectra of the two enantiomers are identical and diagnostic for $C_3$ symmetry while on the NMR time scale the averaged symmetry of the self-assembled edifices appears to be $D_3$. But their CPL spectra are opposite, which enabled the authors to established the absolute configuration as being Λ,Λ for [Eu$_2$(L33R)$_3$]$^{6+}$ and Δ,Δ for [Eu$_2$(L33S)$_3$]$^{6+}$ (Figure 6.20).

Isolation of pure chiral nd-4f helicates was achieved by C. Piguet *et al.* who took advantage of the inertness of Cr$^{III}$ in [LnCr(L25)$_3$]$^{6+}$ (Scheme 6.6) [119]. Indeed, the

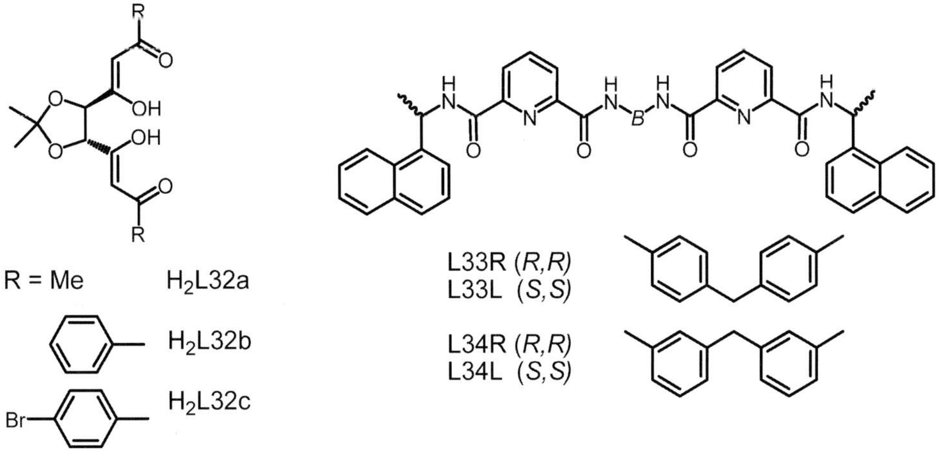

R = Me    H$_2$L32a

H$_2$L32b

Br—◯— H$_2$L32c

L33R (*R,R*)
L33L (*S,S*)

L34R (*R,R*)
L34L (*S,S*)

**Scheme 6.7** *Chiral ligands for the self-assembly of lanthanoid helicates.*

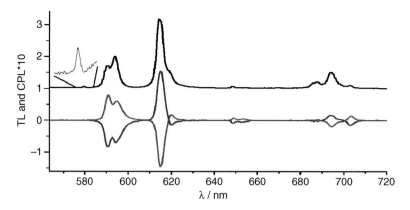

**Figure 6.20** *Top trace (black): luminescence spectrum of Eu₂(L33S)₃]⁶⁺. Bottom traces (coloured): CPL spectra of Eu₂(L33R)₃]⁶⁺ and Eu₂(L33S)₃]⁶⁺ in MeOH. Reproduced with permission from [117]. Copyright 2009 American Chemical Society.*

synthesis requires the availability of facial noncovalent tripodal receptors *P*- or *M*-[$ML_3$]$^{n+}$ with M being an inert transition metal ion to avoid scrambling in solution. First, a racemic LnM$^{II}$ helicate is self-assembled, followed by oxidation into *rac*-LnM$^{III}$. After removal of the Ln$^{III}$ ion with edta, the racemic mixture is loaded onto an ion exchange column and eluted with $Na_2Sb_2[(+)-C_4O_4H_2]_2$ to yield the facial tripodal receptor *fac*-[$M^{III}L_3$]$^{3+}$ with the desired chirality. The latter is finally reacted with Ln$^{III}$. Within the range of investigated nd-4f compounds, both Co$^{III}$ and Cr$^{III}$ qualify, but the difficult oxidation of the former leaves traces of Co$^{II}$ in solution which catalyses the *fac*-[Co(**L36**)₃]$^{3+}$ ⇆ *mer*-[Co(**L36**)₃]$^{3+}$ isomerization, so that Cr$^{III}$ is preferred. Both *P,P*- and *M,M*-[EuCr(L25)₃]$^{6+}$ isomers have been isolated in high purity and their CD and CPL spectra are displayed in Figure 6.21. The two CD spectra are perfect mirrors of each other and the resulting asymmetry factor at 340 nm amounts to 74 M$^{-1}$ cm$^{-1}$. When it comes to luminescence, both Cr($^2$E → $^4$A₂) and Eu($^5$D₀ → $^7$F$_J$), $J = 1$–4, transitions display asymmetry. The luminescence dissymmetry factors $g_{lum}$ are very small for Cr$^{III}$ emission, +0.01 (*M,M*) and −0.01 (*P,P*), as well as for the Eu$^{III}$ transitions to $J = 2$ (±0.07), 3 (∓0.0009) and 4 (±0.033), while they are sizeable for the magnetic dipole transition to $^7$F₁, −0.154 (*M,M*) and +0.163 (*P,P*). Correlation between molecular structure and CPL spectra is presently still not well understood; for instance, in the present case, the signs and magnitudes of the dissymmetry factors do not correlate well with helicity, possibly because the edifice deviates too much from $D_3$ symmetry [120].

Bipyridine ligands derivatized with a chiral pinene unit are strong inductors of chirality in transition metal complexes. Ligands (±)-HL35 (Scheme 6.8) self-assemble with Ln$^{III}$ ions to quantitatively yield trinuclear [Ln₃(±L35)₆(μ-OH)]$^{2+}$ complexes with a propeller-like arrangement of the ligands around the trinuclear metal core inducing helical chirality [121,122]. These trinuclear edifices slowly convert into tetranuclear assemblies with the formula [Ln₄(±L35)₉(μ-OH)]$^{2+}$ and both tri- and tetranuclear complexes display metal-centred luminescence and chirality [123]. Quantum yields for [Ln₃(±L35)₆(μ-OH)]$^{2+}$ in methylene chloride amount to 0.1, 13, 15 and 0.01% for Ln = Nd (−isomer), Eu(+), Tb(+) and Er(−).

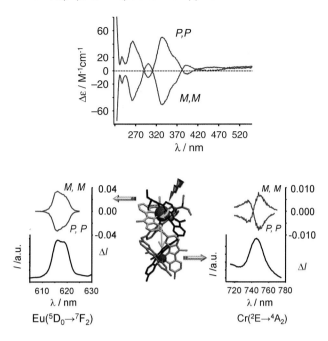

**Figure 6.31** *CD (top) and part of CPL (bottom) spectra of P,P and M,M $[CuCr(L35)_3]^{6+}$ 1 mM in acetonitrile at room temperature.* Redrawn with permission from [119]. Copyright 2004 American Chemical Society.

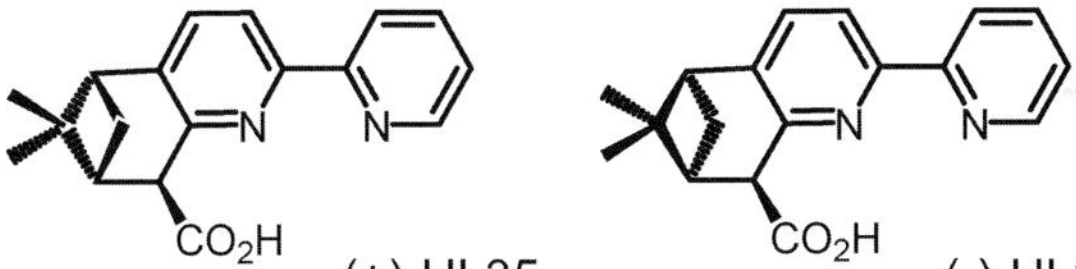

(+)-HL35          (-)-HL35

**Scheme 6.8** *Pinene-derivatized chiral ligands.*

Dissymmetry factors for the Eu(+) and Tb(+) species have been determined for several transitions; for instance, $g_{lum} = -0.088$ for $^5D_0 \rightarrow {}^7F_1$ and $-0.081$ for $^5D_4 \rightarrow {}^7F_5$.

## 6.5 Extended Helical Structures

Numerous examples of lanthanoid [124–128] or lanthanoid–transition metal [129–131] coordination polymers featuring helical one-, two- or three-dimensional chains and/or channels [132] have been engineered with the aim of getting unusual luminescent or magnetic properties [38]. Their description is beyond the scope of the present chapter, so we shall only mention one example in which a dinuclear helicate has been assembled into a one-dimensional coordination polymer. When ligand $H_2L2b$ (Scheme 6.1) is reacted with $Ln^{III}$ ions in the presence of a monodentate ancillary ligand such as water, dimethylformamide or triphenylphosphine oxide, molecular helicates are obtained. However, when

(a)                                    (b)

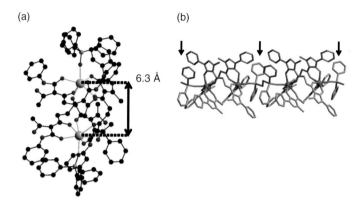

6.3 Å

**Figure 6.22** (a) Crystal structure of [Tb$_2$(L2b)$_3$(dppeO$_2$)$_{1.5}$]$_n$. (b) Fragment of the polymeric one-dimensional chain, with bridging ligands indicated by arrows [133]. Reproduced by permission of The Royal Society of Chemistry.

bidentate diphenylphosphine-ethane dioxide (dppeO$_2$) is used, the bulkier ancillary ligand causes a slightly larger distortion from the idealized $D_3$ symmetry and the helical [Tb$_2$(L2b)$_3$(dppeO$_2$)$_{1.5}$]$_n$ units are connected through Tb$^{III}$ ions by dppeO$_2$ bridges into infinite parallel chains, as shown in Figure 6.22. There are two types of chains corresponding to the two different optical isomers. The length of the repetitive unit along the one-dimensional chain is 16.3 Å with neighbouring inter-chain Tb-Tb distances of 10.1 Å, somewhat longer than in the reference polymeric compound [Ln(NO$_3$)$_3$(dppeO$_2$)$_{1.5}$]$_n$ (9.2 Å). This distance can also be compared with the intramolecular Tb-Tb contacts of 6.5 Å [133]. Both the Tb$^{III}$ molecular helicate [Tb$_2$(L2b)$_3$(H$_2$O)$_2$] and the coordination polymer display sizeable luminescence: quantum yields reach $21 \pm 2$ and $28 \pm 2\%$, respectively [134].

## 6.6 Perspectives

The field of lanthanoid helicates, which started 20 years ago, has produced astonishing supramolecular structures with amazing magnetic and luminescent properties [38]. Chemists have succeeded in programming ligand strands for the selective recognition of homo- and hetero-pairs of lanthanoid ions and/or lanthanoid/transition metal ions. Further extension of the ligand frameworks has led to polynuclear edifices. Theoretical investigation of the recognition process resulted in a satisfying thermodynamic model explaining the site selectivity displayed by these edifices. When it comes to luminescent properties, electronic communication between the metal ions located on the symmetry axis generated unprecedented luminescent properties, including sensitization of the lanthanoid luminescence by d-transition metal ions, lengthening of the excited state lifetime and molecular upconversion. Applications in biology and medicine also highlighted the extraordinary properties and flexibility of these supramolecular edifices. Albeit everything is not said. Chirality of the helicates has not yet been fully taken advantage of, particularly in biological luminescent probes. The rational design of oligomeric structures based on polynuclear helicates is still in its infancy. Moreover, with the exponential development of coordination polymers and nanoparticles, much has still to be discovered when helicates are combined into these materials.

## Acknowledgements

The work of the author was supported by grants from the Swiss National Science Foundation and the Swiss Office for Science and Education within the frame of the COST programme from the European Science Foundation. The author also thanks the World Class University programme (Photovoltaic Materials, Department of Materials Chemistry, Korea University) funded by the Ministry of Education, Science and Technology through the National Research Foundation of Korea (R31-2012-000-10035-0).

## References

1. Albrecht, M. (2001) "Let's twist again" – double-stranded, triple-stranded, and circular helicates. *Chem. Rev.*, **101**, 3457–3497.
2. Constable, E.C., Chotalia, R., and Tocher, D.A. (1992) The 1st example of a mono-helical complex of 2,2′:6′,2″: 6″, 2″:6″,2″:6″,2‴-Sexipyridine – preparation, crystal and molecular structure of bis(nitrato-O,O′) (2,2′:6′,2″:6″, 2″:6″,2″:6″, 2‴-sexipyridine) europium(III) nitrate. *J. Chem. Soc. Chem. Commun.*, **1992**, 771–773.
3. Xing, Y., Li, X., Yan, L., and Yang, R. (1992) Studies on rare-earth coordination compounds. (IX). preparation and characterization of complexes with rare earths and BPMPPD. *Gaodeng Xuexiao Huaxue Xuebao (Chem.J. Chin. Univ.)*, **13**, 14–17.
4. Yang, L. and Yang, R.D. (1995) Synthesis and crystal structure of the dinuclear complex of 1,5-bis(1′-phenyl-3′-methyl-5′-pyrazolone-4′)-1,5-pentanedione with samarium. *Polyhedron*, **14**, 507–510.
5. Bernardinelli, G., Piguet, C., and Williams, A.F. (1992) The 1st self-assembled dinuclear triple-helical lanthanide complex – synthesis and structure. *Angew. Chem, Int, Ed.*, **31**, 1622–1624.
6. Piguet, C., Bünzli, J.-C.G., Bernardinelli, G. *et al.* (1993) Self-assembly and photophysical properties of lanthanide dinuclear triple-helical complexes. *J. Am. Chem. Soc.*, **115**, 8197–8206.
7. Jensen, T.B., Scopelliti, R., and Bünzli, J.-C.G. (2008) Tuning the self-assembly of lanthanide triple stranded heterobimetallic helicates by ligand design. *Dalton Trans.*, **2008**, 1027–1036.
8. Piguet, C., Rivara-Minten, E., Hopfgartner, G., and Bünzli, J.-C.G. (1995) Structural and photophysical properties of pseudo- tricapped trigonal prismatic lanthanide building blocks controlled by Zinc(II) in heterodinuclear d-f complexes. *Helv. Chim. Acta*, **78**, 1541–1566.
9. Bocquet, B., Bernardinelli, G., Ouali, N. *et al.* (2002) The first self-assembled trimetallic lanthanide helicate: different coordination sites in symmetrical molecular architectures. *Chem. Commun.*, **2002**, 930–931.
10. Floquet, S., Borkovec, M., Bernardinelli, G. *et al.* (2004) Programming heterotrimetallic lanthanide helicates: thermodynamic recognition of different metal ions along the strands. *Chem. Eur. J.*, **10**, 1091–1105.
11. Cantuel, M., Gumy, F., Bünzli, J.-C.G., and Piguet, C. (2006) Encapsulation of labile trivalent lanthanides into a homobimetallic chromium(III)-containing triple-stranded helicate.

synthesis, characterization, and divergent intramolecular energy transfers. *Dalton Trans.*, **2006**, 2647–2660.

12. Riis-Johannessen, T., Bernardinelli, G., Filinchuk, Y. *et al.* (2009) Self-assembly of the first discrete 3d-4f-4f triple-stranded helicate. *Inorg. Chem.*, **48**, 5512–5525.

13. Zeckert, K., Hamacek, J., Senegas, J.-M. *et al.* (2005) Predictions, synthetic strategy, and isolation of a linear tetrametallic triple-stranded lanthanide helicate. *Angew. Chem. Int. Ed.*, **44**, 7954–7958.

14. Imbert, D., Cantuel, M., Bünzli, J.-C.G. *et al.* (2003) Extending lifetimes of lanthanide-based NIR emitters (Nd, Yb) in the millisecond range through Cr(III) sensitization in discrete bimetallic edifices. *J. Am. Chem. Soc.*, **125**, 15698–15699.

15. Aboshyan-Sorgho, L., Besnard, C., Pattison, P. *et al.* (2011) Molecular near-infrared to visible light upconversion in a trinuclear d-f-d complex. *Angew. Chem. Int. Ed.*, **50**, 4108–4112.

16. Bünzli, J.-C.G. and Eliseeva, S.V. (2011) Ch. 2, Basics of lanthanide photophysics, in *Springer Series on Fluorescence. Lanthanide Luminescence: Photophysical, Analytical and Biological Aspects, Springer Series on Fluorescence*, vol. **7** (eds P. Hänninen and H. Härmä), Springer, Berlin, pp. 1–45.

17. Ofelt, G.S. (1962) Intensities of crystal spectra of rare earth ions. *J. Chem. Phys.*, **37**, 511–511.

18. Judd, B.R. (1962) Optical absorption intensities of rare earth ions. *Phys. Rev.*, **127**, 750–750.

19. Dieke, G.H. (1968) *Spectra and Energy Levels of Rare Earth Ions in Crystals*, Interscience, New York.

20. Carnall, W.T. (1979) Ch. 24, The absorption and fluorescence spectra of rare earth ions in solution, in *Handbook on the Physics and Chemistry of Rare Earths*, vol. **3** (eds K.A. Gschneidner Jr and L. Eyring), North Holland, Amsterdam, pp. 172–208.

21. Morrison, C.A. and Leavitt, R.P. (1982) Ch. 46, Spectroscopic properties of triply ionized lanthanides in transparent host crystals, in *Handbook on the Physics and Chemistry of Rare Earths*, vol. **5** (ed. K.A. Gschneidner Jr and L. Eyring), North Holland, Amsterdam, pp. 463–692.

22. Hüfner, S. (1983) Ch. 8, Optical spectroscopy of lanthanides in crystalline matrix, in *Systematics and the Properties of the Lanthanides* (ed. S.P. Sinha), D. Reidel, Dordrecht, pp. 303–388.

23. Görller-Walrand, C. and Binnemans, K. (1996) Ch. 155, Rationalization of crystal field parameterization, in *Handbook on the Physics and Chemistry of Rare Earths*, vol. **23** (eds K. A. Gschneidner Jr and L. Eyring), Elsevier Science, Amsterdam, pp. 121–283.

24. Görller-Walrand, C. and Binnemans, K. (1998) Ch. 167, Spectral Intensities of f-f transitions, in *Handbook on the Physics and Chemistry of Rare Earths*, vol. **25** (eds K.A. Gschneidner Jr and L. Eyring), Elsevier Science, Amsterdam, pp. 101.

25. Liu, G.K. and Jacquier, B. (eds) (2005) *Spectroscopic Properties of Rare Earths in Optical Materials*, Tsinghua University Press/Springer, Beijing/Heidelberg.

26. Walsh, B.M. (2006) Judd- Ofelt theory: principles and practices, in *Advances in Spectroscopy for Lasers and Sensing* (eds B. Di Bartolo and O. Forte), Springer, Berlin, pp. 403–433.

27. Hänninen, P. and Härmä, H. (2011) Lanthanide luminescence: photophysical, analytical and biological aspects, in *Springer Series on Fluorescence*, vol. **7** (eds O.S. Wolfbeis and M. Hof), Springer, Berlin.

28. Hüfner, S. (1978) *Optical Spectra of Transparent Rare Earth Compounds*, Academic Press, New York.

29. Liu, G.K. (2005) Ch. 1, Electronic energy level structure, in *Spectroscopic Properties of Rare Earths in Optical Materials, Springer Series in Materials Science*, vol. **83** (eds G.K. Liu and B. Jacquier), Springer, Berlin, pp. 1–94.

30. Bünzli, J.-C.G. (1989) Ch. 7, Luminescent probes, in *Lanthanide Probes in Life, Chemical and Earth Sciences. Theory and Practice* (eds J.-C.G. Bünzli and G.R. Choppin), Elsevier Science, Amsterdam, pp. 219–293.

31. Görller-Walrand, C. and Fluyt, Linda (2010) Ch. 244, Magnetic circular dichroism of lanthanides, in *Handbook on the Physics and Chemistry of Rare Earths*, vol. **40** (eds K.A. Gschneidner Jr, J.-C.G., Bünzli, and V.K. Pecharsky), Elsevier Science, Amsterdam, pp. 1–107.

32. Aebischer, A., Gumy, F., and Bünzli, J.-C.G. (2009) Intrinsic quantum yields and radiative lifetimes of lanthanide tris(dipicolinates). *Phys. Chem. Chem. Phys.*, **11**, 1346–1353.

33. Görller-Walrand, C., Fluyt, L., Ceulemans, A., and Carnall, W.T. (1991) Magnetic dipole transitions as standards for Judd-Ofelt parametrization in lanthanide spectra. *J. Chem. Phys.*, **95**, 3099–3106.

34. Werts, M.H.V., Jukes, R.T.F., and Verhoeven, J.W. (2002) The emission spectrum and the radiative lifetime of $Eu^{3+}$ in luminescent lanthanide complexes. *Phys. Chem. Chem. Phys.*, **4**, 1542–1548.

35. Bünzli, J.-C.G. and Eliseeva, S.V. (2013) Ch. 8.03, Photophysics of lanthanoid coordination compounds, in *Comprehensive Inorganic Chemistry II*, vol. **8** (ed. V.W.-W. Yam), Elsevier, Amsterdam, in press.

36. Comby, S. and Bünzli, J.-C.G. (2007) Ch. 235, Lanthanide near-infrared luminescence in molecular probes and devices, in *Handbook on the Physics and Chemistry of Rare Earths*, vol. **37** (eds K.A. Gschneidner Jr, J.-C.G., Bünzli, and V.K. Pecharsky), Elsevier Science, Amsterdam, pp. 217–470.

37. Burdick, G.W. and Reid, M.F. (2007) Ch. 232, 4f-5d transitions, in *Handbook on the Physics and Chemistry of Rare Earths*, vol. **37** (eds K.A. Gschneidner Jr, J.-C.G., Bünzli, and V.K. Pecharsky), Elsevier Science, Amsterdam, pp. 61–98.

38. Piguet, C. and Bünzli, J.-C.G. (2010) Ch. 247, Self-assembled lanthanide helicates: from basic thermodynamics to applications, in *Handbook on the Physics and Chemistry of Rare Earths*, vol. **40** (eds K.A. Gschneidner Jr, J.-C.G. Bünzli, and V.K. Pecharsky), Elsevier Science, Amsterdam, pp. 303–553.

39. Martin, N., Bünzli, J.-C.G., McKee, V. *et al.* (1998) Self-assembled dinuclear lanthanide helicates: substantial luminescence enhancement upon replacing terminal benzimidazole groups by carboxamide binding units. *Inorg. Chem.*, **37**, 577–589.

40. Elhabiri, M., Scopelliti, R., Bünzli, J.-C.G., and Piguet, C. (1999) Lanthanide helicates self-assembled in water: a new class of highly stable and luminescent dimetallic carboxylates. *J. Am. Chem. Soc.*, **121**, 10747–10762.

41. Piguet, C. and Geraldes, C.F.G.C. (2003) Ch. 215, Paramagnetic NMR lanthanide induced shifts for extracting solution structures, in *Handbook on the Physics and Chemistry and Rare Earths*, vol. **33** (eds K.A. Gschneidner Jr, J.-C.G. Bünzli, and V.K. Pecharsky), Elsevier Science, Amsterdam, pp. 353–463.

42. Platas-Iglesias, C., Elhabiri, M., Hollenstein, M. *et al.* (2000) Effect of a halogenide substituent on the stability and photophysical properties of lanthanide triple-stranded helicates with ditopic ligands derived from bis(benzimidazoyl)pyridine. *J. Chem. Soc. Dalton Trans.*, **2000**, 2031–2043.

43. Bassett, A.P., Magennis, S.W., Glover, P.B. *et al.* (2004) Highly luminescent, triple- and quadruple-stranded, dinuclear Eu, Nd, and Sm(III) lanthanide complexes based on bis- diketonate ligands. *J. Am. Chem. Soc.*, **126**, 9413–9424.

44. Tripier, R., Hollenstein, M., Elhabiri, M. *et al.* (2002) Self-assembled triple-stranded lanthanide dimetallic helicates with a ditopic ligand derived from bis(benzimidazole)pyridine and featuring an (4-isothiocyanatophenyl)ethynyl substituent. *Helv. Chim. Acta*, **85**, 1915–1929.

45. Chauvin, A.-S., Comby, S., Song, B. *et al.* (2007) A polyoxyethylene-substituted bimetallic europium helicate for luminescent staining of living cells. *Chem. Eur. J.*, **13**, 9515–9526.

46. Chauvin, A.-S., Comby, S., Song, B. *et al.* (2008) A versatile ditopic ligand system for sensitizing the luminescence of bimetallic lanthanide bio-imaging probes. *Chem. Eur. J.*, **14**, 1726–1739.

47. Bünzli, J.-C.G., Chauvin, A.-S., Kim, H.K. *et al.* (2010) Lanthanide luminescence efficiency in eight- and nine-coordinate complexes: role of the radiative lifetime. *Coord. Chem. Rev.*, **254**, 2623–2633.

48. Eliseeva, S.V., Song, B., Vandevyver, C.D.B. *et al.* (2010) Increasing the efficiency of lanthanide luminescent bioprobes: bioconjugated silica nanoparticles as markers for cancerous cells. *New J. Chem.*, **34**, 2915–2921.

49. Deiters, E., Song, B., Chauvin, A.-S. *et al.* (2008) Testing the length of the pyridine substituent on cellular uptake in a bimetallic europium luminescent probe. *New J. Chem.*, **32**, 1140–1152.

50. Deiters, E., Song, B., Chauvin, A.-S. *et al.* (2009) Luminescent bimetallic lanthanide bioprobes for cellular imaging with excitation into the visible. *Chem. Eur. J.*, **15**, 885–900.

51. Lessmann, J.J. and Horrocks, W.d. Jr (2000) Supramolecular coordination chemistry in aqueous solution: Lanthanide ion-induced triple helix formation. *Inorg. Chem.*, **39**, 3114–3124.

52. Chauvin, A.-S., Comby, S., Baud, M. *et al.* (2009) Luminescent lanthanide helicates self-assembled from ditopic ligands bearing phosphonic acid or phosphoester units. *Inorg. Chem.*, **48**, 10687–10696.

53. Petoud, S., Bünzli, J.-C.G., Glanzman, T. *et al.* (1999) Influence of charge-transfer states on the Eu(III) luminescence in mononuclear triple helical complexes with tridentate aromatic ligands. *J. Lumin.*, **82**, 69–79.

54. Gonçalves e Silva, F.R., Longo, R.L., Malta, O.L. *et al.* (2000) Theoretical modelling of the low quantum yield observed in an Eu(III) triple helical complex with a tridentate aromatic ligand. *Phys. Chem. Chem. Phys.*, **2**, 5400–5403.

55. Gonçalves e Silva, F.R., Malta, O.L., Reinhard, C. *et al.* (2002) Visible and near-infrared luminescence of lanthanide-containing dimetallic triple-stranded helicates: energy transfer mechanisms in the Sm(III) and Yb(III) molecular edifices. *J. Phys. Chem. A*, **106**, 1670–1677.

56. Eliseeva, S.V. and Bünzli, J.-C.G. (2010) Lanthanide luminescence for functional materials and bio-sciences. *Chem. Soc. Rev.*, **39**, 189–227.

57. Eliseeva, S.V. and Bünzli, J.-C.G. (2011) Rare earths: jewels for functional materials of the future. *New J. Chem.*, **35**, 1165–1176.

58. Khlebtsov, B.N., Khanadeev, V.A., and Khlebtsov, N.G. (2008) Determination of the size, concentration, and refractive index of silica nanoparticles from turbidity spectra. *Langmuir*, **24**, 8964–8970.

59. Jorgensen, C.K. (1962) The nephelauxetic series. *Prog. Inorg. Chem.*, **4**, 73–124.

60. Frey, S.T. and Horrocks, W.d. Jr (1995) On correlating the frequency of the F-7(0)->D-5(0) transition in Eu(3+) complexes with the sum of 'nephelauxetic parameters' for all of the coordinating atoms. *Inorg. Chim. Acta*, **229**, 383–390.

61. Piguet, C., Bünzli, J.-C.G., Bernardinelli, G. *et al.* (1996) Lanthanide podates with predetermined structural and photophysical properties: strongly luminescent self-assembled heterodinuclear d-f complexes with a segmental ligand containing heterocyclic imines and carboxamide binding units. *J. Am. Chem. Soc.*, **118**, 6681–6697.

62. Binnemans, K. and Görller-Walrand, C. (1995) A simple model for crystal field splittings of the 7F1 and 5D1 energy levels of Eu(III). *Chem. Phys. Lett.*, **245**, 75–78.

63. Bünzli, J.-C.G., Chauvin, A.-S., Vandevyver, C.D.B. *et al.* (2008) Lanthanide bimetallic helicates for *in vitro* imaging and sensing. *An. N.Y. Acad. Sci.*, **1130**, 97–105.

64. André, N., Jensen, T.B., Scopelliti, R. *et al.* (2004) Supramolecular recognition of heteropairs of lanthanide ions: a step toward self-assembled bifunctional probes. *Inorg. Chem.*, **43**, 515–529.

65. Jensen, T.B., Scopelliti, R., and Bünzli, J.-C.G. (2007) Thermodynamic parameters governing the self-assembly of head-head-head lanthanide bimetallic helicates. *Chem. Eur. J.*, **13**, 8404–8410.

66. Jensen, T.B., Scopelliti, R., and Bünzli, J.-C.G. (2006) Lanthanide triple-stranded helicates: controlling the yield of the heterobimetallic species. *Inorg. Chem.*, **45**, 7806–7814.

67. Horrocks, W.deW. Jr and Collier, W.E. (1981) Lanthanide ion luminescence probes. Measurement of distances between intrinsic protein fluorophores and bound metal ions: quantitation of enegy transfer between tryptophan and Tb(III) or Eu(III) in the calcium-binding protein parvalbumin. *J. Am. Chem. Soc.*, **103**, 2856–2856.

68. Horrocks, W.deW. Jr, Rhee, M.-J., Snyder, A.P., and Sudnick, D.R. (1980) Laser-induced metal ion luminescence: interlanthanide on energy transfer distance measurements in the calcium-binding proteins parvalbumin and thermolysin. Metalloproteins models address a photophysical problem. *J. Am. Chem. Soc.*, **102**, 3650–3652.

69. Hagan, A.K. and Zuchner, T. (2011) Lanthanide-based time-resolved luminescence immunoassays. *Anal. Bioanal. Chem.*, **400**, 2847–2864.

70. Degorce, F., Card, A., Soh, S. *et al.* (2009) HTRF: a technology tailored for drug discovery – a review of theoretical aspects and recent applications. *Curr. Chem. Genom.*, **3**, 22–32.

71. Bünzli, J.-C.G. (2009) Lanthanide luminescent bioprobes. *Chem. Lett.*, **38**, 104–109.

72. Algar, W.R., Prasuhn, D.E., Stewart, M.H. *et al.* (2011) The controlled display of biomolecules on nanoparticles: a challenge suited to bioorthogonal chemistry. *Bioconjugate Chem.*, **22**, 825–858.

73. Wombacher, R. and Cornish, V.W. (2011) Chemical tags: applications in live cell fluorescence imaging. *J. Biophotonics*, **4**, 391–402.

74. Kobayashi, H. and Choyke, P.L. (2011) Target-cancer-cell-specific activatable fluorescence imaging probes: rational design and *in vivo* applications. *Acc. Chem. Res.*, **44**, 83–90.

75. Bünzli, J.-C.G. (2010) Lanthanide luminescence for biomedical analyses and imaging. *Chem. Rev.*, **110**, 2729–2755.

76. Song, B., Vandevyver, C.D.B., Deiters, E. *et al.* (2008) A versatile method for quantification of DNA and PCR products based on time-resolved Eu$^{III}$ luminescence. *Analyst*, **133**, 1749–1756.

77. Song, B., Vandevyver, C.D.B., Chauvin, A.-S., and Bünzli, J.-C.G. (2008) Time-resolved luminescence microscopy of bimetallic lanthanide helicates in living cells. *Org. Biomol. Chem.*, **6**, 4125–4133.

78. Eliseeva, S.V., Auböck, G., van Mourik, F. *et al.* (2010) Multiphoton-excited luminescent lanthanide bioprobes: two- and three-photon cross sections of dipicolinates derivatives and binuclear helicates. *J. Phys. Chem. B*, **114**, 2932–2937.

79. D'Aléo, A., Picot, A., Baldeck, P.L. *et al.* (2008) Design of dipicolinic acid ligands for the two-photon sensitized luminescence of europium complexes with optimized cross-sections. *Inorg. Chem.*, **47**, 10269–10279.

80. Green, N.M. (1963) Avidin .1. Use of [14C]Biotin for kinetic studies and for assay. *Biochem. J.*, **89**, 585–591.

81. Prat, O., Lopez, E., and Mathis, G. (1991) Europium(III) cryptate: a fluorescent label for the detection of DNSA hybrids on solid support. *Anal. Biochem.*, **195**, 283–289.

82. Claudel-Gillet, S.P., Steibel, J., Weibel, N. *et al.* (2008) Lanthanide-based conjugates as polyvalent probes for biological labeling. *Eur. J. Inorg. Chem.*, **2008**, 2856–2862.

83. Fernandez-Moreira, V., Song, B., Sivagnanam, V. *et al.* (2010) Bioconjugated lanthanide luminescent helicates as multilabels for Lab-on-a-Chip detection of cancer biomarkers. *Analyst*, **135**, 42–52.

84. Bünzli, J.-C.G., Vandevyver, C.D.B., Chauvin, A.-S. *et al.* (2011) Lighting up cancerous cells with lanthanide luminescence. *Chimia*, **65**, 361.

85. Comby, S., Stomeo, F., McCoy, C.P., and Gunnlaugsson, T. (2009) Formation of novel dinuclear lanthanide luminescent Samarium(III), Europium(III), and Terbium(III) triple-stranded helicates from a C-2-Symmetrical Pyridine-2,6-dicarboxamide-Based 1,3-Xylenediyl-linked ligand in MeCN. *Helv. Chim. Acta*, **92**, 2461–2473.

86. Floquet, S., Ouali, N., Bocquet, B. *et al.* (2003) The first self-assembled trimetallic lanthanide helicates driven by positive cooperativity. *Chem. Eur. J.*, **9**, 1860–1875.

87. Dalla-Favera, N., Hamacek, J., Borkovec, M. *et al.* (2008) Linear polynuclear helicates as a link between discrete supramolecular complexes and programmed infinite polymetallic chains. *Chem. Eur. J.*, **14**, 2994–3005.

88. Piguet, C., Rivara-Minten, E., Hopfgartner, G., and Bünzli, J.-C.G. (1995) Molecular magnetism and Iron(II) spin-state equilibrium as structural probes in heterodinuclear d-f complexes. *Helv. Chim. Acta*, **78**, 1651–1672.

89. Piguet, C., Rivara-Minten, E., Bernardinelli, G. *et al.* (1997) Non-covalent lanthanide podates with predetermined physicochemical properties – Iron(II) Spin-state equilibria in self-assembled heterodinuclear d-f supramolecular complexes. *J. Chem. Soc. Dalton Trans.*, **1997**, 421–433.

90. Cantuel, M., Bernardinelli, G., Imbert, D. *et al.* (2002) A kinetically inert and optically active Cr(III) partner in thermodynamically self-assembled heterodimetallic non-covalent d-f podates. *J. Chem. Soc. Dalton Trans.*, **2002**, 1929–1940.

91. Torelli, S., Delahaye, S., Hauser, A. *et al.* (2004) Ruthenium(II) as a novel labile partner in thermodynamic self-assembly of heterobimetallic d-f triple-stranded helicates. *Chem. Eur. J.*, **10**, 3503–3516.

92. Torelli, S., Imbert, D., Cantuel, M. *et al.* (2005) Tuning the decay time of lanthanide-based near infrared luminescence from micro- to milliseconds through d-f energy transfer in discrete heterobimetallic complexes. *Chem. Eur. J.*, **11**, 3228–3242.

93. Riis-Johannessen, T., Dupont, N., Canard, G. *et al.* (2008) Towards inert and preorganized d-block-containing receptors for trivalent lanthanides: The synthesis and characterization of triple-helical monometallic Os(II) and bimetallic Os(II)-Ln(III) complexes. *Dalton Trans.*, **2008**, 3661–3677.

94. Edder, C., Piguet, C., Bünzli, J.-C.G., and Hopfgartner, G. (1997) A water-soluble and strongly luminescent self-assembled non-covalent lanthanide podate. *J. Chem. Soc. Dalton Trans.*, **1997**, 4657–4663.

95. Edder, C., Piguet, C., Bernardinelli, G. *et al.* (2000) Unusual electronic effects of electro-withdrawing sulfonamide groups in optically and magnetically active self-assembled non-covalent heterodimetallic d-f- podates. *Inorg. Chem.*, **39**, 5059–5073.

96. Edder, C., Piguet, C., Bünzli, J.-C.G., and Hopfgartner, G. (2001) High spin iron(II) as inner filter for Eu(III) luminescence in heterodimetallic d-f complexes. *Chem. Eur. J.*, **7**, 3014–3024.

97. Supkowski, R.M. and Horrocks, W.d. Jr (2002) On the determination of the number of water molecules, q, coordinated to europium(III) ions in solution from luminescence decay lifetimes. *Inorg. Chim. Acta*, **340**, 44–48.

98. Xu, H.B., Li, J., Zhang, L.Y. *et al.* (2010) Structures and photophysical properties of homo- and heteronuclear lanthanide(III) complexes with bridging 2-Methyl-8-hydroxylquinoline (HMq) in the μ-phenol mode. *Cryst. Growth Des.*, **10**, 4101–4108.

99. Auzel, F. (2004) Upconversion and Anti-Stokes processes with f and d ions in solids. *Chem. Rev.*, **104**, 139–173.

100. Wang, H.Q., Batentschuk, M., Osvet, A. *et al.* (2011) Rare-earth ion doped up-conversion materials for photovoltaic applications. *Adv. Mater.*, **23**, 2675–2680.

101. Artizzu, F., Mercuri, M.L., Serpe, A., and Deplano, P. (2011) NIR-emissive erbium-quinolinolate complexes. *Coord. Chem. Rev.*, **255**, 2514–2529.

102. Curry, R.J. and Gillin, W.P. (1999) 1.54 μm electroluminescence from erbium (III) tris (8-hydroxyquinoline) (ErQ)-based organic light-emitting diodes. *Appl. Phys. Lett.*, **75**, 1380–1382.

103. Albrecht, M., Osetska, O., Klankermayer, J. *et al.* (2007) Enhancement of near-IR emission by bromine substitution in lanthanide complexes with 2-carboxamide-8-hydroxyquinoline. *Chem. Commun.*, **2007**, 1834–1836.

104. Comby, S., Imbert, D., Vandevyver, C.D.B., and Bünzli, J.-C.G. (2007) A novel strategy for the design of 8-hydroxyquinolinate-based lanthanide bioprobes emitting in the NIR range. *Chem. Eur. J.*, **13**, 936–944.

105. Comby, S., Imbert, D., Chauvin, A.-S., and Bünzli, J.-C.G. (2006) Stable 8-hydroxyquinoline-based podates as efficient sensitizers of lanthanide near-infrared luminescence. *Inorg. Chem.*, **45**, 732–743.

106. Shavaleev, N.M., Scopelliti, R., Gumy, F., and Bünzli, J.-C.G. (2009) Surprisingly bright near-infrared luminescence and short radiative lifetimes of ytterbium in hetero-binuclear Yb-Na chelates. *Inorg. Chem.*, **48**, 7937–7946.

107. Zhong, Y., Si, L., He, H., and Sykes, A.G. (2011) BODIPY chromophores as efficient green light sensitizers for lanthanide-induced near-infrared emission. *Dalton Trans.*, **40**, 11389–11395.

108. Albrecht, M., Fröhlich, R., Bünzli, J.-C.G. *et al.* (2007) Highly efficient near-IR emitting Yb/Yb- and Yb/Al-Helicates. *J. Am. Chem. Soc.*, **129**, 14178–14179.

109. Albrecht, M., Osetska, O., Bünzli, J.-C.G. *et al.* (2009) Homo- and heterodinuclear helicates of lanthanide(III), zinc(II) and aluminium(III) based on 8-hydroxyquinoline ligands. *Chem. Eur. J.*, **15**, 8791–8799.

110. Akine, S., Hotate, S., and Nabeshima, T. (2011) A molecular leverage for helicity control and helix inversion. *J. Am. Chem. Soc.*, **133**, 13868–13871.

111. Akine, S., Taniguchi, T., and Nabeshima, T. (2006) Helical metallohost-guest complexes via site-selective transmetalation of homotrinuclear complexes. *J. Am. Chem. Soc.*, **128**, 15765–15774.

112. Akine, S., Taniguchi, T., Matsumoto, T., and Nabeshima, T. (2006) Guest-dependent inversion rate of a tetranuclear single metallohelicate. *Chem. Commun.*, **2006**, 4961–4963.

113. Riehl, J.P. and Muller, G. (2005) Ch. 220, Circularly polarized luminescence spectroscopy from lanthanide systems, in *Handbook on the Physics and Chemistry of Rare Earths*, vol. 34 (eds K.A. Gschneidner Jr, J.-C.G., Bünzli, and V.K. Pecharsky), Elsevier Science, Amsterdam, pp. 289–357.

114. Do, K., Muller, F.C., and Muller, G. (2008) A promising change in the selection of the circular polarization excitation used in the measurement of Eu(III) circularly polarized luminescence. *J. Phys. Chem. A*, **112**, 6789–6793.

115. Lunkley, J.L., Shirotani, D., Yamanari, K. *et al.* (2008) Extraordinary circularly polarized luminescence activity exhibited by cesium tetrakis(3-heptafluoro-butylryl-(+)-camphorato) Eu(III) complexes in EtOH and CHCl3 solutions. *J. Am. Chem. Soc.*, **130**, 13814–13815.

116. Albrecht, M., Schmid, S., Dehn, S. *et al.* (2007) Diastereoselective formation of luminescent dinuclear lanthanide(III) helicates with enantiomerically pure tartaric acid derived bis (β-diketonate) ligands. *New J. Chem.*, **31**, 1755–1762.

117. Stomeo, F., Lincheneau, C., Leonard, J.P. *et al.* (2009) Metal-directed synthesis of enantiomerially pure dimetallic lanthanide luminescent triple-stranded helicates. *J. Am. Chem. Soc.*, **131**, 9636–9637.

118. Lincheneau, C., Peacock, R.D., and Gunnlaugsson, T. (2010) Europium directed synthesis of enantiomerically pure dimetallic luminescent "squeezed" triple-stranded helicates; solution studies. *Chem. Asian J.*, **5**, 500–504.

119. Cantuel, M., Bernardinelli, G., Muller, G. *et al.* (2004) The first enantiomerically pure helical noncovalent tripod for assembling nine-coordinate lanthanide(III) podates. *Inorg. Chem.*, **43**, 1840–1849.

120. Gawryszewska, P., Legendziewicz, J., Ciunik, Z. *et al.* (2006) On the determination of empirical absolute chiral structures: chiroptical spectrum correlations for D-3 lanthanide (III) complexes. *Chirality*, **18**, 406–412.

121. Mamula, O., Lama, M., Telfer, S.G. *et al.* (2005) A trinuclear Eu(III) array within a diastereoselectively self-assembled helix formed by chiral bipyridine-carboxylate ligands. *Angew. Chem. Int. Ed.*, **44**, 2527–2531.

122. Lama, M., Mamula, O., Kottas, G.S. *et al.* (2007) Lanthanide class of a trinuclear enantiopure helical architecture containing chiral ligands: synthesis, structure, and properties. *Chem. Eur. J.*, **13**, 7358–7373.

123. Lama, M., Mamula, O., Kottas, G.S. *et al.* (2008) Enantiopure, supramolecular helices containing three-dimensional tetranuclear lanthanide(III) arrays: synthesis, structure, properties, and solvent-driven trinuclear/tetranuclear interconversion. *Inorg. Chem.* **47**, 8000–8015.

124. Yue, Q., Yang, J., Li, G.H. *et al.* (2006) Homochiral porous lanthanide phosphonates with 1D triple-strand helical chains: synthesis, photoluminescence, and adsorption properties. *Inorg. Chem.*, **45**, 4431–4439.

125. Bucar, D.K., Papaefstathiou, G.S., Hamilton, T.D., and MacGillivray, L.R. (2008) A lanthanide-based helicate coordination polymer derived from a rigid monodentate organic bridge synthesized in the solid state. *New J. Chem.*, **32**, 797–799.

126. Song, X., Zhou, X., Liu, W. *et al.* (2008) Synthesis, structures, and luminescence properties of lanthanide complexes with structurally related new tetrapodal ligands featuring salicylamide pendant arms. *Inorg. Chem.*, **47**, 11501–11513.

127. Wang, Y.W., Zhang, Y.L., Dou, W. *et al.* (2010) Synthesis, radii dependent self-assembly crystal structures and luminescent properties of rare earth (III) complexes with a tripodal salicylic derivative. *Dalton Trans.*, **39**, 9013–9021.

128. Lucky, M.V., Sivakumar, S., Reddy, M.L.P. *et al.* (2011) Lanthanide luminescent coordination polymer constructed from unsymmetrical dinuclear building blocks based on 4-((1H-benzo[d] imidazol-1-yl)methyl)benzoic acid. *Cryst. Growth Des.*, **11**, 857–864.

129. Gu, X. and Xue, D. (2006) Spontaneously resolved homochiral 3D lanthanide-silver heterometallic coordination framework with extended helical Ln-O-Ag subunits. *Inorg. Chem.*, **45**, 9257–9261.

130. Zhao, B., Zhao, X.Q., Chen, Z. *et al.* (2008) Structures and near-infrared luminescence of unique 4d-4f heterometal-organic frameworks (HMOF). *Crystengcomm*, **10**, 1144–1146.

131. Zhao, X.Q., Zhao, B., Shi, W. *et al.* (2009) Self-assembly of novel 3d-4d-4f heterometalorganic framework based on double-stranded helical motifs. *Dalton Trans.*, **2009**, 2281–2283.

132. Sun, Y.Q., Zhang, J., and Yang, G.Y. (2006) A series of luminescent lanthanide–cadmium–organic frameworks with helical channels and tubes. *Chem. Commun.*, 4700–4702.

133. Semenov, S.N., Rogachev, A.Y., Eliseeva, S.V. *et al.* (2008) First direct assembly of molecular helical complexes into a coordination polymer. *Chem. Commun.*, **2008**, 1992–1994.

134. Semenov, S.N. and Bünzli, J.-C.G. (2008) Photophysical properties of dinuclear helicates and coordination polymers with pyrazolonate ligands, unpublished work.

# 7

# Design of Supramolecular Materials: Liquid-Crystalline Helicates

Raymond Ziessel

*Laboratoire de Chimie Organique et Spectroscopie Avancées (LCOSA)*
*Ecole de Chimie, Polymères, Matériaux, Université de Strasbourg, France*

## 7.1 Introduction

The molecular assemblies formed by the spontaneous association of a large number of components into a specific phase having more or less well defined nanoscopic/microscopic organization and macroscopic characteristics belong to the field of supramolecular chemistry. This holds true for the formation of organized phases such as structured films, layers, membranes, vesicles, micelles, liposomes, nanoparticles, and mesophases. In the last case, the nature of the interactions involved is weak and dominated by hydrophobic interactions and π-π stacking. One of the most challenging research areas in modern chemistry is the control of the information necessary for the process of self-assembly to take place. This information is taken into account during the design and synthesis of multifunctional compounds and materials with desirable or predictable properties, such as luminescence, information transport, catalytic activity, and macroscopic ordering in mesophases [1–4]. In recent years, there has been a tremendous growth in these scientific areas loosely described as supramolecular materials science/chemistry. A plethora of fascinating molecular structures, often involving interlocking of complementary molecular components has raised the possible and future development of innovative miniaturized

*Metallofoldamers: Supramolecular Architectures from Helicates to Biomimetics*, First Edition.
Edited by Galia Maayan and Markus Albrecht.
© 2013 John Wiley & Sons, Ltd. Published 2013 by John Wiley & Sons, Ltd.

devices [5,6]. Numerous examples of supramolecular assemblies based on selective metal ion:ligand interactions have been engineered with the intent of forming symmetrical and esthetically pleasing structures. Elegant strategies have been elaborated in order to build these assemblies that include examples of discrete and infinite helical frameworks [7], hydrogen-bonded [8] or π-π interacting networks [9], three-dimensional supramolecular complexes displaying well defined channels [10], cucurbituril assemblies [11], highly symmetric coordination clusters [12], and organometallic polymers [13].

Many of the new inorganic supermolecules generated over the last few years rely on the coordination of transition metal cations to polypyridine fragments incorporated into oligomeric ribbons or macroscopic loops [14–16]. This work has combined ingenious synthetic strategies with state of the art in characterization, in some cases relying on novel analytical methodologies [17]. By virtue of forming relatively stable and well defined complexes with metal cations, polypyridine ligands have facilitated assembly of molecular double- or triple-stranded helicates [18–21], catenates [22], ladders, and related exotic frameworks [23,24]. Such elaborate molecular architectures are made possible by incorporating several polypyridine units into open-chain or macrocyclic multitopic ligands. Subsequent coordination with cationic metal centers provides the impetus for self-organization into ordered structures; some of these exhibit useful catalytic [25], mesomorphic [26], or electronic [27] properties. Some attempts have been made to identify the driving forces for formation of one structure versus another by systematic variation of the metal and ligand component, but very little is known about the mechanism of formation of most of these metal ion-based self-organized structures [28–33]. Quantitative measures of the thermodynamic stability of such organized scaffoldings are rarely available but are no doubt fundamental to development of the understanding of possible applications of the scaffolds in mesophases active in photoconduction, in catalysis, as chemosensors, or as molecule-based logic gates.

Many long-familiar systems require time-consuming, multistep syntheses from expensive starting materials and tedious chromatographic separations which afford the target ligands in low yields. We recently adopted a different strategy based on addition of coordination sites (imino functions) around a single and central bipyridine coordination shell. The ready availability of these Schiff-base ligands, the simplicity of the purification procedures, the mildness of the reaction conditions, and the high yields allow the preparation of multigram-scale quantities. The addition of particular functional groups to the ligand is a very attractive feature since the presence of two additional donor atoms adjacent to the central bipyridine unit should result in passage from the simple chelating coordination mode (type I) of the bipyridine to a non-conventional mode of coordination (type II) where the ligand bridges two metal centers (Figure 7.1) [34]. An identical situation should be encountered with terpyridine, where again the usual coordination mode (type V) may become non-conventional (type VI). Even where standard coordination modes are adopted by imino group-functionalized polypyridines, they can still be of interest, since ligands of this type were amongst the first to be used in metallomesogen synthesis.

Thus, Schiff-base metal complexes feature among the earliest and most widely studied class of metallomesogens [35]. The advantages of incorporating an imine functionality stems from: (i) as noted above, the ease of preparation and derivatization of the ligands compared to the classical one of type I, (ii) the versatility and

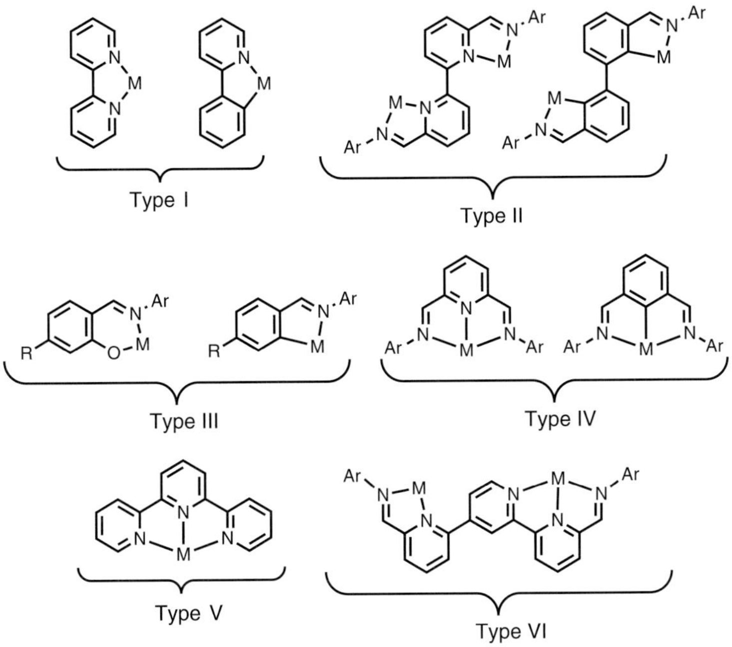

**Figure 7.1** Various coordination modes of imino-polypyridine ligands and corresponding oligopyridine ligands

structural variety of the synthetic protocol, (iii) the excellent donating capabilities of the nitrogen atom of the imine function, and (iv) the strength of the metal–imine bond. Bidentate (type III), tridentate (type IV), tetradendate (type II), and pentaden-date (type VI) Schiff-base ligands are easily obtained by condensation of the corresponding dialdehyde or diketone with aromatic amines under acidic conditions [36]. In particular, 2,6 disubstituted pyridine derivatives have been thoroughly investigated due to the strong coordination abilities of the *N,N,N* donor towards transition metal and lanthanide salts [37]. Several of these complexes are efficient catalysts for olefin polymerization [38] and others have promise as molecular templates for the construction of more sophisticated systems [39].

The aim of this chapter is to give an overview of developments in the construction of supramolecular frameworks which combine the covalent bond-forming capability of metallic centers with imino-functionalized oligopyridine ligands. We review the formation of metallohelicates and the integration of the resulting helicoidal structures into organized assemblies, notably liquid crystals. We also discuss the structure requirements for metallohelicate assembly into mesophases and rigidification of these mesophases by polymerization, together with their possible applications. Different aspects concerning the mechanism of assembly and subsequent dissociation in the presence of a competing ligand, the stabilization of the metallohelicate against oxidative dissociation and its implication for redox catalysis, as well as the internal flexibility of the emerging metallohelicate, will not be discussed in detail.

## 7.2 Imino-Bipyridine and Imino-Phenanthroline Helicates

These metal ion-induced, self-assembled structures were first found with an arene-chromium Schiff-base ligand constructed from arene-chromium *ortho*-toluidine subunits (Figure 7.2) [34]. The high stability constants for the Cu(I) complexes were determined by spectrophotometric titrations that were analyzed in terms of a multistep process. An important caveat of these studies is that positive cooperativity is not mandatory for the formation of a helicate. When reacted with Cu(I), both di-imino-bipyridine and -phenanthroline ligands were found to provide a mixture of 1 : 1 and 1 : 2 (M:L) complexes, the deep violet 1 : 1 species in fact being shown by mass spectrometry to be 2 : 2 complexes to which, for the bipyridine derivative, a helicate structure (Figure 7.2) was assigned. The bipyridine moiety acts as a bridge rather than a chelate and the chelating centres are provided by the imine ($^{Im}N$) and pyridine ($^{Py}N$) donors, with each metal ion coordinated by two of these fragments.

It was surmised at that time that the binuclear complexes were the kinetic products of complexation (formed rapidly under non-equilibrating conditions in non-polar solvents), whereas the mononuclear complexes are the thermodynamic product of complexation when prepared under equilibrating conditions in a coordinative solvent such as acetonitrile [34].

These studies gave rise to interesting questions concerning the possible involvement of an electronic effect inherent to the strong withdrawing character of an arene-chromium tricarbonyl moiety and of possible steric crowding around the potential $^{Py}N,^{Py}N$ chelating center versus the $^{Py}N,^{Im}N$ chelating fragment. It motivated a research program aimed at the construction of imino-oligopyridine frameworks from fluoroaniline, anisidine, and functionalized anilines. A palette of the resulting ligands is given in Figure 7.3 and synthetic details, purification procedures, and characterization may be found in the cited literature. These ligands were synthesized by a standard protocol for Schiff-base synthesis

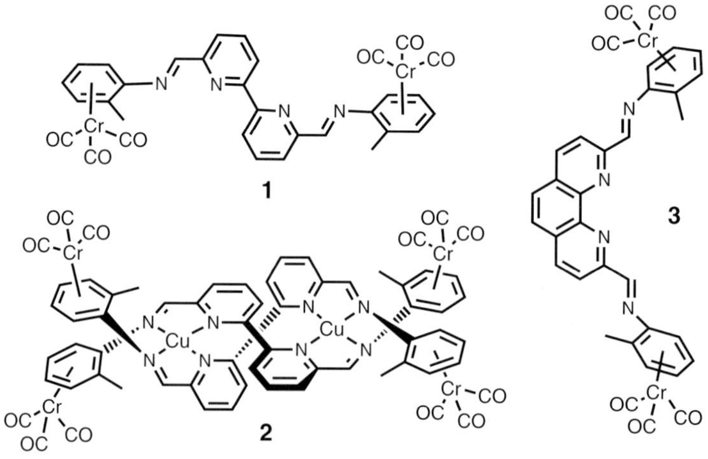

**Figure 7.2** *Arene-chromium imino-bipyridine and imino-phenanthroline ligands and their copper(I) metallohelicate complexes. The charges have been omitted for the sake of clarity.*

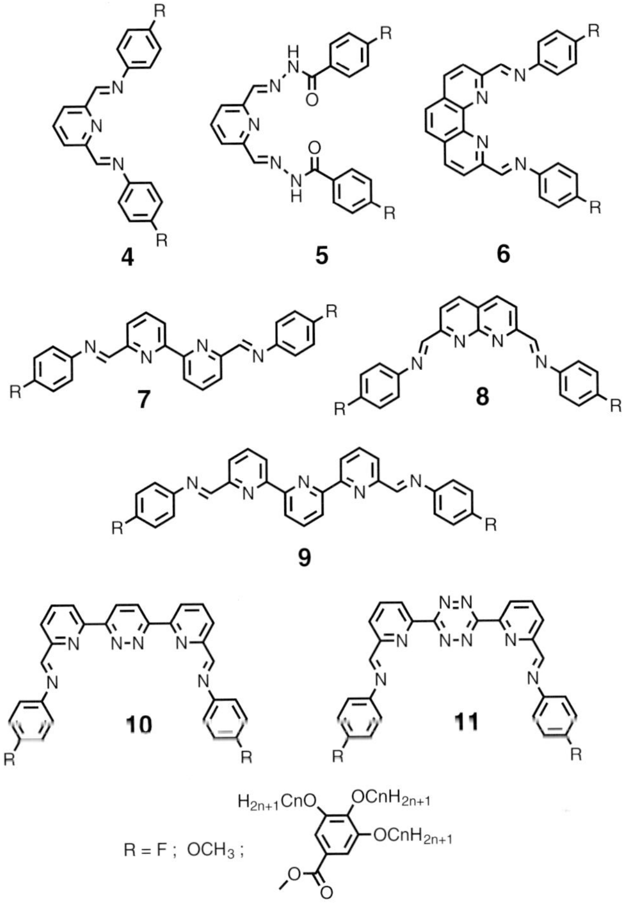

*Figure 7.3* Schematic representation of imino-py ligands bearing various functionalities, including flexible appendages.

employing the condensation of aldehydes with the aniline derivatives in the presence of catalytic amounts of acid.

With Cu(I), ligands **6** and **7** afforded deep-green dinuclear complexes in near-quantitative yields. Here, the Cu(I) cations bind preferentially at the $N_{py},N_{im}$ part of the ligand rather than at the central $N_{py},N_{py}$ fragment, a coordination mode deduced from NMR and FT-IR spectroscopy and confirmed by an X-ray structure determination for each (Figure 7.4).

Indeed, the solid-state structures are best described as double-stranded helicates with the two copper(I) cations having center to center separations of 2.885 and 2.748 Å, respectively, for $[Cu_2(7)_2](ClO_4)_2$ and $[Cu_2(6)_2](ClO_4)_2$ with R = F [33]. These separations are among the shortest metal–metal distances yet found in inorganic helicates [40], being of the same order as found in many copper-containing enzymes [41], but there is no indication of a copper–copper bond. Each ligand wraps around the two copper(I) cations with

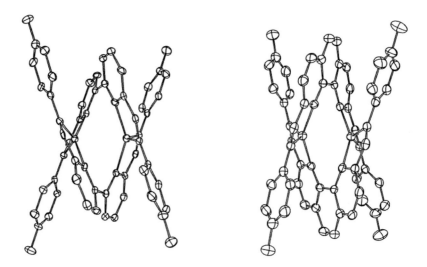

**Figure 7.4** *ORTEP representations of the [Cu$_2$(L)$_2$](ClO$_4$)$_2$ complexes, showing 50% probability ellipsoids, L = **7** (left) and L = **6** (right). The hydrogen atoms are omitted for clarity.*

the bipyridine units being twisted by 33° about the exocyclic central C—C bond. Interestingly, the copper(I) center is significantly closer to the imino N atom than to the pyridine-N. In the phen case, the twist is also clearly evident in the solid state but significantly less pronounced (about 6°; see Figure 7.4 right). The severe internal distortion apparent in the X-ray structures of these helicates is strong testimony to the coordinative ability of the imino function [33].

These crystallographic structures clearly show that, at least in the solid state, individual pyridine groups within a given ligand can be bound to separate cations. Each copper(I) ion is coordinated to four nitrogen atoms, provided by the imine and pyridine functions of two distinct ligands, to give a distorted tetrahedron with a bite angle of 80°. It appears, therefore, that incorporation of imine groups at the α,α′ positions of a bipyridine or phenanthroline ligand can cause the central unit to become non-chelating and, instead, to adopt a bridging role. This is not too surprising for **7**, where individual pyridines can rotate freely, but is unparalleled in the case of the phenanthroline-based ligand **6** where the pyridino N atoms are well sited to form a chelate.

These two unique complexes were the first examples to unveil the ability of bipyridine and phenanthroline framework to act selectively as bridging ligands rather than bidentates. This unusual mode of coordination has further been confirmed in many cases in combinatorial chemistry and dynamic systems [42].

The use of spectrophotometric titrations enabled the evaluation of the stability constants for each step in the helicate formation. These data, as well as the molecular mechanics simulations, support a mechanism in which the first step involves attachment of a copper(I) fragment to a single N donor atom belonging to the imino function. There is apparently positive cooperativity in the binding of bpy-ligand **7** to Cu(I) and the overall stability constant for the metallohelicate is $\log\beta_L = 26$ and compares quite well with the only available literature data for a somewhat comparable process [43]. In the case of the phen-ligand **6**, the helicate formation is not driven by positive cooperativity and the overall stability constant ($\log\beta_L = 16$) reflects the relative difficulty in attaining a suitable geometry. As a matter of fact even for the phen ligand, the mononuclear complex is rather difficult to isolate and the final helicate is the sole compound which is easily isolated in near quantitative yield. The conventional copper(I) bis-chelating $N_{py},N_{py}$ complex has never been observed nor isolated. This is rather surprising considering the plethora of such phenanthroline complexes described in the literature [44].

The systems described with imino-bpy and imino-phen are significantly more rigid than those traditionally used to prepare metallohelicates and an important caveat arising from this study is that positive cooperativity is not a prerequisite for helicate formation. These hybrid ligands do not operate as single chelators towards Cu(I) cations but individual pyridine-N atoms act independently as secondary coordination sites. Metal-induced self-organization of these ligands into helical structures is a multistep process and with **6** a crucial intermediary complex in the formation of the helicate has been isolated. With phenanthroline ligands, however, helicate formation is restricted by thermodynamic considerations rather than by kinetic factors as in certain pentanuclear helicates [45]. These metallohelicates displaying copper(I) centers in close proximity can be broken down under mild conditions by: (i) chemical or electrochemical oxidation, (ii) competitive complexation, or (iii) addition of excess ligand in the case of the phen ligand **6**.

The redox behavior of these unusual helical complexes is also very interesting but has rarely been studied in detail. The copper(I) helicates exhibit four quasi-reversible reduction peaks corresponding to stepwise reduction of each coordinated imino group while in the related free ligands only irreversible reductions are observed. This is interesting because complexation of the imino ligand strongly stabilizes the imine function and favors its reversible reduction. The oxidation of the copper(I) cations is irreversible and results in the removal of one copper center. Poor thermodynamic stability of the resultant copper(II) complex may be due to the fact that a tetrahedral coordination site is unsuitable for binding of Cu(II). Indeed, chemical oxidation of the helicate with Ce(IV) results in the formation of a stable complex with a lime-green color typical of five-coordinate mononuclear copper(II) assembled via complexation to two bpy subunits and a single imino group. Reformation of the dinuclear metallohelicate is quantitative when hydrazine is used as a reducing agent.

Such structural reorganization due to preferential binding of cationic species to secondary complexation sites [translocation processes] have also been observed in related complexes built from ligands possessing additional coordination sites such as the terpyridine ligand **12** (Figure 7.5). The X-ray crystal structure shows that the helicate formed between the imino-terpy **12** and copper(I) cations provides similar environments for the metal centers lying 3.278 Å apart. Each segmented imino-terpy ligand is coordinated to two copper(I) centers in a unsymmetrical fashion. Each ligand provides one bidentate site involving a pyridine-N/imine-N combination and another involving two pyridine-N donors to separate metal ions. Binding of the ligand strands in an antiparallel fashion in the helicate results in each metal ion having an essentially tetrahedral (pyridine-N)$_3$(imine-N) coordination environment (Figures 7.5 and 7.6) [46].

It is noteworthy that the additional imine function is not coordinated (2.629 Å from the copper center), but the fact that a *cis*-conformation of these imine is observed is

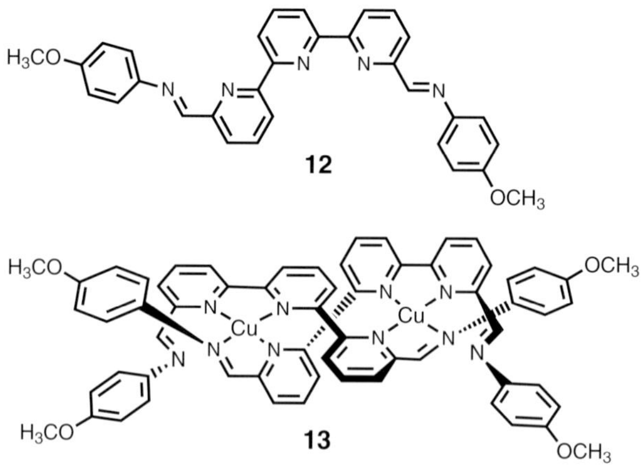

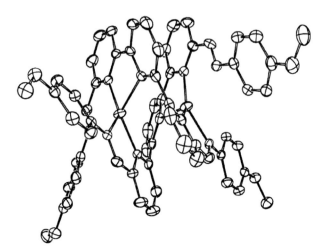

**Figure 7.5** *Schematic representation of the imino-terpyridine ligands and its corresponding copper-helicate.*

auspicious for some electronic interactions as well as for coordination to a metal in a higher oxidation state (Figure 7.6) [46].

Redox cycling in these segmented copper helicates is effective and the helicate is stable in different oxidation states of the metal and also for successive reduction of the ligands. Such stability is derived from the ability of the ligand to provide both 4- and 5-coordinate geometries around the metal centers. These requirements are essential if the metallohelicate is to be used as a redox catalyst or artificial enzyme. There appears to be a limited number of examples of copper helicates or other such related structures that do not undergo large-scale structural reorganization following oxido-reduction reactions.

**Figure 7.6** *ORTEP drawings of the [Cu₂(**12**)₂](ClO₄)₂ complexes, showing 50% probability ellipsoids. The hydrogen atoms are omitted for clarity.*

It is worth pointing out here that redox cycling to produce topological reorganization processes in appropriate scaffoldings is a concept, which has led to the discovery of molecular hysteresis [47,48], molecular memory [49], molecular machines [50], and the precursors of molecular electromechanical motors [51]. In particular, redox-induced movement occurring in multifunctional systems incorporating multistrand arms with distinct (soft or hard donors) coordination sites has been found by changing the oxidation state of iron cation [52,53]. As well, electrochemically induced molecular and ring-gliding motions in pseudo-rotaxane and catenate derivatives have been observed in copper(I) complexes containing different interlocking rings [54].

One clear advantage of these imino-based systems is that very stable complexes can be formed and that the terminal imino group can be readily functionalized without disruption to the helicate structure. This provides the key element for inclusion of the local molecular architecture into an organized macroscopic ensemble, such as calamitic mesophases.

### 7.2.1   Liquid Crystals from Imino-Polypyridine Based Helicates

Metallomesogens (liquid crystals containing metal ions) have become subject to increasing interest, since the introduction of transition metal centers in particular into liquid-crystalline material which may results in the significant modification of physical properties such as color, conduction, magnetism, or redox behavior [55,56]. The ability of transition metal ions to adopt different coordination geometries and to organize elemental synthons around a central core in principle permits the preparation of a wide variety of novel metallomesogens. To date, many different metallic centers and coordination geometries have been incorporated into metallomesogens [57].

Chemical information, as expressed through molecular recognition, provides a means to direct the spontaneous formation of supramolecular species from complementary components. Although control of the supramolecular structure is still a demanding challenge, careful choice of the molecular building blocks makes it possible to predict the nature of the emergent assembly. One of the major challenges remaining in the area of supramolecular chemistry concerns the identification of viable applications (macroscopic function), other than by analytical chemistry, via the integration of individual supramolecular species into an organized network that can be addressed macroscopically [58]. The challenge here is to obtain more functional systems by selective coordination processes around metallic cores, in order to build sophisticated molecular scaffolds (such as helices, grids, ladders, cyclic helicates, numerous polyhedral species) and to provide access to even more sophisticated systems mimicking basic biological functions. Additionally, the conjunction of molecular processes (e.g., metal-induced organization of ligands around metallic cations to produce selectively targeted assemblies) and self-assembly of these units at the macroscopic level to form liquid-crystalline materials offers many interesting features, where unexpected properties could emerge.

Indeed, major goals for the ever-developing field of supramolecular chemistry are: (i) to relate local molecular architecture to macroscopic ordering of the system, (ii) to identify useful applications other than analytical chemistry, and (iii) to integrate individual supramolecular species into organized networks. Such large-scale organization is probably essential for the construction of practical devices from intricate molecular units [59]. Despite this realization little genuine progress has been made with regard to the integration of local order into large-scale multidimensional arrays. In marked contrast, tremendous advances have been made with respect to the construction of exotic super-molecules by stepwise accretion of simple building blocks.

While numerous examples of copper(I) helicates are known it has proved extremely difficult to assemble them into organized assemblies. Special attention has to be given to the number, length and nature of the lipid-like chains and it appears clear that there is a high barrier to arrange self-organized structures into liquid-crystalline material. In this respect, the onset of liquid-crystalline behavior can be seen as a fine balance between order (helix) and chaos (alkyl chains). The net result of optimizing this balance is to produce low-temperature metallomesogens [26]. The initial step in the overall process involves formation of a stable metallohelicate and is itself a challenging proposition in view of the bulky alkyl chains that complicate the gathering of the ligands around the cations. The helix provides essential rigidity that favors subsequent stacking of the aromatic cores.

A simple strategy is to assemble non-mesomorphic but lipid-like organic strands (polycatenary imino-bipyridine ligands) around d-block transition metals to promote formation of a liquid-crystalline state. Our discovery briefly described in the previous section (see above) that a stable copper(I) helicate containing bridging bis-imino-bipyridine subunits is formed selectively and quantitatively by a cooperative process prompted exploration of polycatenary substituted imino-bipyridine ligands as building blocks. These non-discoidal units represent the key component for inclusion of the local molecular architecture (double helix) into an organized macroscopic ensemble. Many kinds of Schiff-base bipy ligands bearing an increasing number of flexible chains and aromatic cycles have been prepared and studied (Figure 7.3). All these new frameworks display well defined melting points (non-mesomorphic material) and formed selectively deep-green dinuclear complexes with copper(I) precursors. Even the presence of very bulky gallate-ether substituents does not significantly perturb the helicoidal arrangement, which was deduced, by analogy with standard ligands, from NMR data. A prototypical example is given in Figure 7.7.

In the case of complex **15**, formed from ligand **14** and copper(I) salts, a well defined columnar mesophase was formed and observed at room temperature [26]. **Cr** accounts for crystal, **Col$_h$** for columnar phase with an hexagonal symmetry and **Iso** for isotropic melt. **ΔH** is the enthalpy changes measured during the phase transition Cr 25 °C (ΔH 132 kJ mol$^{-1}$) Col$_h$ 181 °C (ΔH 3.1 kJ mol$^{-1}$) Iso.

This system, being the first liquid-crystalline metallohelicate, is a rare example of a room temperature metallomesogen and illustrates the exceptional organizational ability of copper(I) cations. It is, in fact, remarkable that this tiny cation is able to induce order at both molecular and supramolecular levels considering the large volume and structural disorder inherent to the non-disklike ligand. This point is well illustrated by comparing molar volumes calculated for the ligand ($V_m = 1356$ cm$^3$ mol$^{-1}$) and the cation ($V_m = 2.2$ cm$^3$ mol$^{-1}$) [26]. The strength of the bonds between the copper(I) cation and the hetero-cyclic amine certainly also contribute to the formation of the final edifice.

The key element of this approach lies with the helix providing rigidity and polarizability to counterbalance the flexibility and non-polarizability of the paraffinic chains. It is this subtle balance between organized and chaotic domains that controls the fate of the mesomorphic material. The ability of imino-bipyridine ligands to form copper(I) helicates

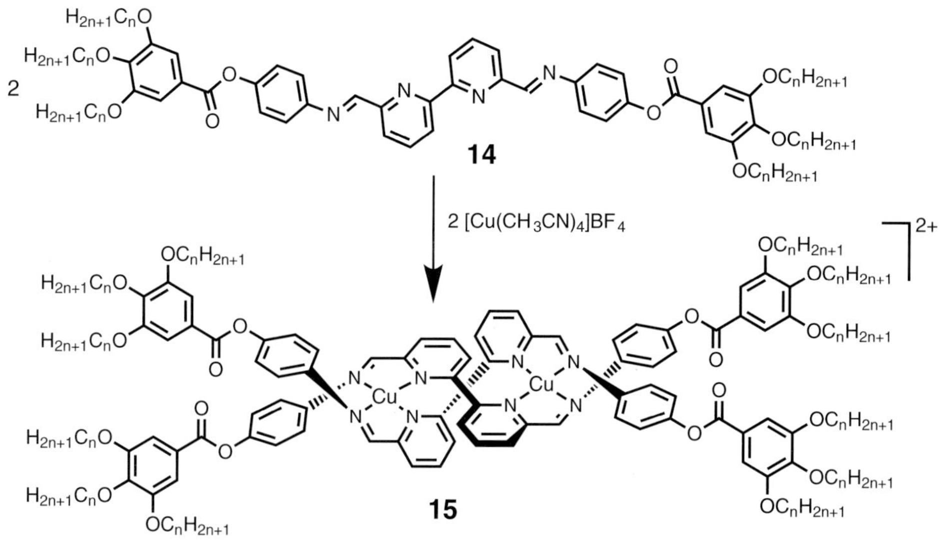

**Figure 7.7** *Schematic representation of the wrapping of the non-mesomorphic ligands **14** around copper(I) cations leading to the metallohelicate **15**. Each copper centre has a single positive charge counterbalanced with an anion (BF₄⁻).*

of non-discoidal shape selectively and quantitatively by a cooperative process is the key element in including the local molecular architecture within an organized macroscopic ensemble. The structure of the liquid-crystalline phase consists of columns of rigid cores surrounded by mobile alkyl chains laterally packed into a two dimensional lattice of rectangular or oblique symmetry (Figure 7.8) [26].

Another significant conclusion reached during this research program is that the introduction of more flexibility into the supermolecular structure does not detract from its ability to assemble into organized networks. It turns out that this delicate operation can be achieved by using a central core constructed from a segmented terpyridine ligand bearing two imino appendages equipped with flexible paraffin chains, so-called phasmidic tails (Figure 7.9) [60].

Complexation to copper(I) gives stable complexes and the first thermotropic terpy-based metallomesogens Cr 115 °C ($\Delta$H 10.1 kJ mol$^{-1}$), Sm$_{\text{biaxial}}$ 187 °C ($\Delta$H 5.6 kJ mol$^{-1}$) Iso.

The liquid-crystalline material displays an unusual type of molecular ordering in the mesophase wherein individual molecules align to form columnar liquid crystals with a lamellar morphology, somewhat similar to that found earlier for charge transfer assemblies [61], macrocyclic aza complexes [62], and orthometallated compounds [63]. This liquid-crystalline phase has a genuine smectic structure and resembles that of linear DNA stretched across a lipid bilayer (Figure 7.10) [64]. The similarity between the smectic liquid-crystalline phase found with this pentadentate ligands and the oriented DNA molecule is striking and of some interest. It is considered that local segregation of the polar helicate from the paraffinic chains creates an interface that facilitates adoption of a non-discoidal liquid-crystalline phase. This seems to be

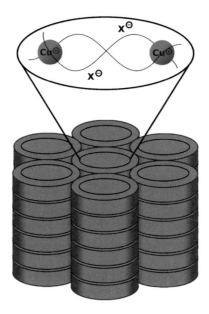

**Figure 7.8** *Idealized packing of the metallohelicate within the columnar mesophase.*

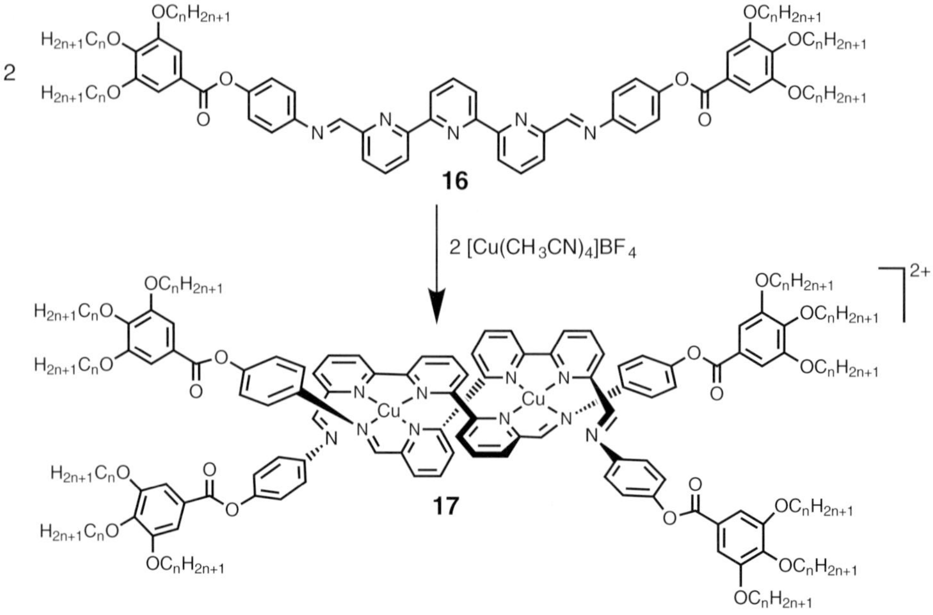

**Figure 7.9** *Schematic representation the mesomorphic helicate **17** formed by a copper(I) induced self-organization process of the segmented terpyridine ligand **16**.*

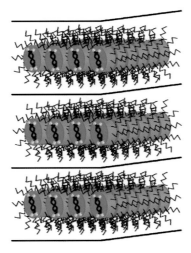

**Figure 7.10**  *Idealized packing of the terpyridine based metallohelicate within the smectic mesophase. The yellow signs represent the uncoordinated imine functions.*

a viable method by which to obtain new types of self-organized assemblies of nanoscopic dimensions.

Nevertheless, the stability and processability of metallomesogens is lower than that of similar organic compounds and this drawback has restricted the development of these materials. One strategy to improve their processability with the aim of fabricating smart devices is the incorporation of metals into liquid-crystalline polymeric assemblies. The design of such metallomesogenic polymers generally employs similar methodologies to those used in organic liquid-crystalline polymers [65]. In the design of these reactive mesogens, moieties that undergo radical polymerization (mainly acrylates or methacrylates) have generally been chosen because radical polymerization is more compatible with organic functional groups than other polymerization techniques. *In situ* polymerization of lyotropic metallomesogens has been studied in regard to biomembrane models, drug-delivery systems, and templates for nanocomposites [66]. Along these lines a series of reactive amphiphiles that contain transition metal or lanthanide ions chelated to car boxylate moieties were prepared and polymerized by subsequent irradiation in the presence of a radical photo-initiator. Selection of an appropriate metal ion enables both the dimensions and the properties of the nanostructured polymeric network to be controlled [67]. The confinement of metal ions within the nanometric channels of the polymeric matrix provides the possibility of using these materials as nanoreactors for catalytic transformations [68]. Also interesting is the generation of CdS nanoparticles in nanometric channels by exposing a polymerized network of the $Cd^{II}$ metallomesogen to $H_2S$ vapor [69]. More recently *in situ* polymerization of thermotropic metallomesogens has proven to be feasible, affording anisotropic materials for application in optical technologies [70].

Bipy–Schiff base metallohelicates bearing between four and 12 acrylate substituents form self-organized structures, some of which are shown in Figure 7.11. In all cases, the material remains mesomorphic at temperatures between −15 and 140 °C as deduced from differential scanning calorimetry (DSC). Photochemically initiated polymerization was

**18**  R₁ = R₂ = R₃ = $\sim OC_{12}H_{24}O$

**19**  R₁ = $OC_{12}H_{25}$ ;  R₂ = R₃ = $\sim OC_{12}H_{24}O$

**20**  R₁ = R₂ = $OC_{12}H_{25}$ ;  R₃ = $\sim OC_{12}H_{24}O$

**Figure 7.11** *Schematic representation of the bipyridine based metallohelicate bearing various numbers of acrylate fragments at the periphery.*

unsuccessful due to severe auto-filtration of the light by the strongly absorbing helicates. However, thermal polymerization using AIBN as initiator gave rise to a thermally stable polymeric network in which the columnar organization of the helicate was maintained (Figures 7.12 and 7.13). Close examination of the polymerized sample by X-ray diffraction revealed a mesomorphic state over a very large temperature range, as would be expected for a rigidified structure. The diffraction pattern revealed a single narrow

**Figure 7.12** *Left hand side: photographs of the polymerized ligand precursor of helicate **20** (white polymer), duramer of the metallohelicate **20** (dark green), duramer obtained from the dark green polymer after treatment with KCN in hot DMF. Right hand side: crème powder of the non-polymerized ligand, precursor of metallohelicate **20** (non-mesophorphic ligand), green powder metallohelicate **20** before polymerization (mesomorphic complex). The polymerized copper-helicate sits in the middle of the glass slides.*

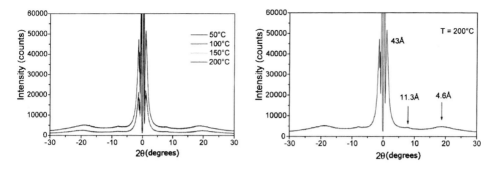

**Figure 7.13** *X-ray diffraction pattern obtained on micro slides of the duramer obtained by polymerization of the metallohelicate* **20**.

diffraction peak at 43 Å in the small angle domain, whereas two broad, diffuse bands were observed at 4.6 and 11.3 Å, corresponding respectively to the molten alkyl chains and the periodicity of the pillaring of the helicate subunits inside the column. Notice that for the monomer **20** the peaks were found at similar values (42, 4.5 and 10 Å), confirming that the columnar order is maintained during the polymerization procedure. Furthermore, the narrowing of the diffraction peak in the small angle regime by increasing the temperature is in keeping with a better organization of the column inside the polymeric network (Figure 7.13).

It appear that the polymerization kinetics depend on the ordering effect of the polymerizable groups, anisotropic mobility, and diffusion in the liquid-crystalline medium. Helicates **18** and **19** were less well suited to the formation of reproducible material (*duramer*) than **20**, which seems the best starting material to generate homogeneous polymeric networks. A schematic representation of the overall process is shown in Figure 7.14.

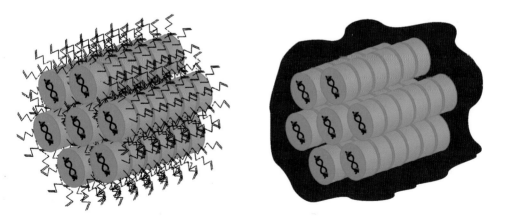

**Figure 7.14** *Schematic representation of the columnar arrangement of metallohelicate* **20** *before polymerization as deduced from X-ray power diffraction (left part) and after polymerization (right part) as deduced from X-ray diffraction on microslides of the polymer.*

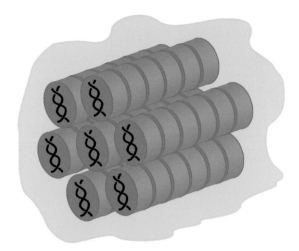

***Figure 7.15*** *Schematic representation of the columnar arrangement of the metal free ligand after cyanide extraction from the duramer as deduced from X-ray diffraction on micro-slides of the polymer.*

Interestingly, the Cu(I) can be extracted from the green duramer using KCN in DMF, leaving the duramer morphology unchanged while the materials becomes pale yellow (Figures 7.15 and 7.12c). Unfortunately, this process is very slow probably because it is limited by the diffusion of cyanide anions inside the polymeric matrix. X-ray diffraction of the metal-free polymers shows a columnar organization of the ligand with cell parameters close to the one measured with the helicate (Figure 7.13 right). This confirms that immobilization of the helicate due to reticulation of the acrylate function is an interesting method of access to well defined nano-structured networks.

Further developments include the use of these ligands to complex metals in a octahedral environment. As a first step, Fe(II) and Ni(II) complexes have been prepared and characterized by X-ray crystallography (Figure 7.16). In the Ni case with bipy ligand **7**, the six-coordinate nickel atom is surrounded by four N atoms belonging to bipy subunits and two N atoms of imino groups belonging to each ligand. The additional two imino functions (which are not ligated to the metal center), lies at 4.4 and 3.4 Å in a *trans* conformation for the former and in a *cis* conformation for the latter. In the latter case some electronic interaction wit the metal center are effective. Here the mode of coordination of the ligand is of Type **VII** (Figure 7.17). In the iron(II) complex with ligand **8**, the metal is surrounded by two imino-naphthyridine ligands in a chelating fashion. The eight N donor atoms define a distorted square antiprism, which is composed of two sets of trapezoidal arrangements in perpendicular orientations (Figure 17.16 right).

The structure of this complex exhibits some striking similarities with that of $[Fe(1,8\text{-naphthyridine})_4]^{2+}$ [71], which also shows a dodecahedral arrangement of the ligand and the coordination mode of the ligand is of Type VIII shown in Figure 7.17. It displays interesting mesomorphic behavior over a large temperature range, depending on the nature of the counter-anion. Most of these salts display a columnar mesophase with octahedral symmetry.

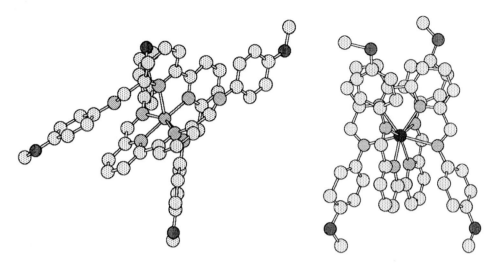

**Figure 7.16** *(Left) ORTEP drawings of [Ni(**7**)₂](ClO₄)₂ complexes. (Right) ORTEP drawings of [Fe(**8**)₂](ClO₄)₂ complexes, The hydrogen atoms are omitted for clarity and 50% probability thermal ellipsoids are used in the drawings.*

The success achieved with the previous complexation reactions with copper(I) salts prompted us to investigate the construction of helicates with the naphthyridine ligand **8** and larger supramolecular systems with the diazine-based building blocks **10** and **11**. The first helicate formed with a naphthyridine-based ligand was obtained and characterized by an X-ray structure determination (Figure 17.18 left), showing a Type IX coordination mode. With ligand **11**, a tetranuclear complex, again characterized by X-ray crystallography (Figure 17.18 right), was selectively obtained. Further illustration of the versatility

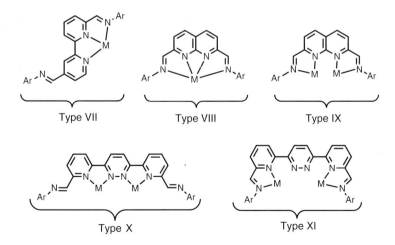

**Figure 7.17** *Schematic representation of the coordination mode of imino-py ligands based on naphtyridine and pyridine-pyridazine ligands.*

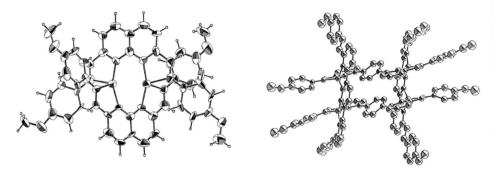

**Figure 7.18** *(a) ORTEP representations of the [Cu(9)₂](ClO₄)₂ complexes. (b) ORTEP representations of the [Cu(11)]₄(ClO₄)₄ complexes, The hydrogen atoms are omitted for clarity and 50% probability ellipsoids are shown.*

and ambidentate coordination modes of these kinds of Schiff-base ligands is shown here, where the same ligand within the same complex adopts two different coordination modes, Type X and Type XI (Figure 7.17).

## 7.3 Conclusions

General trends can be discerned from these studies of several series of metallomesogens constructed from imino-oligopyridinic scaffoldings forming metallohelicates. While the mesophase obtained with the Cu-terpy helicate is smectic, that of the corresponding bpy complex is columnar and that obtained with a mononuclear copper(I) imino-pyridine complex is columnar with hexagonal symmetry [72,73]. There are two main structural differences between these polypyridine complexes that might account for this variation in mesogenic morphology. First, the extended length of the central aromatic core is expected to stretch the molecule into a calamitic shape that favors a smectogenic arrangement. Second, the fluxional motion inherent to the terpy-based metallohelicate increases the entropy of the system and favors the formation of the mesophase. It is this local fluctuation that establishes microdomains sustaining liquid-crystalline behavior at room temperature. Furthermore, the Cu-terpy helicate forms a highly unusual smectic mesophase comprising layers of metallohelicates arranged in equidistant columns but without 3-D correlation of the layers. Such structures are made possible by combining: (i) internal flexibility of the coordinated polytopic ligands, (ii) ancillary coordination sites to stabilize emerging redox centers, and (iii) multiple flexible sidechains.

The realization that such cationic complexes are liquid crystals at room temperature opens the door to several broad and fruitful areas of supramolecular chemistry. With this information at hand, it might be expected that the design and construction of nanoscale molecular assemblies and supramolecular arrays intended to store, transfer or display information could be reached in the near future. Although control of the supramolecular structure is still a demanding challenge, the careful choice of molecular building blocks makes it possible to predict the nature of the emergent supramolecular structure.

This point has been further substantiated by the qualitative comparison of the pyridine, bipyridine and terpyridine Schiff-based ligands bearing phasmidic tails.

A key element of this work involves the use of segmented polytopic ligands bearing a vacant coordination site. This facilitates the formation of symmetrical metallohelicates that remain in labile conformations. The fluctional motion of the ligands gliding across the metal centers provides for the low-temperature mesogenic phase and for the stabilization of higher valence states. This suggests ways to generate new metallomesogens displaying highly desirable properties. It is clear from the few examples of liquid-crystalline materials obtained from these artificial molecular architectures that the list of novel assemblies will grow rapidly in the near future, despite the difficulty in ensuring the targeted properties and properly characterizing the emerging mesophases. The self-assembly process is very versatile and allows the preparation of a great number of discrete supramolecular species with well defined, pre-designed shapes and geometries. The most important advantages of this approach are its wide applicability and the large and different number of potentially suitable transition metal complexes and multidentate nitrogen- or oxygen-containing ligands available as building blocks. Excellent product yields and the high thermodynamic stability constants of the formed complexes that are inherent to such self-assembly processes have been observed in many cases. These results demonstrate two key points with respect to the design of self-assembling systems. First, it is possible to combine imino functions with bipyridine and terpyridine fragments to create stable supramolecular complexes in a productive fashion. Second, judicious tailoring of the ligand with flexible alkyl chains can produce macroscopically organized phases with textures strongly dependent on the architecture of the rigid frame.

## 7.4 Outlook and Perspectives

Up to date, there has been a steady and progressive interest in the synthesis of oligopyridines and an impressive number of such substances are now known. These compounds are not especially appealing on their own, although there are some interesting synthetic facets, and certain compounds form stable complexes with many cations. It is this interest for the synthesis of metal complexes that has driven research into finding new and improved oligopyridines, and such complexes figure prominently in historical accounts of both coordination chemistry and analysis. More recently, metal oligopyridines have been used to construct exotic molecular architectures, to sensitize photoelectrochemical cells, and to label biomaterials as the basis for selective sensors and as magnetic materials. In the future oligopyridines will find even more applications, especially as materials scientists seek to develop miniaturized devices. We have argued the case that Schiff-base templates endow certain metal complexes with special properties and this claim is the result of a determined and sustained effort to produce a comprehensive catalogue of suitably functionalized compounds. In the present survey, we have outlined a synthetic strategy for grafting imino groups to a variety of different oligopyridines. The synthetic routes have been adapted to ensure that various terminal groups (from one to three paraffinic chains) can be attached and to provide means by which to tune the flexibility and microsegregation behavior of the emerging material. Throughout this work, we have become evermore convinced that these hybrid ligands should not be considered as classical bipy, phen, or terpy frameworks but should be regarded as a new type of ligands. We are only just beginning to explore the multitudinous opportunities provided by these ligands and, while they might not be a cure-all for every problem, their versatility and scope are such that they will become fundamental building blocks in emerging molecular machinery. The greatest challenge facing this field is how far it may be possible to correlate the microscopic architecture with the morphology of the emerging mesomorphic material.

The next phase of this work is to construct larger molecules bearing targeted functions that facilitate the formation of mesomorphic materials or molecules able to undergo self-segregation. Here, we foresee this kind of Schiff-base oligopyridines will control the local geometry and electronic properties at the molecular level, while ancillary functions favor self-association into organized arrays. An elegant avenue would be to nanostructure these mesophases into polymeric matrices where the memory of the initial supramolecular arrangement is preserved.

## Acknowledgements

R.Z. wishes to express his sincere appreciation to Professor Jack Harrowfield for commenting on the manuscript before publication and to Delphine Hablot for producing all the artwork in this contribution.

## References

1. (a) Ziessel, R., Hissler, M., El-ghayoury, A., and Harriman, A. (1998) Multifunctional transition metal complexes Information transfer at the molecular level. *Coord. Chem. Rev.*, **178**, 1251–1298; (b) Ramamurthy, V. and Schanze, K.S. (1999) *Multimetallic and Macromolecular Inorganic Photochemistry*, Marcel Dekker, New York.
2. Lehn, J.-M. (1995) *Supramolecular Chemistry*, VCH, Weinheim.
3. Parshall, G.W. and Ittel, S.D. (1992) *Homogeneous Catalysis: The Applications and Chemistry of Catalysis by Soluble Transition Metal Complexes*, Wiley-Interscience, New York, pp. 151–261.
4. (a) Guillon, D. (1999) Columnar order in thermotropic mesophases. *Struct. Bond. (Berlin)*, **95**, 41–82; (b) Donnio, B. and Bruce, D.W. (1999) Metallomesogen. *Struct. Bond. (Berlin)*, **95**, 193–216.
5. Munakata, M., Wu, L.P., and Kuroda-Sowa, T. (1999) Silver(I) coordination complexes of 1,10-Phenanthroline-5,6-dione with 1D chain and 2D network structure. *Adv. Inorg. Chem.*, **46**, 173–303.
6. (a) Balzani, V. Gomez-Lopez, M., and Stoddart, J.F. (1998) Molecular machines. *Acc. Chem. Res.*, **31**, 405–414; (b) Balzani, V., Credi, A., Raymo, F.M., and Stoddart, J.F. (2000) Artificial molecular machines. *Angew. Chem. Int. Ed.*, **39**, 3348–3391; (c) Boyle, M.M., Smaldone, R.A., Whalley, A.C. *et al.* (2011) Mechanised materials. *Chem. Sci.*, **2**, 204–210.
7. (a) Lehn, J.-M., Rigault, A., Siegel, J. *et al.* (1987) Spontaneous assembly of double-stranded helicates from oligobipyridine ligands and copper(I) cations: structure of an inorganic double helix. *Proc. Natl Acad. Sci. USA*, **84**, 2565–2569; (b) Piguet, C., Hopfgartner, G., Bocquet, B. *et al.* (1994) Self-assembly of heteronuclear helical complexes with segmental ligands. *J. Am. Chem. Soc.*, **116**, 9092–9102; (c) Suzuki, T., Kotsuki, H., Isobe, K. *et al.* (1995) Drastic

structural change in Silver(1) complexes with alteration of the optical activity of a pyridine derivative ligand: helical arrays with extended structure and an optically inactive dinuclear complex. *Inorg. Chem.*, **34**, 530–531; (d) Psillakis, E., Jeffery, J.C., McCleverty, J.A., and Ward, M.D. (1997) Synthesis and co-ordination chemistry of the tetradentate chelating ligand 1,3-bis[3-(2-pyridyl)pyrazol-1-yl]propane: crystal structures of complexes with FeII, CuII, ZnII, AgI and PbII. *J. Chem. Soc. Dalton Trans.*, **1997**, 1645–1648; (e) Albrecht, M. and Fröhlich, R. (1997) Controlling the orientation of sequential ligands in the self-assembly of binuclear coordination compounds. *J. Am. Chem. Soc.*, **119**, 1656–1661.

8. (a) Chowdhry, M.M., Mingos, D.M.P., White, A.J.P., and Williams, D.J. (1996) Novel supra-molecular self-assembly of a transition-metal–organo network based on simultaneous coordi-nate- and hydrogen-bond interactions. *Chem. Commun.*, **1996**, 899–900; (b) Smith, G., Reddy, A.N., Byriel, K.A., and Kennard, C.H.L. (1995) Preparation and crystal structures of the silver (I) carboxylates [Ag$_2${C$_6$H$_4$(CO$_2$)$_2$}(NH$_3$)$_2$], [NH$_4$][Ag$_5${C$_6$H$_3$(CO$_2$)$_3$}$_2$(NH$_3$)$_2$(H$_2$O)$_2$]·H$_2$O and [NH$_4$][Ag{C$_4$H$_2$N$_2$(CO$_2$)$_2$}]. *J. Chem. Soc. Dalton Trans.*, **1995**, 3565–3570.

9. (a) Dai, J., Yamamoto, M., Kuroda-Sowa, T. *et al.* (1997) Double hydrogen bond- and π-π-stacking-assembled two-dimensional copper(I) complex of 2-hydroxyquinoxaline. *Inorg. Chem.*, **36**, 2688–2690; (b) Munakata, M., Wu, L.P., Kuroda-Sowa, T. *et al.* (1997) Construc-tion and Conductivity of W-Type Sandwich Silver(I) Polymers with Pyrene and Perylene. *Inorg. Chem.*, **36**, 4903–4905.

10. (a) Batsanov, A.S., Begley, M.J., Hubberstey, P., and Stroud, J. (1997) Discrete dinuclear and polymeric copper(I) cations bridged by 4,4′-bipyridine, *trans*-1,2-bis(4-pyridyl)ethene or bis (4-pyridyl) disulfide. *J. Chem. Soc. Dalton Trans.*, **1997**, 1947–1957; (b) Munakata, M., Wu, L. P., Yamamoto, M. *et al.* (1996) Construction of three-dimensional supramolecular coordination copper(I) compounds with channel structures hosting a variety of anions by changing the hydro-gen-bonding mode and distances. *J. Am. Chem. Soc.*, **118**, 3117–3124.

11. (a) Whang, D., Jeon, Y.-M., Heo, J., and Kim, K. (1996) Self-assembly of a polyrotaxane. *J. Am. Chem. Soc.*, **118**, 11333–11334; (b) Whang, D. and Kim, K. (1997) Polycatenated two-dimensional polyrotaxane net. *J. Am. Chem. Soc.*, **119**, 451–452. (c) Whang, D., Park, K.-M., Heo, J., Ashton, P., and Kim, K. (1998) Molecular necklace: quantitative self-assembly of a cyclic oligorotaxane from nine molecules. *J. Am. Chem. Soc.*, **120**, 4899–4900; (d) Roh, S.-G., Park, K.-M., Sakamoto, S., Yamaguchi, K., and Kim, K. (1999) Synthesis of a five-membered molecular necklace: A 2+2 approach. *Angew. Chem. Int. Ed.*, **38**, 637–643.

12. Caulder, D.L., and Raymond, K.N. (1999) The rational design of high symmetry coordination clusters. *J. Chem. Soc. Dalton Trans.*, **1999**, 1185–1200.

13. (a) Black, J.R., Champness, N.R., Levason, W., and Reid, G. (1996) Self-assembly of ribbons and frameworks containing large channels based upon methylene-bridged dithio-, diseleno , and ditelluroethers. *Inorg. Chem.*, **35**, 4432–4438; (b) Pfitzner, A. and Zimmerer, S. (1997) (CuI)$_3$Cu$_2$TeS$_3$: layers of Cu$_2$TeS$_3$ in Copper(I) iodide. *Angew. Chem. Int. Ed.*, **36**, 982–984.

14. Piguet, C. (1999) Helicates and related metallosupramolecular assemblies: Toward structurally controlled and functional devices. *J. Incl. Phenom. Macrocyclic Chem.*, **34**, 361–391.

15. Olenyuk, B., Fechtenkötter, A., and Stang, P.J. (1998) Molecular architecture of cyclic nano-structures: use of co-ordination chemistry in the building of supermolecules with predefined geometric shapes. *J. Chem. Soc. Dalton Trans.*, **1998**, 1707–1728.

16. Mamula, O., Von Zelewsky, A., and Bernardinelli, G. (1998) Completely stereospecific self-assembly of a circular helicate. *Angew. Chem. Int. Ed.*, **37**, 289–293.

17. Hopfgartner, G., Piguet, C., and Henion, J.D. (1994) Ion spray-tandem mass spectrometry of supramolecular coordination complexes. *J. Am. Soc. Mass Spectrom.*, **5**, 748–756.

18. (a) Lindsey, J.S. (1991) Self-assembly in synthetic routes to molecular devices. Biological prin-ciples and chemical perspectives: a review. *New J. Chem.*, **15**, 153–180; (b) Lawrence, D.S. Jiang, T., and Levett, M. (1995) Self-assembling supramolecular complexes. *Chem. Rev.*, **95**,

2229–2260; (c) Phip, D. and Stoddart, J.F. Self-assembly in natural and unnatural systems. *Angew. Chem. Int. Ed.*, **35**, 1154–1196. (1996).

19. (a) Hasenknopf, B., Lehn, J.M., Kneisel, B.O. *et al.* (1996) Self-assembly of a circular double helicate. *Angew. Chem. Int. Ed.*, **35**, 1838–1840; (b) Schwach, M., Hausen, H.D., and Kaim, W. (1996) Pulled molecular strings and stacked molecular decks. Chelate-ring formation vs. metal-metal-bridging in Dicopper(I) Complexes of 2,2′-Bipyrimidine with diphosphine ligands of variable polymethylene chain length. *Chem. Eur. J.*, **2**, 446–456. (c) Woods, C.R., Benaglia, M., Cozzi, F., and Siegel, J.S. (1996) Enantioselective synthesis of Copper(I) bipyridine based helicates by chiral templating of secondary structure: transmission of stereochemistry on the nanometer scale. *Angew. Chem. Int. Ed.*, **35**, 1830–1833.

20. (a) Piguet, C. Bernardinelli, G., and Hopfgartner, G. (1997) Helicates as versatile supra-molecular complexes. *Chem. Rev.*, **97**, 2005–2062; (b) Piguet, C., Borkovec, M., Hamacek, J., and Zeckert, K. (2005) Strict self-assembly of polymetallic helicates: the concepts behind the semantics. *Coord. Chem. Rev.*, **249**, 705–726.

21. (a) Albrecht, M. (1998) Dicatechol ligands: novel building-blocks for metallo-supramolecular chemistry. *Chem. Soc. Rev.*, **27**, 281–288; (b) Albrecht, M. (2001) Lets twist again – Double-stranded, triple-stranded and circular Helicates. *Chem. Rev.*, **101**, 3457–3498; (c) Albrecht, M. (2004) Supramolecular templating in the formation of helicates. *Top. Curr. Chem.*, **248**, 105–139.

22. (a) Chambron, J.-C. Dietrich-Buchecker, C.O., and Sauvage, J.-P. (1993) From classical chiral-ity to topologically chiral catenands and knots. *Top. Curr. Chem.*, **165**, 131–162; (b) Chambron, J.-C., Dietrich-Buchecker, C.O., Heitz, V. *et al.* (1993) *Transition Metals in Supramolecular Chemistry, NATO ASI Series*, **448** (eds L. Fabbrizzi and A. Poggi), Kluwer, Dordrecht, pp. 371–385.

23. Piguet, C., Hopfgartner, G., Williams, A.F., and Bünzli, J.-C.G. (1995) Self-assembly of the first heterodinuclear d-f triple helix in solution. *Chem. Commun.*, **1995**, 491–493.

24. (a) Ziessel, R. Matt, D., and Toupet, L. (1995) Construction of a phosphane-based metallo-syn-thon suitable for the selective formation of a tetranuclear $Ru_2–Cu_2$ macrocycle. *Chem. Comm.*, **1995**, 2033–2035; (b) Romero, F.M., Ziessel, R., Dupont-Gervais, A., and Van Dorsselaer, A. (1996) Monitoring the iron(II)-induced self-assembly of preorganized tritopic ligands by electrospray mass spectrometry: unique formation of metallomacrocycles. *Chem. Comm.*, **1996**, 551–553.

25. Hill, C.L. and Zhang, X. (1995) A 'smart' catalyst that self-assembles under turnover condi-tions. *Nature*, **373**, 324–326.

26. El-ghayoury, A., Douce, L., Skoulios, A., and Ziessel, R. (1998) Cation-induced macroscopic ordering of non-mesomorphic modules – a new application for metallohelicates. *Angew. Chem. Int. Ed.*, **37**, 2205–2208.

27. Collier, C.P., Wong, E.W., Belohradsky, M. *et al.* (1999) Electronically configurable molecular-based logic gates. *Science*, **285**, 391–394.

28. Marquis-Rigault, A., Dupont-Gervais, A., Van Dorsselaer, A., and Lehn, J.M. (1996) Investiga-tion of the self-assembly pathway of pentanuclear helicates by electrospray mass spectrometry. *Chem. Eur. J.*, **2**, 1395–1398.

29. (a) Kersting, B., Telford, J.R., Meyer, M., and Raymond, K.N. (1996) Gallium(III) catecholate complexes as probes for the kinetics and mechanism of inversion and isomerization of sidero-phore complexes. *J. Am. Chem. Soc.*, **118**, 5712–5721; (b) Kersting, M. Meyer, R.E., and Pow-ers, K.N. (1996) Raymond, dinuclear catecholate helicates: their inversion mechanim. *J. Am. Chem. Soc.*, **118**, 7221–7222; (c) Meyer, M., Kersting, B., Powers, R.E., and Raymond, K.N. (1997) Rearrangement reactions in dinuclear triple helicates. *Inorg. Chem.*, **36**, 5179–5191.

30. (a) Charbonnière, L.J., Williams, A.F., Frey, U. *et al.* (1997) A comparison of the lability of mononuclear octahedral and dinuclear triple-helical complexes of cobalt(II). *J. Am. Chem.*

*Soc.*, **119**, 2488–2496; (b) Charbonnière, L.J., Williams, A.F., Piguet, C. *et al.* (1998) Structural, magnetic, and electrochemical properties of dinuclear triple helices: Comparison with their mononuclear analogues. *Chem. Eur. J.*, **4**, 485–496.

31. Meyer, M., Albrecht-Gary, A.-M., Dietrich-Buchecker, C.O., and Sauvage, J.-P. (1997) Dicopper(I) trefoil knots: topological and structural effects on the demetalation rates and mechanism. *J. Am. Chem. Soc.*, **119**, 4599–4607.

32. Blanc, S., Yakirevitch, P., Leize, E. *et al.* (1997) Allosteric effects in polynuclear triple stranded ferric complexes. *J. Am. Chem. Soc.*, **119**, 4934–4944.

33. Ziessel, R., Harriman, A., Suffert, J. *et al.* (1997) Copper(I) helicates containing bridging but nonchelating polypyridine fragments. *Angew. Chem. Int. Ed.*, **36**, 2509–2511.

34. Ziessel, R. and Suffert, J. (1990) Unexpected copper(I) complexation behaviour observed in the synthesis of novel polynuclear chromium(0)–copper(I) complexes. *Chem. Commun.*, **1990**, 1105–1107.

35. (a) Serrano, J.L. (ed.) (1996) *Metallomesogens, Synthesis, Properties and Applications*, VCH, Weinheim; (b) Neve, F. (1996) Transition metal based ionic mesogens. *Adv. Mater.*, **8**, 277–289; (c) Collinson, S.R. and Bruce, D.W. (1999) Chapter 7, in *Transition Metals in Supramolecular Chemistry* (ed. J.P. Sauvage), John Wiley & Sons, Inc., New York, pp. 285–369; (d) Hoshino, N. (1998) Liquid crystal properties of metal–salicylaldimine complexes: chemical modifications towards lower symmetry. *Coord. Chem. Rev.*, **174**, 77–108.

36. (a) Alyea, E.C., Ferguson, G., and Restivo, R.J. (1975) Crystallographic study of dinitrato[2,6-diacetylpyridine bis(anil)]nickel(II). *Inorg. Chem.*, **14**, 2491–2495, and references therein; (b) Blake, A.J., Lavery, A.J., Hyde, T.I., and Schröder, M. (1989) Precursor catenand complexes: synthesis, structure, and electrochemistry of bis(2,6-di-iminopyridyl) complexes of nickel(II). The single-crystal X-ray structure of [NiL$^4_2$][BF$_4$]$_2$. *J. Chem. Soc. Dalton Trans.*, **1989**, 965–970.

37. (a) Alyea, E.C., Ferguson, G., Restivo, R.J., and Merrell, P.H. (1975) Structural investigation of a nickel(II) complex containing a new type of terdentate ligand with a planar NNN donor set and both uni and bi dentate nitrato groups. X-Ray analysis of 2,6-bis-(phenyliminoethyl)pyridine(dinitrato)nickel(II). *J. Chem. Soc. Chem. Commun.*, **8**, 269–270; (b) Ferguson, G. and Restivo, R.J. (1976) Structural characterization of metal complexes of 2,6-diacetylpyridine-bis (imines). Crystal and molecular structure of dinitrato{2,6-bis[1-(phenylimino)ethyl]pyridine} copper(II). *J. Chem. Soc. Dalton Trans.*, **1970**, 518–521; (c) Alyea, E.C. and Merrell, P.H. (1978) Synthesis and characterization of metal complexes of terdentate NNN donor.txt ligands derived fromo 2,6-diacetylpyridine. Nickel(II) complexes with 2,6-diacetylpyridinebis(imines). *Inorg. Chim. Acta*, **28**, 91–97; (d) Edwards, D.A., Mahon, M.F., Martin, W.R. *et al.* (1990) Manganese(II) complexes containing the tridentate ligands 2,6 bis[1 (phenylimino)ethyl]pyridine, L$^1$, or 2,6-bis[1-(4-methoxyphenylimino)ethyl]pyridine, L$^2$. The molecular structures of five-co-ordinate [MnBr$_2$L$^1$] and the zinc analogue [ZnCl$_2$L$^1$]. *J. Chem. Soc. Dalton Trans.*, **1990**, 3161–3168; (e) Edwards, D.A., Edwards, S.D., Martin, W.R. *et al.* (1992) Cobalt(II) complexes containing nitrogen donor tridentate ligands derived from 2,6-diacetylpyridine and aniline or 4-methoxyaniline. *Polyhedron*, **11**, 1569–1573.

38. (a) Small, B.L., Brookhart, M., and Bennett, A.M.A. (1998) Highly active iron and cobalt catalysts for the polymerization of ethylene. *J. Am. Chem. Soc.*, **120**, 4049–4050; (b) Griffiths, E.A. H., Britovsek, G.J.P., Gibson, V.C., and Gould, I.R. (1999) Highly active ethylene polymerisation catalysts based on iron: an *ab initio* study. *Chem. Commun.*, **1999**, 1333–1334; (c) Britovsek, G.J.P., Bruce, M., Gibson, V.C. *et al.* (1999) Iron and cobalt ethylene polymerization catalysts bearing 2,6-Bis(Imino)Pyridyl ligands: synthesis, structures, and polymerization studies. *J. Am. Chem. Soc.*, **121**, 8728–8740; (d) de Bruin, B., Bill, E., Bothe, E. *et al.* (2000) Molecular and electronic structures of bis(pyridine-2,6-diimine)metal complexes [ML$_2$](PF$_6$)$_n$ (n=0, 1, 2, 3; M=Mn, Fe, Co, Ni, Cu, Zn). *Inorg. Chem.*, **39**, 2936–2947.

39. Andrew, L.V., Alcock, N.W., Heppert, J.A., and Busch, D.H. (1998) An octahedral template based on a new molecular turn: synthesis and structure of a model complex and a reactive, diphenolic ligand and its metal complexes. *Inorg. Chem.*, **37**, 6912–6920.

40. (a) Drew, M.G.B., Lavery, A., McKee, V., and Nelson, S.M. (1985) The structure of a dinuclear copper(I) complex of a Schiff-base ligand containing a copper–copper bond. *J. Chem. Soc., Dalton Trans.*, **1985**, 1771–1774; (b) Piguet, C. Bernardinelli, G., and Williams, A.F. (1989) Preparation and crystal structure of the unusual double-helical copper(I) complex bis(2,6-bis(1-methylbenzimidazol-2-yl)pyridine)dicopper(I) naphthalene-1,5-disulfonate. *Inorg. Chem.*, **28**, 2920–2925.

41. Kaim, W. and Rall, J. (1996) Copper – a "modern" bioelement. *Angew. Chem. Int. Ed.*, **35**, 43–60.

42. (a) Hutin, M., Schalley, C.A., Bernardinelli, G., and Nitschke, J.R. (2006) Helicate, Macro-cycle, or catenate: dynamic topological control over subcomponent self-assembly. *Chem. Eur. J.*, **12**, 4069–4070; (b) Sarma, R.J. and Nitschke, J.R. (2008) Self-assembly in systems of sub-components: simple rules, subtle consequences. *Angew. Chem. Int. Ed.*, **47**, 377–380.

43. Pfeil, A. and Lehn, J.M. (1992) Helicate self-organisation: positive cooperativity in the self-assembly of double-helical metal complexes. *Chem. Commun.*, **1992**, 838–840.

44. (a) Dietrich-Buchecker, C.O. and Sauvage, J.-P. (1991) *Bioorganic Chemistry Frontiers*, vol. **2**, Springer, Berlin, pp. 197–248; (b) Harriman, A. and Sauvage, J.-P. (1996) *Chem. Soc. Rev.*, **41** and references cited therein.

45. Marquis-Rigault, A., Dupont-Gervais, A., Baxter, P.N.W. *et al.* (1996) Self-assembly of an 11-component cylindrical inorganic architecture: electrospray mass spectrometry and thermo-dynamic studies. *Inorg. Chem.*, **35**, 2307–2310.

46. El-ghayoury, A., Harriman, A., De Cian, A. *et al.* (1998) Redox cycling of segmented copper helicates. *J. Am. Chem. Soc.*, **120**, 9973–9974.

47. (a) Sano, M. and Taube, H. (1991) Molecular hysteresis. *J. Am. Chem. Soc.*, **113**, 2327–2328; Tomita, A. and Sano, M. (1994) Linkage isomerizations of (Sulfoxide)ammineruthenium com-plexes induced by electrochemical processes. *Inorg. Chem.*, **33**, 5825–5830;Sano, M. and Taube, H. (1996) Determination of memory life of a molecular hysteresis molecule by thin layer CV 981. *Chem. Lett.* and references cited therein.

48. Sessoli, R., Gatteschi, D., Caneschi, A., and Novak, M.A. (1993) Magnetic bistability in a metal-ion cluster. *Nature*, **365**, 141–143.

49. Yashima, E., Maeda, K., and Okamoto, Y. (1999) Memory of macromolecular helicity assisted by interaction with achiral small molecules. *Nature*, **399**, 449–451.

50. Drexler, K.E. (1992) *Nanosystems, Molecular Machinery, Manufacturing and Computation*, John Wiley & Sons, Inc., New York.

51. Sauvage, J.-P. (1991) Transition metal-containing rotaxanes and catenanes in motion: toward molecular machines and motors. *Acc. Chem. Res.*, **31**, 611–619.

52. (a) Zelikovich, L. Libman, J., and Shanzer, A. (1995) Molecular redox switches based on chem-ical triggering of iron translocation in triple-stranded helical complexes. *Nature*, **374**, 790–792; (b) Canevet, C. Libman, J., and Shanzer, A. (1996) Molecular redox-switches by ligand exchange. *Angew. Chem. Int. Ed.*, **35**, 2657–2660.

53. Ward, T.R., Lutz, A., Parel, S.P. *et al.* (1999) An iron-based molecular redox switch as a model for iron release from enterobactin via the salicylate binding mode. *Inorg. Chem.*, **38**, 5007–5017.

54. (a) Livoreil, A. Dietrich-Buchecker, C.O., and Sauvage, J.P. (1994) Electrochemically triggered swinging of a [2]-Catenate. *J. Am. Chem. Soc.*, **116**, 9399–9400; (b) Collin, J.P. Gavina, P., and Sauvage, J.P. (1996) Monitoring the iron(II)-induced self-assembly of preorganized tritopic ligands by electrospray mass spectrometry: unique formation of metallomacrocycles. *Chem. Commun.*, **1996**, 551–553.

55. (a) Polishchuk, A.P. and Timofeeva, T.V. (1993) Metal-containing liquid-crystal phases. *Russ. Chem. Rev.*, **62**, 291–321; (b) Deschenaux, R. and Goodby, J.W. (1995) Ch. 9, in *Ferrocenes* (eds A. Togni and T. Hayashi), VCH, Weinheim, pp. 471–495; (c) Bruce, D.W. (1996) *Inorganic Materials*, 2nd edn (eds D.W. Bruce and D. O'Hare), John Wiley & Sons, Inc., New York, pp. 429–522; (c) Serrano, J.L. (1996) *Metallomesogens*, VCH, Weinheim; (d) Giroud-Godquin, A.M. (1998) My 20 years of research in the chemistry of metal containing liquid crystals. *Coord. Chem. Rev.*, **171/180**, 1485–1499.

56. (a) Tschierske, C. (1996) Molecular self-organization of amphotropic liquid crystals. *Prog. Polym. Sci.*, **21**, 775–852; (b) Serrano, J.L. (1997) The state of the art in metal-lomesogenic polymers. *Prog. Polym. Sci.*, **21**, 873–911; (c) Stebani, U., Lattermann, G., Wittenberg, M. and Wendorff, J.H. (1996) Metallomesogens with branched, dendrimeric amino ligands. *Angew. Chem. Int. Ed.*, **35**, 1858–1861; (d) Tschierske, C. (1998) Non-conventional liquid crystals-the importance of micro-segregation for self-organisation. *J. Mater. Chem.*, **8**, 1485–1508.

57. Donnio, B. and Bruce, D.W. (1999) Metallomesogens. *Struct. Bond. (Berlin)*, **95**, 193.

58. (a) Stupp, S.I., LeBonheur, V., Walker, K. *et al.* (1997) Supramolecular materials: self-organized nanostructures. *Science*, **276**, 384–389; (b) Klepppinger, R. Lillya, C.P., and Yang, C. (1997) Discotic liquid crystals through molecular self-assembly. *J. Am. Chem. Soc.*, **119**, 4097–4102.

59. Reek, J.N.H., Priem, A.H., Engelkamp, H. *et al.* (1997) Binding features of molecular clips. Separation of the effects of hydrogen bonding and π-π interactions. *J. Am. Chem. Soc.*, **119**, 9956–9964.

60. Ziessel, R., Douce, L., El-ghayoury, A. *et al.* (2000) Unusual smectic ordering of unlocked copper bis-(terpyridine) complexes. *Angew. Chem. Int. Ed.*, **39**, 1489–1493.

61. Davidson, P., Levelut, A.-M., Strzelczka, H., and Gionis, V. (1983) Nature of the mesophase of a conducting charge transfer complex: neither discotic nor calamitic. *J. Phys. Fr.*, **44**, L823–L828.

62. Lattermann, G., Schmidt, S., Kleppinger, R., and Heiz Wendorff, J. (1992) The first example of a tridentate azamacrocyclic metallomesogen. *Adv. Mater.*, **1**, 30–33.

63. El-ghayoury, A., Douce, L., Skoulios, A., and Ziessel, R. (1998) π-Stacked *ortho*-palladated bipyridine complexes exhibiting unusual liquid crystalline behavior. *Angew. Chem. Int. Ed.*, **37**, 1255–1257.

64. Salditt, T., Koltover, I., Rädler, J.O., and Safinya, C.R. (1997) Two-dimensional smectic ordering of linear DNA chains in self-assembled DNA-cationic liposome mixtures. *Phys. Rev. Lett.*, **79**, 2582–2585.

65. Chisholm, M.H. (2000) One-dimensional polymers and mesogens incorporating multiple bonds between metal atoms. *Acc. Chem. Res.*, **33**, 53–61 and references therein.

66. (a) Mueller, A. and O'Brien, D.F. (2002) Supramolecular materials via polymerization of mesophases of hydrated amphiphiles. *Chem. Rev.*, **102**, 727–757; (b) Hentze, H.P. and Kaler, E.W. (2003) Polymerization of and within self-organized media. *Curr. Opin. Colloid Interface Sci.*, **8**, 164–178.

67. (a) Deng, H., Gin, D.L., and Smith, R.C. (1998) Polymerizable lyotropic liquid crystals containing transition-metal and lanthanide ions: Architectural control and introduction of new properties into nanostructured polymers. *J. Am. Chem. Soc.*, **120**, 3522–3533; (b) Smith, R.C., Fischer, W.M., and Gin, D.L. (1997) Ordered poly(p-phenylenevinylene) matrix nanocomposites via lyotropic liquid-crystalline monomers. *J. Am. Chem. Soc.*, **119**, 4092–4093.

68. (a) Ding, J.H. and Gin, D.L. (2000) Catalytic Pd nanoparticles synthesized using a lyotropic liquid crystal polymer template. *Chem. Mater.*, **12**, 22–24; (b) Gu, W. Zhou, W.J., and Gin, D. L. (2001) A nanostructured, scandium containing polymer for Lewis acid catalysis in water. *Chem. Mater.*, **13**, 1949–1951.

69. Gray, D.H. and Gin, D.L. (1998) Polymerizable lyotropic liquid crystals containing transition-metal ions as building blocks for nanostructured polymers and composites. *Chem. Mater.*, **10**, 1827–1832.

70. (a) Robbie, K. Broer, D.J., and Brett, M.J. (1999) Chiral nematic order in liquid crystals imposed by an engineered inorganic nanostructure. *Nature*, **399**, 764–766; (b) Theissen, U., Zilker, S.J., Pfeuffer, T., and Strohriegl, P. (2000) Photopolymerizable cholesteric liquid crystals—new materials for holographic applications. *Adv. Mater.*, **12**, 1698–1700; (c) Penterman, R., Klink, S.I., de Koning, H. *et al.* (2002) All-metallic three-dimensional photonic crystals with a large infrared bandgap. *Nature*, **417**, 55–58.

71. Ziessel, R., Harriman, A., El-ghayoury, A. *et al.* (2000) First assembly of copper(I) naphthyridine-based helicates. *New J. Chem.*, **24**, 729–732.

72. Douce, L., El-ghayoury, A., Skoulios, A., and Ziessel, R. (1999) Columnar mesophases from tetrahedral copper(I) cores and Schiff-base derived polycatenar ligands. *Chem. Commun.*, **1999**, 2033–2034.

73. Douce, L., Diep, T.H., Ziessel, R. *et al.* (2003) Columnar liquid crystals from wedge-shaped tetrahedral copper(I) complexes. *J. Mater. Chem.*, **13**, 1533–1539.

# 8

# Helicates, Peptide-Helicates and Metal-Assisted Stabilization of Peptide Microstructures

*Markus Albrecht*

*Institut für Organische Chemie, RWTH Aachen University, Germany*

## 8.1   Introduction

Amino acids and their oligomers, peptides and proteins are of essential biological importance. In proteins, structure and function is combined in an exceptionally elegant way.
The specific features of the single amino acid components, which depend on the nature of
the side chain functionalities (polarity, etc.), are crucial for the structural arrangement –
the folding. Specific properties are introduced by the action and interaction of those simple groups on the periphery of the peptide [1].

Due to the high number of donor atoms hidden in peptides, they are excellent chelating
ligands for metal ions [2,3]. Coordinated metals are often involved in reaction centers or
might just act as connecting units to stabilize specific structures. In nature, examples are
found for both of those situations:

1. A representative example for proteins in which the metal is the reactive moiety, which
   promotes chemical reactivity, is the class of carboxypeptidases. In this group of
   enzymes zinc(II) is the active metal center which facilitates the cleavage of peptidic
   bonds [4].
2. The same metal, zinc(II), is present in the zinc finger proteins. Here the metal coordinates to histidine and cysteine residues and acts as a connecting unit that arranges the

*Metallofoldamers: Supramolecular Architectures from Helicates to Biomimetics*, First Edition.
Edited by Galia Maayan and Markus Albrecht.
© 2013 John Wiley & Sons, Ltd. Published 2013 by John Wiley & Sons, Ltd.

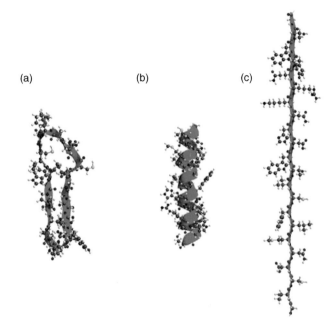

(a)          (b)          (c)

**Scheme 8.1**  *Different kinds of peptide structures: (a) a loop/turn structure, (b) an α-helix, and (c) a single peptide strand with β-sheet conformation.*

peptide chain in a specific geometry. Both α-helical and β-sheet folding are induced within segments of a larger loop type structure. This conformationally fixed metallo-protein is able to bind to DNA and to regulate its function [5].

The present chapter introduces the chemistry of peptide metal–ligand conjugates and their coordination compounds. The binding of metal ions to the ligand donor sites on peptides induces special conformations. Initially, selected examples of peptide metal-ligand conjugates are described to introduce the field. Then, peptide-based ligands are described which are used in the self-assembly of dinuclear double- or triple-stranded helicates. This topic directly leads to ligand substituted peptides which, upon metal coordination, adopt specific structures: loops/turns, α-helices, or β-sheets (Scheme 8.1).

## 8.2   Selected Examples of Metal Peptide Conjugates

Metal peptide conjugates are based on a "direct" interaction of the peptide backbone with metal ions or on the coordination of metal ions to donors at peptide side chains. In non-natural systems metal-binding sites can be attached to the peptide either at the side chain or at the terminus [6,7].

The coordination of metal ions to the peptide backbone occurs either by binding to the carboxylate (Figure 8.1, site a) or to the amine terminus (Figure 8.1, site b), to the carbonyl oxygen (Figure 8.1, site c) or to deprotonated amide functionalities (Figure 8.1,

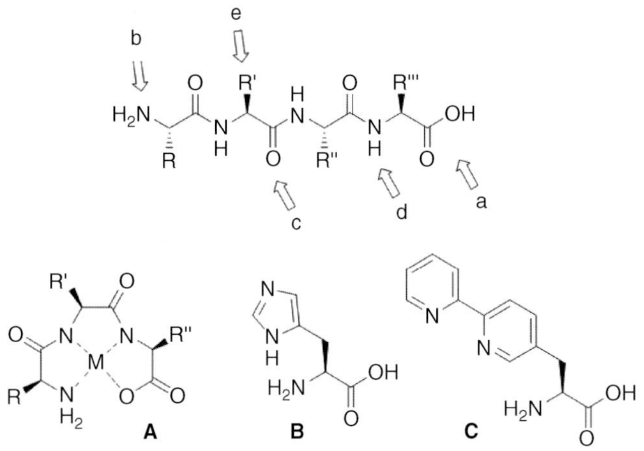

**Figure 8.1** *A short peptide possessing different metal binding sites a–e and selected examples of natural (A, B) and non-natural (C) metal-binding sites.*

site d). Direct backbone coordination is exemplified in **A**, showing binding of the terminal carboxylate and amine as well as deprotonated amide units. Related structures are frequently found with copper(II) ions as the respective metal center [8].

An ideal anchoring unit for metals can be the side chain of the amino acid building blocks of the peptide. In this context, the natural occurring amino acid histidine **B** is the raw model for the development of artificial metal coordinating amino acids as building blocks for peptides. Modification of the imidazole nitrogen donor-binding site results in amino acids like the 2,2′-bipyridine derivative **C**[9]. In addition to **C,** the corresponding terpyridine derivatives have also been prepared [10]. Although the nitrogen donor systems are the most widely used ones, significant work has been devoted towards the introduction of phosphorous [11] or oxygen donors [12] as amino acid side chains.

Figure 8.2 shows selected examples of bipyridine-substituted amino acid or peptide metal conjugates **1–3.** In **1,** a bipyridine unit is incorporated into a short peptide by the attachment of Val-Val units to the 3- and 5-position of the ligand unit. Bipyridine is able to bind ruthenium(II) bis bipyridine, while cobalt(III)pentaamine can be coordinated to the carboxylate terminus. This arrangement allows investigation of electron transfer processes between the metal centers along the short Val-Val β-sheet [13].

In compound **2,** a central trisamide-linked bis(bipyridine) coordinates the redox- and photoactive ruthenium(II) bipyridine fragment [14]. In a similar fashion, cobalt(II) fixes a Gly-Gly-Gly-Gly loop in **3.** It coordinates to two bipyridine units, forming a tetrahedral complex. The amino acid termini of the organic ligand are brought together by the corresponding conformational change and the through space interaction of those units by luminescence quenching has been investigated [15].

An alternative to the side chain functionalization of peptidic amino acid building blocks is the attachment of ligand units to the N- or C-terminus of peptides. This will be discussed in more detail below. However, Figures 8.3 and 8.4 show examples **4, 6** and **7** with metal complexes attached to the terminus of peptide strands.

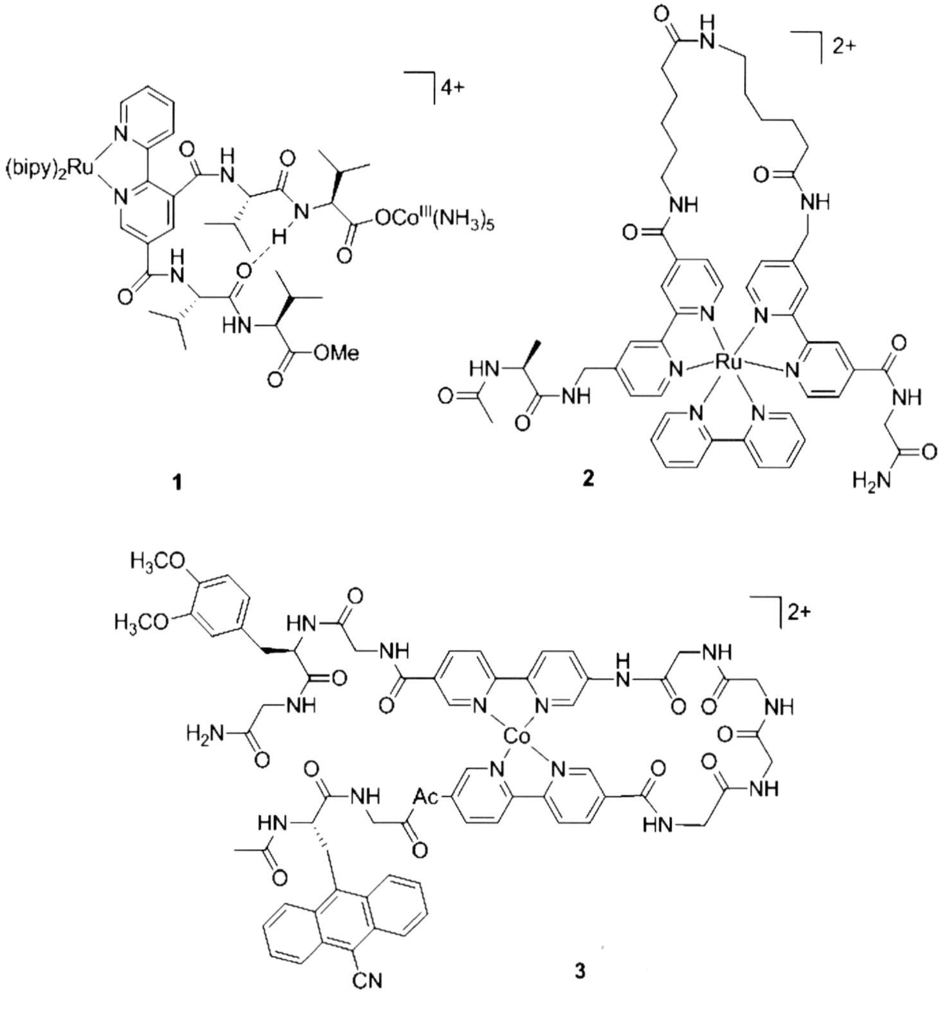

**Figure 8.2**   *Bipyridine-based metal amino acid or metal peptide conjugates.*

A series of organometallic ferrocene derivatives was studied in which the organo-metallic unit is incorporated into peptide structures. Examples are the dimeric ferrocenyl Gly-Cys derivative **4** [16] or the doubly amino acid bridged bisferrocene **5** [17].

Compounds **6** and **7** represent classical coordination compounds, which are attached to the N-terminus of peptides. Especially complex **7** possesses a high potential for applications in radio therapeutics and diagnostics [18].

The examples shown in Figures 8.2–8.4 are only a few representatives of a huge class of compounds which are prepared in order to investigate the mutual interplay between a peptide and a metal center. Those studies are mainly directed towards the development of novel sensing and diagnostic materials, to the search for new reactivities and selectivities, or to new bioactive coordination compounds [19].

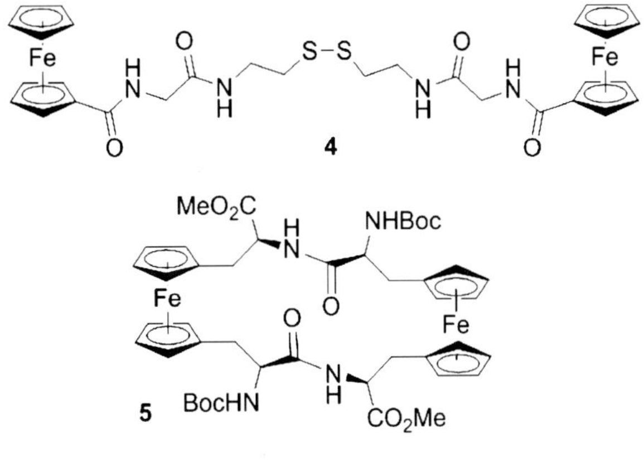

**Figure 8.3** *Ferrocene amino acid/peptide conjugates.*

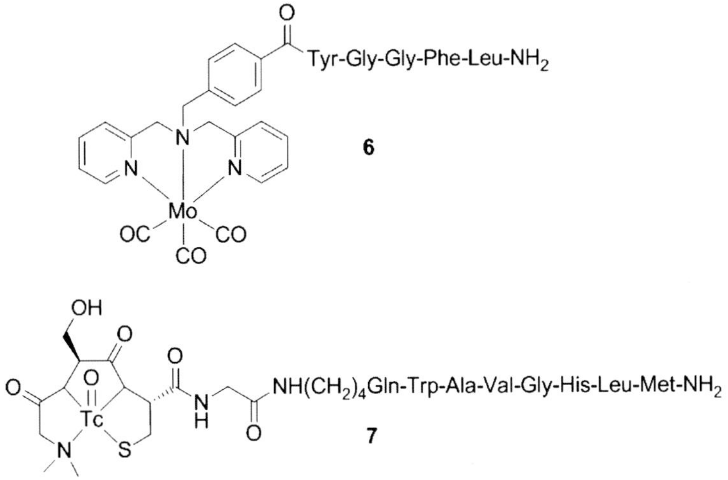

**Figure 8.4** *Peptides bearing terminal complex units.*

## 8.3 Helicates and Peptide-Helicates

In this section, helicates will be briefly introduced and peptide-helicates incorporating amino acids or short peptides as spacers will be described in detail.

### 8.3.1 Helicates

In 1987, the term "helicate" was introduced by Lehn for oligonuclear coordination compounds in which two or more linear ligand strands wrap around two or more metal centers (Figure 8.5) [20]. The complexes are formed in a "spontaneous" self-assembly process by a simple mixing of ligands and metal ions [21,22].

**Figure 8.5** *Schematic representation of a double- and a triple-stranded dinuclear helicate.*

Figure 8.6 shows representative examples of oligodentate ligands used in helicate chemistry [23–26].

Starting in the 1990s, catechol ligands like **12** were intensely studied. Upon deprotonation of the catechol unit, an effective ligand system for the formation of helicates is obtained. **12** and related derivatives were used to study the mechanisms of the self-assembly, structure, ability to form host–guest complexes, and stereochemical features of the oligonuclear helicates [27]. In the latter case it was found that appropriate design of the spacer enables the diastereoselective formation of either the "classical" helicates or the meso-form. In the meso-form one metal complex unit is configured $\Lambda$ and the other $\Delta$. Those non-chiral coordination compounds with two helical domains possessing opposite configuration are termed side by side complexes, meso-helicates, or mesocates [28].

Today the chemistry of helicates still represents an evolving field of research, due to the model character of the relatively simple metallosupramolecular species. Intense study of

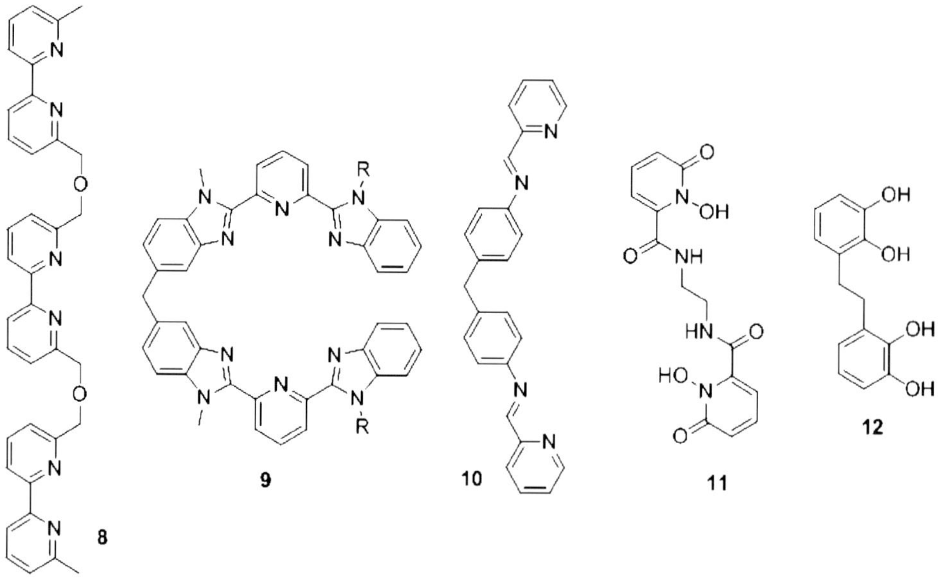

**9**   **10**   **11**   **12**   **8**

**Figure 8.6** *Representative examples for ligands forming helicates in the presence of appropriate metal ions.*

the helicates allows a deeper insight into mechanisms which are also active for much more complicated systems. In addition, "function" (e.g., metal–metal communication) has become important. As a further special aspect of helicate chemistry the structural organization of linear ligand strands upon metal coordination has to be mentioned.

### 8.3.2 Peptide-Helicates

In helicates, two or more metal ions are connected through oligodentate ligands bearing connecting bridges. In this context, it became of interest whether simple amino acids or short peptides can be used as spacers of the helicating ligands. Those would allow the investigation of the mutual interplay of structural information at the metal and at the peptide. The self-assembly of well defined complexes should be controlled by a subtle cooperation of the different effects.

Thus, amino acid bridged dicatechol ligands **13**-$H_4$ were prepared by standard peptide coupling reactions using common protecting group strategies (Figure 8.7) [29].

Complexation of the ligands **13**-$H_4$ with titanium(IV) ions under basic conditions did not lead to the formation of triple-stranded helicates [30]. ESI MS showed that double-stranded coordination compounds of the type [(**13**)$_2$(OR′)$_2$Ti$_2$]$^{2-}$ with hydroxyl or alcoholate coligands are formed. However, mass spectrometry cannot distinguish between different isomers. Proton NMR, in contrast, shows a collection of several isomers to be present. This is due to different relative orientations of the amino acid bridged ligands (parallel or antiparallel) and to different configurations at the metal complex units ($\Lambda$ or $\Delta$). In total, seven different isomers can be formed (Figure 8.8).

Figure 8.9 shows representative example proton NMR resonances of one of the diastereomeric benzyl protons in the NMR spectrum of [(**13d**)$_2$(OMe)$_2$Ti$_2$]$^{2-}$. Initially (15 h after preparation of the sample), seven different species can be observed in the spectrum. After 14 days several of the signals nearly disappeared and one species is highly dominant. This means that initial complex formation proceeds unspecifically under kinetic

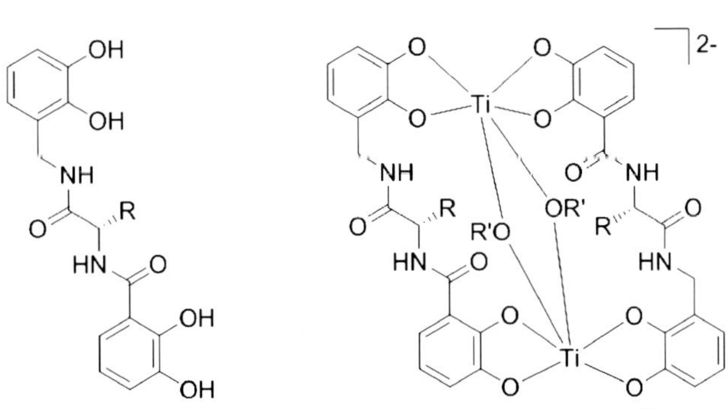

**13**-$H_4$

[(**13**)$_2$(OR′)$_2$Ti$_2$]$^{2-}$

R = H (**a**), Me (**b**), iPr (**c**), Bz (**d**) ...
R′ = H, Me

**Figure 8.7** *Amino acid bridged dicatechol ligands and their respective dinuclear double-stranded titanium(IV) complexes (only the isomer with parallel orientation of the ligand strands is shown).*

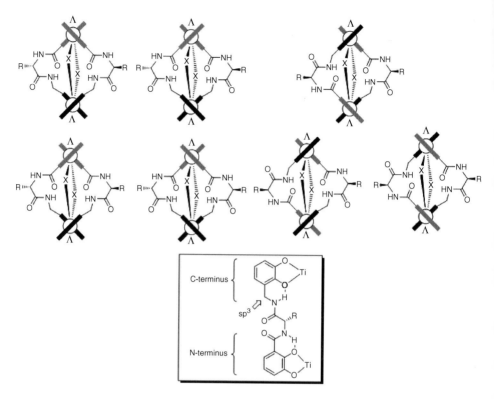

**Figure 8.8** *Schematic representation of the seven isomers which can be formed from ligands* **13** *and titanium(IV) ions. Catechol units are only shown as bars (grey: N-terminal; black: C-terminal); X = OR. The inset shows the preferred conformation of the ligand due to intramolecular NH–O hydrogen bonding.*

control and the thermodynamically favored product is only obtained after a couple of days at room temperature.

It was found that an initially high optical rotation of the solution of complexes decreases over a period of some time. This effect indicates a "loss of stereochemical information" which can be explained by the preferred formation of a $\Lambda\Delta$ heterochiral dinuclear titanium(IV) complex. This assumption is supported by computational considerations showing that four stabilizing intramolecular hydrogen bonds are present in the heterochiral complex. In the corresponding homochiral compound only two can be formed.

A series of crystal structures of $Li_2[(\mathbf{13})_2(OR)_2Ti_2]$ (R = Me, H) complexes with different amino acids were obtained which allowed a conformational analysis. The final considerations followed a series of sequential steps:

1. From the observations of the optical rotation and from computational results it was assumed that the dominating species adopts a heterochiral $\Lambda\Delta$ configuration at the metal complexes (see above).

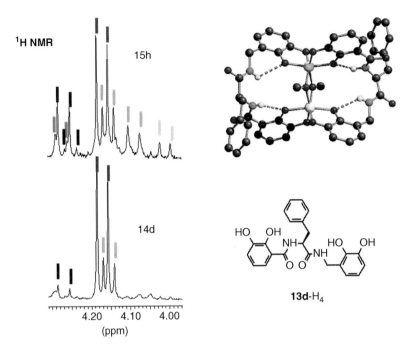

**Figure 8.9** *Part of the NMR spectrum of the dinuclear titanium(IV) complex of ligand **13d** showing the signal of one of the two diastereomeric benzylic protons after 15 h and 14 days and structure of the dinuclear dianionic complex [(**13d**)₂(OMe)₂Ti₂]²⁻ as found in the crystal.*

2. The conformation at the amino acid backbone can be analyzed based on the X-ray structural data using Ramachandran's method. Therefore the dihedral angles Φ and Ψ at the amino acid spacers were set into correlation.
   The analysis (Figure 8.10) reveals that the amino acid residues prefer to adopt the conformation of either a right-handed α-helix or a right-handed twisted β-sheet.
3. Considerations regarding the ligand backbone reveal that two different linkages are present between the catechol units and the amino acids. Those linkages should be responsible for stereochemical induction. On the one hand there is the benzyl amide at the C-terminus with an sp³ carbon atom within the six membered ring formed by NH—O hydrogen bonding. Due to the sp³ carbon this unit is rather flexible and is able to level out strains, which are built up upon metal coordination. The catecholamide (N-terminus), on the other hand, forms a planar six-membered ring by hydrogen bonding. This unit is rigid and transfers stereochemical information from the amino acid to the metal complex. From the experimental results it can be deduced that S-configuration at the amino acid induces Λ at the N-terminal catechol complex.

All the considerations described under 1–3 yield only one possible preferred isomer of Li₂[(**13**)₂(OR)₂Ti₂]. The titanium complexes have to have different configurations, one Λ and one Δ. In order to adopt a right-handed helical twist at the amino acid, both N-termini have to bind to the Λ-configured metal center.

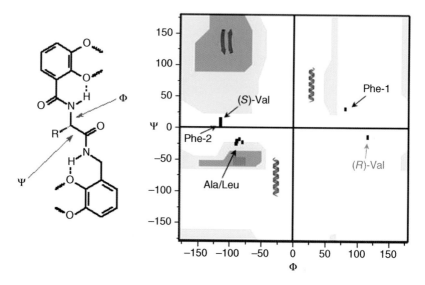

**Figure 8.10** *Ramachandran plot for dihedral angles at amino acid residues of a series of complexes $Li_2[(13)_2(OR)_2Ti_2]$.*

In conclusion, the thermodynamically most favored isomer is the one with a parallel orientation of the ligands with both N-termini binding to a $\Lambda$ and the C-termini binding to a $\Delta$-configured titanium center. Those studies impressively show how the interplay of experiment and theory together with conformational considerations can help to understand relatively simple stereochemical processes [31].

Substitution of the amino acid in the spacer of the dicatechol ligands **13** by short peptides results in ligands **14** and **15**. Now the binding sites are far separated and no binding of small bridging coligands can occur. The formation of triple-stranded helicates should be possible with titanium(IV).

Accordingly, ligand **14**-$H_4$ coordinates to titanium(IV) ions and forms the triple-stranded helicate $[\mathbf{14}_3Ti_2]^{4-}$ which has been characterized by ESI MS. $^1$H NMR spectroscopy shows that initially a mixture of isomers is formed. In DMSO it slowly transforms into only one thermodynamically favored species with all three ligand strands orientated parallel (Figure 8.11).

Use of the longer tripeptide ligand **15**-$H_4$ in coordination studies with titanium(IV) affords a mixture of compounds. Triple-stranded helicates are only one component of the mixture. Due to the flexibility of the relatively big ligand, loop type structures are additionally formed in which both catechols of a ligand bind to the same metal center.

This is also observed, if dipeptides with less sterically demanding side chains are used instead of the Val-Val spacer of **14**. It is obvious that ligand **14** represents a special case in which the ligand has to adopt a linear arrangement, enforced by the short chain length in concert with repulsion of the bulky groups. Reduction of the constraints by using longer peptides or less bulky groups enables back-binding of the chelating units to bind to only one metal center [32].

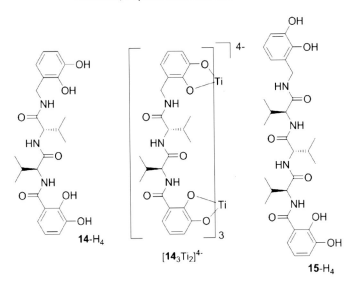

**Figure 8.11** *Helicating ligands bearing di- or tripeptides as spacers and the thermo-dynamically preferred titanium complex [14₃Ti₂]⁴⁻ of ligand 14 with parallel arrangement of all three ligand strands.*

For the preparation of well defined peptide helicates, metals with different coordination preferences in combination with different ligand units can be used to gain some structural control over the outcome of a coordination study.

The principle is schematically presented in Figure 8.12. If sequential ligands are used, the strands can be orientated either parallel or antiparallel. The addition of one kind of metal ion preferably results in an oligonuclear complex with an antiparallel orientation of the ligand strands. The similar metal ions are approaching a coordination environment as similar as possible. Nature tries to avoid charge separation between similar centers. If the same (or a related) ligand strand reacts with two different kinds of metal ions, one will prefer the "first", the other the "second" coordination site. A parallel orientation of the ligand strands will result in a heteronuclear complex [33].

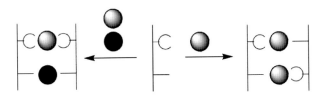

**Figure 8.12** *Control of the relative orientation (parallel versus antiparallel) of sequential ligands (presented as straight line bearing two different binding sites for metals). They are able to react with two different metals (grey and black) to result in a heterodinuclear complex with parallel orientation of the ligands. Reaction with only one kind of metal (grey) results in the formation of a complex with antiparallel orientation of the strands.*

The group of Lehn used a similar strategy for the formation of heteronuclear and heteroleptic helicates containing a high level of information.

Two strands of the bis-bipyridine/mono-terpyridine **16** bind two copper(I) ions and one iron(II) ion to form a heterotrinuclear double-stranded helicate. The denticity of the coordination site controls the position of the different metals. The two copper(I) coordinate two bipyridine units, adopting tetrahedral coordination geometries, while iron(II) prefers an octahedron formed by two terpyridine units. Ligand **17** leads to the opposite situation in the helicate with two iron bis-terpyridine moieties and only one copper bis-bipyridine moiety. This principle also works with the unsymmetrical ligand **18**, leading to a bis-copper(I) mono-iron(II) complex.

The series of ligands **16–18** results in the formation of heterotrinuclear iron(II)–copper(I) complexes with the sequence of the metals (Cu-Fe-Cu for **16**, Fe-Cu-Fe for **17**, Fe-Cu-Cu for **18**) programmed into the ligand strand (Figure 8.13). In addition, a mixture of ligands **16** and **17** forms heteroleptic trinuclear double stranded helicates $[(16)(17)Cu_3]^{6+}$ if copper(II) ions are added. Here the metal ions prefer a coordination number higher than four, resulting in the binding of one bipyridine and one terpyridine to the same metal

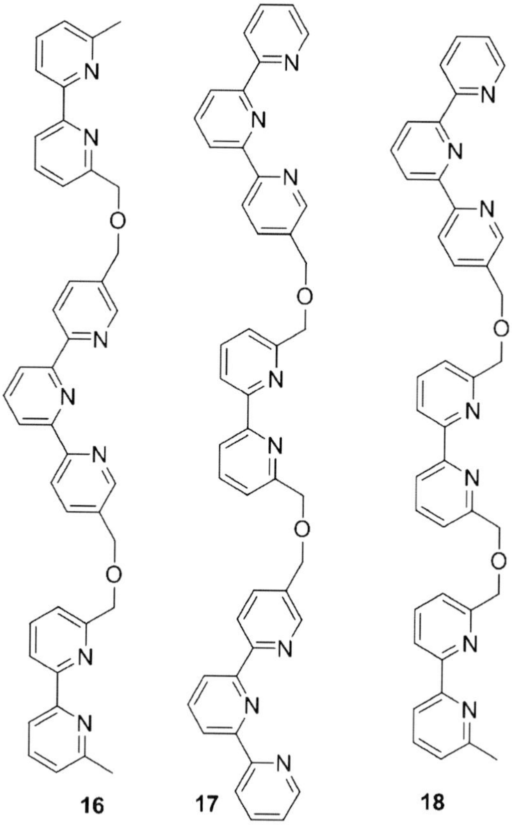

**16**          **17**          **18**

*Figure 8.13*   *Sequential bipyridine/terpyridine ligands.*

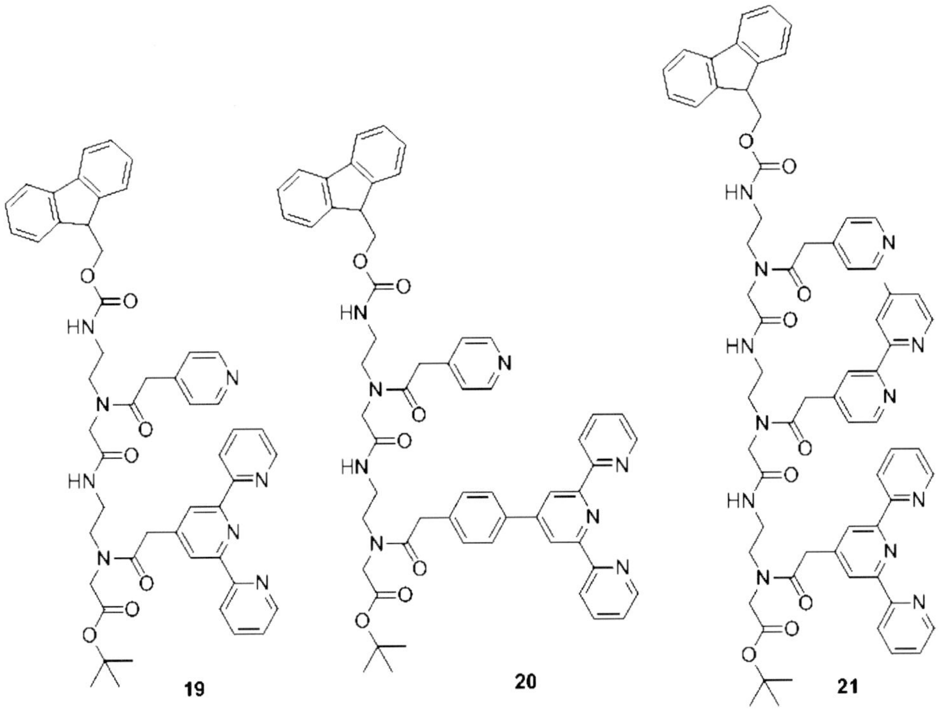

**Figure 8.14** *Sequential peptide-type ligands bearing pyridine, bipyridine and terpyridine as metal coordination sites.*

center. This enforces the formation of the hetero-stranded complex [34]. Recent studies by Williams *et al.* transferred the Lehns concept to related peptide derivatives.

The aminoethylglycine derivatives **19–21** were prepared (Figure 8.14). These bear as side chain functionalities mono- (pyridine), bi- (bipyridine), and tridentate (terpyridine) units. The addition of copper(II) salts results in the formation of double-stranded helicates with an antiparallel orientation of the two strands. The self-assembly is followed by spectrophotometry and the formed coordination compounds are characterized by positive ESI-MS spectra.

In the case of **19** and **20** a dinuclear helicate is formed [35] while ligand **21** results in a trinuclear complex with two copper(II) ions coordinating to terpyridine and pyridine and one copper(II) binding to two bipyridines. Thus each of the copper(II) ions is tetra-coordinated. Due to the high oxidation state at the metal, a square planar geometry is preferred (Figure 8.15) [36].

This describes how helicate-type coordination complexes can self-assemble from ligands, which are based on peptidic units. Here different motifs are used. The metal coordinating units are either attached to the termini of the peptide or are introduced as side chains. In addition, it has been shown that the incorporation of coligands can influence the outcome of a coordination study and that it is possible to control the relative orientation (parallel or antiparallel) of sequential/directional ligands. The ligands themselves are

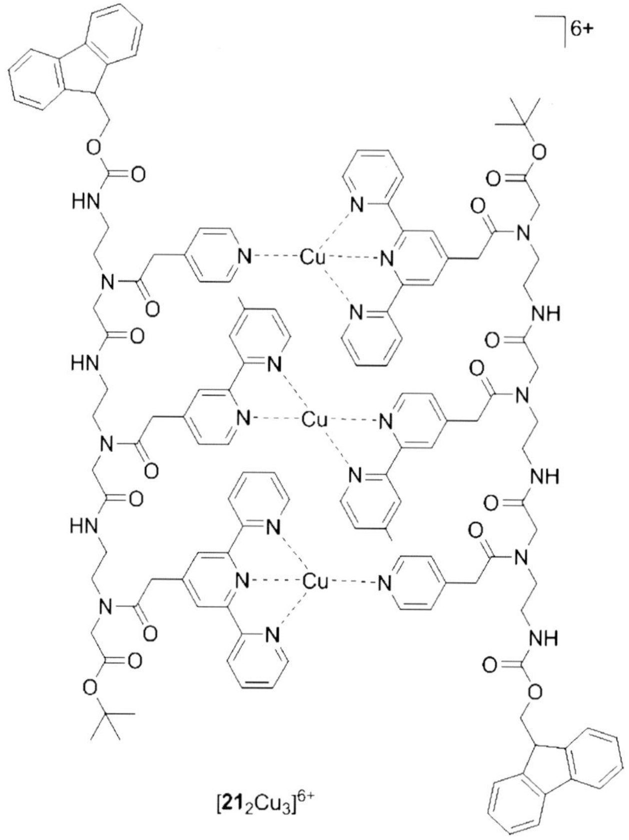

**Figure 8.15** *The trinuclear double-stranded helicate-type complex [21₂Cu₃]⁶⁺ showing anti-parallel orientation of the peptide type-ligands.*

usually highly flexible. However, in helical coordination compounds they have to adopt a conformationally fixed structure.

## 8.4 Metal-Assisted Stabilization of Peptide Microstructures

The investigation of peptide-helicates showed that metal coordination to peptide/metal–ligand hybrids is a tool to fix the conformation at the peptidic moiety. Viewing this from the point of the peptide it should be a possibility to stabilize well defined secondary structures by metal coordination (Scheme 8.2). Corresponding strategies to fix loops/turns, helices or sheets will be described in the following section [37].

### 8.4.1 Loops and Turns

As stated above, upon titanium coordination ligand **15**-H$_4$ forms the triple-stranded helicate [**15**$_3$Ti$_2$]$^{4-}$ along with complexes in which the ligand units fold back and both catechol units bind to the same metal. The latter is of special interest because here a loop type conformation is induced at the peptide [38]. This structural motif can be obtained as the

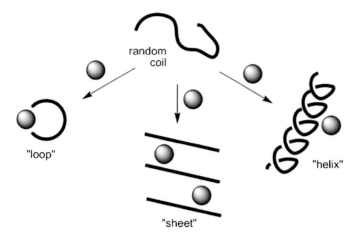

**Scheme 8.2** *Schematic representation of metal controlled fixation of secondary peptide structures.*

dominating/only species, either if a mixture of ligand **15** and catechol (1 : 1) is used in the coordination studies, or if the titanium(IV) unit is substituted by molybdenum(VI)dioxo (Figure 8.16). In the latter case the metal binds two catechols, while the third coordination site is already blocked by the two oxo ligands.

The Leu-Ala-Leu substituted dipyridine ligand **22** was prepared to form a metallamacrocycle by coordination to the Pd(en)$^{2+}$ (en = ethylene diamine) unit. Indeed the 1 : 1 complex is obtained as the dominant species. However, the corresponding 2 : 2 and 3 : 3 complexes are also observed in significant amounts [39]. The same concept is used with the bisterpyridine ligand **23**, which in the presence of iron(II) cleanly produces the metallamacrocycle [**23**Fe]$^{2+}$ (Figure 8.17) [40].

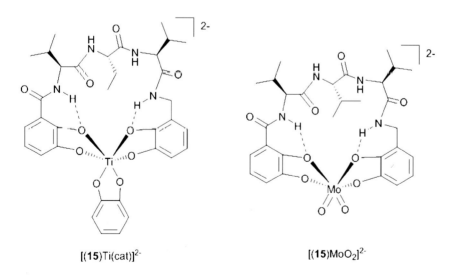

$$[(15)Ti(cat)]^{2-} \qquad\qquad [(15)MoO_2]^{2-}$$

**Figure 8.16** *Metal complexes in which a well defined Val-Val-Val loop-type structure is enforced by the binding of the metal ion.*

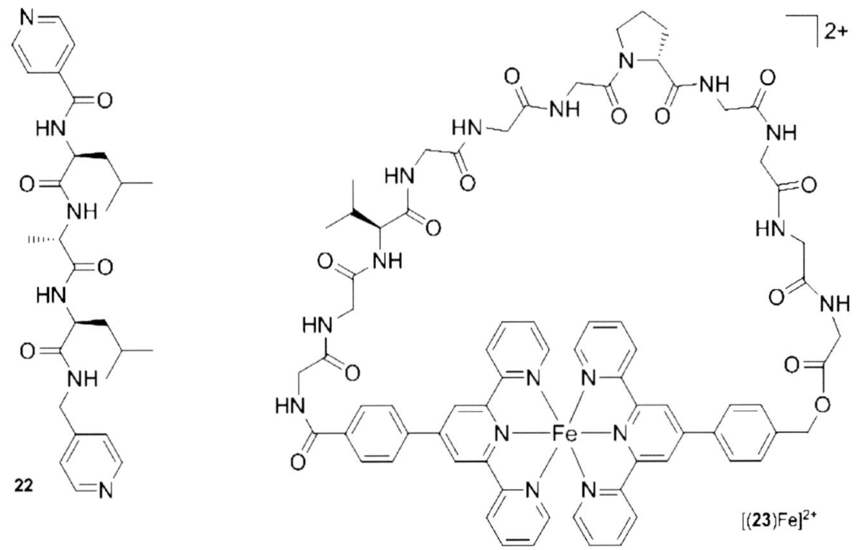

**Figure 8.17**   *Pyridine/terpyridine terminated peptides for the formation of metallomacrocyclic structures.*

The formation of conformationally well defined metallomacrocycles from random coil peptides terminated by metal-binding sites is of interest. However, the control of structure is often related to the control of a specific function of the molecule. One important property of peptides is their bioactivity. It would be desirable to control the bioactivity (e.g., on/off) of simple short peptides by effective processes like metal coordination. The following describes some examples of short catechol terminated peptides which, upon metal coordination, mimic the active parts of loop type peptides found in nature.

1. *Segetalins* are a class of naturally occurring cyclopeptides, isolated from the seeds of *Vaccaria segetalis*. Those seeds are used in Chinese home medicine for the treatment of breast infections, to promote milk secretion, to activate blood flow, and to treat amenorrhea. A comparison of different *Segetalins* (as an example *Segetalin A* is shown in Figure 8.18) makes it most probable that the active part is the Trp-Ala-Gly or the Trp-Ala-Gly-Val peptide sequence [41]. Thus, short peptides with catecholamide as well as dihydroxybenzyl termini were prepared and submitted to coordination studies with cis-molybdenum(VI)dioxo. The macrocyclic $1:1$ complexes [ $24MoO_2]^{2-}$ and $[25MoO_2]^{2-}$ are easily obtained. NMR spectroscopy of the complexes reveals sharp and well separated resonances indicating a well defined conformation of the ring. In the case of random coil free ligands, broader and less resolved spectra are observed [42,43]. The study shows that the desired metallomacrocycles are quantitatively formed with different oligopeptides. This provides a way to stabilize bioactive peptide turns.
2. *Urotensin II* was isolated from the goby fish *Gillichthys mirabilis*. It is one of the most potent vasoconstrictors known [44]. A structure–function analysis shows that the Trp-Lys-Tyr part of the cysteine-linked macrocycle is the bioactive part of the molecule. Incorporation of this front as a spacer into the dicatechol ligand **26** and

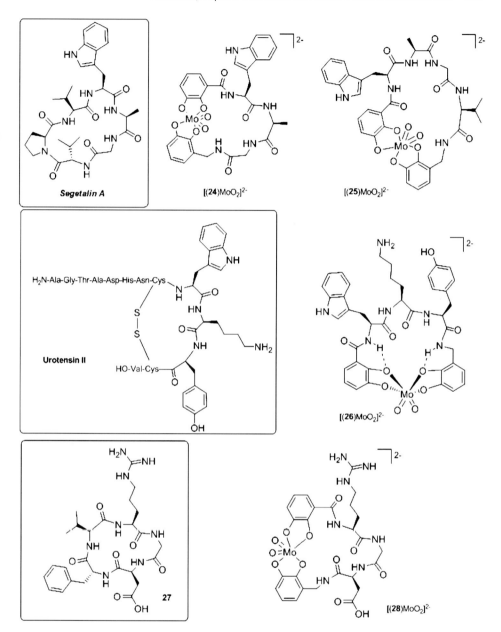

**Figure 8.18** *Naturally occurring bioactive loop structures and their macrocyclic molybdenum(VI)dioxo analogous.*

coordination to molybdenum(VI)dioxo results in the formation of the metallomacrocycle $[\mathbf{26}MoO_2]^{2-}$ with the active part fixed in a loop type conformation [45].

3. The arginine glycine aspartic acid (RGD = Arg-Gly-Asp) sequence is termed the universal cell recognition sequence, because it binds to integrins which are located as

membrane proteins on the surfaces of cells. RGD binding blocks cell–cell interaction. This is for example the active mode of the highly toxic protein *Echistatin* found in the venom of the viper *Echis carinatus* [46]. The cell recognition can be used for a novel concept of cancer therapy. From structure–activity studies it is well known that the RGD sequence is most active if it is placed in a loop type region of a peptide, as observed in *Echistatin*. Thus, many macrocyclic RGD derivatives were prepared, with **27** being the most prominent [47]. $[\mathbf{28}MoO_2]^{2-}$ is easy to obtain by either solution or solid phase preparation of the ligand followed by complexation of the molybdenum(VI)dioxo unit. The macrocycle containing the RGD sequence in the ring is obtained quantitatively.

The described examples show that it is possible to prepare simple linear peptide derivatives adopting in their free state a random coil structure. Metal coordination to terminal binding sites fixes a loop/turn structure, which might possess enhanced bioactivity.

### 8.4.2   α-Helices

The previous section shows that metal coordination is able to fix peptide units in a specific structure. α-Helices represent a prominent motif in peptide chemistry which is important for the interaction of peptides. The α-helix is formed due to intramolecular hydrogen bonding between amide NH units and carbonyl oxygen atoms. This hydrogen bonding leads to the helical motif if three amino acid residues separate the hydrogen bond donor from the corresponding acceptor [48]. A trick for the stabilization of an α-helix is to introduce attractive interactions on one front of the helical column. This can be done by non-polar amino acids, which are orientated to only one face of the helical "column" and thus preferably interact with other derivatives (e.g., other α-helices) by hydrophobic interactions. Another approach is to tether two residues which are located in the *i* and *i* + *4* positions of the peptide strand (Figure 8.19).

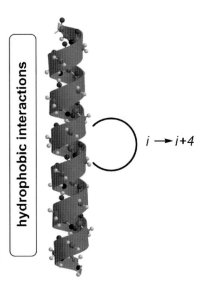

**Figure 8.19**   *Two approaches to stabilize α-helical peptide structures.*

In order to stabilize the α-helical structure by interaction of hydrophobic parts, two (or more) helices with hydrophobic faces have to be brought together in hydrophilic media. Metal templates are appropriate to fix the helices in close proximity to each other.

Scheme 8.3 shows a representative example for this principle. A bipyridine-terminated peptide **29** has been prepared. This 15mer adopts a random coil structure. Coordination of three molecules of **29** to an appropriate metal ion leads to an equilibrium between the anti and syn coordination isomers. In the anti isomer the random coil structure is still preferred. In the syn isomer, however, the α-helical peptides interact through their hydrophobic faces. Thus, a cooperative templating effect

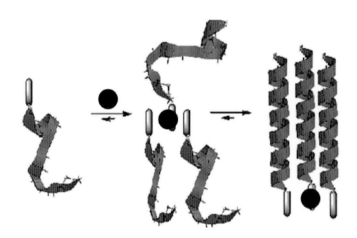

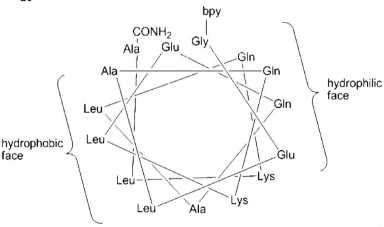

**Scheme 8.3** *Stabilization of α-helices by hydrophobic interactions between three peptide strands. The strands are brought together by metal coordination of three terminal bipyridine units to one metal ion.*

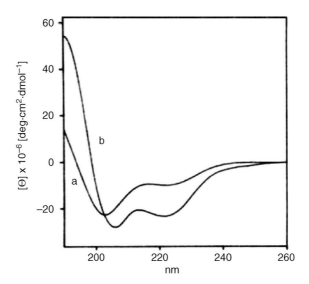

**Figure 8.20** *CD spectra of* **29** *in the absence (a) and presence (b) of NiCl₂.* Reprinted with permission from Ref. [49]. Copyright 1992 American Chemical Society.

stabilizes, on the one hand, the α-helical structure at the peptide (for CD spectroscopic evidence, see Figure 8.20) and, on the other hand, the statistically disfavored syn conformation at the metal complex unit [49].

As an alternative to this, two amino acids which are located in $i$ and $i + 4$ position of the peptide sequence can be tethered in order to stabilize the α-helix (Scheme 8.4). This can be done by the introduction of a covalent linkage [50]. However, an effective and synthetically much easier alternative is non-covalent attraction between amino acid side chains in the corresponding positions. The interaction can take place through H-bonding, electrostatics, or metal coordination.

To utilize metal coordination to the side chains of short peptides for the stabilization of the secondary structure, a His-Ala-Ala-Ala-His derivative has been prepared as a ligand of the palladium(II)ethylene diamine complex **30**. Complex formation proceeded very well and the obtained coordination compound was thoroughly investigated by different NMR techniques. It was shown that the short peptide of the metal complex adopts a single helical turn of an α-helix. In uncoordinated form the short sequence shows a random coil structure [51,52].

Introduction of a unit like **30** into a longer peptide induces the α-helical turn, which progresses through the whole peptide chain. Thus, the α-helical structure is not only induced at the metal coordination site but is transferred along the amino acid sequence. This was shown with peptide **31** which, upon coordination of ruthenium(II) tetraamine to the two histidine residues in the 7- ($i$) and 11-position ($i + 4$) of the strand, adopts an α-helical structure [53].

A further peptide that has been used to test this concept is **32** (Figure 8.21). In this case, two amino diacetates were attached to the side chains of the peptide with variation of

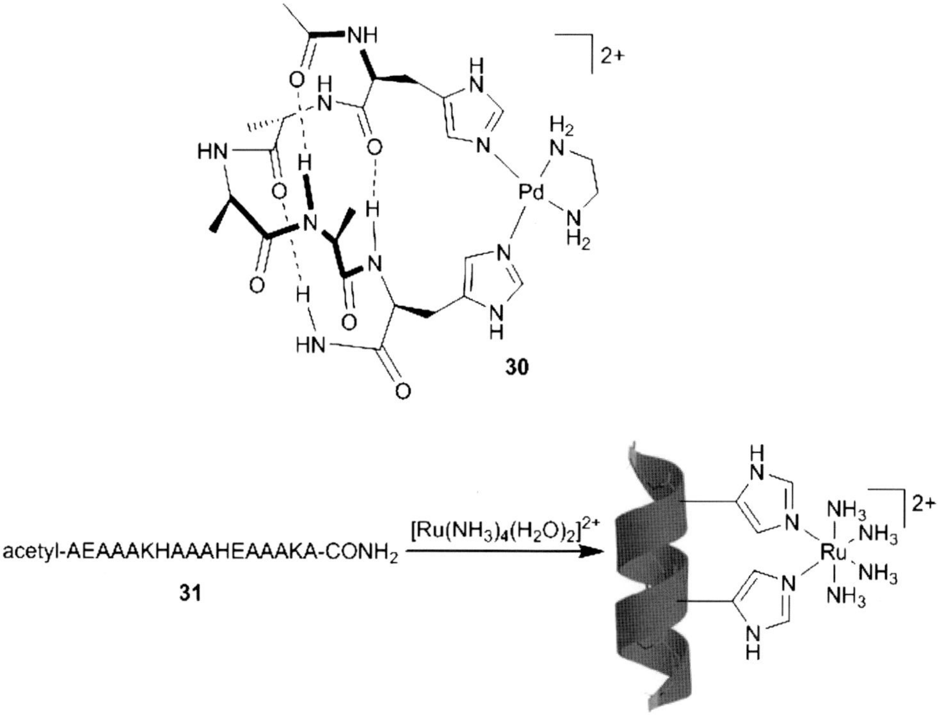

**30**

acetyl-AEAAAKHAAAHEAAAKA-CONH$_2$

**31**

$\xrightarrow{[Ru(NH_3)_4(H_2O)_2]^{2+}}$

**Scheme 8.4** *Stabilization of α-helical peptide structures through coordination to histidine residues located in i and i + 4 position of the peptide strand.*

the length of the connecting linkage. The introduction of two such units into the sequence affords a compound with a high analogy to the ethylene diamine tetraacetate (EDTA) chelator. The ethylene backbone of the parent EDTA is substituted by the short peptide. Coordination of **32** to cobalt(II), nickel(II), copper(II), zinc(II), or cadmium(II) results in the formation of coordination compounds with a preferred α-helical structure. CD spectroscopy reveals that especially cadmium(II) is highly effective in stabilizing the helix [54].

Other metal-binding amino acids for incorporation into peptides are the crown ether **A** and the cyclic triamine **B**. With peptides possessing two units of **A** no α-helix stabilization was observed. However, Boc-Ala$_3$-A-Ala$_2$-A-Ala$_3$-NHC$_3$H$_7$ adopts a turn structure in the presence of cesium cations [55].

In contrast, introduction of the triazanonane amino acid **B** into a peptide afforded a helix-loop-helix motif upon addition of zinc(II) ions. The obtained complex is an active catalyst for transesterification of phosphate esters [56].

Peptide **33** bears two peripheral phosphane units (Figure 8.21). Upon coordination of rhodium(I) to the two diphenylphosphane units the α-helix structure of the peptide is supported [57]. Derivatives similar to **33** were used as catalysts in hydrogenation reactions [58].

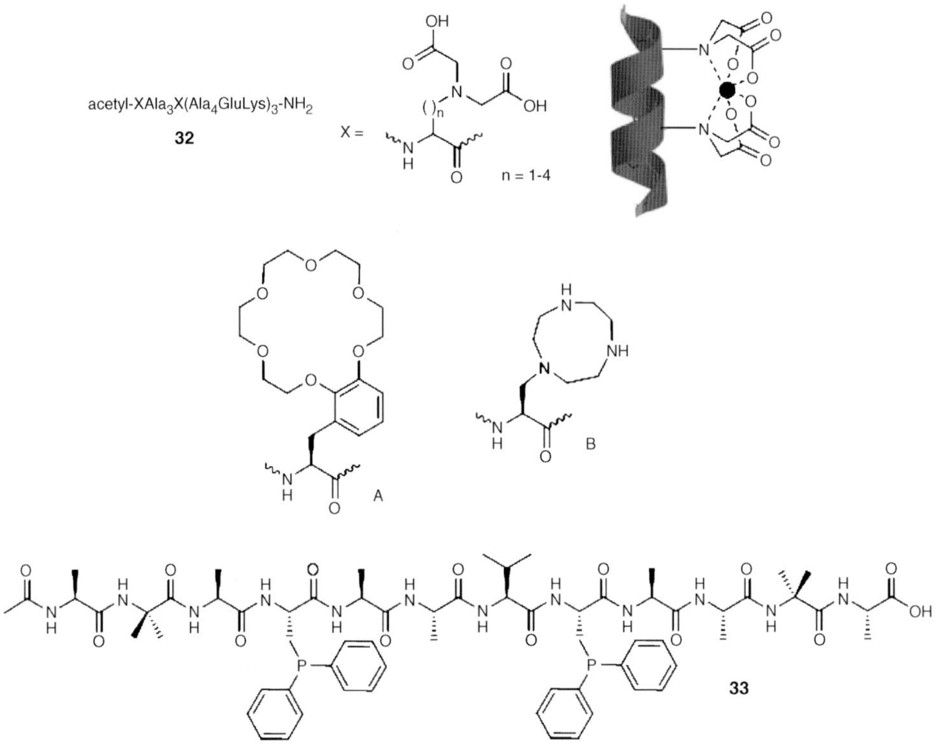

**Figure 8.21** *Two peptide strands **32** and **33** with non-natural binding sites (e.g., X) in i and i + 4 position of a peptide which upon metal coordination induce α-helical structures and unnatural amino acids **A** and **B** appropriate as building blocks to be incorporated into peptides.*

The metallopeptide **34** represents an achiral peptide (Figure 8.22) in which a short $3_{10}$-helix is stabilized by the coordination of platinum(II) or palladium(II). Metallopeptide **34** shows some decelerated chirality interconversion upon coordination of the metals to the side chain pyridines [59].

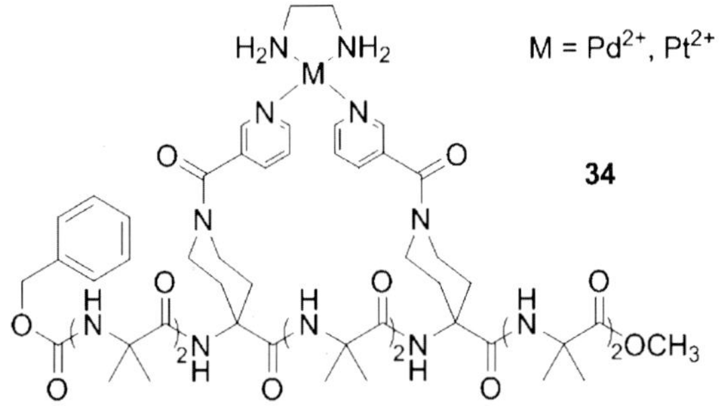

**Figure 8.22** *An optically inactive $3_{10}$-helical peptide, stabilized by palladium(II) or platinum (II) coordination.*

In this section different approaches for the stabilization of the α-helical motif are described. One is the metal templated generation of α-helix bundles, which cooperatively interact by the hydrophobic "fronts" of the α-helices. An alternative approach is to introduce amino acids with coordination sites in *i* and *i*+*4* position of the peptide strand. Upon metal binding both units have to point to the same face of the peptide and thus induce a short α-helical segment, which transfers the helical information along the attached peptide chain.

### 8.4.3  β-Sheets

For the stabilization of peptide β-sheet structures, two (or more) peptide strands have to be orientated close to each other in a parallel (or anti-parallel) fashion in order to support the NH—O hydrogen bonding between the amino acid residues. Two different approaches will be described to show how metal ions can lead to a stabilization of this secondary peptide structure.

2,2′-Bipyridine is a unit that acts as a hinge. In the uncoordinated form, the two nitrogen atoms are orientated anti to each other while upon metal coordination rotation around the central aryl–aryl bond has to occur. Consequently, in the free form substituents in the 6 and 6′ positions of the bipyridine are pointing away from each other, while in the metal complex they are located close to each other. In the bispeptide substituted bipyridine **35** this principle has been used to stabilize a β-sheet structure (Scheme 8.5).

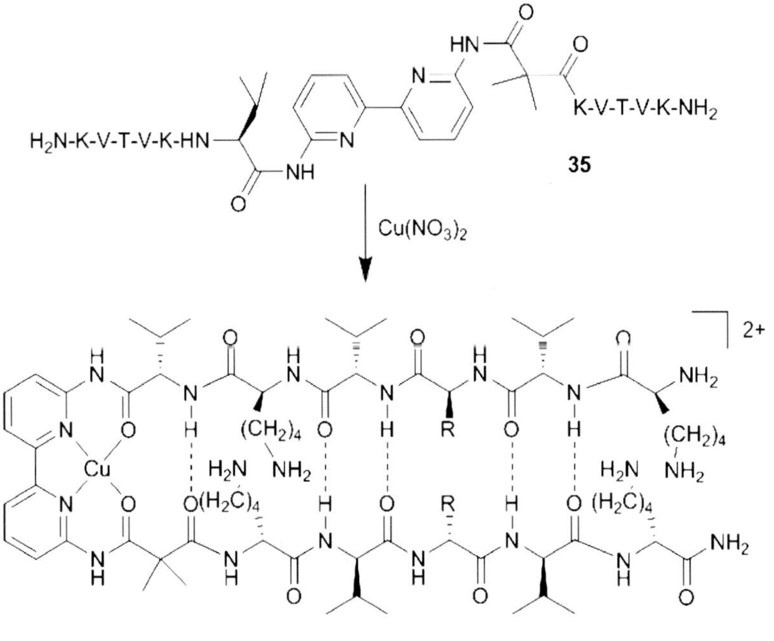

**Scheme 8.5**  *Induction of a β-sheet structure by coordination of copper(II) to a bipyridine bearing two peptide units in 6 and 6′ position.*

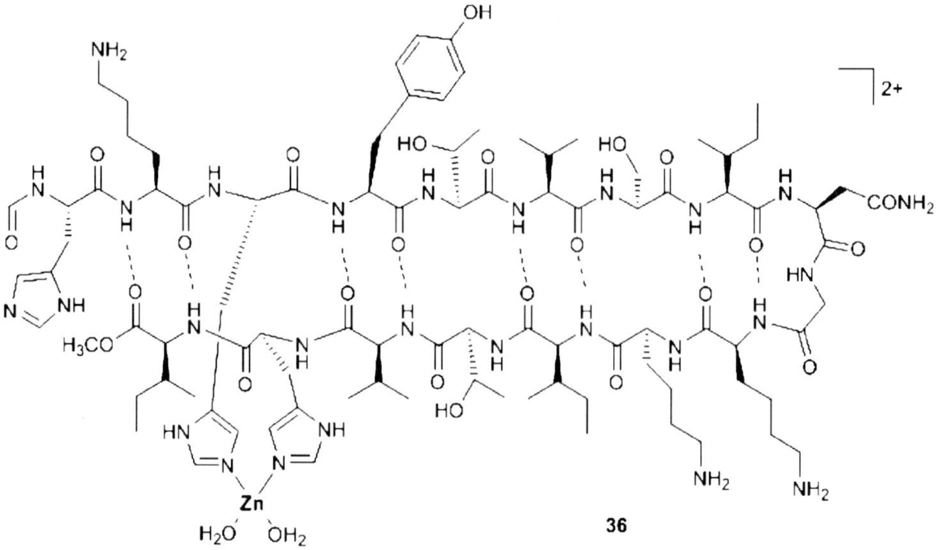

**Figure 8.23** *A zinc(II) stabilized β-hairpin.*

The addition of copper(II) ions to **35** results in the formation of a 1 : 1 complex in which the short peptidic moieties are orientated parallel to each other. This is supported by an additional metal coordination of carbonyl units neighboring the coordination site, resulting in the favored square planar coordination at the metal. The peptide units now easily undergo hydrogen bonding and an anti-parallel β-sheet with five intramolecular hydrogen bonds is formed. In principle, the bipyridine copper complex of **35** mimics a turn leading to the preorganization of the two peptidic segments for non-covalent interaction [60].

An alternative for the metal-assisted β-sheet formation of **35** can be found in the β-hairpin structure **36**. Two histidine residues in the peptide are ideally located in the strand in order to bind zinc(II) ions (Figure 8.23). This results in a huge macrocycle, which by hydrogen bonding adopts a β-sheet structure. Here, the turn unit, composed of the Asn-Gly dipeptide, is located far from the metal coordination sites [61].

In the two discussed examples shown for the stabilization of β-sheet structures different approaches are used. The strands are forced close to each other either by the introduction of a turn mimic or by tethering the termini of the peptide by metal coordination. In either case, the stabilization of the structures relies on cooperative effects between the binding of the metal ions and the formation of hydrogen bonds.

## 8.5 Conclusion

Metal coordination is an important tool to assemble and stabilize well defined structures and networks. It is utilized for the self-assembly of metallosupramolecular structures which can reach high complexity. The introduction of amino acids or peptidic units allows hybrids to be built, which connects structural supramolecular chemistry with the functional biochemical world.

In this chapter a connection was drawn between helicates (as supramolecular entities) and peptide-helicates (which already show some simple structural features of proteins). This finally led to the metal-assisted stabilization of peptide secondary structures: loops/turns, α-helix, and β-sheet. Thus, simple metal-centered coordination was discussed as a basis for the chemistry of metalloproteins in which single metal ions can play a crucial role in order to stabilize well defined structures. This chapter finally will end with a biochemical motif already mentioned in the introduction: the zinc finger proteins. They represent an exceptionally beautiful example of how the binding of a single metal can affect the structure and biological function of an otherwise structurally not defined and inactive protein (Figure 8.24) [62].

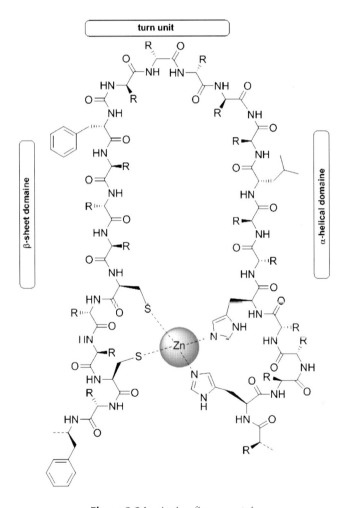

**Figure 8.24** *A zinc finger protein.*

# References

1. Sewald, N. and Jakubke, H.-D. (2002) *Peptides: Chemistry and Biology*, Wiley-VCH, Weinheim.
2. Voyer, N. (1997) The development of peptide nanostructures, in *Top. Curr. Chem*, vol. **184** (ed. F.P. Schmidtchen), Springer, Heidelberg, pp. 1–38.
3. Lombardi, A., Summa, C.M., Geremia, S. *et al.* (2000) Retrostrutural analysis of metalloproteins: application to the design of a minimal model for diiron proteins. *Proc. Natl Acad. Sci. U.S.A.*, **97**, 6298–6305.
4. Christianson, D.W. and Lipscomb, W.N. (1989) Carboxypeptidase A. *Acc. Chem. Res.*, **22**, 62–69.
5. Berg, J.M. (1990) Zinc fingers and other metal-binding domains. Elements for interactions between macromolecules. *J. Biol. Chem.*, **265**, 6513–6516.
6. Albrecht, M., Stortz, P., and Nolting, R. (2003) Peptide/metal-ligand hybrids for the metal-assisted stabilization of peptide-microstructures. *Synthesis*, **2003**, 1307–1320.
7. Sovago, I. and Ösz, K. (2006) Metal ion selectivity of oligopeptides. *Dalton Trans.*, **2006**, 3841–3854.
8. Freeman, H.C. and Taylor, M.R. (1965) Crystallographic studies of metal-peptide complexes. III. Disodium glyclglyclglyclglycinocuprate(II) decahydrate. *Acta Crystallogr*, **18**, 939–952.
9. Imperiali, B., Prins, T.J., and Fisher, S.L. (1993) Chemoenzymic synthesis of 2-amino-3-(2,2'-bipyridinyl)propanoic acids. *J. Org. Chem.*, **58**, 1613–1616.
10. Khatyr, A. and Ziessel, R. (2001) New ligands bearing chiral bioactive fragments. *Org. Lett.*, **3**, 1857–1860.
11. Gilbertson, S.R., Collibee, S.E., and Agarkov, A. (2000) Asymmetric catalysis with libraries of palladium β-turn phosphine complexes. *J. Am. Chem. Soc.*, **122**, 6522–6523.
12. Pattus, F. and Abdallah, M.A. (2000) Siderophores and iron-transport in microorganisms. *J. Chin. Chem. Soc.*, **47**, 1–20.
13. Gretchikhine, A.B. and Ogawa, M.Y. (1996) Photoinduced electron transfer along a β-sheet mimic. *J. Am. Chem. Soc.*, **118**, 1543–1544.
14. Bishop, B.M., McCafferty, D.G., and Erickson, B.W. (2000) 4'-Aminomethyl-2,2'-bipyridyl-4-carboxylic acid (Abc) and related derivatives: Novel bipyridine amino acids for the solid-phase incorporation of a metal coordination site within a peptide backbone. *Tetrahedron*, **56**, 4629–4638.
15. Torrado, A. and Imperiali, B. (1996) New synthetic amino acids for the design and synthesis of peptide-based metal ion sensors. *J. Org. Chem.*, **61**, 8940–8948.
16. Bediako-Amoa, I., Silerova, R., and Kraatz, H.-B. (2002) Ferrocenoyl glycylcystamine: organization into a supramolecular helicate structure. *Chem. Commun.*, **2002**, 2430–2431.
17. Maricic, S. and Frejd, T. (2002) Synthesis and conformational analysis of optically active ferrocene containing macrocyclic peptides. *J. Org. Chem.*, **67**, 7600–7606.
18. Dirscherl, G. and König, B. (2008) The use of solid-phase synthesis techniques for the preparation of peptide–metal complex conjugates. *Eur. J. Org. Chem.*, **2008**, 597–634.
19. Ming, L.J. (2010) Metallopeptides - from drug discovery to catalysis. *J. Chin. Chem. Soc.*, **57**, 285–299.
20. Lehn, J.-M., Rigault, A., Siegel, J. *et al.* (1987) Spontaneous assembly of double-stranded helicates from oligobipyridine ligands and copper(I) cations: structure of an inorganic double helix. *Proc. Natl Acad. Sci. USA*, **84**, 2565–2569.
21. Piguet, C., Bernardinelli, G., and Bünzli, J.-C.G. (1997) Helicates as versatile supramolecular complexes. *Chem. Rev.*, **97**, 2005–2062.
22. Albrecht, M. (2001) "Lets twist again" – Double-stranded, triple-stranded and circular helicates. *Chem. Rev.*, **101**, 3457–3498.

23. Piguet, C. and Bünzli, J.-C.G. (1999) Mono- and polymetallic lanthanide-containing functional assemblies: a field between tradition and novelty. *Chem. Soc. Rev.*, **28**, 347–358.

24. Bünzli, J.-C.G. and Piguet, C. (2005) Taking advantage of luminescent lanthanide ions. *Chem. Soc. Rev.*, **34**, 1048–1077.

25. Meistermann, I., Morenao, V., Prieto, M.J. *et al.* (2002) Intramolecular DNA coiling mediated by metallo-supramolecular cylinders: differential binding of P and M helical enantiomers. *Proc. Natl Acad. Sci. USA*, **99**, 5069–5074.

26. Scarrow, R.C., White, D.L., and Raymond, K.N. (1985) Ferric ion sequestering agents. 14. 1-Hydroxy-2(1H)-pyridinone complexes: properties and structure of a novel iron-iron dimer. *J. Am. Chem. Soc.*, **107**, 6540–6546.

27. Albrecht, M. (1998) Dicatechol ligands: novel building-blocks for metallo-supramolecular chemistry. *Chem. Soc. Rev.*, **27**, 281–288.

28. Albrecht, M. (2000) How do they know? Influencing the relative stereochemistry of the complex units of dinuclear triple-stranded helicate-type complexes. *Chem. Eur. J.*, **6**, 3485–3489.

29. Albrecht, M., Napp, M., and Schneider, M. (2001) The synthesis of amino acid bridged dicatechol derivatives. *Synthesis*, **2001**, 468–472.

30. Albrecht, M., Napp, M., Schneider, M. *et al.* (2001) Kinetik versus thermodynamic control of the self-assembly of isomeric double-stranded dinuclear titanium(IV) complexes from phenylalanine-bridged dicatechol ligands. *Chem. Commun.*, **2001**, 409–410.

31. Albrecht, M., Napp, M., Schneider, M. *et al.* (2001) Dinuclear titanium(IV) complexes from amino acid-bridged dicatechol ligands: Formation, structure, and conformational analysis. *Chem. Eur. J.*, **7**, 3966–3975.

32. Albrecht, M., Spieß, O., Schneider, M., and Weis, P. (2002) The fixation of linear versus loop-type peptidic structures by metal coordination: The coordination chemistry of Val-Val- and Val-Val-Val-bridged dicatechol ligands. *Chem. Commun.*, **2002**, 786–787.

33. Albrecht, M. and Fröhlich, R. (1997) Controlling the orientation of sequential ligands in the self-assembly of binuclear coordination compounds. *J. Am. Chem. Soc.*, **119**, 1656–1661.

34. Smith, V.C. and Lehn, J.-M. (1996) Helicate self-assembly from heterotopic ligand strands of specific binding site sequence. *Chem. Commun.*, **1996**, 2733–2734.

35. Coppock, M.B., Kapelewski, M.T., Youm, H.W. *et al.* (2010) Cu[II] cross-linked antiparallel dipeptide duplexes using heterofunctional ligand-substituted aminoethylglycine. *Inorg. Chem.*, **49**, 5126–5133.

36. Coppock, M.B., Miller, J.R., and Williams, M.E. (2011) Assembly of a trifunctional artificial peptide into an anti-parallel duplex with three Cu(II) cross-links. *Inorg. Chem.*, **50**, 949–955.

37. Albrecht, M., Stortz, P., and Nolting, R. (2003) Peptide/metal-ligand hybrids for the metal-assisted stabilization of peptide-microstructures. *Synthesis*, **2003**, 1307–1320.

38. Albrecht, M. and Stortz, P. (2005) Metallacyclopeptides: Artificial analogues of naturally occurring peptides. *Chem. Soc. Rev.*, **2005**, 496–506.

39. Albrecht, M., Stortz, P., Engeser, M., and Schalley, C.A. (2004) Solid phase synthesis of a double 4-pyridinyl terminated Leu-Ala-Leu tripeptide and macrocyclization by palladium(II) coordination. *Synlett*, **2004**, 2821–2823.

40. Constable, E.C., Housecroft, C.E., and Mundwiler, S. (2003) Metal-directed assembly of cyclo-metallopeptides. *Dalton Trans.*, **2003**, 2112.

41. Morita, H., Yun, Y.S., Takeya, K. *et al.* (1995) A cyclic heptapeptide from Vaccaria segetalis. *Tetrahedron*, **51**, 5987–6170.

42. Albrecht, M., Stortz, P., and Weis, P. (2003) Mimicing the biologically active part of the cyclopeptides Segetalin A and B by clipping of a linear tripeptide-derivative by metal coordination. *Supramol. Chem.*, **15**, 477–483.

43. Albrecht, M., Stortz, P., and Weis, P. (2003) Facile solid-phase synthesis of the WAGV-tetrapeptide front of the cyclopeptides Segetalin A and B terminated by catechol moieties and formation of a metalla-cyclopeptide. *Synlett*, **2003**, 867–869.

44. Ames, R.S., Sarau, H.M., Chambers, J.K. *et al.* (1999). Human urotensin-II is a potent vaso-constrictor and agonist for the orphan receptor GPR14. *Nature*, **401**, 282–286.

45. Albrecht, M., Stortz, P., Runsink, J., and Weis, P. (2004) Preparation of tripeptide-bridged dica-techol ligands and their macrocyclic molybdenum(VI) complexes. Fixation of the RGD-sequence and the WKY-sequence of Urotensin II in a cyclic conformation. *Chem. Eur. J.*, **10**, 3657–3666.

46. Gan, Z.R., Gould, R.J., Jacobs, J.W. *et al.* (1988) Echistatin. A potent platelet aggregation inhibitor from the venom of the viper, Echis carinatus. *J. Biol. Chem.*, **263**, 19827–19832.

47. Haubner, R., Finsinger, D., and Kessler, H. (1997) Stereoisomeric peptide libraries and peptido-mimetics for designing selective inhibitors of the $\alpha_v\beta_3$ integrin for a new cancer therapy. *Angew. Chem. Int. Ed.*, **36**, 1374–1389.

48. Fersht, A. (1998) *Structure and Mechanism in Protein Science*, Freeman, New York.

49. Ghadiri, M.R., Soares, C., and Choi, C. (1992) A convergent approach to protein design. Metal ion-assisted spontaneous self-assembly of a polypeptide into a triple-helix bundle protein. *J. Am. Chem. Soc.*, **114**, 825–831.

50. Kapurniotu, A., Buck, A., Weber, M. *et al.* (2003) Conformational restriction via cyclization in $\beta$-amyloid peptide A$\beta$(1-28) leads to an inhibitor of A$\beta$(1-28) amyloidogenesis and cyto-toxicity. *Chem. Biol.*, **10**, 149–159.

51. Kelso, M.J., Hoang, H.N., Appleton, T.G., and Fairlie, D.P. (2000) The first solution stucture of a single $\alpha$-helical turn. A pentapeptide $\alpha$-helix stabilised by a metal clip. *J. Am. Chem. Soc.*, **122**, 10488–10489.

52. Kelso, M.J., Hoang, H.N., Oliver, W. *et al.* (2003) A cyclic metallopeptide induces $\alpha$ helicity in short peptide fragments of thermolysin. *Angew. Chem. Int. Ed.*, **42**, 421–424.

53. Ghadiri, M.R. and Fernholz, A.K. (1990) Peptide architecture. Design of stable $\alpha$-helical metal-lopeptides via a novel exchange-inert ruthenium(III) complex. *J. Am. Chem. Soc.*, **112**, 9633–9635.

54. Ruan, F., Chen, Y., and Hopkins, P.B. (1990) Metal ion-enhanced helicity in synthetic peptides containing unnatural, metal-ligating residues. *J. Am. Chem. Soc.*, **112**, 9403–9404.

55. Voyer, N., Roby, J., Deschenes, D., and Bernier, J. (1995) Chiral recognition of carboxylic acids by bis-crown ether peptides. *Supramol. Chem.*, **5**, 61–69.

56. Rossi, P., Tecilla, P., Baltzer, L., and Scrimin, P. (2004) De-novo metallonucleases based on helix-loop-helix motifs. *Chem. Eur. J.*, **10**, 4163–4170.

57. Gilbertson, S.R., Chen, G., and McLoughlin, M. (1994) Versatile building block for the synthe-sis of phosphine-containing ceptides: The sulfide of diphenylphosphinoserine. *J. Am. Chem. Soc.*, **116**, 4481–4482.

58. Gilbertson, S.R., Wang, X., Hoge, G.S. *et al.* (1996) Synthesis of phosphine–rhodium com-plexes attached to a standard peptide synthesis resin. *Organometallics*, **15**, 4678–4680.

59. Ousaka, N., Tani, N., Sekiya, R., and Kuroda, R. (2008) Decelerated chirality interconversion of an optically inactive $3_{10}$-helical peptide by metal chelation. *Chem. Commun.*, **2008**, 2894–2896.

60. Schneider, J.P. and Kelly, J.W. (1995) Synthesis and efficacy of square planar copper complexes designed to nucleate $\beta$-sheet structure. *J. Am. Chem. Soc.*, **117**, 2533–2546.

61. Platt, G., Chung, C.-W., and Searle, M.S. (2001) Design of histidine-$Zn^{2+}$ binding sites within a $\beta$-hairpin peptide: enhancement of $\beta$-sheet stability through metal complexation. *Chem. Com-mun.*, **2001**, 1162–1163.

62. Quinlan, K.G.R., Verger, A., Yaswen, P., and Crossley, M. (2007) Amplification of zinc finger gene 217 (ZNF217) and cancer: When good fingers go bad. *Biochim. Biophys. Acta*, **1775**, 333–340.

# 9

# Artificial DNA Directed toward Synthetic Metallofoldamers

*Guido H. Clever*[1] *and Mitsuhiko Shionoya*[2]

[1]*Institute for Inorganic Chemistry, Georg-August University Göttingen, Germany*
[2]*Department of Chemistry, University of Tokyo, Japan*

## 9.1 Introduction

### 9.1.1 Oligonucleotides are Natural Foldamers

Most biopolymers such as proteins and oligosaccharides possess the ability to fold into secondary (and higher) structural motifs in a predetermined way. Whereas it is generally acknowledged that polypeptides can fold into a variety of structures such as β-sheets, barrels and α-helices depending on the underlying amino acid sequence, deoxy-ribonucleic acid (DNA) is most commonly known for adopting a right-handed double helical structure termed B-DNA. Indeed, this B-DNA conformation is believed to be the predominant form adopted by DNA in the natural environment, but it is by far not the only type of secondary structure known in the context of oligonucleotides [1,2]. Figure 9.1 shows the structures of the three most prominent DNA secondary structures known today and summarizes a few of their main characteristic properties. A multitude of further structures, parameters and information on the conditions when to expect a certain type of secondary structure are described elsewhere [3] and will not be the focus of this section.

The molecular structures of the oligonucleotide double-strands reveal that all forms show a helical twist along the axis rather than adopting the shape of a straight ladder.

*Metallofoldamers: Supramolecular Architectures from Helicates to Biomimetics*, First Edition.
Edited by Galia Maayan and Markus Albrecht.
© 2013 John Wiley & Sons, Ltd. Published 2013 by John Wiley & Sons, Ltd.

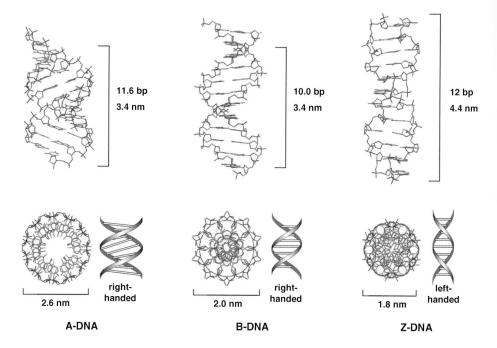

**Figure 9.1** *The three most important DNA secondary structure families in side view (top) and along the helix axis (bottom). A-DNA features a right-handed double helix with the strongly tilted base pairs aligned around a central hollow cavity. It is found for DNA at low humidity and the most common RNA duplex structure. B-DNA is the predominant conformation of double-stranded DNA. It is a right-handed helix of parallel-stacked Watson–Crick base pairs with one full helical twist every ten base pairs. Z-DNA features a left-handed helix with pair wise clustered base pairs and is mainly found in CpG alternating sequences. The less common C-, D- and E-DNA families, triple helical structures and quadruplexes are not presented here.*

This twisting is an inherent feature of the double-strand having its origins in the shape, connectivity, conformation and electronic properties of the phosphate backbone, the 2′-deoxyribose sugar and the hydrogen-bonded and π-stacked nucleobases [4]. External influences such as temperature and ionic strength can have a tremendous effect on the adopted helix structure, as does the base pair sequence. Interestingly, not only double-stranded oligonucleotides show a pronounced secondary structure; single-stranded DNA was also found to show a helical ordering based on its sugar–phosphate backbone structure and π-stacking between neighboring bases [5]. Forming linear double-strands, however, is not the end of the line when regarding the degree of structural complexity implemented by oligonucleotides. DNA can form structures termed hairpins and bulges as well as three- or even four-way junctions, which were shown to have tremendous biological significance in the process of homologous recombination of paternal and maternal genes in the process of inheritance [2]. Even more so, RNA is able to adopt a great variety of secondary and tertiary structures with implications in the regulation of gene expression. Furthermore, complex folded RNA structures in the form of tRNA and ribosomal subunits are the basis of gene translation [2]. The control and nanotechnological exploitation of these structural features of DNA (and to a lesser extent RNA) are discussed in Section 9.1.3. Before that, however, we shall discuss the biological function of DNA and then see how the highly developed properties of oligonucleotides may find application in bio-inspired artificial systems and materials.

### 9.1.2 Biological Functions and Beyond

The main biological functions of DNA are the storage and propagation of genetic information. All the information encoding the physical manifestation and life of an organism is stored in the form of a DNA-based library in each cell. DNA therefore makes use of a four-letter code of the bases, adenine (A), thymine (T), guanine (G) and cytosine (C). Triplet sequences of these bases encode for the natural 20 amino acids that are built into the polypeptide programmed by the DNA sequence via an RNA transcript as the information mediator [2]. In double-stranded DNA, each nucleobase of the sequence pairs with its counter part (A with T and G with C), making the double-strand containing the same sequence twice (once in a "sense" and once in an "antisense" fashion). This feature is not only the fundamental basis for the processes of DNA replication and transcription but also ensures the integrity of the code in the common case where one side of the redundant double sequence suffers physical damage and the remaining copy serves as a template for the repair process. A crucial parameter for an efficient and error-free copy process performed by the corresponding DNA replication, transcription and repair enzymes is the fidelity of correct base pairing. Nature has evolved the base pairs A-T and G-C (termed "Watson–Crick" base pairs named after the discoverers of the DNA double helical structure) in a way that A pairs exclusively with T and G exclusively with C in order to conserve the integrity of the genetic code. Interestingly, this high degree of orthogonality of base pairing is not only manifested on the level of (sequential) single base pair formation events as it is the case in the processes of DNA replication, transcription and repair, but also leads to a very high sequence specificity for the formation of longer double strands from pairs of single strands with matching sequences [6]. In practice this means that even a larger number (hundreds) of individual single-stranded sequences mixed with their matching counterstrands in the same (aqueous) solution leads to the "self-sorting" of the system to give a product mixture containing only double-strands composed of matching, error-free sequences. A prerequisite to the success of this experiment, however, is that the sequences are not too similar, not too short and that the system is given enough time and thermal energy to equilibrate from an initial chaos of kinetically formed, mispaired structures to the thermodynamic minimum being the sum of all perfectly paired combinations. This is usually best achieved by heating the mixture of single-strands in a buffered, aqueous solution containing a rather high concentration of NaCl (and sometimes MgCl$_2$ and other salts) to a temperature above which all oligonucleotides should exist as unpaired single strands (see the discussion of the DNA melting temperature in Section 9.4.2) and subsequently allowing the system to slowly cool down to (usually) room temperature, thereby allowing all possible combinations to find their thermodynamic sink ("hybridization").

Although this experiment surely describes a process of utmost biological significance, we may already have noticed that this self-assembling puzzle game in which all pieces find their designated counter parts just depending on their intrinsic, preprogrammed sequence information may find applications far away from its original biological meaning. But not only this feature of sequence-specific self-assembly makes DNA a highly attractive molecule for bio-inspired technological developments. Likewise, the aforementioned understanding of – and control over – the types of double helical and higher-order structures such as junctions and bulges that has been collected and discovered by numerous scientists over the last 50 years or so sets us in the comfortable position today of being able to create new DNA structures by design. Furthermore, oligonucleotides of lengths of up to about 100 bases are routinely synthesized on autonomously working DNA synthesizer machines based on solid-phase synthesis and the phosphoramidite protocol with total control of the desired sequence (see Section 9.3.4) [7]. This absolutely non-biological process of DNA synthesis starting from (commercially available) building blocks and yielding sequences containing a programmed order of the bases A, T, G and C is fast, reliable and comparatively cheap and is offered as a service by a number of companies around the world today. Since automated DNA synthesis is nothing else than a sequence of carefully developed chemical reaction steps, it opens the possibility of incorporating artificial nucleobases inside an oligonucleotide that may either be slight modifications of the natural bases or totally artificial structures having nothing in common with natural bases. Although the practicality of chemical DNA synthesis is limited to the preparation of strands of not much more than about 100 bases, systems with much longer sequences are not out of reach when combining the aforementioned techniques with methods from the molecular biology toolbox, such as sequence-specific strand cleavage by restriction enzymes, strand joining by ligase enzymes and strand elongation and replication by the polymerase chain reaction (PCR) [2,8].

These features of DNA, along with some additional properties (inherent chirality, amphiphilic distribution of hydrophilic phosphates and hydrophobic bases) render it a highly interesting construction material with possible applications in future nano-technology, as we shall see in this section [9].

### 9.1.3   DNA Nanotechnology

As a first example for the use of chemically synthesized, yet unmodified (as compared to natural DNA) oligonucleotides pursuing an absolutely non-biological purpose, we will discuss the direction of nanotechnology dealing with the programmed self-assembly of 2- and 3-D structures from DNA as the building material [10].

Regarding again the experiment discussed in Section 9.1.2, we can imagine that such a puzzle game might not only be designed in a way that strands and counterstrands specifically combine in a 1 : 1 fashion, but one long single strand can also bind a number of shorter single strands all matching with certain stretches of the sequence along the long strand (from now on termed the template strand). If we now further consider the possibility of implementing structural motifs such as hairpins and junctions, we might imagine that the shorter strands not only bind the long strand once by making use of their entire sequence, but rather bind with one end to one designated part of the template strand and with their other end to another stretch along the sequence of the template strand, thereby folding it into a structure that is determined by the distance and angle between the two binding sites. We shall call these shorter strands binding the template strand twice (or even more times) and thereby spatially arranging it "staple strands". Figure 9.2a shows

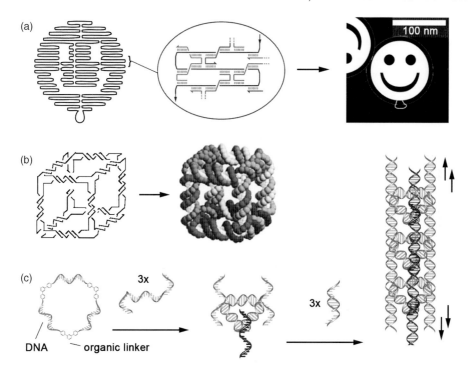

**Figure 9.2**  *Examples of DNA nanotechnology. (a) Two-dimensional DNA origami self-assembles from a long templating single-strand and a carefully designed mix of short single-stranded "staples" into a predetermined pattern; Reprinted with permission from Ref. [11b]. Copyright 2009 WILEY-VCH Verlag GmbH & Co. KGaA, Weinheim (b) likewise, three-dimensional constructs like Seeman's cube can be assembled from a set of suitable sequences; Reprinted with permission from Ref. [13a]. Copyright 1998 WILEY-VCH Verlag GmbH & Co. KGaA, Weinheim (c) circular single-stranded constructs of a triangular shape containing three organic corners are hybridized with single-strands creating shape-persistent building blocks that can polymerize into long tubes via hybridization of their sticky ends.*

the result of numerous of such staple strands folding the template strand into a predetermined 2-D pattern, here a smiley face, as designed before by a computer program based on the length and sequence information of the used template strand and a carefully chosen set of staple strands. Mixing the template and all staple strands together in the right stoichiometry and allowing the mixture to slowly cool down so that all matching sequences find each others then results in the formation of nanoscopic smileys consisting solely of double-stranded DNA, as can be seen in the subsequent AFM analysis [11].

Since the staple strands are nothing more than relatively short single-stranded DNA sequences that can be ordered from a DNA synthesis company nowadays via the internet, obtaining them just requires to provide the sequence information for each desired strand.

The only remaining question is where to get the long, single-stranded template sequence, since automated DNA synthesis can provide us with single-stranded DNA but not such long strands, and the usual PCR process can give us long strands but not single-stranded. The answer to this problem was found by looking at the few biological systems which store their information in the form of single-stranded oligonucleotides: certain types of viruses. Based on how these viruses replicate their genome, researchers found a way to obtain long, single-stranded DNA using techniques from molecular biology that will not be further discussed here. Most appealingly, this direction of DNA nanotechnology has been termed DNA origami by its founders [12].

Now it should be a simple task for us to imagine porting this building principle from the second to the third dimension in order to create objects such as the cube depicted in Figure 9.2b. Following this seminal work of DNA-based 3-D nanoconstruction by Seeman, others subsequently described the self-assembly of even larger and more complex 3-D shapes such as boxes with a controllable lid to open and close constructed of only DNA [13].

Figure 9.2c shows an example of how artificial structures such as 1,1′:3′,1″-terphenyl moieties can be incorporated into large circular DNA single strands in order to preorganize them in the shape of a triangle. Subsequently, three matching single strands, comprising overhangs on both ends, wrap around the preformed triangle and thereby equip it with six single-stranded studs, two each protruding from every corner in opposite directions. After another strand has been added, these overhangs are joined together and long polymeric tubes of triangular cross-section are formed which again can be visualized on surfaces by techniques such as AFM [14]. It was further shown that guest compounds such as gold nanoparticles can be loaded and released from the prism-shaped compartments of the polymeric structure, which might give new perspectives for DNA-based drug delivery [15].

Although the sense of playing puzzle games with DNA on the nanoscale might be questioned, a couple of systems may serve as examples for possible future applications: (1) the precise molecular positioning of reaction partners on top of DNA origami tiles allows the elucidation of reaction mechanisms on the single molecule scale, independent of substrate orientation and distance [12], (2) constructs consisting of DNA and other materials act as specific sensors for small molecules [9], (3) DNA-based molecular machines based on dynamic hybridization/dehybridization processes such as legs walking along a track help to understand related biological systems such as kinesin walking on microtubuli [16], (4) self-replicating systems help to gain understanding concerning the origin of life on earth [17] and (5) the self-assembly of electron-conducting and electronic information processing DNA structures might enable the bottom-up development of computer chips based on DNA and organic molecules instead of silicon wafers [18]. It is especially this latter idea which has spurred the investigation of electron conductance through DNA double strands. After realizing the low (but not zero!) intrinsic conductivity of unmodified DNA duplexes [19], the focus has now turned upon double-stranded DNA containing metal ions coordinated in the middle of its helix structure [20].

### 9.1.4 Interactions of DNA with Metal Ions

Metal ions can interact with oligonucleotides in a number of ways of which several have direct biological importance and others have therapeutic or technological relevance. Figure 9.3 summarizes the main interactions found between oligonucleotides and metal ions or coordination compounds. DNA is almost always associated with mono- and divalent ions such as Na(I) and Mg(II) ions (Figure 9.3a) which are closely associated

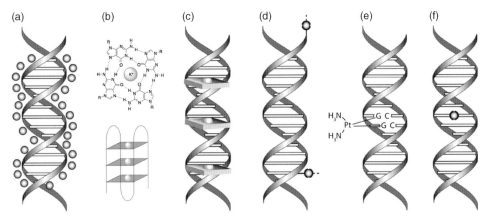

**Figure 9.3** *Types of DNA metal interactions. (a) Unspecific backbone/groove binding for charge compensation; (b) binding of K(I) in the centre of G-quadruplex structures; (c) groove binding/intercalation of metal complexes; (d) specific coordination of metals to artificial ligands attached to the ends or sides of DNA strands; (e) inter- or intrastrand crosslinking of natural nucleobases by coordination compounds such as cis-platin; (f) metal base pairing in the centre of the DNA double helix.*

with the negatively charged phosphate groups along the backbone for reasons of charge compensation and hence structural stabilization (although it must be noted that ammonium ions or related biological compounds such as spermine serve a similar job) [2,21]. The G-quadruplex motif found in the telomeric G-rich repeats was found to be stabilized by K(I) ions sitting in the central cavity between each of the $G_4$ squares (Figure 9.3b) [22]. The intercalation (and/or groove binding) of metal complexes such as $[Ru(bpy)_3]^{2+}$ and related structures inside the base stack of double-stranded DNA has relevance for the research on electron conductance through DNA, the use of DNA as chiral ligand in enantioselective catalysis [23] and the development of DNA targeting therapeutics (Figure 9.3c) [24]. Systems capable of binding metal ions selectively by ligands attached covalently to the ends or sides of DNA strands have been investigated for new ways of metal-mediated DNA nanoconstruction and their ability to enhance the conductivity of DNA (Figure 9.3d) [25]. Metal complexes directly binding to the natural nucleobases by coordination bonds are most prominently exemplified by the complex formed between the cytostatic drug *cis*-platin and two neighboring guanine bases via the coordination of both of their N7-nitrogen atoms (Figure 9.3e) [26]. Finally, metal ions can be coordinated right inside the core of the DNA double helix either by replacing the protons in the hydrogen bonds holding together the pairing bases in natural Watson–Crick base pairs or between the donor atoms of artificial ligands built inside the DNA [20].

## 9.2 The Quest for Alternative Base Pairing Systems

Enhancing the intrinsically low electron conductivity of natural DNA by doping the strands with metal ions is one motivation that led to the development of metal base pairs, as has already been exposed in the preceding sections. Another motivation that should not be left out here is the topic of finding alternative base pairing schemes which are orthogonal to the natural Watson–Crick base pairing and therefore might be used to expand the natural genetic code [27]. The emerging field of synthetic biology might benefit from such a development in future.

### 9.2.1 Modifications of the Hydrogen Bonding Pattern

The first approaches aiming at expanding the genetic code by using artificial base pairs were based on modifications of the hydrogen bonding patterns between the nucleobases of strand and counterstrand. Therefore, numerous heterocyclic nucleobase-like compounds were synthesized having hydrogen bond donor and acceptor sites similar to the natural nucleobases but arranged in other positions. Pairs of these artificial nucleobases were then tested for their propensity to undergo heterodimerization and for their orthogonality to the natural nucleobases, both in the form of their monomers and in an oligonucleotide context. From these works, important knowledge was derived about the role of hydrogen bonding for the sequence-specific hybridization of single strands [28].

### 9.2.2 Shape Complementarity

Later studies concentrated on the role of the nucleobases' shape and their ability to interact by π-stacking along the double helical structure. By using flat, aromatic nucleobase analogs devoid of any hydrogen donor and acceptor sites (called "hydrophobic bases") it was found that the role of hydrogen bonding should not be overestimated, since already the shape complementarity of the pairing bases in strand and counterstrand allowed specific base pairing in short duplexes. Furthermore, using DNA replicating and transcribing enzymes (polymerases), it was found that the shape of the bases seems to play a very important role in the base recognition in enzyme-catalyzed processes of templated oligonucleotide polymerization [29].

### 9.2.3 Metal Coordination

Early experiments to introduce metal ions into the center of the double helix made use of unmodified DNA strands at high pH (thereby facilitating the deprotonation of the nucleobases) and metal ions such as Zn(II), Co(II) and Ni(II) [30]. The structure and conductive properties of this so-called M-DNA were, however, controversially discussed and, despite some structural proposals, the exact positions of the metal ions inside or around the DNA strands remain unclear [20]. In contrast, it was unambiguously shown that Hg(II) ions coordinate strongly between two thymine bases oppositely arranged in a DNA double strand [31]. Whereas such T-T mismatches lead to a significant destabilization of the duplex structure in the absence of Hg(II), the addition of one equivalent of Hg(II) leads to the formation of a duplex stabilizing T-Hg(II)-T base pair in which the imide protons are removed. Likewise it was found that two oppositely arranged cytosine bases can bind Ag (I) to form the C-Ag(I)-C base pair (Figure 9.4) [32]. Since these two coordination events were found to proceed largely in an orthogonal manner, systems containing both C-C mismatches and T-T mismatches can be individually addressed by Ag(I) or Hg(II), respectively. This principle has been used to generate DNA-based logical switches that react on the input of Ag(I) and/or Hg(II) (see Section 9.6.2).

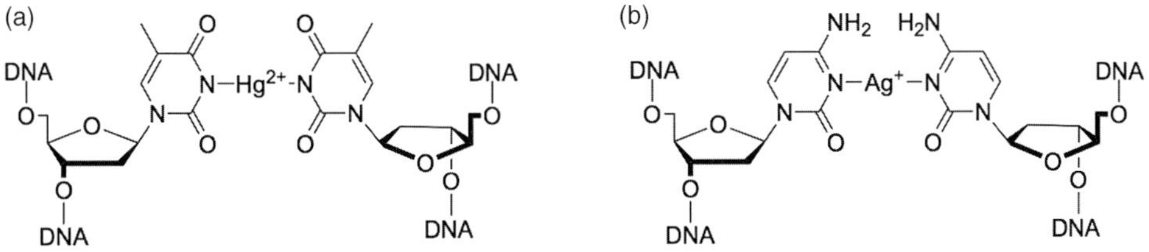

(a)                                                        (b)

**Figure 9.4**   (a) The thymine-Hg(II) (T-Hg(II)-T) base pair and (b) the cytosine-Ag(I) (C-Ag(I)-C) base pair.

## 9.3   Design and Synthesis of Metal Base Pairs

### 9.3.1   Rational Design of Metal Base Pairs

Although both the T-Hg(II)-T and the C-Ag(I)-C base pairs have proven to be very useful and have the advantage that no artificial ligands have to be introduced into the DNA, the latter circumstance also means that mixed sequences with Watson–Crick base pairs have to be carefully designed in order to prevent undesired interactions between the metal ions and parts of the sequence that were not supposed to be metallated.

Another approach to introduce metal ions into specific positions of a DNA double helix was therefore developed by Shionoya and others using artificial ligand-type nucleobases that are specifically designed to coordinate metal ions without undergoing any interactions with the natural base pairs [20].

The basic design principles that have to be considered for the development of these so-called metal base pairing systems are summarized in Figure 9.5. In brief, one has to consider mainly geometric and electronic features. The former ones include the dimensions of the metal complex in terms of the distance between the connection points to the sugar moieties (Figure 9.5a), the steric bulk when viewed along the base stack (Figure 9.5b) and the angle between the metal base pairs' long axis and the bonds to the C1′ position of the sugars (Figure 9.5c). Furthermore, the coordination number and geometry favored by the used metal ion (Figure 9.5d) and provided by the donor atom(s) of the ligands (Figure 9.5e) and the net charge of the resulting metal base pairs must be carefully adjusted. The electronic interactions between the metal ion and the ligand's donor atoms must be chosen to enable tight binding of the metal ion to the ligands alone (and not to the

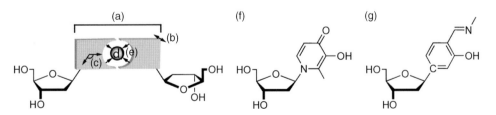

**Figure 9.5**   (a–e) Features to be considered in the design of metal base pairs. Ligands consisting of (f) an N-glycosidic base (hydroxypyridone) and (g) a C-glycosidic base (salicylaldimine).

natural bases). Therefore, the HSAB principle should be considered and the inorganic chemistry literature should be consulted for the binding constant and geometry of corresponding metal complexes, if available. The redox and pH stabilities of the metal complex in the aqueous media used in DNA chemistry should also be considered.

Figure 9.5f and g show for two ligands, which were extensively studied in DNA metal base pairing, that the connection of the ligand moiety to the C1′ position of the backbone sugar may be achieved via an *N*-glycosidic bond (likewise to the natural nucleobases) or via a *C*-glycosidic bond (which requires an sp$^3$-sp$^2$ *C-C* coupling reaction as a key synthetic step). Furthermore, the synthesis and use of the building blocks in automated DNA synthesis may require the implementation of a suitable protecting group strategy. These and other synthetic issues are discussed again in Section 9.3.3.

### 9.3.2 Model Studies

In case where the metal base pair design does not follow the example of a literature known metal complex or the planned coordination chemistry is not trusted to work out in the context of DNA chemistry, model studies and test experiments should be carried out before the decision is made to synthesize a new ligand-functionalized nucleobase for the incorporation into duplex DNA [33].

The required test experiments using simple model compounds based on the same ligand design or even the monomeric ligand-functionalized nucleoside might include: (1) determination of the p$K_a$ value(s) of the donor site(s) of the ligand as a measure for the prospective metal binding ability, (2) metal ion binding in aqueous buffer solution at high ionic strength or at various pH values as monitored by NMR, UV and mass spectrometry, (3) thereupon calculation of binding stoichiometry (Job's plot) and binding constant, (4) preparation of single-crystalline samples of the coordination compound and comparison of the obtained X-ray structure with the structural features of (B-type) DNA and (5) control experiments evaluating the influence of the used metal ions on unmodified DNA (non-specific binding?) and likewise the compatibility of the prepared model complexes in the presence of DNA in the solution.

Computer-aided model studies may include: (1) comparing the structures and energies of alternative metal binding patterns by quantum mechanical modeling, (2) the generation of a simple molecular mechanics (or even physical plastic) model in order to estimate the impact of the metal base pair incorporation on the DNA secondary structure and (3) a more sophisticated atomistic molecular mechanics or molecular dynamics (or even higher theory level) model study considering the base stack integrity, helical twist, deviation from linear duplex structure and other parameters affected by the metal base pair and the surrounding solvent and counter ions.

### 9.3.3 Synthesis of Modified Nucleosides

Following the design of a new metal base pairing system, test experiments and modeling studies, a synthetic strategy has to be developed for the successful incorporation of the artificial nucleoside into DNA double strands. Figure 9.6 depicts the general strategy for the synthesis of an *N*- or *C*-glycosidic nucleoside based on the standard DNA sugar 2′-deoxyribose (note, however, that combinations of metal base pairing with backbone modifications such as open chain carbohydrates and peptides have also been described) [34].

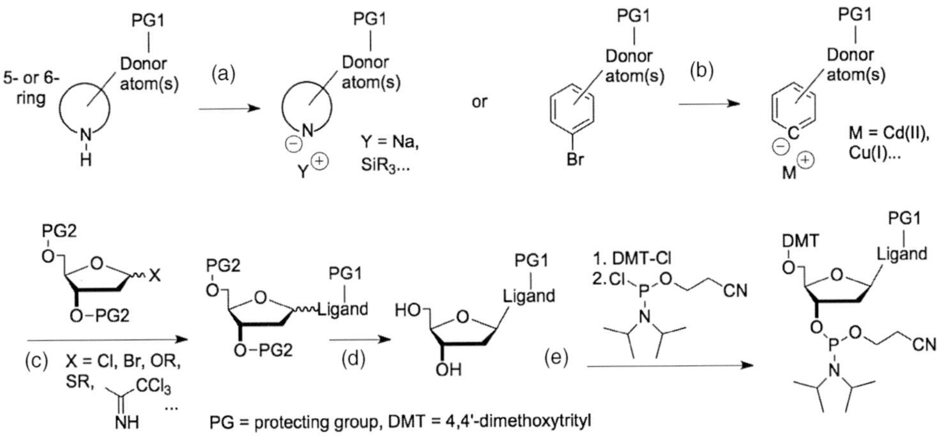

**Figure 9.6** *General overview of the synthesis of ligand-modified nucleosides. (a) activation for N-glycosylation; (b) activation for C-glycosilation; (c) reaction with glycosyl donors; (d) separation of anomers and deprotection; (e) attachment of protecting and activating groups for automated DNA synthesis.*

In a convergent synthetic approach, an *N*-nucleophilic (Figure 9.6a) or *C*-nucleophilic (Figure 9.6b) ligand-modified nucleobase carrying suitable protecting groups must be prepared and subsequently brought into reaction with a glycosyl donor (a $3'O$-/$5'O$-protected, $1'$-activated $2'$-deoxyribosyl electrophile) such as Hofmann's α-$1'$-deoxyribosylchloride to yield the nucleoside, usually as a mixture of α- and β-anomers (Figure 9.6c). Afterwards, the desired β-anomer has to be separated from the undesired α-anomer (usually by chromatography or recrystallization) and the protecting groups are removed from the $3'O$- and $5'O$-positions (Figure 9.6d). Care has to be taken in the unambiguous characterization of the two anomeric isomers (e.g., by NOESY-NMR spectroscopy or single-crystal structure determination) to make sure that indeed the desired anomer ends up in the final DNA product.

Finally, the $5'O$-position is equipped with the 4,4′-dimethoxytrityl (DMT) protecting group and the $3'O$-position is phosphorylated with 2-cyanoethyl-*N*,*N*-diisopropylchloro phosphoramidite to yield the phosphoramidite building block needed for the standard protocol of the automated solid-phase DNA synthesis (Figure 9.6e) [7]. Both latter reagents have been specifically developed for automated DNA synthesis and we shall see later (Section 9.3.4) the reason for this. Since the product contains an acid labile DMT group and a phosphorus(III) center very prone to hydrolysis and oxidation to a phosphorus(V) by dioxygen, special care has to be taken during synthesis and purification and the final product should be checked by $^1$H- and $^{31}$P-NMR spectroscopies.

The commercially available phosphoramidite building blocks for the natural nucleobases A, G and C (but not T) carry protecting groups on nucleophilic positions of the bases that survive the conditions of DNA synthesis but are smoothly cleaved after DNA synthesis to release the fully unprotected oligonucleotide. Similar considerations have to be made for any protecting groups attached to the artificial nucleobase: whereas their main purpose is to protect the nucleophilic (or electrophilic) sites on the ligand from interference with the automated DNA synthesis process, their full removal from the final oligonucleotide, preferably in aqueous medium and without harming the DNA molecule, must be tested in control experiments beforehand.

### 9.3.4 Automated Oligonucleotide Synthesis

Having the artificial phosphoramidite building block in hand, incorporation into oligonu-cleotides is achieved by automated DNA synthesis on a commercially available machine along with the natural nucleotides [7]. The machine, termed an oligonucleotide synthe-sizer (or DNA synthesizer), is merely a computer-controled system of pumps, tubes and valves working strictly anhydrously under a protecting gas in order to deliver all needed reagents in the right amounts and temporal sequence. The underlying chemical principle is solid-phase synthesis using porous glass or polystyrene beads as the solid support on which the first nucleotide is usually already bound (Figure 9.7). The cyclic process lead-ing to a stepwise growing of the single-stranded oligonucleotide sequence can be divided into the following four steps:

1. In the *coupling* step, the free 5'-OH group of the last nucleotide of the bound nascent strand reacts in a nucleophilic substitution reaction with the added phosphoramidite (activated at the phosphorous center) under formation of a phosphorus(III) ester.
2. In the following *capping* step, unreacted oligonucleotide strands are inactivated for further elongation (leading to false sequences) by blocking the free 5'-OH groups via esterification with acetic anhydride from further reaction.
3. Subsequently, the unstable phosphorus(III) center is *oxidized* to a phosphorus(V) ester using an iodine containing reagent.
4. In the following *deprotection* step, the 5'$O$-DMT group is cleaved from the newly coupled nucleoside by the use of acid.

The remainder of the DMT protecting group, the 4,4'-dimethoxytrityl cation, gives an intense red color to the solution leaving the solid support cartridge. This is collected in an integrated photometer cell in a modern DNA synthesizer machine in order to estimate its molar amount, which is proportional to the amount of deprotected oligonucleotide and can be taken as a comparative measure for the coupling efficiency in each cycle. The

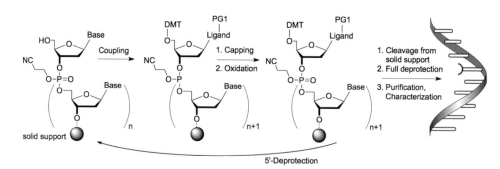

**Figure 9.7** *The cycle of automated DNA synthesis based on the phosphoramidite protocol.*

deprotection step is then followed by the coupling step for the next base until the end of the cyclic synthesis process is reached. Subsequently, the oligonucleotide is released from the solid support by the treatment with aqueous ammonia which not only cleaves the ester bonds through which the strands are attached to the solid phase, but also quantitatively removes all protecting groups sitting on the (natural) nucleobases and the phosphate backbone. Depending on the chosen protecting group strategy, the artificial nucleotides may require special operations of deprotection to obtain the fully deprotected single strand. This step is followed by a purification using chromatography on disposable reverse-phase cartridges or by HPLC or gel electrophoresis. After an optional desalting step, the products should be thoroughly characterized by MALDI- or ESI-mass spectrometry and HPLC to confirm sequence identity, purity and full removal of protecting groups. The concentration of the single-strands in solution is then estimated by UV spectroscopy (requiring one to know the molar extinction of all incorporated nucleobases, including the artificial one).

### 9.3.5 Enzymatic Oligonucleotides Synthesis

Although artificial nucleosides can be incorporated into short oligonucleotides by chemical synthesis as described above, the enzymatic incorporation into long DNA strands following the concept of the polymerase chain reaction (PCR) is highly desirable to obtain even longer strands containing artificial nucleosides and to investigate the implication of nucleoside modifications on biological molecules and living systems. Using very tolerant or engineered polymerases, indeed this has been shown to be possible. Therefore, the artificial nucleosides have to be turned into 5'-triphosphates, the natural substrates for polymerases, instead of 5'-DMT-protected 3'-phosphoramidites [35].

## 9.4 Assembly and Analysis of Metal Base Pairs Inside the DNA Double Helix

### 9.4.1 Strategies for Metal Incorporation

The synthesis, purification and determination of DNA concentration yield solutions of the matching single strands containing the ligand-modified nucleosides in either strand. Prior to the addition of the metal ions forming the metal base pair, the double strand may usually be preformed by allowing both complementary single strands to hybridize. This is achieved by slowly cooling a hot solution of both strands combined in a 1 : 1 ratio in a buffered aqueous solution at high ionic strength. The chosen buffer should provide a suitable pH for metal complexation by the artificial nucleosides (and must not contain EDTA or any other chelating agents!), the electrolyte is usually NaCl but may be changed to $NaClO_4$ or $NaNO_3$, when Ag(I) or Hg(II) are the metal ions of choice. The DNA concentration is usually in the micromolar range.

Subsequently, the metal ion under investigation is added as an aqueous solution (usually prepared from a salt containing a weakly coordinating anion such as sulfate, perchlorate or nitrate), often followed by various spectroscopic methods in the form of a titration experiment [20].

### 9.4.2 Analytical Characterization in Solution

Several analytical methods are routinely applied for the characterization of metal base pair containing DNA double-strands, while the application of other methods, such as NMR and EPR, have only been demonstrated in a few cases so far.

One of the most useful analytical techniques in the characterization of DNA modifications is the UV-based melting curve analysis (also termed a thermal de- and renaturation study) [36]. The aim of this method is the determination of the so-called melting point $(T_M)$ of the DNA double strand under investigation, which is defined as the temperature at the transition point in the melting curve (absorption against temperature). It is a measure for the characteristic temperature at which a specific double strand under given conditions (concentration, pH, ionic strength) dehybridizes to yield a solution containing the corresponding single strands (Figure 9.8a). The basis for obtaining a temperature-dependent change in UV absorption (usually measured at the absorption maximum of DNA at 260 nm) can be found in the different molar absorptivity of double-stranded DNA compared to the absorbance sum of the corresponding two single strands. Whereas double-stranded DNA has a relatively low molar absorptivity due to quenching effects in the duplex structure, the corresponding single strands have a higher extinction (an effect termed "hyperchromicity") which can be monitored in order to follow the thermal denaturation of double to single strands and vice versa. Different DNA sequences of the

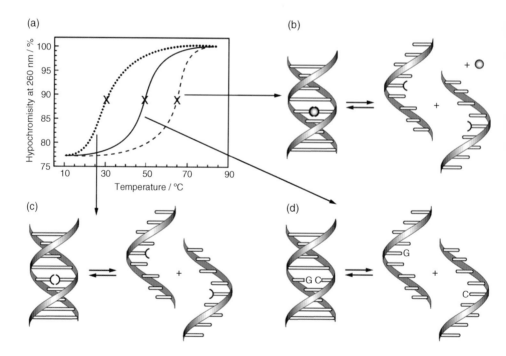

**Figure 9.8** *(a) Typical DNA melting curves for (b) metal-containing ligand-modified DNA; (c) the same strand in the absence of metal ions and (d) the same sequence containing a G-C base pair instead of the ligands.*

same length give different melting points. In particular, GC-rich duplexes have higher melting points than AT-rich sequences due to the higher stability of the triply hydrogen-bonded G-C base pairs. The same may be observed for DNA sequences containing one or several Watson–Crick base pairs replaced by pairs of artificial nucleobases. When the modified base pairs have a higher stability than the natural ones, the melting point rises, whereas when they contribute less to the attractive interaction between the strands, the melting point decreases. Concerning the topic of metal base pairing, it is instructive to compare the melting points of a metal base pair containing a double strand (Figure 9.8b), a ligand containing a double strand in the absence of metal ions (Figure 9.8c) and a double strand containing a Watson–Crick base pair at the same position (Figure 9.8d). Whereas the oppositely arranged ligands usually form a destabilizing mismatch in the absence of metal ions (with a melting point lower than the duplex containing the Watson–Crick base pair), the addition of a suitable metal ion results in a significant increase in melting temperature, often exceeding the value obtained from the unmodified DNA sequence.

In cases where the coordination of the metal ions to the ligand-containing DNA duplexes gives rise to other spectroscopically observable features such as color change, fluorescence or changes in circular dichroism, the respective spectroscopic changes may be monitored in dependence of the amount of metal added and/or change in temperature. Furthermore, the interaction of the double strands with the metal ions may be studied by HPLC, gel electrophoresis, EPR and NMR spectroscopic methods.

### 9.4.3  X-Ray Structure Determination

Single-crystal X-ray structure determination usually yields the best structural description of a coordination compound in terms of connectivity, bond distances, angles and stereo-chemical relationships. In the case of metal-containing oligonucleotide duplexes, however, the growing of suitable single crystals and the interpretation of diffraction data is often a time- and material-consuming process requiring a good bit of luck. So far, only a limited number of crystal structures of oligonucleotides containing metal base pairs have been reported. Figure 9.9 shows two structures reported in the literature. The first structure contains two isolated asymmetric metal base pairs consisting of one tridentate pyridine-2,6-dicarboxylate (Dipic) ligand and one monodentate pyridine ligand coordinating a Cu(II) ion in a square-planar geometry (Figure 9.9a and b) [37]. Interestingly, the palindromic dodecamer forms a Z-type, left-handed helical structure, a fact that seems to be controlled by the requirement of the Cu(II) ions adopting a Jahn–Teller distorted octahedral coordination environment in which the O6 of the neighboring guanine base on one side of the metal base pair and the ribose O4′ of a thymine on the other side coordinate the Cu(II) ion in its apical positions (dashed lines in Figure 9.9a). Especially the latter coordination of the backbone sugar to the two metal centers contained in the dodecamer sequence seems to have such a strong influence on the whole duplex conformation, leading to a nearly perfect Z-DNA-type structure.

Figure 9.9c shows another example, where two Cu(II)-mediated base pairs based on the hydroxypyridone ligand are incorporated into an artificial duplex consisting of an open chain propylene glycol backbone [38]. The main intention of incorporating the metal base pairs into this construct actually was their positive effect on duplex stability in order

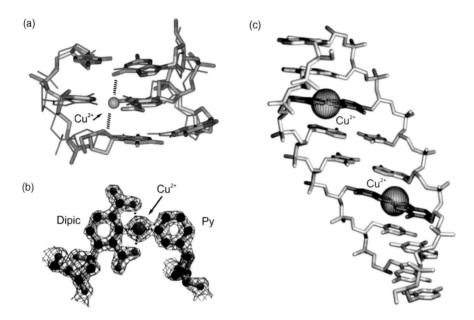

**Figure 9.9** *X-ray crystal structures of the Dipic-Cu(II)-Pyridine base pair in a palindromic dodecamer sequence: (a) side view (the contacts to the upper base oxygen and lower sugar oxygen are represented by dashed lines, Reproduced with permission from [37c]. Copyright 2001 American Chemical Society); (b) top view. Reproduced with permission from [37c]. Copyright 2001 American Chemical Society. (c) X-ray structure of a backbone-modified GNA duplex containing two hydroxypyridone-Cu(II) base pairs. Reproduced with permission from [38a]. Copyright 2001 American Chemical Society.*

to elevate the melting temperature of this short double strand significantly above room temperature. Therefore, the incorporation of metal base pairs into this duplex can be seen as an early application of the concept of metal base pairing.

The X-ray structure of a salen-Cu(II)-containing DNA strand in complex with a fragment of DNA polymerase I from *Bacillus stearothermophilus* was recently reported [39]. The structure shows that the metal base pair fits perfectly inside the B-DNA-like helical structure. Furthermore, it was shown that the salen metal base pair can be enzymatically incorporated into oligonucleotides using the triphosphate of a salicylic aldehyde nucleoside in the presence of ethylenediamine as a reversible cross-linker.

## 9.5 Artificial DNA for Synthetic Metallofoldamers

### 9.5.1 Overview

Since the pioneering work by Shionoya, Tanaka [40] and Schultz [37], a number of further metal base pairing systems have been developed, a selection of which are depicted in Figure 9.10 [20,41]. Simple monodentate nitrogen ligands such as imidazole (Figure 9.10a) [33,42] and pyridine (Figure 9.10b) [43] were found suitable to coordinate

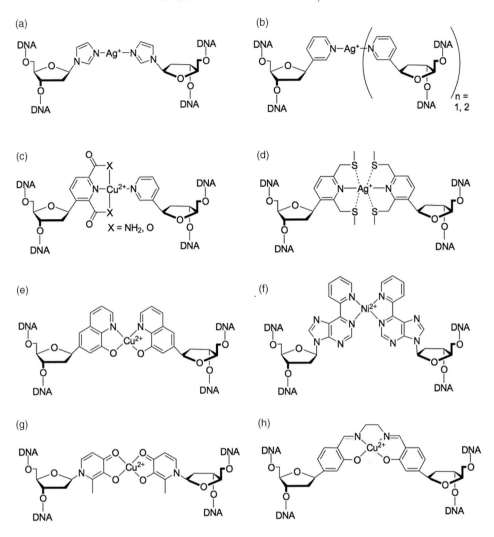

**Figure 9.10** *Overview of a selection of metal base pairs.*

Ag(I) ions in DNA duplexes. Depending on the sequence design, the latter ligand was even able to form metal-mediated triple helices containing trigonal-planar coordinated Ag(I). The asymmetric base pair based on pyridine-2,6-dicarboxylate (or derivatives) and pyridine (Figure 9.10c) was already discussed in Section 9.4.3. A symmetric variant containing soft sulfur atoms as coordinating atoms with the ability to coordinate Ag(I) is shown in Figure 9.10d [44]. Figure 9.10e shows a Cu(II)-mediated base pair based on the hydroxyquinoline ligand [45], whereas Figure 9.10f shows a nice example of a ligand system based upon the natural purine base adenine from which it can be synthesized in a few steps [46]. Figure 9.10g and h depict the hydroxypyridone and the salen metal base pairs which will be discussed in more detail below.

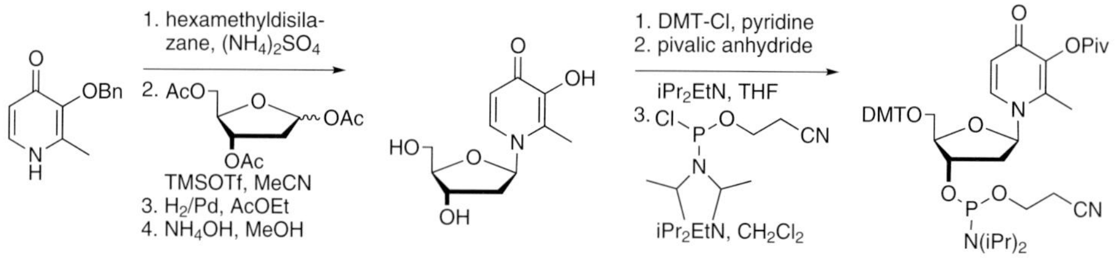

**Figure 9.11** *Synthesis of the N-glycosidic hydroxypyridone base and its phosphoramidite.*

### 9.5.2 The Hydroxypyridone Base Pair

One of the most widely used metal base pairing systems is the hydroxypyridone metal base pair (Figure 9.10g) which can be employed for the coordination of Cu(II) (and also Fe(III), see Section 9.6.1) inside DNA [47]. The synthesis of the *N*-glycosidic ligand is briefly shown in Figure 9.11, the incorporation into artificial DNA strands according to Section 9.3.4 is followed by a cleavage and deprotection step using aqueous ammonium hydroxide in order to cleave the pivaloyl protecting group alongside all other protecting groups of the synthesized oligonucleotide.

Using this ligand, it was shown for the first time that not only one but up to five directly stacked Cu(II)-mediated base pairs could be incorporated into a DNA duplex without compromising the integrity of the double helical structure [48]. Most interestingly, EPR measurements showed a ferromagnetic coupling between the stacked paramagnetic copper complexes (Figure 9.12 and Section 9.6.3). Like the salen metal base pair discussed in the following section, the hydroxypyridone base pair was successfully used in the programmed assembly of heterobimetallic stacks inside DNA (see Figure 9.15) [49].

### 9.5.3 The Salen Base Pair

Whereas all metal base pairing systems discussed so far are composed of two individual mono- or bidentate ligands reacting with the metal ion in a 2:1 fashion, the following example effectively forms one tetradentate ligand inside the DNA that covalently cross-links a strand and its counterstrand and thereby leads to a tremendous thermal stabilization of the double helical structure. This system is based on the well known *N,N'*-ethylene-bis-salicylimine ("salen") ligand which is widely used in catalysis and binds a

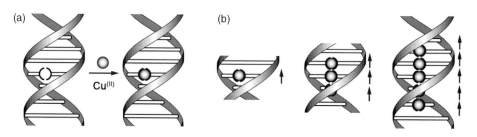

**Figure 9.12** *(a) Assembly and (b) stacking of the Cu(II)-mediated hydroxypyridone base pair.*

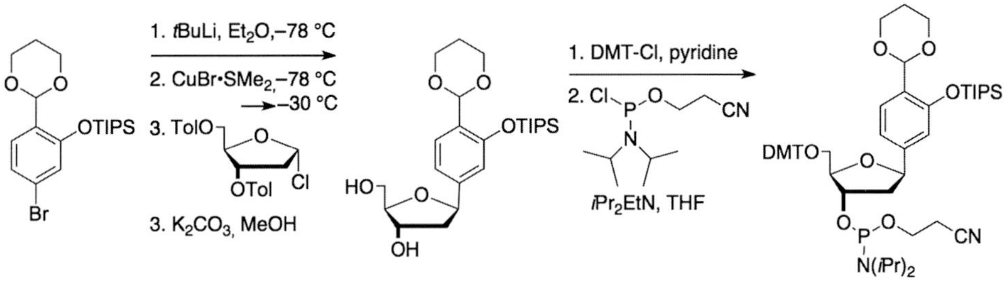

**Figure 9.13** Synthesis of the C-glycosidic salicylic aldehyde base (a precursor to the salen ligand) and its phosphoramidite.

large number of transition metal ions in a square-planar, square-pyramidal or octahedral fashion [50]. The direct precursor of the ligand is a salicylic aldehyde, which reacts with ethylenediamine to the tetradentate ligand. Such a salicylic aldehyde, intended to be incorporated into DNA as a C-nucleoside, was prepared according to Figure 9.13 carrying a silyl protecting group on the phenolic hydroxyl group and having its aldehyde functionality masked as a cyclic acetal.

After DNA synthesis and the removal of all protecting groups, salicylic aldehyde containing single strands of matching sequence were hybridized and ethylenediamine was added in excess in order to form the salen ligand as a covalent cross-link between the strands. Since the formed imine bonds are not stable in the aqueous medium used, the salen ligand forms and reopens again in a dynamic equilibrium which can be seen by the relatively small increase in melting temperature ($+5$ K) upon the addition of ethylenediamine to the duplex. This situation changes dramatically when one equivalent of transition metal ions is added: the metal coordinates inside the tetradentate salen ligand with a high affinity and thereby stabilizes the ligand's imine bonds against hydrolytic cleavage. Consequently, the covalent cross-link conveys a tremendous thermal stability to the double helix. In the case of Cu(II), the increase in melting temperature was found to be larger than 40 K, which is the largest duplex stabilization achieved by metal base pairing so far [50].

In accordance with the metal stacking experiments using the hydroxypyridone base pair discussed in Section 9.5.2, the salen base pair was also used to produce arrays of stacked metal ions inside the DNA double helix (Figure 9.14). By incorporating ten consecutive ligands, metal stacks spanning one entire helical twist (considering a B-DNA type structure) were generated using transition metal ions such as Cu(II) and Mn(II) (which is oxidized to Mn(III) upon complexation) [51]. Interestingly, the weaker bound Mn(II) ions gave better results in terms of uniformity and purity of the formed metallated double helices, most likely attributed to the "self-healing" of preliminary formed, kinetically misfolded products into the perfectly stacked array of ten metal base pairs as the thermodynamically most stable product (Figure 9.15a).

In an attempt to introduce even more complexity into the system, a second metal base pairing system orthogonal to the first one was introduced into the same double helix in order to create mixed metal arrays of predetermined sequence inside the double helix (Figure 9.15b) [49]. Therefore, the salen-Cu(II) base pair was combined with the non-

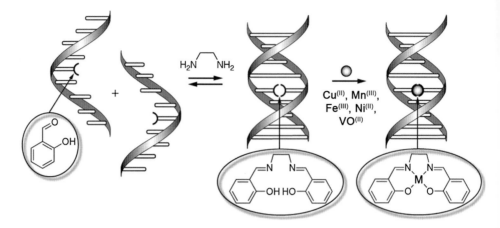

**Figure 9.14** *Stepwise assembly of salen ligand and the salen metal base pair inside the DNA double helix.*

interfering T-Hg(II)-T base pair as shown in Figure 9.15b. Likewise, it was demonstrated that the hydroxypyridone-Cu(II) base pair could be combined with the pyridine-Hg(II) base pair in order to create mixed metal arrays (not shown) [49]. Since the sequential nature of the solid-phase oligonucleotide synthesis process unambiguously determines the order of the incorporated ligands and titration and mass spectrometry experiments showed the binding of the anticipated number and type of metal ions to these ligands, the systems can be regarded as programmed assemblies of linear chains of two kinds of metal ions that might find future application in DNA-based molecular electronics. In a related work, Ono *et al.* were recently able to incorporate T-Hg(II)-T and C-Ag(I)-C base pairs into the same double strand [52].

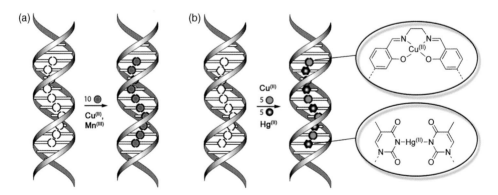

**Figure 9.15** *(a) Stacking of ten consecutive salen metal base pairs and (b) assembly of mixed metal arrays from strands containing a programmed sequence of salen ligands and T-T mismatches using Cu(II) and Hg(II) ions.*

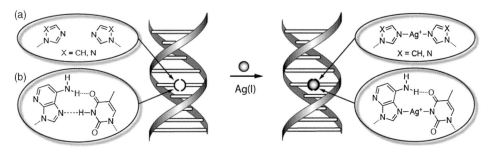

**Figure 9.16** *(a) The imidazole and the 1,2,4-triazole base pairs as well as (b) the 1-deazaadenine-thymine base pair are all suited to incorporate Ag(I) ions into DNA double helices.*

### 9.5.4 The Imidazole, Triazole and 1-Deazaadenine-Thymine Base Pairs

A series of simple *N*-glycosidic metal base pairs mainly coordinating Ag(I) ions were reported by Müller and coworkers [33,42]. Figure 9.16a shows the structure of the imidazole (X = CH) and 1,2,4-triazole (X = N) base pairs and Figure 9.16b shows a hydrogen-bonded base pair between thymine and 1-deazaadenine (in the so-called Hoogsteen bonding mode) in which one of the two hydrogen bonds can be replaced by a Ag(I) ion.

Using three consecutive imidazole base pairs having the nitrogen donor positions enriched in $^{15}$N, NMR studies yielded the secondary structure depicted in Figure 9.17 [53]. The analysis revealed that the duplex adopts a right-handed helical structure showing only minor deviations from the canonical B-type DNA conformation. Furthermore, the $^{15}$N-enrichement allowed an unambiguous allocation of the three Ag(I) ions coordinated between the imidazole ligands, as seen by the observation of $^1J(^{15}\text{N}, \, ^{107/109}\text{Ag})$ couplings.

Further noteworthy is the fact that, by using the 1-deazaadenine-Ag(I)-T base pair, the longest stretch of stacked metal base pairs realized so far includes 19 Ag(I) ions, thus reaching over almost two complete helical turns [54].

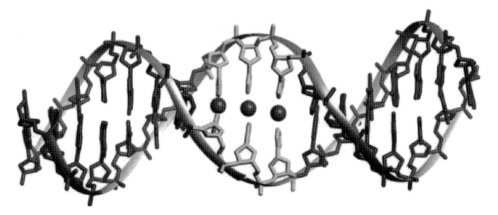

**Figure 9.17** *NMR structure of a B-DNA-like duplex containing three stacked Ag(I)-imidazole base pairs.*

## 9.6    Functions, Applications and Future Directions

### 9.6.1    Duplex Stabilization and Conformational Switching

The first two decades of research on metal base pairing mainly focused on the development and characterization of new metal-mediated base pairing systems in order to elucidate the critical parameters and scope of metal complexation. Since a number of systems have been proven robust and reliable in the context of various sequences, current research efforts are shifting towards the application of metal base pairs in fields such as nanotechnology and medical diagnostics.

The increase in duplex stabilization by metal base pairing has been applied to the stabilization of short double strands, as explained in Section 9.4.3. The process depicted in Figure 9.18a goes even further: a palindromic single-strand containing a central stretch of three 1,2,4-triazole ligands forms a hairpin structure in the absence of Ag(I) ions [42]. The addition of Ag(I) to the sample results in a structural conversion into a double-stranded helix containing three consecutive Ag(I)-mediated base pairs. Such a metal-triggered conformational change might find use in future DNA-based nanomechanical devices. Even more intriguing is the reaction shown in Figure 9.18b: a tetrameric oligonucleotide consisting of hydroxypyridone ligands was found to form triple helical, tetranuclear complexes upon addition of Fe(III) since the Fe(III) ions strictly demand a six-coordinate environment [55]. Keeping it in mind that hydroxypyridone-containing strands form double strands with Cu(II) as the coordinated metal, a structural conversion between double-stranded and triple-stranded constructs upon a change of metal ion is imaginable.

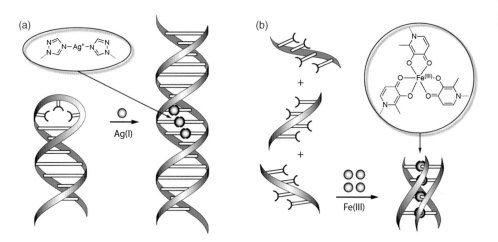

**Figure 9.18**  *Examples of structural transitions triggered by the addition of metal ions to ligand-containing DNA sequences. (a) Hairpin-duplex transformation upon binding of Ag(I). (b) Formation of triple helices by incorporation of hexacoordinate Fe(III) into hydroxypyridone-containing sequences that otherwise give a double-stranded DNA with Cu(II).*

### 9.6.2 Sensor Applications

Examples that are probably nearer to the real world application of metal base pairing systems may be found in the field of molecular sensing. In this direction, Willner *et al.* reported a system based on Hg(II)-selective, T-rich- or Ag(I)-selective, C-rich-DNA single strands bound to luminescent CdSe/ZnS quantum dots (Figure 9.19) [56]. In the absence of the metal ions, irradiation of the nanoscopic core-shell particles results in the emission of light with a wavelength specific for the size of the quantum dot. After the addition of the appropriate transition metal ions that bind to the DNA sequences, this fluorescence is quenched, thereby signaling the binding event. The two systems depicted in Figure 9.19a and b were found to behave independently, which was used to realize the functions of logic gates with "AND" as well as "OR" signal processing behavior.

### 9.6.3 Magnetism and Electrical Conductance

Towards the application of metal base pairs in the emerging fields of molecular magnetism and electronics, a few steps have been undertaken so far. As mentioned in Section 9.5.2, linear arrays of directly stacked hydroxypyridone-Cu(II) base pairs were found to result in an ferromagnetic coupling of all involved Cu(II) centers [48]. Similar EPR measurements were performed for stacks of two salen-Cu(II) base pairs in DNA which showed that, in this case, anti-ferromagnetic coupling was predominant [57]. These experimental studies were supported and interpreted by high-level DFT quantum mechanical modeling studies, which showed that the stacked hydroxypyridone-Cu(II) complexes are tightly linked by two of the ligand's oxygen atoms acting as bridging ligands (Figure 9.20a). This results in an orthogonal arrangement of the single occupied molecular orbitals [58]. Consequently, the absence of a continuous exchange path along the bridging ligands results in ferromagnetic coupling according to the Goodenough–Kanamori rules of magnetic superexchange. In contrast, the salen-Cu(II) base pairs adopt an arrangement in which the energetically preferred anti-ferromagnetic coupling is the result (Figure 9.20b).

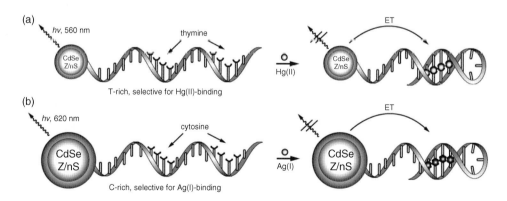

**Figure 9.19** *Quantum dot-DNA hybrid structures with (a) specificity for Hg(II) and (b) specificity for Ag(I) deliver a change in fluorescence intensity upon metal addition as a readout signal.*

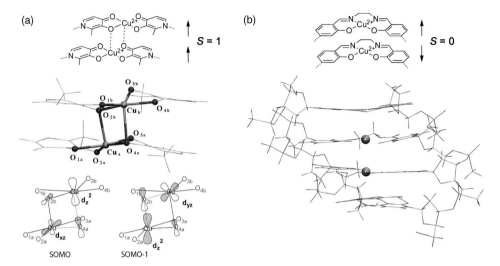

**Figure 9.20** *Different magnetic behaviours of stacked metal base pairs containing paramagnetic Cu(II). (a) The stacking of hydroxypyridone-Cu(II) base pairs leads to a ferromagnetic interaction between the Cu(II) centers due to the formation of square [Cu₂O₂] arrangements resulting in an orthogonal arrangement of the single occupied molecular orbitals. (b) In contrast, the stacking of salen-Cu(II) base pairs leads to an antiferromagnetic interaction between the Cu(II) centers. Reprinted with permission from Ref. [58]. Copyright 2009 WILEY-VCH Verlag GmbH & Co. KGaA, Weinheim.*

Motivated by the vision of future DNA-based molecular electronics [20,59], first experiments on the electrical conductivity of metal base pairing DNA were undertaken recently. To this end, short DNA double strands containing no, one or three metal base pairs were covalently attached between the two ends of single-walled carbon nanotubes, as depicted in Figure 9.21a [60]. Microcontacting of the nanotubes on either side allowed the measurement of the electrical current through the devices in response to the applied voltage. It was found that the measured conductivity significantly responded to the presence or absence of metal ion(s) coordinated by the metal base pair(s), which was demonstrated by the cyclic switching of the conductivity after treating the system in turns with EDTA and the metal ion, respectively (Figure 9.21b). This effect was only observed when using the double strands containing the hydroxypyridone ligands, and no effect of metal addition was seen for the conductivity of the unmodified control strand. Furthermore, the duplex containing three consecutive metal base pairs indicated a somewhat higher conductivity than the one with only one metal base pair.

By employing a different approach, other researchers recently investigated the mobility of charge carriers in metal base pair containing DNA strands using microwave spectroscopy, thus coming to similar results [61].

### 9.6.4 Future Directions

The topic of metal base pairing is still a pretty young one, with important milestones already left behind, such as the proof of synthetic feasibility, the specific binding of various

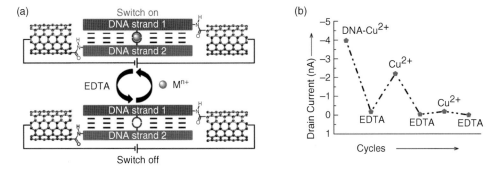

**Figure 9.21** *Direct measurement of electrical conductance through metal base pair containing DNA duplexes. (a) Experimental setup consisting of a metal-DNA duplex covalently bound between the ends of two carbon nanotubes. (b) The development of the measured current in cyclic presence or absence of the metal ions shows a clear dependence of the conductivity on the binding of the metal inside the DNA (although each cycle leads to a gradual device degradation). Reprinted with permission from Ref. [60]. Copyright 2011 WILEY-VCH Verlag GmbH & Co. KGaA, Weinheim.*

transition metal ions, the stacking and mixing of metal ions inside the same duplex and the resulting implications on properties such as duplex stability, magnetism and electrical conductivity. Especially the latter two topics, however, require both a deeper understanding of the underlying mechanisms and the examination of more metal base pair-containing systems by additional independent experimental techniques. With regard to DNA-based molecular electronics, the combination of metal base paired sequences with the toolbox of DNA origami may allow the construction and self-assembly of electrofunctional devices in the near future. Combining the principles of electrical conductivity with the spin-based properties of paramagnetic metal ions along the path of electric conductance through the metallo-DNA duplex may contribute to the development of the emerging field of molecular spintronics [62] which encodes information using not only a difference in electrical potential ("on", "off") but also the spin of the conducting electrons.

Further, the development of metal base pairing may also contribute to the field of analytical chemistry, in particular medical or environmental diagnosis. Not only is the direct sensing of transition metal ions specifically binding to the offered DNA-based ligands of interest here, but also more complex sensing and signaling cascades based on DNA constructs may benefit from the incorporation of metal-triggered duplex stabilization or conformational switching events.

# References

1. Watson, J.D. and Crick, F.H.C. (1953) Molecular structure of nucleic acids: a structure for deoxyribose nucleic acid. *Nature*, **171**, 737–738.
2. Lodish, H., Berk, A., Zipursky, S.L. *et al.* (2007) *Molecular Cell Biology*, 6th edn, Freeman, New York.
3. Saenger, W. (1984) *Principles of Nucleic Acid Structure*, Springer, New York.

4. Guckian, K.M., Schweitzer, B.A., Ren, R.X.-F. *et al.* (2000) Factors contributing to aromatic stacking in water: evaluation in the context of DNA. *J. Am. Chem. Soc.*, **122**, 2213–2222.

5. Isaksson, J., Acharya, S., Barman, J. *et al.* (2004) Single-stranded adenine-rich DNA and RNA retain structural characteristics of their respective double-stranded conformations and show directional differences in stacking pattern. *Biochemistry*, **43**, 15996–16010.

6. Chalikian, T., Völker, J., Plum, G., and Breslauer, K. (1999) A more unified picture for the thermodynamics of nucleic acid duplex melting: A characterization by calorimetric and volumetric techniques. *Proc. Natl Acad. Sci. USA*, **96**, 7853–7858.

7. Gait, M.J. (ed.) (1990) Oligonucleotide Synthesis: A Practical Approach, IRL, New York.

8. Keren, K., Berman, R.S., and Braun, E. (2004) Patterned DNA metallization by sequence-specific localization of a reducing agent. *Nano Lett.*, **4**, 323–326.

9. (a) Niemeyer, C.M. (2001) Nanoparticles, proteins, and nucleic acids: biotechnology meets materials science. *Angew. Chem. Int. Ed.*, **40**, 4128–4158; (b) Wengel, J. (2004) Nucleic acid nanotechnology-towards Ångström-scale engineering. *Org. Biomol. Chem.*, **2**, 277–280; (c) Storhoff, J.J. and Mirkin, C.A. (1999) Programmed materials synthesis with DNA. *Chem. Rev.*, **99**, 1849–1862; (d) Bandy, T.J., Brewer, A., Burns, J.R. *et al.* (2010) DNA as supramolecular scaffold for functional molecules: progress in DNA nanotechnology. *Chem. Soc. Rev.*, **40**, 138–148; (e) Sorensen, M.D., Petersen, M., and Wengel, J. (2003) Functionalized LNA (locked nucleic acid): high-affinity hybridization of oligonucleotides containing N-acylated and N-alkylated 2-amino-LNA monomers. *Chem. Commun.*, **2003**, 2130–2131.

10. (a) Aldaye, F.A., Palmer, A., and Sleiman, H.F. (2008) Assembling materials with DNA as the guide. *Science*, **321**, 1795–1799; (b) Feldkamp, U. and Niemeyer, C.M. (2006) Rational design of DNA nanoarchitectures. *Angew. Chem. Int. Ed.*, **45**, 1856–1876; (c) Endo, M. and Sugiyama, H. (2009) Chemical approaches to DNA nanotechnology. *ChemBioChem*, **10**, 2420–2443; (d) Simmel, F.C. (2008) Three-dimensional nanoconstruction with DNA. *Angew. Chem. Int. Ed.*, **47**, 5884–5887; (e) Heckel, A. and Famulok, M. (2008) Building objects from nucleic acids for a nanometer world. *Biochimie*, **90**, 1096–1107.

11. (a) Rothemund, P.W.K. (2006) Folding DNA to create nanoscale shapes and patterns. *Nature*, **440**, 297–302; (b) Endo, M., and Sugiyama, H. (2009) Chemical approaches to DNA nanotechnology. *ChemBioChem*, **10**, 2420–2443.

12. (a) Maune, H.T., Han, S., Barish, R.D. *et al.* (2010) Self-assembly of carbon nanotubes into two-dimensional geometries using DNA origami templates. *Nat. Nanotechnol.*, **5**, 61–66; (b) Yan, H. (2004) Nucleic Acid Nanotechnology. *Science*, **306**, 2048–2049; (c) Somoza, A. (2009) Evolution of DNA Origami. *Angew. Chem. Int. Ed.*, **48**, 9406–9408.

13. (a) Seeman, N.C. (1998) Nucleic acid nanostructures and topology. *Angew. Chem. Int. Ed.*, **37**, 3220–3238; (b) Seeman, N.C. (2003) DNA in a material world. *Nature*, **421**, 427–431; (c) Seeman, N.C. (2005) The challenge of structural control on the nanoscale: bottom-up self-assembly of nucleic acids in 3D. *Int. J. Nanotechnol.*, **2**, 348; (d) Douglas, S.M., Dietz, H., Liedl, T. *et al.* (2009) Self-assembly of DNA into nanoscale three-dimensional shapes. *Nature*, **459**, 414–418; (e) Andersen, E.S., Dong, M., Nielsen, M.M. *et al.* (2009). Self-assembly of a nanoscale DNA box with a controllable lid. *Nature*, **459**, 73–76.

14. Aldaye, F.A., Lo, P.K., Karam, P. *et al.* (2009) Modular construction of DNA nanotubes of tunable geometry and single- or double-stranded character. *Nat. Nanotechnol.*, **4**, 349–352.

15. Lo, P., Karam, P., Aldaye, F.A. *et al.* (2010) Loading and selective release of cargo in DNA nanotubes with longitudinal variation. *Nat. Chem.*, **2**, 319–328.

16. (a) Seeman, N.C. (2005) From genes to machines: DNA nanomechanical devices. *Trends Biochem. Sci.*, **30**, 119–125; (b) Sherman, W.B. and Seeman, N.C. (2004) A precisely controlled DNA biped walking device. *Nano Lett.*, **4**, 1203–1207; (c) Shin, J.-S. and Pierce, N.A. (2004) A synthetic DNA walker for molecular transport. *J. Am. Chem. Soc.*, **126**, 10834–10835.

17. Sievers, D. and von Kiedrowski, G. (1994) Self-replication of complementary nucleotide-based oligomers. *Nature*, **369**, 221–224.

18. (a) Carroll, R.L. and Gorman, C.B. (2002) The genesis of molecular electronics. *Angew. Chem. Int. Ed.*, **41**, 4378–4400; (b) Wassel, R.A. and Gorman, C.B. (2004) Establishing the molecular basis for molecular electronics. *Angew. Chem. Int. Ed.*, **43**, 5120–5123.

19. (a) Guo, X., Gorodetsky, A.A., Hone, J. *et al.* (2008) Conductivity of a single DNA duplex bridging a carbon nanotube gap. *Nat. Nanotechnol.*, **3**, 163–167; (b) Mas-Balleste, R., Castillo, O., Sanz Miguel, P.J., *et al.* (2009) Towards molecular wires based on metal-organic frameworks. *Eur. J. Inorg. Chem.*, **2009**, 2885–2896.

20. (a) Clever, G.H., Kaul, C., and Carell, T. (2007) DNA–metal base pairs. *Angew. Chem. Int. Ed.*, **46**, 6226–6236; (b) Clever, G.H. and Shionoya, M. (2010) Metal–base pairing in DNA. *Coord. Chem. Rev.*, **254**, 2391–2402; (c) Clever, G.H. and Shionoya, M. (2012) *Interplay Between Metal Ions and Nucleic Acids. Metal Ions in Life Sciences, vol. 10.* (eds A. Sigel, H. Sigel, and R.K.O. Sigel), Springer, Dordrecht.

21. (a) Egli, M. (2002) DNA-cation interactions: quo vadis? *Chem. Biol.*, **9**, 277–286; (b) Hud, N.V. and Polak, M. (2001) DNA–cation interactions: the major and minor grooves are flexible ion-ophores. *Curr. Opin. Struct. Biol.*, **11**, 293–301.

22. (a) Williamson, J.R., Raghuraman, M.K., and Cech, T.R. (1989) Monovalent cation-induced structure of telomeric DNA: The G-quartet model. *Cell*, **59**, 871–880; (b) Smith, F.W. and Feigon, J. (1992) Quadruplex structure of Oxytricha telomeric DNA oligonucleotides. *Nature*, **356**, 164–168.

23. Roelfes, G. and Feringa, B.L. (2005) DNA-based asymmetric catalysis. *Angew. Chem. Int. Ed.*, **44**, 3230–3232.

24. (a) Sundquist, W.I. and Lippard, S. (1990) The coordination chemistry of platinum anticancer drugs and related compounds with DNA. *Coord. Chem. Rev.*, **100**, 293–322; (b) Sato, K., Chikira, M., Fujii, Y., and Komatsu, A. (1994) *Chem. Commun.*, **1994**, 625; (c) Mandal, S.S., Varshney, U., and Bhattacharya, S. (1997) Role of the central metal ion and ligand charge in the DNA binding and modification by metallosalen complexes. *Bioconjugate Chem.*, **8**, 798–812; (d) Rokita, S.E., and Burrows, C.J. (2003) in *Salen-Metal Complexes, Small Molecule DNA and RNA Binders* (eds M. Demeunynck, C. Bailly, and W.D. Wilson), Wiley-VCH, Weinheim.

25. (a) Gothelf, K.V. and LaBean, T.H. (2005) DNA-programmed assembly of nanostructures. *Org. Biomol. Chem.*, **3**, 4023–4037; (b) Bannwarth, W., Pfleiderer, W., and Müller, F. (1991) Energy transfer within oligonucleotides from a lumazine (= Pteridine-2,4(1*H*,3*H*)-dione) chromophore to bathophenanthroline-ruthenium(II) complexes. *Helv. Chim. Acta*, **74**, 1991–1999; (c) Hurley, D.J. and Tor, Y. (1998) Metal-containing oligonucleotides: solid-phase synthesis and lumines-cence properties. *J. Am. Chem. Soc.*, **120**, 2194–2195; (d) Nguyen, T., Brewer, A., and Stulz, E. (2009) Duplex stabilization and energy transfer in zipper porphyrin DNA. *Angew. Chem. Int. Ed.*, **48**, 1974–1977.

26. (a) Kelland, L. (2007) The resurgence of platinum-based cancer chemotherapy. *Nat. Rev. Cancer*, **7**, 573–584; (b) Schliepe, J., Berghoff, U., Lippert, B., and Cech, D. (1996) Automated solid phase synthesis of platinated oligonucleotides via nucleoside phosphonates. *Angew. Chem. Int. Ed. Engl.*, **35**, 646–648.

27. Krueger, A.T. and Kool, E.T. (2009) Redesigning the architecture of the base pair: toward biochemical and biological function of new genetic sets. *Chem. Biol.*, **16**, 242–248.

28. (a) Seela, F. and He, Y. (2003) 6-Aza-2′-deoxyisocytidine: synthesis, properties of oligonucleo-tides, and base-pair stability adjustment of DNA with parallel strand orientation. *J. Org. Chem.*, **68**, 367–377; (b) Benner, S.A. (2004) Understanding nucleic acids using synthetic chemistry. *Acc. Chem. Res.*, **37**, 784–797; (c) Switzer, C., Moroney, S.E., and Benner, S.A. (1989) Enzymatic incorporation of a new base pair into DNA and RNA. *J. Am. Chem. Soc.*, **111**, 8322–8323; (d) Piccirilli, J.A., Krauch, T., Moroney, S.E., and Benner, S.A. (1990) Enzymatic

incorporation of a new base pair into DNA and RNA extends the genetic alphabet. *Nature*, **343**, 33–37; (e) Switzer, C., Moroney, S.E., and Benner, S.A. (1993) Enzymic recognition of the base pair between isocytidine and isoguanosine. *Biochemistry*, **32**, 10489–10496; (f) Horlacher, J., Hottiger, M., Podust, V.N., Huebscher, U., and Benner, S.A. (1995) Recognition by viral and cellular DNA polymerases of nucleosides bearing bases with nonstandard hydrogen bonding patterns. *Proc. Natl Acad. Sci. USA*, **92**, 6329–6333.

29. (a) Kool, E.T. (1998) Replication of non-hydrogen bonded bases by DNA polymerases: A mechanism for steric matching. *Biopolymers*, **48**, 3–17; (b) Guckian, K.M., Krugh, T.R., and Kool, E.T. (1998) Solution structure of a DNA duplex containing a replicable difluorotoluene–adenine pair. *Nat. Struct. Biol.*, **11**, 954–959; (c) Kool, E.T., Morales, J.C., and Guckian, K.M. (2000) Mimicking the structure and function of DNA: insights into DNA stability and replication. *Angew. Chem. Int. Ed.*, **39**, 990–1009; (d) Ogawa, A.K., Wu, Y., McMinn *et al.* (2000) Efforts toward the expansion of the genetic alphabet: information storage and replication with unnatural hydrophobic base pairs. *J. Am. Chem. Soc.*, **122**, 3274–3287; (e) Wu, Y., Ogawa, A.K., Berger, M. *et al.* (2000) Efforts toward expansion of the genetic alphabet: optimization of interbase hydrophobic interactions. *J. Am. Chem. Soc.*, **122**, 7621–7632.

30. (a) Rakitin, A., Aich, P., Papadopoulos, C. *et al.* (2001) Metallic conduction through engineered DNA: DNA nanoelectronic building blocks. *Phys. Rev. Lett.*, **86**, 3670–3673; (b) Aich, P., Labiuk, S.L., Tari, L.W., *et al.* (1999) M-DNA: a complex between divalent metal ions and DNA which behaves as a molecular wire. *J. Mol. Biol.*, **294**, 477–485; (c) Nokhrin, S., Baru, M., and Lee, J.S. (2007) A field-effect transistor from M-DNA. *Nanotechnology*, **18**, 095205.

31. (a) Katz, S. (1952) The reversible reaction of sodium thymonucleate and mercuric chloride. *J. Am. Chem. Soc.*, **74**, 2238–2245; (b) Katz, S. (1963) The reversible reaction of Hg(II) and double-stranded polynucleotides a step-function theory and its significance. *Biochim. Biophys. Acta*, **68**, 240–253; (c) Buncel, E., Boone, C., Joly, H. *et al.* (1985) Metal ion-biomolecule interactions. XII. $^1$H and $^{13}$C NMR evidence for the preferred reaction of thymidine over guanosine in exchange and competition reactions with Mercury(II) and Methylmercury(II). *Inorg. Biochem.*, **25**, 61–73; (d) Kuklenyik, Z., and Marzilli, L.G. (1996) Mercury(II) site-selective binding to a DNA hairpin. Relationship of sequence-dependent intra- and interstrand cross-linking to the hairpin–duplex conformational transition. *Inorg. Chem.*, **35**, 5654–5662; (e) Ono, A., and Togashi, H. (2004) Highly selective oligonucleotide-based sensor for mercury(II) in aqueous solutions. *Angew. Chem. Int. Ed.*, **43**, 4300–4302; (f) Miyake, Y., Togashi, H., Tashiro, M. *et al.* (2006) Mercury(II)-mediated formation of thymine–Hg(II)–thymine base pairs in DNA duplexes. *J. Am. Chem. Soc.*, **128**, 2172–2173; (g) Tanaka, Y., Oda, S., Yamaguchi, H. *et al.* (2007) Mercury(II)-mediated formation of thymine–Hg(II)–thymine base pairs in DNA duplexes. *J. Am. Chem. Soc.*, **129**, 244.

32. (a) Ono, A., Cao, S., Togashi, H. *et al.* (2008) Specific interactions between silver(I) ions and cytosine–cytosine pairs in DNA duplexes. *Chem. Commun.*, **2008**, 4825–4827; (b) Megger, D.A. and Müller, J. (2010) Silver(I)-mediated cytosine self-pairing is preferred over hoogsteen-type base pairs with the artificial nucleobase 1,3-dideaza-6-nitropurine. *Nucleosides, Nucleotides Nucleic Acids*, **29**, 27–38.

33. (a) Müller, J., Böhme, D., Düpre, N. *et al.* (2007) Differential reactivity of α and β-2′-deoxyribonucleosides towards protonation and metalation. *J. Inorg. Biochem.*, **101**, 470–476; (b) Müller, J., Böhme, D., Lax, P. *et al.* (2005) Metal ion coordination to azole nucleosides. *Chem. Eur. J.*, **11**, 6246–6253.

34. (a) For *N*-glycosidation methods see: Vorbrüggen, H. and Ruh-Polenz, C. (2001) *Handbook of Nucleoside Synthesis*, John Wiley & Sons, Inc., New York;(b) For C-glycosidation methods see: Stambaský, J., Hocek, M., and Kocovský, P. (2009) C-nucleosides: synthetic strategies and biological applications. *Chem. Rev.*, **109**, 6729–6764; (c) Bihovsky, R., Selick, C., and Guisti, I.

(1988) Synthesis of C-glucosides by reactions of glucosyl halides with organocuprates. *J. Org. Chem.*, **53**, 4026–4031; (d) Griesang, N. and Richert, C. (2002) Oligonucleotides containing a nucleotide analog with an ethynylfluorobenzene as nucleobase surrogate. *Tetrahedron Lett.*, **43**, 8755–8758.

35. (a) Jäger, S., Rasched, G., Kornreich-Leshem, H. *et al.* (2005) A versatile toolbox for variable DNA functionalization at high density. *J. Am. Chem. Soc.*, **127**, 15071–15082; (b) Gramlich, P.M.E., Wirges, C.T., Gierlich, J., and Carell, T. (2008) Synthesis of modified DNA by PCR with alkyne-bearing purines followed by a click reaction. *Org. Lett.*, **10**, 249–251.

36. Wartell, R.M. and Benight, A.S. (1985) Thermal denaturation of DNA molecules: A comparison of theory with experiment. *Physics Reports*, **126**, 67–107.

37. (a) Meggers, E., Holland, P.L., Tolman, W.B. *et al.* (2000) A novel copper-mediated DNA base pair. *J. Am. Chem. Soc.*, **122**, 10714–10715; (b) Zimmermann, N., Meggers, E., and Schultz, P.G. (2004) A second-generation copper(II)-mediated metallo-DNA-base pair. *Bioorg. Chem.*, **32**, 13–25; (c) Atwell, S., Meggers, E., Spraggon, G., and Schultz, P.G. (2001) Structure of a copper-mediated base pair in DNA. *J. Am. Chem. Soc.*, **123**, 12364–12367.

38. (a) Schlegel, M.K., Essen, L.-O., and Meggers, E. (2008) Duplex structure of a minimal nucleic acid. *J. Am. Chem. Soc.*, **130**, 8158–8159; (b) Schlegel, M.K., Zhang, L., Pagano, N., and Meggers, E. (2009) Metal-mediated base pairing within the simplified nucleic acid GNA. *Org. Biomol. Chem.*, **7**, 476–482.

39. Kaul, C., Müller, M., Wagner, M. *et al.* (2011) Reversible bond formation enables the replication and amplification of a crosslinking salen complex as an orthogonal base pair. *Nat. Chem.*, **3**, 794–800.

40. Tanaka, K. and Shionoya, M. (1999) Synthesis of a novel nucleoside for alternative DNA base pairing through metal complexation. *J. Org. Chem.*, **64**, 5002–5003.

41. (a) Tanaka, K. and Shionoya, M. (2007) Programmable metal assembly on bio-inspired templates. *Coord. Chem. Rev.*, **251**, 2732–2742; (b) Müller, J. (2008) Metal-ion-mediated base pairs in nucleic acids. *Eur. J. Inorg. Chem.*, 3749–3763.

42. Böhme, D., Düpre, N., Megger, D.A., and Müller, J. (2007) Conformational change induced by metal-ion-binding to DNA containing the artificial 1,2,4-triazole nucleoside. *Inorg. Chem.*, **46**, 10144–10119.

43. Tanaka, K., Yamada, Y., and Shionoya, M. (2002) Formation of silver(I)-mediated DNA duplex and triplex through an alternative base pair of pyridine nucleobases. *J. Am. Chem. Soc.*, **124**, 8802–8803.

44. Zimmermann, N., Meggers, E., and Schultz, P.G. (2002) A novel silver(I)-mediated DNA base pair. *J. Am. Chem. Soc.*, **124**, 13684–13685.

45. Zhang, L. and Meggers, E. (2005) An extremely stable and orthogonal DNA base pair with a simplified three-carbon backbone. *J. Am. Chem. Soc.*, **127**, 74–75.

46. (a) Switzer, C., Sinha, S., Kim, P.H., and Heuberger, B.D. (2005) A purine-like nickel(II) base pair for DNA. *Angew. Chem. Int. Ed.*, **44**, 1529–1532; (b) Switzer, C., and Shin, D. (2005) A pyrimidine-like nickel(II) DNA base pair. *Chem. Commun.*, **2005**, 1342–1344.

47. Tanaka, K., Tengeiji, A., Kato, T. *et al.* (2002) Efficient incorporation of a copper hydroxypyridone base pair in DNA. *J. Am. Chem. Soc.*, **124**, 12494.

48. Tanaka, K., Tengeiji, A., Kato, T. *et al.* (2003) A discrete self-assembled metal array in artificial DNA. *Science*, **299**, 1212–1213.

49. Tanaka, K., Clever, G.H., Takezawa, Y. *et al.* (2006) Programmable self-assembly of metal ions inside artificial DNA duplexes. *Nat. Nanotechnol.*, **1**, 190–194.

50. (a) Clever, G.H., Polborn, K., and Carell, T. (2005) A highly DNA-duplex-stabilizing metal–salen base pair. *Angew. Chem. Int. Ed.*, **44**, 7204–7208; (b) Clever, G.H., Söltl, Y., Burks, H., Spahl, W., and Carell, T. (2006) Metal–salen-base-pair complexes inside DNA: complexation overrides sequence information. *Chem. Eur. J.*, **12**, 8708–8718.

51. Clever, G.H. and Carell, T. (2007) Controlled stacking of 10 transition-metal ions inside a DNA duplex. *Angew. Chem. Int. Ed.*, **46**, 250–253.
52. Yanagida, K., Hamochi, N., Sasano, K. *et al.* (2007) Preparation of DNA double helical structures containing both mercury and silver ions. *Nucleic Acids Symp. Ser.*, **51**, 179–180.
53. Johannsen, S., Megger, N., Böhme, D. *et al.* (2010) Solution structure of a DNA double helix with consecutive metal-mediated base pairs. *Nat. Chem.*, **2**, 229–234.
54. Polonius, F.-A. and Müller, J. (2007) An artificial base pair, mediated by hydrogen bonding and metal-ion binding. *Angew. Chem. Int. Ed.*, **46**, 5602–5604.
55. Takezawa, Y., Maeda, W., Tanaka, K., and Shionoya, M. (2009) Discrete self-assembly of iron (III) Ions inside triple-stranded artificial DNA. *Angew. Chem. Int. Ed.*, **48**, 1081–1084.
56. Freeman, R., Finder, T., and Willner, I. (2009) Multiplexed analysis of $Hg^{2+}$ and $Ag^+$ ions by nucleic acid functionalized CdSe/ZnS quantum dots and their use for logic gate operations. *Angew. Chem. Int. Ed.*, **42**, 7818–7821.
57. Clever, G.H., Reitmeier, S.J., Carell, T., and Schiemann, O. (2010) Antiferromagnetic coupling of stacked Cu(II)–salen complexes in DNA. *Angew. Chem. Int. Ed.*, **49**, 4927–4929.
58. Mallajosyula, S.S. and Pati, S.K. (2009) Conformational tuning of magnetic interactions in metal–DNA complexes. *Angew. Chem. Int. Ed.*, **48**, 4977–4981.
59. (a) Carell, T., Behrens, C., and Gierlich, J. (2003) Electrontransfer through DNA and metal-containing DNA. *Org. Biomol. Chem.*, **1**, 2221–2228; (b) Liu, S.-P., Weisbrod, S.-H., Tang, Z. *et al.* (2010) Direct measurement of electrical transport through G-quadruplex DNA with mechanically controllable break junction electrodes. *Angew. Chem. Int. Ed.*, **49**, 3313–3316.
60. Liu, S., Clever, G.H., Takezawa, Y. *et al.* (2011) Direct conductance measurement of individual metallo-DNA duplexes within single-molecule break junctions. *Angew. Chem. Int. Ed.*, **50**, 8886–8890.
61. (a) Kawai, K., Osakada, Y., Fujitsuka, M., and Majima, T. (2008) Charge separation in acridine- and phenothiazine-modified DNA. *J. Phys. Chem. B*, **112**, 2144–2149; (b) Yamagami, R., Kobayashi, K., Saeki, A. *et al.* (2006) Photogenerated hole mobility in DNA measured by time-resolved microwave conductivity. *J. Am. Chem. Soc.*, **128**, 2212–2213; (c) Isobe, H., Yamazaki, N., Asano, A. *et al.* (2011) Electron mobility in a mercury-mediated duplex of triazole-linked DNA. *Chem. Lett.*, **40**, 318.
62. (a) Sanvito, S. and Rocha, A.R. (2006) Molecular-spintronics: the art of driving spin through molecules. *J. Comput. Theor. Nanosci.*, **3**, 624–642; (b) Kwok, K.S. and Ellenberger, J.C. (2002) Moletronics: future electronics. *Materials Today*, **5**, 28–37; (c) Shinwari, M.W., Deen, M.J., Starikov, E.B., and Cuniberti, G. (2010) Electrical conductance in biological molecules. *Adv. Funct. Mater.*, **20**, 1865–1883.

# 10

# Metal Complexes as Alternative Base Pairs or Triplets in Natural and Synthetic Nucleic Acid Structures

*Arnie De Leon, Jing Kong, and Catalina Achim*
*Department of Chemistry, Carnegie Mellon University, USA*

## 10.1  Introduction

Advances made in recent decades in understanding biological processes at the molecular level have stimulated progress in the quest for biomimetic approaches for the synthesis of artificial devices that perform various functions, such as electron or ion transfer, catalysis, controlled changes in magnetic properties, or mechanical motions in response to an electrochemical, pH, or light signal. A fundamental issue in the synthesis of the devices is the rational and precise assembly by covalent and/or non-covalent interactions of a discrete number of molecular components in a manner in which they can sense and communicate with each other when a change in their environment takes place. In this context, hybrid inorganic–nucleic acid molecules that contain one or more transition metal ions have been pursued recently. The coexistence of metal complexes and nucleobase pairs leads to molecules with a relatively large set of properties that can be adjusted to achieve the desired molecular architecture and function. This chapter presents an overview of the research in the area of metal-containing, ligand-modified nucleic acids that emerged and has developed in the last decade and a half.

The typical synthesis of molecules containing several metal ions is based on either: (a) the use of polytopic ligands, whose synthesis requires several, possibly different, organic reactions and, consequently, has a low yield (Figure 10.1a), or (b) the self-assembly of

*Metallofoldamers: Supramolecular Architectures from Helicates to Biomimetics*, First Edition.
Edited by Galia Maayan and Markus Albrecht.
© 2013 John Wiley & Sons, Ltd. Published 2013 by John Wiley & Sons, Ltd.

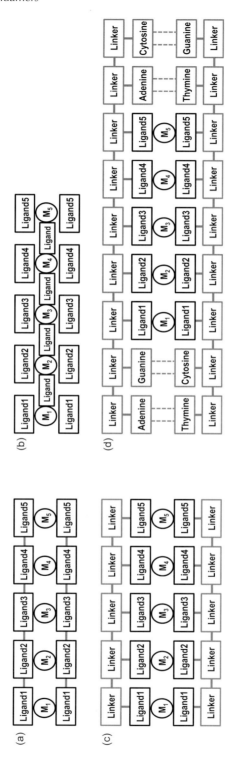

**Figure 10.1** *Cartoon representation of strategies for the synthesis of heteroarrays of transition metal ions. The metal arrays are based on: (a) polytopic ligands, (b) self-assembly using small ligands, (c) polytopic ligands in which different ligands are attached to the same backbone made of monomers termed "linker" in the figure, (d) polytopic ligands in which different ligands and nucleobases are attached to the same nucleic acid backbone.*

small ligands and metal ions (Figure 10.1b), which may generate many isomers from a set of different ligands and metal ions. A solution to this problem is offered by a strategy for creating homo- and heterometallic molecules based on the use of ligand-containing nucleic acid oligomers synthesized by only one, very high yield reaction between backbone linkers to which various ligands are attached (Figure 10.1c). Using this strategy, oligomers with different ligands arranged in any sequence can be obtained in high yield. The difficulty that remains in the case of oligomers composed exclusively of ligand-containing monomers is the presence of multiple, different coordination sites that may have comparable affinity for metal ions and thus, upon metal coordination, it can lead to several super molecules with different arrangements of the oligomers. This problem can be alleviated or avoided if the oligomers contain besides ligands another recognition site that drives the assembly of oligomers in a unique way independent of the metal ion(s) (Figure 10.1d). Ligand-modified nucleic acid duplexes such as those generically represented in Figure 10.1d are possible outcomes of this strategy. The formation of the duplex is driven by the information stored in the sequence of nucleobases of the two strands that interact through Watson–Crick base pairing. Subsequent to the duplex hybridization, metal coordination takes place at the ligand sites rather than at the nucleobases because, under a wide range of conditions, these sites have higher affinity for the metal ions than the nucleobases. Conversely, it is also possible to first self-assemble the ligand-modified nucleic acid oligomers through metal coordination and then create the conditions for Watson–Crick hybridization, for example by a change in temperature or by the addition of a complementary oligomer.

The use of ligand-modified nucleic acids to create hybrid inorganic–nucleic acid structures ensures that the metal binding is site specific rather than uniform, as is the case when the metal ions bind to either the backbone or the nucleobases of the non-modified nucleic acid. In DNA, the G-N7 and A-N7 positions are the most common coordination sites for transition metal ions, with the G-N7 being preferred over the A-N7 site [1]. The factors that affect metal binding to DNA include: (a) the nucleophilicity of the phosphates and nucleobases and their hard and soft Lewis base character, respectively, (b) the accessibility of these groups to metal ions, (c) the molecular electrostatic potential of DNA, (d) the ability of the groups neighboring the metal coordination site to form hydrogen bonds with the water molecules coordinated to the metal, and, (e) in the case of binding of multiple metal ions to the same DNA duplex, the changes in the secondary structure of the DNA and in the nucleophilicity of the phosphate and nucleobases induced by the first metal ions that bind to the DNA [2].

Although not site specific, metal binding to non-modified DNA was successfully exploited to create nano-size, hybrid inorganic–nucleic acid structures with interesting conduction or charge transfer properties. For example, a conductive silver wire was obtained by treatment of a DNA scaffold with $Ag^+$, followed by reduction of $Ag^+$ to metallic silver, and subsequent uniform deposition of Ag or Au on the initial Ag clusters formed on DNA [3,4]. Sequence specificity of the Ag deposition has been achieved by using RecA proteins for sequence-specific protection of the DNA [5]. While this strategy offered sequence specificity, the metal ion localization was with cluster/nanometer resolution rather than atom/angstrom resolution, respectively. A different strategy for the synthesis of hybrid inorganic–nucleic acid structures in which the metal ions are uniformly distributed on DNA was to coordinate 3d transition metal ions to non-modified DNA at

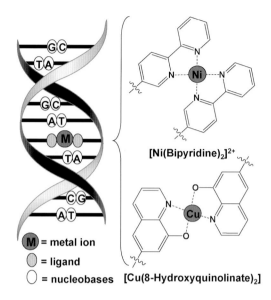

*M* = metal ion
= ligand
= nucleobases

[Ni(Bipyridine)₂]²⁺

[Cu(8-Hydroxyquinolinate)₂]

**Figure 10.2** *Cartoon representation of a nucleic acid duplex containing one metal-based alternative base pair and examples of complexes formed with the two ligands from the two single strands of the duplex.*

high pH, which led to the coordination of one metal ion per nucleobase pair to form M-DNA [6]. The structure, stability, and electronic properties of M-DNA have also been evaluated by theoretical methods [7].

In 1999 Shionoya and collaborators published a paper containing the first mention of the possibility to substitute nucleobase pairs in DNA with pairs of ligands and binding of metal ions to these ligands to create metal-based alternative base pairs (bp) in duplexes that contain both Watson–Crick and coordination bonds (Figure 10.2) [8]. They synthesized several nucleosides and characterized metal complexes of these nucleosides [9]. In 2000 Schultz published the first DNA duplex containing a metal-mediated alternative base pair based on a monodentate and a tridentate ligand coordinated to Cu²⁺ [10], shortly followed by Tor [11] and Shionoya [12], who synthesized DNA duplexes modified with a pair of two bidentate ligands that bound 3d transition metal ions. Since then, several other groups including those of Switzer [13], Müller [14], and Carell [15] have applied this method to synthesize DNA duplexes containing one or several transition metal ions. The same strategy was applied to the successful modification of peptide nucleic acids (PNAs) [16], glycol nucleic acids (GNAs) [17], and locked nucleic acids (LNAs) [18].

To date, several strategies have been used to create systems based on synergetic formation of hydrogen and coordination bonds, and π-stacking interactions (Figure 10.2). Ligand incorporation in the center of nucleic acid oligomers (Figure 10.3a) was used to create metal-containing, ligand-modified duplexes. In this approach, the coordination complex formed between the metal ion and the ligands plays the role of an artificial base pair and will be referred to as a metal–ligand alternative base pair or metal-mediated base pair. Ligands have also been used as connectors of single-stranded (ss) nucleic acids

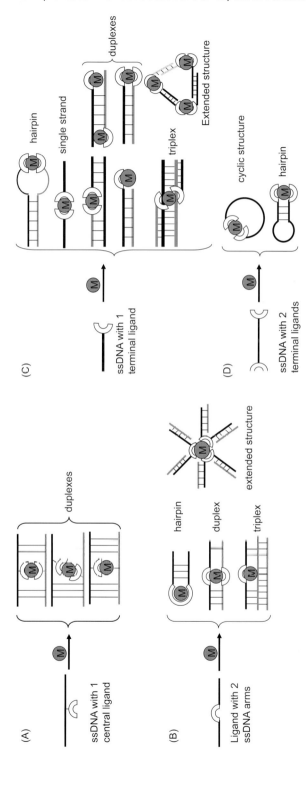

**Figure 10.3** *Hybrid inorganic–nucleic acid structures based on four types (a–d) of ligand-modified nucleic acid strands and transition metal ions.*

(Figure 10.3b). Terminal modification of nucleic acid oligomers with ligands was used to create a broad range of structures, including metal-containing, single strands, hairpins, duplexes, and triplexes (Figure 10.3c, d). Within these hybrid inorganic–nucleic acid structures, there are synergetic relationships between: (a) metal coordination and oligonucleotide hybridization, and (b) the properties of the metal complexes and those of the nucleic acid that form the hybrid inorganic-nucleic acid ensemble. For example, upon hybridization of ligand-modified oligonucleotides to form a duplex that contains ligands in complementary positions, the binding constant for a metal ion to the ligands brought in close proximity of each other by the duplex is generally larger than that to the same free ligand [12b,16c]. Conversely, ligand-modified oligonucleotides that may form two alternate structures such as a duplex or a hairpin (e.g., Figure 10.3b) will adopt preferably one of these two structures in which the metal ion has its most common coordination number and geometry [19].

## 10.2 Brief Overview of Synthetic Analogues of DNA: PNA, LNA, UNA, and GNA

Metal-based alternative base pairs have been incorporated to date in DNA, RNA, and several of their synthetic analogues, including peptide nucleic acid (PNA), locked nucleic acid (LNA), unlocked nucleic acid (UNA), and glycol nucleic acid (GNA). These analogues can form duplexes by Watson–Crick base pairing with DNA and RNA, and thus have potential for antisense and antigene applications. Their backbones are different from that of DNA, which makes them chemically and biochemically more stable than DNA, and thus more appropriate in certain nanotechnology and biology applications.

PNA, which was discovered in 1991, has a pseudo-peptide backbone originally based on N-(2-aminoethylglycine) (Aeg), which is neutral and achiral (Figure 10.4) [20]. The thermal stability of PNA·PNA duplexes is higher than that of PNA·DNA or DNA·DNA duplexes because the backbone is neutral. Hence, mismatches and in general chemical modifications that affect the base pairing, have a larger effect on the thermal stability of

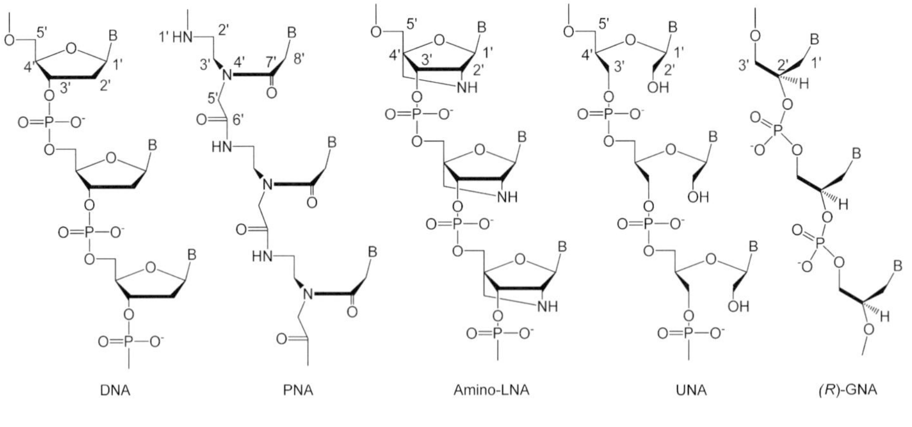

**Figure 10.4** *Chemical structure of single strands of DNA, PNA, LNA, and GNA.*

PNA·PNA duplexes than on PNA·DNA or DNA·DNA duplexes. The synthesis of PNA monomers is relatively simple [21]. PNA monomers containing the A, G, C, and T bases are commercially available from several suppliers. Numerous PNA monomers with different pseudo-peptide backbone or with alternative nucleobases have been prepared [22]. For example, alanyl PNA is a peptide-based nucleic acid structure with an alanyl-derived backbone having the nucleobases attached to the β-carbon [23]. PNA oligomers are prepared simply by modified solid phase synthesis using either a Boc or a Fmoc protection strategy [24]. The attachment of amino acids or carboxylic acid derivatives of ligands or metal complexes to the amino end of the PNA oligomers is straightforward. Positively charged amino acids such as lysine and arginine, and hydrophilic groups such as poly-ethylene glycol (PEG) have been introduced into PNA oligomers to improve their solubility, enabling the preparation of millimolar solutions of PNA [24]. Cysteine has also been introduced in PNA to form self-assembled monolayers (SAMs) on gold surfaces [25].

In contrast to the chiral sugar–phosphate backbone of DNA, the amino-ethylglycine backbone of PNA does not contain stereocenters. A preferred helical handedness can be induced in the PNA·PNA duplexes by the attachment of a D- or L-amino acid at the carboxy or C-terminus of one or both strands of a duplex [26]. X-ray crystallography showed that the PNA·PNA duplexes, which adopt a distinct "p-form" helix in both crystals and solution, exist in crystals as a 1:1 mixture of right- and left-handed duplexes even if they contain a L-lysine, which indicates that the chiral induction effect exerted by the lysine is comparable in strength to the crystal packing forces [27]. A preferred handedness can also be induced in PNA duplexes by a stereogenic center at the α- or γ-position of the PNA backbone [22o,28]. The incorporation of an (S)-Me stereogenic center at the γ-backbone position induces the formation of a right-handed structure in duplexes as well as in ss PNA [22o,29].

Several metal complexes have been attached to PNA because they can be used as IR [e.g., (benzene)chromiumtricarbonyl] or electrochemistry probes {e.g., ferrocene, cobaltocene, or $[Ru(bipy)_3]^{2+}$}, or to quantify the cellular uptake of PNA. These complexes have been linked to PNA by using metal complex-containing PNA monomers or alkyne-containing PNA monomers to which the metal complex could be linked by click chemistry (Figure 10.5) [30]. Studies of the thermal stability of a DNA·PNA duplex have shown that the duplex was destabilized by the ferrocene "clicked" at the middle of the PNA strand of the duplex.

Only LNA is an RNA analogue developed in 1998 [32], in which the ribose group of the sugar phosphate backbone contains a methylene bridge between the 2'-oxygen atom of the ribose and the 4'-carbon atom [33]. Unlocked nucleic acid (UNA) lacks the C2'-C3' bond normally found in ribonucleosides (Figure 10.4). Several C-glycoside analogues of LNA have been prepared including an amino-LNA that has a 2'-NH group instead of the 2'-oxygen (Figure 10.4). This chemical modification preorganizes the LNA monomer into a locked C3'-endo conformation similar to that observed in A-type DNA, and reduces the entropic penalty for the formation of duplexes containing LNA. Consequently, LNA·LNA and hetero LNA·DNA and LNA·RNA duplexes are more stable than DNA·DNA or DNA·RNA ones. The NMR structures of partially- or completely-modified LNA·DNA and LNA·RNA hybrids showed that the LNA nucleotides induce a preference for an A-type, right-handed duplex structure because they adopt a C3'-endo conformation and because they steer the sugar conformations of neighboring nucleotides into similar

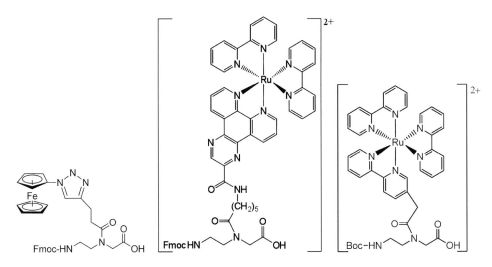

**Figure 10.5** *PNA monomers used for the labeling of PNA-containing duplexes with transition metal complexes [30,31]. These complexes do not function as alternative nucleobase pairs.*

C3′-endo conformations [34]. We note that it is common to refer to DNA that contains only several LNA (or GNA, see p. 341) residues as LNA (or GNA).

The nucleobases of LNA are preorganized for good stacking interactions, which causes very high thermal stability of LNA-modified duplexes [33b]. For example, the introduction of LNA into the DNA strand of a DNA·RNA duplex has been shown to increase the $T_m$ by 1–8 °C per LNA monomer relative to the unmodified duplex. In contrast, the high flexibility of UNA causes a destabilization of UNA-containing duplexes when compared to the ones lacking UNA. For example, the $T_m$ of a UNA·RNA duplex decreased by up to 10 °C per UNA monomer incorporated in the DNA strand.

Glycol nucleic acid GNA has an acyclic three-carbon, propylene glycol-phosphodiester backbone that includes a stereocenter (Figure 10.4) [17b,35]. GNA oligomers form antiparallel, helical duplexes based on Watson–Crick base pairing, which are more stable than the analogous homo DNA or RNA duplexes. Crystallographic studies of ds GNA that contained a brominated cytidine or uracil, or a complex of copper with hydroxypyridone ligands showed the existence of two forms of backbone conformations, namely a condensed N-type and an elongated M-type, respectively [36]. Molecular dynamics simulations of a non-modified GNA duplex showed an all-gauche conformation characteristic of M-type GNA but with a higher helical twist common than N-type GNA structures [37]. GNA and LNA oligomers can be synthesized by standard solid phase DNA synthetic methods because the backbone of both molecules are formed by phosphodiester bonds.

## 10.3 Metal-Containing, Ligand-Modified Nucleic Acid Duplexes

An interesting approach to incorporate transition metal ions in nucleic acid duplexes is the formation of artificial metallo-base pairs (Figure 10.2) [38]. This approach was first described by Tanaka *et al.* and it involves the substitution of nucleobase pairs with ligands that have a high affinity for metal ions [8]. Ligands juxtaposed at complementary positions within a nucleic acid duplex form a high affinity binding site for metal ions. The DNA monomers in which the nucleobase was replaced with a ligand were termed ligandosides by Tor *et al.* [11].

This novel class of metal–DNA hybrid structures is interesting for several reasons. First, it adds a novel dimension to the growing field of artificial base pairs, which were previously based on alternative hydrogen bonding patterns or on hydrophobic interactions [39]. The higher strength of coordinative bonds when compared to hydrogen bonds bodes well for the formation of alternative base pairs whose stability surpasses that of the natural, Watson–Crick base pairs and may be of interest for applications in nano- or biotechnology. Second, metal-based alternative base pairs could be used in the extension of the genetic code [10]. Recent work showed that indeed, metal-containing alternative base pairs can be recognized and extended by polymerases [15i,40]. Third, for the inorganic chemist, the interest in such structures is related to the possibility of using the nucleic acid duplex as a scaffold in which one or multiple metal ions are incorporated in a rational and controlled manner. The bridging of the gap between extended metal clusters produced by materials science, and the metal clusters synthesized by supramolecular inorganic chemistry, makes the efforts aimed at using nucleic acids as scaffold for transition metal ions worthwhile.

### 10.3.1   Design Strategy

To make possible metal ion incorporation at specific locations in nucleic acid duplexes, ligands are included in nucleic acid oligomers, typically in complementary positions. Most ligands used to date to form metal–ligand alternative base pairs contain aromatic rings, and can participate in π-π stacking interactions with adjacent bases. Generally these ligands cannot form hydrogen bonds to bridge the two complementary oligonucleotides forming the duplex. In all studies published so far (most of which are cited in this and other chapters of this book), the stability of the ligand-modified duplexes was similar to or lower than that of the duplexes that have an A·T or G·C base pair instead of the pair of ligands.

The ligands chosen to create metal-containing nucleic acid duplexes form typically square planar metal complexes, can participate in π-stacking interactions (with the adjacent base pairs), and have been placed usually in complementary positions within nucleic acid duplexes. Most of the ligands have 1–3 metal binding sites that are part of aromatic rings. To ensure that the metal ions are incorporated only at the positions where the duplex has been modified, the affinity of the ligands for the metal ions must be higher than that of the natural nucleobases, in particular that of the G bases, which coordinate to metal ions in a monodentate fashion through GN7. Preferably the metal ions should have the ability to form square-planar complexes; octahedral complexes with the ligands in equatorial positions and the axial positions occupied by donor atoms from the adjacent nucleobases or solvent molecules may also play the role of alternative base pairs.

An important condition that artificial base pairs must fulfill to be useful for the extension of the genetic code is to be orthogonal to natural base pairs. Therefore, the metal ion should form complexes only with the extraneous ligands and not mixed-ligand complexes with the natural bases. This condition was verified when polymerases were used to extend ss DNA containing either thymine or salicylaldehyde as ligands in the presence of $Hg^{2+}$ or ethylene diamine and $Cu^{2+}$, respectively [15i,40]. Bipyridine may be another candidate for genetic code expansion because metal coordination to bipyridine ligands introduced opposite a natural nucleobase in DNA (as well as PNA) duplexes had no effect on- or caused a decrease in- the thermal stability of the duplexes [16a,41]. This result suggests that the metal ion does not form an alternative base pair with mixed bipyridine-nucleobase coordination. The decrease in the melting temperature upon metal addition may have been caused by metal coordination to bipyridine only, which in turn may cause a distortion of the duplex and/or a loss of stacking interactions.

Incorporation of multiple metal ions in adjacent positions within duplexes is likely to depend on the overall charge of the metal complex and of the nucleic acid. In DNA duplexes, the phosphodiester backbone can act as intrinsic counter anion for positively charged metal complexes incorporated in the duplex and thus can mitigate electrostatic repulsion between adjacent metal–ligand complexes. In DNA analogues that are neutral, the electrostatic repulsion is weak if neutral metal complexes are incorporated in the duplex.

## 10.3.2    Duplexes Containing One Alternative Metal–Ligand Base Pair with Identical Ligands

Incorporation of a pair of ligands in complementary positions in the middle of DNA duplexes creates a metal-binding site of $[n + m]$ type, where $n$ and $m$ are the number of donor atoms through which the two ligands coordinate to the metal ion. Metal binding sites of the $[n + n]$ type are simpler to create because the same ligand-containing phosphoramidite can be used to incorporate the $n$-dentate ligand in both complementary strands of the duplex. In the following paragraph, we describe the nucleic acid duplexes that contain metal-mediated alternative base pairs reported to date according to their $[n + m]$ coordination schemes.

### *10.3.2.1    [1 + 1] Coordination*

$Ag^+$ and $Hg^{2+}$ ions can form linear, two-coordinate complexes with monodentate ligands, such as the nitrogen-coordinating thymine (**T**), pyridine (**Py**) and imidazole (**Imidazole**). Both $Ag^+$ and $Hg^{2+}$ have been shown to interact with nucleic acids duplexes. For example, $Hg^{2+}$ forms [**T**·$Hg^{2+}$·**T**] complexes in which $Hg^{2+}$ coordinates to the N3 atom of the thymine nucleobase. This property of $Ag^+$ and $Hg^{2+}$ has been exploited in recent years to create [**T**·$Hg^{2+}$·**T**] and [**C**·$Ag^+$·**C**] alternative base pairs within DNA duplexes. For a review of research on metal incorporation in DNA using as ligands natural nucleobases or purine and pyrimidine derivatives, we refer the reader to reference [42].

β-Carbon ligandosides containing imidazole, 1,2,4-triazole (**Triazole**), or tetrazole ligands had been synthesized [43]. The basicity and the related, metal-binding affinity of these ligandosides decreases from imidazole, to 1,2,4-triazole, to tetrazole. In keeping with this relationship, the imidazole ligandoside formed 2 : 1 complexes with both $Ag^+$ and $Hg^{2+}$ but the tetrazole did not form complexes with these metal ions (Figure 10.6). The **Triazole**[44] and **Imidazole**[45] ligandosides have been incorporated in DNA duplexes. These duplexes contained three pairs of ligands in the middle of their sequence

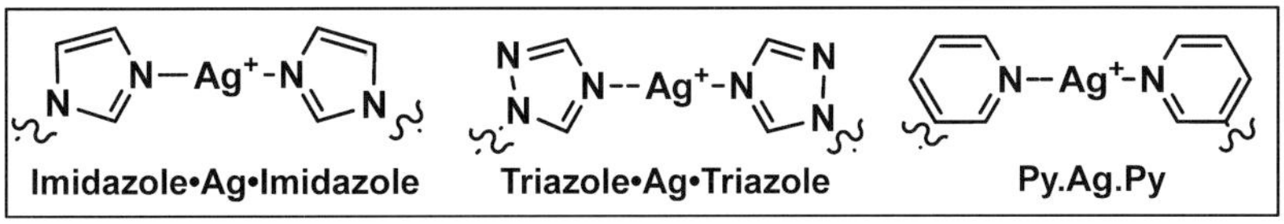

| Imidazole•Ag•Imidazole | Triazole•Ag•Triazole | Py.Ag.Py |

**Figure 10.6** *Alternative base pairs based on two-coordinate metal complexes.*

(see also Section 10.4). In the absence of metal ions, the 17-bp oligonucleotide d(A$_7$**Tri-azole**$_3$T$_7$) formed a hairpin structure. In the presence of Ag$^+$, the same oligonucleotide formed a duplex with a palindromic sequence, in which three central metal-containing, alternative base pairs [**Triazole**.Ag.**Triazole**] were flanked by **AT** base pairs [44]. The structural change from hairpin to duplex induced by Ag$^+$ was demonstrated by using a combination of Ultraviolet (UV), circular dichroism (CD), and fluorescence spectroscopy, MALDI-ToF mass spectrometry, and dynamic light scattering. The **Imidazole**-containing duplex has been characterized by X-ray crystallography (see Section 10.4) [45].

The stability constant of complexes formed between Ag$^+$ and pyridine is such that a [Ag(**Py**)$_2$]$^+$ complex does not form in aqueous solutions containing micromolar concentrations of Ag$^+$ and **Py** [46]. Nevertheless, in the same concentration range, a [Ag(**Py**)$_2$]$^+$ complex formed within DNA duplexes that contained a pair of **Py**s in complementary positions, which can be attributed to a duplex-induced, supramolecular chelate effect (Table 10.1, entry 1) [12b,47]. The coordination of one equivalent of Ag$^+$ to the pair of **Py** was confirmed by $^1$H-NMR titrations, which showed that the Ag$^+$ binding affected exclusively the protons of **Py**s and not those of the natural nucleobases. Ag$^+$ coordination to **Py** had a stabilization effect on the modified DNA duplexes that depended on the sequence and length of the DNA duplex. For example, Ag$^+$ had no effect on the stability of a **Py**-modified DNA duplex that had 15 bp [47] but increased the stability of the same type of duplex that had a different sequence and 21 bps (Table 10.1, entry 1) [12b]. The stabilization effect of Ag$^+$ on the latter duplex increased with the Ag$^+$ concentration and was largest in the presence of three equivalents of Ag$^+$.

### 10.3.2.2 [2 + 2] Coordination

Several groups have developed metal-based alternative base pairs with [2 + 2] coordination by using the bidentate ligands such as bipyridine, 8-hydroxyquinoline, hydroxypyridone, or salicylaldehyde. **Salen**, a four-dentate ligand that coordinates to metal ions through two nitrogen and two oxygen atoms, is also included in this section because of the similarity between its coordination complexes and those with bis(salicylaldehyde) coordination. This ligand was created by metal-templated reaction of salicyladehyde ligands incorporated in DNA with bis-amines.

***N,N-Coordinating Ligands: Bipyridine and Related Aromatic Bis-Imine Ligands.*** 2,2'-Bipyridine (**Bpy**) has a high affinity for a variety of transition metal ions [46,48] and can participate in π-stacking interactions. Hence, this ligand was one of the first ones used to create metal-binding sites in nucleic acids (Figure 10.7). For incorporation in ss DNA, **Bpy** has been connected at its 5-position to the 1' position of the 2'-deoxy-D-ribose, either directly [49] or through a methylene linker [11b]. (Table 10.1,

**Table 10.1** Melting Temperatures of Duplexes Containing Metal Based Alternative Base Pairs. If present during $T_m$ measurement, the transition metal ion and number of equivalents with respect to the NA duplex are specified after the $T_m$ value.

| Ligand Name (Abbr.) | Nucleic Acid | Sequence | $T_m$ (°C) | Conc.[a] (μM) | Ref. |
|---|---|---|---|---|---|
| [1 + 1] Pairs of ligands | | | | | |
| 1   Pyridine (Py) | DNA | 5'-CACATTALTGTTGTA-3'<br>3'-GTGTAATLACAACAT-5'<br>Control: AT/TA<br>CG/GC | 25<br>25 (1 eq Ag$^+$)<br>39/39<br>41/39 | 1 | 47 |
| | DNA | 5'-T$_{10}$LT$_{10}$-3'<br>3'-A$_{10}$LA$_{10}$-5' | 34<br>38 (1 eq Ag$^+$)<br>41 (3 eq Ag$^+$)<br>34 (1 eq Cu$^{2+}$)<br>34 (1 eq Ni$^{2+}$)<br>34 (1 eq Pd$^{2+}$)<br>34 (1 eq Hg$^{2+}$) | 1.2 | 12b |
| | | Control: TA | 47 | | |
| [2 + 2] Pairs of ligands | | | | | |
| 2   Bipyridine (Bpy) | DNA | 5'-AGTCGLCGACT-3'<br>3'-TCAGCLGCTGA-5'<br>Control: CG | 57<br>64 (1 eq Cu$^{2+}$)<br>57 | 1 | 11b |
| 3   5'-Methyl-bipyridine (MeBpy) | PNA | H-GTAGLTCACT-Lys<br>Lys-CATCLAGTGA-H<br>Control: AT | 48<br>59 (1 eq Ni$^{2+}$)<br>67 | 5 | 16a,50 |
| 4   4-(2'-Pyridyl)-pyrimidinone (Pyr$^P$) | DNA | 5'-CTTTCTLTCCCT-3'<br>3'-GAAAGALAGGGA-5' | 23<br>41 (4 eq Ni$^{2+}$)<br>30 (4 eq Co$^{2+}$)<br>27 (4 eq Cu$^{2+}$)<br>24 (4 eq Zn$^{2+}$)<br>24 (4 eq Fe$^{2+}$)<br>24 (4 eq Mn$^{2+}$) | 2.5 | 13a |
| | | Control: TA<br>CG | 37<br>40 | | |

| | | | | | | |
|---|---|---|---|---|---|---|
| 5 | 6-(2′-Pyridyl)-purine (Pur$^P$) | DNA | 5′-CTTTCTLTCCCT-3′<br>3′-GAAAGALAGGGA-5′ | 29<br>46 (2 eq Ni$^{2+}$)<br>39 (2 eq Co$^{2+}$)<br>31 (2 eq Cu$^{2+}$)<br>31 (2 eq Zn$^{2+}$)<br>31 (2 eq Ag$^+$)<br>29 (2 eq Fe$^{2+}$)<br>29 (2 eq Mn$^{2+}$)<br>29 (2 eq Eu$^{3+}$)<br>27 (2 eq Pd$^{2+}$) | 2.5 | 13b |
| | | GNA | Control: TA<br>CG<br>3′-AATATTA_TATTTTA-2′<br>2′-TTATAATL_ATAAAAT-3′ | 37<br>40<br>35<br>53 (2 eq Ni$^{2+}$)<br>48 (2 eq Ag$^{2+}$)<br>42 (2 eq Cu$^{2+}$)<br>42 (2 eq Co$^{2+}$)<br>37 (2 eq Cd$^{2+}$)<br>36 (2 eq Zn$^{2+}$)<br>35 (2 eq Au$^{3+}$)<br>33 (2 eq Pd$^{2+}$) | 2 | 53 |
| 6 | 2-Aminophenol (AP) | DNA | Control: AT<br>5′-A$_{10}$LA$_{10}$-3′<br>3′-T$_{10}$LT$_{10}$-5′<br>Control: AT | 51<br>43<br>40 (1 eq Cu$^{2+}$)<br>47 | 1.19 | 54 |
| 7 | 8-Hydroxyquinoline (Q) | DNA | 5′-CACATTALTGTTGTA-3′<br>3′-GTGTAATLACAACAT-5′<br>Control: AT<br>CC | 36<br>65 (1 eq Cu$^{2+}$)<br>41<br>45 | 2 | 35b |
| | | PNA | H-GTAGLTCACT-Lys<br>Lys-CATCLAGTGA-H<br>Control: AT | 46<br>>79 (1 eq Cu$^{2+}$)<br>67 | 5 | 16b |
| | | GNA | 5′-CACATTALTGTTGTA-3′<br>3′-GTGTAATLACAACAT-5′<br>Control: AT/GC | n.a.<br>71 (1 eq Cu$^{2+}$)<br><30 | 2 | 35b |

*(continued)*

**Table 10.1** *(Continued)*

| | Ligand Name (Abbr.) | Nucleic Acid | Sequence | $T_m$ (°C) | Conc.[a] (μM) | Ref. |
|---|---|---|---|---|---|---|
| 8 | Hydroxypyridone (H) | DNA | 5'-CACATTAL̲TGTTGTA-3'<br>3'-GTGTAAT̲L̲ACAACAT-5'<br>Control: AT | 37<br>50 (1 eq Cu$^{2+}$)<br>44 | 2 | [12a] |
| | | GNA | 3'-AATATTAL̲TATTTTA-2'<br>2'-TTATAAT̲L̲ATAAAAT-3' | 37<br>70 (2 eq Cu$^{2+}$)<br>53 (2 eq Zn$^{2+}$)<br>42 (2 eq Co$^{2+}$)<br>39 (2 eq Ni$^{2+}$)<br>38 (2 eq Cd$^{2+}$)<br>37 (2 eq Au$^{3+}$)<br>37 (2 eq Ag$^{+}$)<br>37 (2 eq Pd$^{2+}$)<br>51 | 2 | [53] |
| 9 | Salicylaldehyde (Sal) | DNA | Control: AT<br>5'-CACATTAL̲TGTTGTA-3'<br>3'-GTGTAAT̲L̲ACAACAT-5'<br><br>Control: AT | 40 (pH 9.0 Ches)<br>55 (1.3 eq Cu$^{2+}$ Ches)<br>41 (pH 9.0 Hepes)<br>41 (2 eq Mn$^{2+}$ Hepes)<br>41 (pH 7.4 Tris)<br>50 (pH 7.4 Tris) | 3.0 | [15a] |
| 10 | N,N'-bis-(salicylidene)-ethylendiamine (Salen) | DNA | 5'-CACATTAL̲TGTTGTA-3'<br>3'-GTGTAAT̲L̲ACAACAT-5'<br>exces ethylenediamine<br><br>Control: AT | 46 (pH 9.0 Hepes)<br>82 (1 eq Cu$^{2+}$)<br>55 (1.3 eq Cu$^{2+}$)<br>82 (2 eq Cu$^{2+}$)<br>69 (1 eq Mn$^{2+}$)<br>49 (ex. Zn$^{2+}$)<br>37 (ex. Ni$^{2+}$)<br>50 (pH 7.4 Tris) | 3.0 | [15a] |

| # | Ligand | Type | Sequence | Values | | Ref |
|---|--------|------|----------|--------|---|-----|
| **[3 + 3] Metal complexes** | | | | | | |
| 11 | 2,6-(2'-Pyridyl)-purine (Pur$^{2,6\text{-py}}$) | DNA | 5'-CTTTCTLTCCCT-3'<br>3'-GAAACGALAGGGA-5'<br>Control: TA/GC | 38<br>50 (2 eq Ag+)<br>See 5, above | 2.5 | 13d,61a |
| 12 | 2,6-(2'-Bispyridyl)-purine (Pur$^{6\text{-bpy}}$) | DNA | 5'-CTTTCTLTCCCT-3'<br>3'-GAAACGALAGGGA-5'<br>Control: TA/GC | 35<br>42 (2 eq Ag+)<br>See 5, above | 2.5 | 13d,61a |
| 13 | 4-(2'-Bipyridyl) pyrimidinone (Pyr$^{\text{bipy}}$) | DNA | 5'-CTTTCTLTCCCT-3'<br>3'-GAAACGALAGGGA-5'<br>Control: TA/GC | 32<br>31 (2 eq Ag+)<br>See 5, above | 2.5 | 13c |
| 14 | Terpyridine (Tpy) | DNA | 5'-GTGALATGC-3'<br>3'-CACLA$^-$ACG-5'<br><br>Control:<br>5'-GTGATATGC-3'<br>3'-CACTATACG-5' | 14<br>38 (1 eq Ni$^{2+}$)<br>35 (1 eq Cu$^{2+}$)<br>35 (1 eq Zn$^{2+}$) | 1 | 61b |
| 15 | 2,6-bis-(Ethylthiomethyl) pyridine (Spy) | DNA | 5'-CACATTALTGTTGTA-3'<br>3'-GTGTAATLACAACAT-5' | 28<br>23<br>43 (1 eq Ag+) | 1 | 47 |
| **[2 + 2] Metal complexes with different ligands** | | | | | | |
| 16 | Hydroxypyridinone-6-(2'-Pyridyl)-purine (H·Pur$^{P}$) | RNA | 3'-AATAT ALTATTTTA-2'<br>2'-TTATAATL ATAAAAT-3'<br><br>Control: AT | 37<br>74 (2 eq Cu$^{2+}$)<br>51 (2 eq Zn$^{2+}$)<br>50 (2 eq Ni$^{2+}$)<br>51 | 2 | 53 |
| **[3 + 1] Metal complexes** | | | | | | |
| 17 | 2,6-(2'-Pyridyl)-purine·Pyridine (Pur$^{2,6\text{-py}}$·Py) | DNA | 5'-CTTTCTLTCCCT-3'<br>3'-GAAAGAL'AGGGA-5'<br>L = 3-Py, L' = Pur$^{2,6\text{-py}}$<br><br>L = 4-Py, L' = Pur$^{2,6\text{-py}}$<br>Control: TA<br>CG | 24<br>40 (2 eq Ag+)<br><br>24<br>32 (2 eq Ag+)<br>39, 34 (2 eq Ag+)<br>38, 41 (2 eq Ag+) | 2.5 | 13d,61a |

*(continued)*

**Table 10.1** (*Continued*)

| | Ligand Name (Abbr.) | Nucleic Acid | Sequence | $T_m$ (°C) | Conc.[a] (μM) | Ref. |
|---|---|---|---|---|---|---|
| 18 | 2,6-(2'-Bispyridyl)-purine·Pyrdine (Pur^{6-bpy}·Py) | DNA | 5'-CTTTCTLTCCCT-3' <br> 3'-GAAAGAL'AGGGA-5' <br> L = 3-Py, L' = Pur^{6-bpy} <br><br> L = 4-Py, L' = Pur^{6-bpy} <br> Control: TA/GC | 29 <br> 32 (2 eq Ag$^+$) <br><br> 28 <br> 39 (2 eq Ag$^+$) <br> See 5, above | 2.5 | 13d,61a |
| 19 | 4-(2'-Bipyridyl) pyrimidinone·Pyridine (Pyr^{bpy}·Py) | DNA | 5'-CTTTCTLTCCCT-3' <br> 3'-GAAAGAL'AGGGA-5' <br> L = 4-Py, L' = Pyr^{bpy} <br><br><br><br><br><br><br><br><br> L = 3-Py, L' = Pyr^{bpy} <br><br> Control: TA/GC | 22 <br> 35 (2 eq Ag$^+$) <br> 25 (2 eq Co$^{2+}$) <br> 25 (2 eq Zn$^{2+}$) <br> 23 (2 eq Cu$^{2+}$) <br> 23 (2 eq Ni$^{2+}$) <br> 21 (2 eq Fe$^{2+}$) <br> 21 (2 eq Tl$^+$) <br> 22 (2 eq Mn$^{2+}$) <br> 22 (2 eq Mg$^{2+}$) <br> 22 <br> 27 (2 eq Ag$^+$) <br> See 5, above | 2.5 | 13c |
| 20 | Pyridine-2,6-dicarboxylate·Pyridyl (Dipic·Py) | DNA | 5'-CACATTALTGTTGTA-3' <br> 3'-GTGTAATL'ACAACAT-5' <br> L = Py, L' = Dipic <br><br> Control: AT <br> L = Dipic, L' = Py <br> Control: AT/TA <br> GC/CG | n.a. <br> 39 (1 eq Cu$^{2+}$) <br> 40 (2 eq Cu$^{2+}$) <br> 40 (5 eq Cu$^{2+}$) <br> 40 (15 eq Cu$^{2+}$) <br> 41 (15 eq Cu$^{2+}$) <br> 37 (ex. Cu$^{2+}$) <br> 39/39 (ex. Cu$^{2+}$) <br> 40/40 (ex. Cu$^{2+}$) | 2 | 10,63 |
| 21 | Pyridine-2,6-dimethylcarboxamide·Pyridyl (Dipam·Py) | DNA | 5'-CACATTALTGTTGTA-3' <br> 3'-GTGTAATL'ACAACAT-5' <br> L = Dipam, L' = Py <br> Control: AT/TA <br> GC/CG | 28 <br> 43 (ex. Cu$^{2+}$) <br><br> See 20, above | 1 | 63 |

| | | | | | | |
|---|---|---|---|---|---|---|
| 22 | Pyridine-2,6-dicarboxamide-Pyridyl (MeDipam·Py) | DNA | 5'-CACATTALTGTTGTA-3'<br>3'-GTGTAATL'ACAACAT-5'<br>L = MeDipam, L' = Py<br>Control: AT/TA<br>GC/CG | 27<br>27 (ex. Cu$^{2+}$)<br><br>See 20, above | 1 | 63 |
| 23 | 2,6-bis-(Ethylthiomethyl) pyridine-pyridyl (Spy·Py) | DNA | 5'-CACATTALTGTTGTA-3'<br>3'-GTGTAATL'ACAACAT-5'<br><br>L = Spy, L' = Py<br>Control: AT/TA<br>GC/CG | 24<br>35 (1 eq Ag$^+$)<br><br>39/39<br>39/41 | 1 | 47 |
| 24 | 2,2'-Dipicolylamine·PO$_4^{3-}$ (DPA·PO$_4$) | LNA | 5'-GTGALATGC-3'<br>3'-CACLATACG-5'<br><br><br>5'-GTGALATGC-3'<br>3'-CACTALACG-5' | 34<br>53/24 (1/10 eq Ni$^{2+}$)<br>40/20 (1/10 eq Cu$^{2+}$)<br>36/31 (1/10 eq Zn$^{2+}$)<br>39<br>47/50 (1/10 eq Ni$^{2+}$)<br>42/39 (1/10 eq Cu$^{2+}$)<br>46/47 (1/10 eq Zn$^{2+}$)<br>28 | 1 | 18a |
| 25 | 2,2'-Dipicolylamine·Imidazcle (DPA·Imidazole) | DNA·GNA | Control: TA<br>5'-GAGGGAAGALAAG-3'<br>3'-CT CCC T TCT L'TTC-5'<br>L = DPA, L' = Imidazole<br>Control: | 24<br>30 (1 eq Ag$^+$)<br>27 (1 eq Au$^{3+}$)<br><br>40 | 0.5 | 17a |
| 26 | 2,2'-Dipicolylamine-1,2,4-triazole (DPA·Triazole) | DNA·GNA | 5'-GAGGGAAGAAAG-3'<br>3'-CTCCCTTCTTTC-5'<br>5'-GAGGGAAGALAAG-3'<br>3'-CTCCCTTCTL'TTC-5'<br>L = DPA, L' = Triazole<br>Control: See 20, above | 24<br>27 (1 eq Ag$^+$)<br>27 (1 eq Au$^{3+}$)<br>See 20, above | 0.5 | 17a |
| 27 | 2,2'-Dipicolylamine·Tetrazole (DPA·Tetrazole) | DNA·GNA | 5'-GAGGGAAGALAAG-3'<br>3'-CTCCCTTCTL'TTC-5'<br>L = DPA, L' = Tetrazole<br>Control: See 20, above | 24<br>26 (1 eq Ag$^+$)<br>28 (1 eq Au$^{3+}$)<br>See 20, above | 0.5 | 17a |

ND stands for "not detected".

$^a$Concentration is of the duplex.

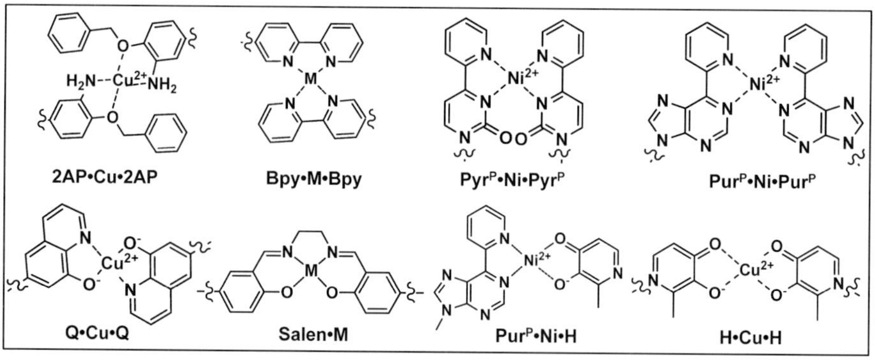

**Figure 10.7** *Alternative base pairs based on metal complexes that contain a pair of bidentate ligands.*

entry 2). For incorporation in ss PNA, **Bpy** (or 5-methyl-2,2'-bipyridine) has been connected at its 5'-position to the secondary NH group in the Aeg backbone through a CH$_2$-CO linker (Table 10.1, entry 3) [16a,c,50]. We note that the properties of dsPNAs containing 2,2'-bipyridine or 5-methyl-2,2'-bipyridine were similar [16c]. Two other ligands that have the same cis-diimine metal binding site as 2,2'-bipyridine have been used also to form metal complexes that function as alternative base pairs in DNA duplexes [13a,b]. 4-(2'-pyridyl)-pyrimidinone (**Pyr$^P$**; Table 10.1, entry 4) and 6-(2'-pyridyl)-purine (**Pur$^P$**; Table 10.1, entry 5) are formally obtained by substitution with pyridine of the 4-amino group of cytosine and of the 6-amino group of adenine, respectively (Figure 10.7).

Substitution of a nucleobase pair in the middle of a DNA or PNA duplex with a pair of **Bpy**, **Pyr$^P$** or **Pur$^P$** ligands reduced the thermal stability of the duplexes below that of the corresponding, non-modified duplex, which can be related to the fact that these ligands do not form H-bonds within the pair [13a,b,16a,c]. Nevertheless, there are cases of both DNA and PNA duplexes in which **Bpy** did not affect the melting temperature [27b,c,49]. The decrease in the melting temperature of 12-bp DNA duplexes containing **Pur$^P$** ligands was smaller than that for duplexes with the same sequence with **Pyr$^P$** ligands, which could be due to better $\pi$-stacking of the larger **Pur$^P$** ligands [13a,b].

The metal-based alternative base pairs influence the properties of the nucleic acid duplexes and the duplex context influences the properties of the metal complexes. It is difficult to identify simple cause–effect relationships between these factors and the properties of nucleic acid duplexes because of the limited number and diversity of systems studied this far. Nevertheless we identify below several correlations between the properties of the metal complex and of the duplexes. These correlations can be considered in the design of new metal-containing duplexes for specific synthetic or functional goals.

1. The way in which a given ligand is attached to the nucleic acid backbone, that is, the position in the ligand where the attachment is made, and the chemical nature of the linker between the ligand and the backbone of the nucleic acid can influence the effect of the ligand and metal complex on the duplex.

An increased conformational flexibility of the ligand can increase the entropic penalty for the formation of the ligand- and metal complex-modified duplex. However, an inflexible linker may represent a steric barrier to the formation of the metal-containing duplex.

In experiments by Brotschi *et al.*, a 19-bp DNA duplex with a central **Bpy·Bpy** pair in which the two **Bpy** were directly attached to the ribose was as stable as a 19-bp duplex with a G·C pair in the place of the **Bpy·Bpy** pair, although the two **Bpy**s cannot form hydrogen bonds and are not complementary in shape [49]. In experiments by Weizman and Tor, an 11-bp DNA duplex with a central **Bpy·Bpy** pair in which the two **Bpy** were attached to the ribose through a methylene linker was as stable as a 10-bp duplex that had the same sequence but no **Bpy·Bpy** pair, although in general extension of a DNA duplex by a base pair leads to an increase in the duplex stability (Table 10.1, entry 2) [11b]. These results suggest that the **Bpy·Bpy** pair connected through the methylene linkers to the duplex backbone has a lower stabilization effect than a Watson–Crick base pair, and thus than a **Bpy·Bpy** pair directly attached to the DNA backbone. It must nevertheless be noted that the effect of bipyridine substitution can be modulated by the sequence and/or the length of the nucleic acid duplex as these parameters were different for the DNA duplexes used in the experiments by Weizman *et al.* and by Brotschi *et al.* [11b,49].

2. The chemical nature of the nucleic acid influences the effect of a given ligand and metal complex on the modified nucleic duplex.

For example, in experiments by Brotschi *et al.* and by Weizman *et al.* [11b,49], the effect of a central **Bpy·Bpy** pair on the DNA duplex stability was very small (see above). In contrast, the effect of a central pair of the same ligands on a 10-bp Aeg-based PNA duplex destabilized the duplex by more than 10 °C [16a]. This difference between the effect of **Bpy** in DNA and PNA correlates with the fact that central mismatches cause a significantly larger destabilization of PNA duplexes than of DNA ones. In turn, this effect can be linked to the fact that DNA duplexes, which contain strands with negatively charged backbones, are less stable than PNA duplexes with the same sequence because the Aeg-based PNA has neutral strands (see Section 10.2).

3. The position of the ligand modification in the duplex, for example, in the center versus close to the end of the duplex, can modulate the effect of the ligand substitution because the relative contribution of central base pairs to duplex stabilization is higher than that of terminal base pairs that are subject to fraying.

For example, the melting temperature for a 10-bp PNA duplex that contained a **Bpy·Bpy** pair close to the end of the duplex was almost the same as that of the duplex that contained an AT base pair instead of the **Bpy·Bpy** pair [16c]. In contrast, the melting temperature of the same duplex in which the **Bpy·Bpy** pair was in the center of the duplex was significantly lower than that of the non-modified duplex (Table 10.1, entry 3) [16c].

4. The effect on duplex stability of a metal-containing, nucleic acid duplex can be affected by the sequence and length of the nucleic acid.

It is well known that the melting temperature of nucleic acid duplexes and triplexes increases with the length of these systems, which is due to cooperativity effects between the nucleobase pairs, and with the increasing GC content of the duplexes. For example, we observed that a palindromic, 8-bp PNA duplex was slightly more stable than a 9-bp PNA duplex with higher GC content [16e]. Given these differences, one and the same ligand substitution can have a different effect on the stability of duplexes with different length and sequence. Indeed, incorporation of a **Bpy·Bpy** pair in the middle of a 9-bp PNA duplex or an 8-bp palindromic PNA duplex had significantly different effect on the thermal stability of the duplexes, destabilizing the former and affecting very little the stability of the latter [27c].

5. The stability constant of the complex formed by a metal ion with the free ligand in solution may influence the change in duplex thermal stability caused by metal binding to the same ligands within a duplex.

For example, examination of the literature on metal-containing nucleic acid duplexes leads to the observation that most common metal ions introduced in the duplexes are $Ni^{2+}$ and $Cu^{2+}$, which are also the metal ions that usually form complexes with the highest stability constants in the series of 3d transition metals.

Addition of $Fe^{2+}$ or $Zn^{2+}$ did not affect the thermal stability of a 10-bp PNA duplex that contained a central **Bpy·Bpy** ligand pair [16c]. The stabilization of the duplex increased in the order $Co^{2+} < Cu^{2+} < Ni^{2+}$, which coincides with the order of increasing stability constants of the complexes formed by these ions with the free bipyridine [16c]. Addition of one equivalent of $Ni^{2+}$ to 10-bp PNA duplexes with a central pair of bipyridines partially restored the loss in thermal stability caused by substitution of an AT base pair with the bipyridine ligands [16a]. We note that the lack of a stabilization effect of the metal ion on ligand-containing nucleic acid duplexes does not exclude an interaction between the metal ion and the ligands. For example, UV-vis titrations of **Bpy**-modified PNA duplex monitored at 320 nm, the wavelength where the $M^{2+}$-coordinated bipyridine has $\pi$-$\pi^*$ transitions, demonstrated that: (a) one $Ni^{2+}$ ion binds to the bipyridine ligands within the 10-bp PNA duplex [16a], and (b) $Zn^{2+}$ interacts with the bipyridines although it does not affect the thermal stability of the duplex [51]. CD spectra of the bipyridine-modified PNA duplexes in the absence or presence of $Ni^{2+}$ showed a pattern characteristic for a left-handed PNA helix [16a]. This handedness is induced by an L-lysine situated at the C-terminus of PNA strands that form the duplex. Two additional spectral features at 290–320 nm appeared in the CD spectrum upon $Ni^{2+}$ binding, suggesting that the duplex may exert a chiral induction effect onto the pro-chiral $[Ni(\mathbf{Bpy})_2]^{2+}$ complex [16a].

6. The geometry of the metal complex can affect the properties of the duplexes in which they act as alternative nucleobase pairs.

For example, the stabilization conferred by $Ni^{2+}$ binding to Bpy-modified PNA duplexes did not make the duplexes more stable than the corresponding duplexes that had an AT base pair instead of a pair of bipyridine ligands [16a]. This is despite the fact that Ni-N coordination bonds are stronger than hydrogen bonds [26,28]. The apparent dichotomy may be due to the fact that $[\mathrm{MBpy}_2]$ complexes adopt a distorted square planar geometry [52], which destabilizes the adjacent base pairs. In contrast, the stabilization induced by $Cu^{2+}$ in a PNA duplex with the same sequence of nucleobases that contained a pair of 8-hydroxyquinoline (Q) ligands was higher than that for the non-modified duplex, that is, the PNA duplex that had a central AT base pair instead of the pair of ligands [16b]. Indeed $[\mathrm{CuQ}_2]$ has a square planar geometry while the two bipyridine ligands coordinated to $Ni^{2+}$ (or another metal ion) cannot be coplanar. Nevertheless we note that the difference in the effect of metal ions on the thermal stability of ligand-modified PNA duplexes may also be due to the fact that not only the geometry but also the stability constants of complexes of $Cu^{2+}$ with **5Q** and of $Ni^{2+}$ with **Bpy** ligands are also different.

$Ni^{2+}$ binding to the ligand-modified duplexes increased the melting temperature of the **Pyr$^P$**- and **Pur$^P$**-modified DNA duplexes by the same number of degrees. This is surprising to some degree given the fact that there are geometric differences between the **Pyr$^P$·Ni$^{2+}$·Pyr$^P$** and **Pur$^P$·Ni$^{2+}$·Pur$^P$** complexes; the size of the former complex (estimated as distance between corresponding atoms bound to the backbone, i.e., $\sim$5 Å) is significantly smaller than that of a natural base pair ($\sim$9 Å; Figure 10.1).

The stabilization conferred to the DNA or PNA duplexes by the [ML$_2$] complexes with **Bpy**, **Pur$^P$**, or **Pyr$^P$** ligands is at most equal to that of a GC base pair, despite the fact that the complexes contain coordination bonds, which are stronger than hydrogen bonds. This discrepancy may be due to the fact that in [ML$_2$] complexes, the two aromatic cis-diimine ligands cannot be strictly coplanar, because of the steric clash between the $3,3'$-protons of the carbon atoms adjacent to the nitrogens. As a consequence, these complexes undergo either a tetrahedral or a bow-step distortion [52]. In general, $\pi$-stacking interactions between the [M(**Bpy**)$_2$] complexes in solid state favor the latter distortion. DFT calculations for the [Ni(**Pur$^P$**)$_2$]$^{2+}$ also predict a "bow-step" square planar geometry, which may be further reinforced by stacking interactions between the **Pur$^P$** ligands and adjacent nucleobases [13b].

The change in melting temperature induced by DNA binding of a 3d divalent, transition metal ion to the **Pyr$^P$** or **Pur$^P$** ligands in modified duplexes decreased in the order $Ni^{2+} > Co^{2+} > Cu^{2+} > Fe^{2+} \sim Zn^{2+} \sim Mn^{2+}$ (Table 10.1, entry 4–5). This trend is similar to that observed for the 10-bp PNA duplexes that contained a pair of bipyridine ligands [16c] and it correlates with that of the increasing stability constants for metal complexes of bipyridine. $Cu^{2+}$ represents an exception, which could be due to the relationship between steric constraints exerted by the duplex and Jahn Teller distortion at the $Cu^{2+}$.

A bis(pyridopurine) ligand site has been also introduced in GNA duplexes [53]. $Ni^{2+}$ binding to the ligand-modified GNA duplex caused a large increase in the $T_m$ of the duplex (larger than that caused by $Cu^{2+}$ and several other transition metal ions). In contrast, a GNA duplex with the same sequence but containing a pair of hydroxypyridone ligands was stabilized to the largest extent by $Cu^{2+}$ (see below p. 357).

### *O,N-Coordinating Ligands: 2-Aminophenol, 8-Hydroxyquinoline, and Salen.*

Ortho-aminophenol [9b–d] (**AP**) in its benzyl-protected form was incorporated in a ligandoside, which in turn was used to synthesize two complementary DNA strands with a central pair of **AP** ligands (Table 10.1, entry 6 and Figure 10.7). The thermal stability of the 21-bp DNA duplex formed from these two strands was slightly lower than that of a duplex with an AT base pair instead of the pair of **AP** ligands, both in the absence and presence of $Cu^{2+}$. The fact that $Cu^{2+}$ binding did not affect the stability of the AP-modified DNA duplex can be attributed to the fact that the benzyl protection group renders the ligand significantly bulkier than the A or T bases and reduces the binding affinity of the ligand for metal ions [54].

8-Hydroxyquinoline (**Q**) has been used for the efficient incorporation of metal ions into DNA, PNA and GNA duplexes. In this section, we understand by GNA duplex, a DNA duplex that contains one base pair with a C3 glycol backbone (Figure 10.4). The phenolic group and the aromatic nitrogen of **Q** are in an optimal 1,4 relationship for coordinating metal ions and, consequently, the affinity of **Q** for metal ions is very high. Furthermore, this ligand is particularly amenable for creating alternative base pairs because it forms square planar $[ML_2]$ complexes. The fact that the $[MQ_2]$ complexes with divalent metal ions are neutral bodes well for the incorporation of multiple, adjacent $[MQ_2]$ complexes within a nucleic acid duplex. The extended hydrophobic and aromatic surface of the ligand is ideal for $\pi$-stacking.

Substitution of an AT base pair with a pair of **Q** ligands in the middle of DNA, PNA, or GNA duplexes led to a significant loss of stability for the duplex, which indicates that the two ligands are not hydrogen bonded at pH 7 (Table 10.1, entry 7) [16b,35b]. Addition of $Cu^{2+}$ increased the thermal stability of **Q**-DNA, **Q**-PNA, and **Q**-GNA duplexes by >29 °C. This large increase suggests that the strong coordinative bonds in $[MQ_2]$ alternative base pair are the dominant factor for the stability of the duplexes. In the presence of $Cu^{2+}$, the thermal stability of DNA or GNA duplexes that contain a natural base across from **Q** was significantly lower than that of duplexes containing $[CuQ_2]$ [35b]. This observation indicates that the $[CuQ_2]$ alternative base pair is highly specific irrespective of the nucleic acid backbone.

The formation of metal complexes with high binding stability within the PNA duplex can lead to high mismatch tolerance, that is, the thermal stability of metal-containing, ligand-modified duplexes is not affected by the existence of several mismatches. For example, in the presence of $Cu^{2+}$, the partly self-complementary ssPNAs that contained a **Q** ligand H-GTAG**Q**TCACT-Lys-NH$_2$ or NH$_2$-Lys-CATC**Q**AGTGA-H formed duplexes with melting temperatures close to those of corresponding $[CuQ_2]$-containing PNA duplexes formed from fully complementary oligomers [16b]. The high duplex stability is likely due to the fact that $[CuQ_2]$ complexes bridge the nucleic acid strands at all temperatures, reducing the loss of translational entropy associated with the duplex formation. Indeed, titrations at 25 and 95 °C with $Cu^{2+}$ of PNA duplexes that contained a pair of **Q** ligands in complementary positions showed that $[CuQ_2]$ complexes form with the PNA at a $Cu^{2+}$:**Q**-PNA oligomer ratio in solution of 1 : 2 [16b]. As a result, the gain in enthalpy from $\pi$-stacking of several but not all base pairs is a sufficient driving force for duplex formation, even in the absence of hydrogen bonding between several base pairs in the duplex. In the duplexes that contained mismatches, the nucleobases not involved in Watson–Crick base pairing can bulge out of the duplex and could be used potentially to construct higher-dimensionality, hybrid inorganic PNA structures.

An interesting observation made when one compares the titrations of **Q**-containing PNA with $Cu^{2+}$ at 25 and 95 °C is that $[CuQ]$ complexes can form at high temperature (at which the PNA strands are not hold together by Watson Crick interactions) but cannot form at low temperature (at which the PNA exists in duplex form) [16b]. The hypothesis that this difference is due to a supramolecular chelate effect exerted by the duplex was verified by isothermal calorimetry (ITC) titrations [16e]. The calorimetric studies showed quantitatively that the Watson–Crick hybridization of two PNA strands into a duplex that contains a pair of ligands in complementary positions can significantly increase the stability constants of the metal complexes formed by the ligands within the PNA by a few orders of magnitude when compared to the corresponding stability constants for the same type of complexes with free ligands.

This effect is similar to that in which polytopic ligands form complexes more stable than the corresponding complexes with individual ligands.

On the other hand, steric constraints such as those exerted on the ligand position by the duplex or by the way in which the ligand is attached to the PNA backbone limit the supramolecular effect. In other words, the relative orientation of the ligands within the PNA duplex may disfavor the formation of a metal complex. Indeed, a PNA duplex that contained **Q** ligands attached to the backbone through the ortho position to the imine nitrogen of the ligand formed [CuQ$_2$] complexes with stability constant lower than that of [CuQ$_2$] complexes with Q ligands in solution [16e].

The structure of GNA or fully complementary PNA duplexes as probed by CD spectroscopy was not affected by the presence in the duplex of a [CuQ$_2$] complex [16b,17b]. Partly complementary PNA duplexes did not show a CD spectrum in the absence or in the presence of Cu$^{2+}$, suggesting that the mismatches prevent the transmission of the chiral induction effect of the L-lysines situated at the C-terminus of the PNA strands. Nevertheless, the EPR spectra of the **Q**-PNA duplexes formed from partly complementary strands were similar to each other and to the spectrum of the fully complementary **Q**-PNA duplex [16b], and confirmed the coordination of two **Q** ligands to Cu$^{2+}$. This observation suggests a similarity in the environment of the duplexes around the [CuQ$_2$] complexes irrespective of the degree of complementarity of the two strands that form the duplex.

Schiff bases are very versatile ligands and have high affinity for metal ions. The N,N'-bis(salicylidene)ethylendiamine or salen (**Salen**) version is widely used for its catalytic properties. **Salen** formation from salicylaldehyde (**Sal**) and ethylenediamine is a reversible process and the coordination of metal ions is generally required as a driving force for the reaction to occur in aqueous environment. The preorganization of an aldehyde and an amine close to each other onto a template entropically favors the formation of the condensation product in dilute samples, a strategy used early on in Inorganic Chemistry to form **Salen** complexes. In 1997, a ss DNA template was used to support the condensation of a 3'-end aldehyde group with the 5'-end amino group of two different, shorter ss DNA that were complementary to adjacent portions of the DNA template [55]. In a related approach, Czlapinski and Sheppard showed the formation in presence of ethylenediamine but in the absence of a metal ion of a non-metallated **Salen** between two salicylaldehydes situated at the end of opposite strands in a 15-bp DNA duplex [56]. The two salicylaldehyde ligands were linked directly to the terminal phosphate group of each strand rather than to a sugar moiety and were likely to π-stack with the nucleobase pairs in a manner similar to that in which overhang nucleobases stack at the end of a nucleic acid duplex. The reversible formation of a non-metallated form of **Salen** was invoked to explain the hysteresis observed in the presence of ethylenediamine in the melting of a DNA duplex that contained two salicyladehydes in central, complementary positions [15d].

Usually, the two aromatic rings of **Salen** in coordination complexes are coplanar. Hence **Salen** bears structural similarity to a nucleobase pair and it can participate in π-π interactions with adjacent nucleobase pairs. This compatibility between the Salen ligands and their metal complexes with the base pairs in the DNA duplex is supported by the results of CD studies. Room temperature CD spectra of DNA duplexes that contained a central of terminal pair of salicylaldehyde ligands and formed a M$^{2+}$–**Salen** complex in the presence of excess ethylenediamine and $Cu^{2+}$ or $Mn^{2+}$ showed the features character-istic for a B-DNA structure [15a,56]. The small shifts in the position and intensity of CD bands upon the formation of the metal–Salen complexes can be interpreted as indicative of some structural differences induced by the metal complex in the duplex.

The substitution in a 20-bp DNA duplex of one AT base pair with a pair of salicylalde-hydes lowered the stability of the duplex (Table 10.1, entry 9) [15a]. At pH 9.0, the stabil-ity of the salicylaldehyde-containing DNA duplex increased in the presence of $Cu^{2+}$ or of $Cu^{2+}$ and methyl amine, a gain that was attributed to the formation of an interstrand square-planar $[Cu(Sal)_2]$ complex or $[Cu^{2+}\text{-bis(hydroxo-imine)}]$ complex, respectively. Comparison of the stabilization effect exerted on the duplex by either $Cu^{2+}$ ($\Delta T_m \sim 15\,°C$) or ethylenediamine ($\Delta T_m \sim 6\,°C$) with the effect of the combination of $Cu^{2+}$ and ethylenediamine ($\Delta T_m > 40\,°C$) indicated a cooperative stabilization effect exerted by the metal coordination and ethylene crosslinking [15a]. The stabilization of the duplex upon formation of $[Cu(Salen)]$ complexes depended on the identity of the metal ion. A stoichiometric amount of $Mn^{2+}$ produced a large duplex stabilization, simi-lar to that of $Cu^{2+}$, but $Zn^{2+}$ and $Ni^{2+}$ caused a relatively small stabilization and only if added in excess to duplex solutions containing ethylenediamine.

The mechanical properties of the mixed-sequence, 20-bp DNA duplex containing a central pair of salicylaldehyde ligands have significantly been affected by the coordina-tion of $Cu^{2+}$ to the two ligands [15g]. The rupture force of the ds DNA was significantly higher in the presence of $Cu^{2+}$. The analysis of the rate-dependent separation of the two strands led to the conclusion that thermal off-rates for the non-modified and salicylalde-hyde-modified DNA was about $10^{-13}\,s^{-1}$, while the on- and off- rates for the $Cu^{2+}$-con-taining salicylaldehyde-modified DNA were about $10^4\,M^{-1}\,s^{-1}$ and $10^{-3}\,s^{-1}$, respectively. The mechanism underlying the mechanical properties of the metal-contain-ing DNA duplex depended on the rate with which the mechanical stress was applied. Spe-cifically, the mechanism was dominated by the Watson–Crick part of the duplex at low loading rates and by the metal complex at high loading rates.

Two DNA hairpins containing a **Salen-M** complex have been also reported. The melt-ing temperature of the hairpin in which **Salen-M** played the role of loop was only 2–3 °C higher than the melting temperature of the control hairpins that had a loop of three nucle-obases, GAA, TAA or TTT [56]. In contrast, the melting temperature of the hairpin in which **Salen-M** played the role of an alternative base pair in the stem of the hairpin was higher by several tens of degrees than that of the hairpin which contained two **Sal** ligands. This difference between the effect of the **Salen-M** complex on the thermal stability of the hairpin depending on its position is similar to the observed effect of the position of a metal complex alternative base pair on the thermal stability of the duplex in which the metal complex is incorporated (see above p. 351, for bipyridine-based alternative base pairs).

*O,O-Coordinating Ligands: Hydroxypyridone.* Hydroxypyridone (**H**) can form square planar $[ML_2]$ complexes with metal ions (Figure 10.7) [57]. Tanaka *et al.* [12a] and Schlegel *et al.* [53] synthesized **H**-containing phosphoramidites and used them to incor-porate **H** ligands into DNA and GNA oligomers, respectively (Table 10.1, entry 8). UV titrations [12a,57,58] and mass spectrometry [12a] confirmed the formation of a $[CuH_2]$ complex in solutions of the hydroxypyridone DNA ligandoside with $Cu^{2+}$. As observed for other ligand-modified duplexes, replacement of an AT base pair with a pair of **H** ligands in the middle of 15-bp DNA or GNA duplexes decreased the stability of the duplexes. Binding of $Cu^{2+}$ to the **H** ligands led to a stabilization of the duplexes with respect to both the non-modified and ligand-modified duplexes (Table 10.1, entry 8). $Zn^{2+}$ and $Co^{2+}$ determined also an increase in the $T_m$ of the **H**-containing GNA duplexes, while $Ni^{2+}$, $Pd^{2+}$, $Ag^+$, and $Au^{3+}$ did not affect the thermal stability of the duplexes.

The structure of two GNA duplexes that did not contain alternative base pairs [36,37] and one GNA duplex containing two isolated $[CuH_2]$ alternative base pairs [59] have been determined by X-ray crystallography. The high flexibility of the simplified backbone of GNA underscores the relatively large structural differences observed between these duplexes. The GNA duplexes have a helical pitch of 26 Å with 10 bp per turn while the metal-containing GNA duplex has a helical pitch of 60 Å with 16 bp per turn. The $[CuH_2]$ complex has a square planar geometry with a very small propeller twist between the planes of the two **H** ligands. The complex has a $C_1' - C_1'$ backbone–backbone distance larger by 2 Å than a Watson–Crick base pair. This difference may be the reason why the **H** ligandotides have an anti conformation while the nucleotides in Watson–Crick base pairs have a gauche conformation with respect to the O3'-C3'-C2'-O2' torsion angle. In contrast, in the backbone of each of two GNA duplexes that did not contain metal-based alternative base pairs, the vicinal O3'-C3' and C2'-O2' bonds had a strictly alternating gauche and anti conformation with each base pair containing one nucleotide in the gauche and one in the anti conformation.

A potential application of metal-containing nucleic acid structures is in molecular electronics. As a step towards building molecular devices based on duplexes that contain metal-based alternative base pairs, Liu *et al.* assessed charge transport in DNA duplexes containing one pair of **H** ligands in the presence of $Cu^{2+}$ [60], and Wierzbinski *et al.* studied the charge transfer of **Bpy**-modified PNA duplexes in the absence and presence of $Zn^{2+}$ [51]. Liu *et al.* conducted single molecule break junction experiments using single-wall carbon nanotubes as the electric contact for the DNA duplexes. Wierzbinski *et al.* measured the electrochemical rate constants for charge transfer by cyclic voltammetry and the conductance by conducting mode atomic force microscopy for PNA duplexes. Both groups found that the charge transport properties of the nucleic acid duplexes that contained one ligand pair in the presence of transition metal ions are similar to those of non-modified duplexes of similar length. The conductance of the DNA duplexes increased when multiple $[CuH_2]$ complexes were incorporated into the DNA duplexes [60].

### 10.3.2.3 *[3 + 3] Coordination*

The octahedral geometry of six-coordinate metal complexes is such that one can anticipate unfavorable steric interactions between an octahedral complex used as an alternative base pair and the adjacent base pairs. Hence octahedral complexes have been pursued much less than lower coordination number complexes as alternative base pairs. Nevertheless in several cases in which pairs of tridentate ligands **L** have been introduced in DNA duplexes, it was found that the thermal stability of the duplexes increased significantly in the presence of transition metal ions. This experimental finding is indicative of the formation of $[ML_2]$ complexes that contribute to the stabilization of the duplex. The finding does not necessarily mean that the metal ion achieves coordination number six as the tridentate ligands can also bind through fewer than three atoms. Hence, the title of this section "[3 + 3] Coordination" defines the presence of six donor atoms in the ligand environment created in a duplex but does not identify necessarily the coordination number of the metal in the complex with the ligands.

A pair of tridentate ligands **Pur**$^{2,6\text{-py}}$, **Pur**$^{6\text{-bpy}}$, **Pyr**$^{bpy}$, **Tpy**, or **Spy** was introduced in complementary positions in the center of DNA duplexes (Table 10.1, entries 11–14) [13c, d,47,61]. These ligands were originally intended for the purpose of creating a [3 + 1] coordination site in which a three-dentate ligand would be positioned in a complementary position to a monodentate one, e.g. **Py** (Figure 10.8). Surprisingly it was found that, in the presence of transition metal ions, the DNA duplexes containing a pair of tridentate ligands were usually more stable than those containing a [3 + 1] pair of ligands (compare entries 11 and 17, or 12 and 18, or 13 and 19 in Table 10.1). This finding can be attributed to the fact that the stability constants of complexes of [3 + 3] type are typically very high due to the chelate effect.

Given the high stability constants of metal complexes with polydentate ligands, complexes that had a pair of tridentate ligands have been used also to create interstrand crosslinks in DNA, PNA, LNA, and UNA. Several of these complexes are mentioned in this section, although they cannot function as alternative base pairs from a structural point of view because the two ligands cannot be coplanar.

Yang *et al.* synthesized DNA duplexes with a terminal pair of terpyridine (**Tpy**) ligands [62]. The thermal stability of the **Tpy**-modified DNA duplex was higher than that of the unmodified DNA duplex, an effect similar to that exerted by nucleobase overghangs on the stability of duplexes. The addition of Fe(II) and Co(II) to the **Tpy**-modified DNA duplex further increased the melting temperature of the duplex ($\Delta T_m > 17\,^{\circ}$C). In contrast,

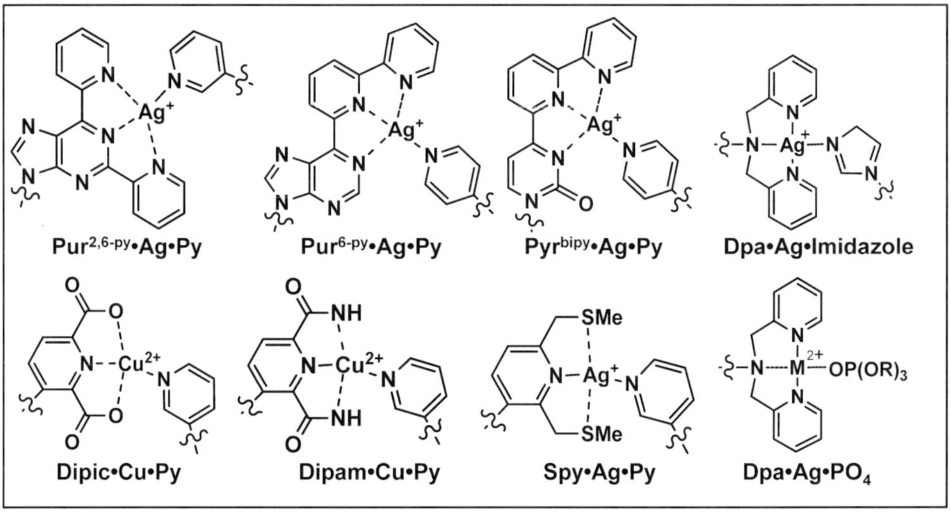

**Figure 10.8** *Alternative base pairs based on metal complexes that contain one tridentate and one monodentate ligand.*

$Zn^{2+}$, $Ag^+$, and $Cu^{2+}$ did not affect the $T_m$ of the ligand-modified DNA duplex. Interesting from an inorganic chemistry point of view, is the fact that Yang *et al.* also used a pair of bidentate ligands or a pair of one bidentate and one tridentate ligands at the end of the duplex and found that $Cu^{2+}$ showed good binding with the mixed ligand environment while $Cu^+$ showed good binding with the pair of tridentate ligands. Hence the three ligand environments, bis-tridentate is-bidentate, and tridentate-bidentate, showed selectivity for specific metal ions.

Bezer *et al.* synthesized a chiral PNA monomer containing **Tpy** and used it to create PNA duplexes with a pair of **Tpy** ligands in the center or at the end of the PNA duplex [16d]. They showed that one $Cu^{2+}$ ion coordinates to the pair of ligands to form a [Cu (**Tpy**)$_2$]$^{2+}$ complex. Although bulky, the complex did not prevent the formation of PNA duplexes. As expected, the PNA duplex with a terminal metal complex was much more stable than that with a central complex. A similar [Cu(**Tpy**)$_2$]$^{2+}$ was formed in 2'-amino-LNA and UNA duplexes by Kalek *et al.* [18b] and Karlsen *et al.* [61b], respectively. The effect of the interstrand metal complexes on the thermal stability of LNA or UNA duplexes depended on the relative position of the two **Tpy** ligands with respect to each other. The effect was smaller in the case of **Tpy**-containing UNA duplexes than LNA duplexes because UNA is more flexible than LNA.

### 10.3.3 Duplexes Containing One Alternative Metal–Ligand Base Pair with Different Ligands

The metal complexes discussed in the previous section have a pair of identical ligands. Hence, when they are introduced in two complementary nucleic acid strands, the respective strands are self-complementary at the level of the metal-containing alternative base pairs. Metal complexes with two different ligands can function as non self-complementary, alternative hetero nucleobase pairs. Such complexes with [2 + 2] or [3 + 1] coordination have been incorporated in DNA homoduplexes and in DNA·GNA heteroduplexes.

*10.3.3.1 [2 + 2] Coordination*

Schlegel *et al.* have used the bidentate hydroxypyridone and pyridopurine ligands to create a $Cu^{2+}$-containing alternative hetero base pair in a GNA duplex (Table 10.1, entry 16) [53]. $Cu^{2+}$ coordination to the hydroxypyridone–pyridopurine hetero base pair caused an increase of 24 °C in the melting temperature of the GNA duplex when compared to the duplex with the same sequence that had a GNA A·T base pair instead of the ligand pair.

*10.3.3.2 [3 + 1] Coordination*

A pair of a neutral tridentate (**Pur$^{2,6-py}$**, **Pur$^{6-bpy}$**, or **Pyr$^{bpy}$**) ligand and a neutral monodentate pyridine ligand was introduced in the middle of 12-bp DNA duplexes by Switzer [13d,61a]. The ligand-containing duplexes had a lower melting temperature than the corresponding non-modified duplexes (Table 10.1, entries 17–19). $Ag^+$ but not other transition metal ions restored the stability of the duplexes completely or partly.

The first metal–ligand alternative base pair incorporated in DNA duplexes was of [3 + 1] type and was based on tridentate pyridine-2,6-dicarboxylate (**Dipic**) and monodentate pyridine (**Py**) ligands, which were introduced in complementary positions in the middle of the duplex (Table 10.1, entry 20) [10]. Subsequently, three other tridentate ligands that bear structural similarity to **Dipic** have been used to create [3 + 1] metal bind-ing sites with **Py** in DNA, namely pyridine-2,6-dicarboxamide (**Dipam**; Table 10.1, entry 21) [63], pyridine-2,6-(N-methyl-)dicarboxamide (**Me-Dipam**; Table 10.1, entry 22) [63], and 2,6-bis(ethylthiomethyl)pyridine (**SPy**; Table 10.1, entry 23) [47]. In the absence of a metal ion, incorporation in DNA duplexes of any of the four tridentate ligands opposite a pyridine destabilized the duplexes in a manner similar to that of a mismatch or prevented completely the formation of a duplex.

Carboxamide ions, of which **Dipam** is an example, are the conjugate bases of amides. They form readily in aqueous solution at neutral pH and coordinate to metal ions through the amide nitrogen. For example, in neutral solutions of picolinamide and $Cu^{2+}$ or $Ni^{2+}$, a bis chelate complex forms in which the ligand is deprotonated and coordinated to the metal ion through the pyridine and amide nitrogens [64]. The methyl carboxamide group in **Me-Dipam** has a higher $pK_a$ than the carboxamide group in **Dipam**, and thus the protonation equilibrium at neutral pH is shifted towards the neutral form of the ligand and reduces the ligand's metal-binding affinity. This reduction is further accentuated by the steric effect of the methyl group. These considerations explain the relative stability of DNA duplexes containing one of the first three tridentate ligands across pyridine in the presence of hard metal ions, which decreases in the order [M(**Dipam**)**Py**] > [M(**Dipic**)**Py**] > [M(**Me-Dipam**)**Py**] (Table 10.1, entries 20–22). In the presence of $Cu^{2+}$, the stabil-ity of the duplexes that contained a central **Dipic·Py** or a **Dipam·Py** pair of ligands was similar to or higher than that of duplexes with all-natural base pairs [10,63].

Interestingly, to achieve stabilization of the Dipam.Py modified double helix duplex, a minimum amount of 15 equivalents of $Cu^{2+}$ per duplex was required although for the apparently less stable **Dipic·Py** duplex, one equivalent of $Cu^{2+}$ was sufficient. Possible explanations for this observation could be that the binding constant of $Cu^{2+}$ to **Dipic·Py** is higher than that to **Dipam·Py**, and that either: (a) the binding constant does not corre-late directly with the stabilization of the duplex, or (b) small steric (or electronic) factors related to the presence of the proton on the carboxamide of **Dipam** have a subtle influence on the duplex stability.

EPR spectroscopy has been used to demonstrate quantitative binding of $Cu^{2+}$ to DNA duplexes that contained a central **Dipic·Py** pair of ligands and to obtain information on the coordination environment of $Cu^{2+}$ [10]. The EPR spectrum of the ligand-modified duplex in the presence of $Cu^{2+}$ was rhombic, which is characteristic of a complex with square planar geometry but does not exclude. Weak coordination of ligands to the axial position of the complex. The $g$ and hyperfine $A_{Cu}$ values for $Cu^{2+}$ and the superhyperfine $A_N$ values determined by simulation of the spectrum were close to those measured for the synthetic [Cu(**Dipic**)(**Py**)] complex [65]. The square planar geometry of the [3 + 1] $Cu^{2+}$ complex was further confirmed by a crystal structure of a related DNA duplex, which contained two isolated **Dipic**-$Cu^{2+}$-**Py** alternative base pairs [66] (see Section 10.4).

By virtue of its coordination through sulfur, the **Spy** ligand is a soft Lewis base and has a high affinity for soft metal ions. The addition of one equivalent of $Ag^+$ to the duplex that contained a **Spy·Py** pair led to an increase in duplex stability compared to the metal-free, ligand-modified duplex (Table 10.1, entry 23). Although this ligand pair was designed specifically to bind a soft metal ion, addition of other soft Lewis acids such as $Pd^{2+}$, $Pt^{2+}$, or $Au^{3+}$ had no effect on the duplex stability. The same selectivity for one metal ion has been observed, for example, for DNA duplexes that contained a **Dipic·Py** pair of ligands.

While $Cu^{2+}$ coordination increased the **Dipic·Py** duplex stability, the addition of $Ce^{3+}$, $Mn^{2+}$, $Fe^{2+}$, $Co^{2+}$, $Ni^{2+}$, $Zn^{2+}$, $Pd^{2+}$, or $Pt^{2+}$ had no effect.

2,2′-Dipicolylamine (**DPA**) is a tridentate ligand that has been introduced in ss LNA (attached to the 2′-NH backbone group), ss PNA (attached to the $NH_2$ end of the PNA) or ss GNA (attached to the backbone methylene group) (Figure 10.4) [17a,18a,67]. In the GNA·DNA duplexes, **DPA** was situated across a monodentate ligand **Imidazole, Triazole**, or **Tetrazole** [17a]. In the absence of transition metal ions, the **DPA**-modified GNA·DNA duplexes had a thermal stability lower than that of the non-modified GNA·DNA duplexes and independent of the nature of the monodentate ligand [17a]. In the **DPA**-modified LNA·DNA duplexes, **DPA** did not have an extraneous group in the complementary position. These duplexes had a thermal stability higher than that of the non-modified LNA·DNA duplexes [18a]. This effect is opposite to that observed for ligand-modified DNA duplexes, which are typically less stable than their non-modified counterparts, but is similar to that observed for LNA·DNA duplexes containing aromatic substituents at the 2′-NH group [68].

**DPA** was shown to form [3 + 1] metal-based alternative base pairs in three types of nucleic acid duplexes. When the **DPA**-containing ss LNA and ss PNA formed hetero-duplexes with DNA, a transition metal ion could coordinate the tridentate ligand together with a monodentate phosphate group from the opposite DNA strand. $Ni^{2+}$, $Cu^{2+}$, and $Zn^{2+}$ were shown to coordinate to one or two **DPA** ligands situated within an LNA·LNA duplex and stabilize the duplexes (Table 10.1, entry 24). In most cases, the addition of more than one equivalent of metal ion per pair of adjacent **DPA** ligands destabilized the duplex, suggesting that steric and electrostatic repulsion occur when each of the **DPA** ligands binds a metal ion (Table 10.1, entry 24). In the GNA·DNA duplexes, **DPA** was situated across a monodentate ligand **Imidazole, Triazole**, or **Tetrazole** [17a]. In the presence of $Ag^+$, the melting temperature of these duplexes increased when compared to the $T_m$ of the ligand-modified duplexes in the absence of the metal ions, but not with respect to the non-modified duplexes (Table 10.1, entries 25–27). The increase in $T_m$ correlated with the basicity of the mononuclear azole group.

Molecular modeling and DFT calculations on **DPA**-modified LNA·DNA duplexes supported the fact that a [3 + 1] **DPA**-$M^{2+}$-$PO_4^{3-}$ coordination mode can be realized by coordination of a **DPA** ligand from one strand and a phosphodiester from either the same strand or from the opposite strand of the LNA·DNA duplex [18a]. This coordination model is compatible with the data obtained for **DPA**-modified PNA·DNA duplexes, namely that: (a) the stability of a $Ni^{2+}$-containing, **DPA**-modified PNA·DNA duplex did not depend on the duplex sequence, and (b) the effect of $Ni^{2+}$ on the duplex stability depended on the ionic strength of the solution [67]. In the case of LNA·DNA duplexes that contained two or more **DPA** ligands, theoretical studies showed that one transition metal ion can coordinate two **DPA** ligands from opposite strands to form either a mononuclear or dinuclear complex that acted as a crosslink between the two strands [69].

## 10.4 Duplexes Containing Multiple Metal Complexes

Duplexes or triplexes that contain several metal-containing alternative base pairs entry point in the synthesis of homo- and hetero-metal polynuclear clusters within nucleic acid structures. Such complexes can have a variety of interesting electronic, magnetic and chemical properties, including the behavior of single molecule magnets, valence delocalization, directional charge transfer, and catalysis. The formation of polynuclear clusters within nucleic acid duplexes requires the presence of multiple ligands, which can lead to the formation of several isomers. Hence, the synthesis and characterization of duplexes containing multiple metal-based alternative base pairs are more challenging than those of duplexes containing one metal complex and require use of a combination of spectroscopic and structural characterization methods.

Table 10.2 shows the melting temperatures of the nucleic acid duplexes that contain several isolated or adjacent metal binding sites. Examination of the data in the Table 10.2 shows that typically:

1. The melting temperature of the ligand-containing DNA duplexes is higher in the presence of the transition metal ion that coordinates to the ligands (entries 3,7,8,9,10). This property, which is similar to that of duplexes containing one metal binding site, is due to the formation of relatively strong coordination bonds between the ligands and metal ion.
2. The melting temperature of the DNA duplexes with the same sequence of nucleobases increases with the increase in the number of metal-based alternative base pairs, irrespective of the metal complexes being isolated or adjacent (entries 2,4,5). There are exceptions from this trend, for example, the DNA duplex that contains **Spy.Py** pairs is an exception from this trend (entry 1). In another example, a duplex that contained a single **SPy·SPy** pair was more stable than a duplex with two **SPy·SPy** pairs, but less stable than a duplex containing three **SPy·SPy** pairs [47].
3. The melting temperature increases with the number of equivalents of metal ion added to the ligand-modified DNA duplexes (entries 3 and 6).

An interesting observation regarding the thermal stability of DNA duplexes with adjacent metal complexes was made by Megger *et al.*, who synthesized two palindromic 18-bp DNA duplexes that did not contain any natural nucleobase pair (Table 10.2, entry 10). Each of the 18 pairs consisted of one 1,3-dideaza-2′-deoxyadenosine (**Dda**) and one thymine and could coordinate two $Ag^+$ ions to form a Hoogsteen-type alternative base pair. In one of the duplexes, the self-complementary ss DNA contained a stretch of nine thymine ligands followed by a stretch of Dda. In the other duplex, the strand had alternating thymines and Dda. The melting temperature measured in the presence of two equivalents of $Ag^+$ for the former duplex was 20 °C lower than that of the latter duplex. This observation was found in agreement with the fact that Hoogsteen-type double helices are preferentially formed from alternating AT sequences.

Multiple bidentate bisimine ligands have been introduced in DNA and PNA duplexes. These studies made possible a comparison of the effect of the chemical nature of the duplex on incorporation of multiple metal complexes in duplexes, with the caveat that the nucleobase sequence of the duplexes was not the same. Multiple, adjacent pairs of bipyridine ligands introduced in both DNA or PNA duplexes caused a systematic decrease of the duplex stability, as did the pairing of **Bpy** against either a natural base or an abasic site [16c,70]. Switzer *et al.* [13b] and Franzini *et al.* [16c] examined the effect of $Ni^{2+}$ on **Pur**[P]-modified DNA and **Bipy**-modified PNA duplexes, respectively, containing three adjacent pairs of bidentate ligands. The increase in the melting temperature of the DNA

**Table 10.2**  *Melting Temperatures $T_m$ of DNA Duplexes with Multiple Metal-Based Alternative Base Pairs [ML$_n$].*

| Relative position of [ML$_n$] | # M sites | Ligands | Sequence | $T_m$ (°C) | [dsDNA] (μM) | Ref. |
|---|---|---|---|---|---|---|
| Isolated [ML$_n$] | 2 | 2,6-bis-(Ethy thiomethyl)pyridine·pyridyl (SPy·Py) | 5'-CACALTACTGLTGTA-3'<br>3'-GTGTL'ATGACL'ACAT-5' | 37 (2 eq Ag$^+$) | 1 | 47 |
| | 3 | L = Spy, L' = Py | 5'-CACL̲TTAL'TGTL̲GTA-3'<br>3'-GTGL'AATL̲ACAL'CAT-5' | 26 (3 eq Ag$^+$) | | |
| | 2 | 2,6-bis-(Ethy thiomethyl)pyridine (Spy·Spy) | 5'-CACALTACTGLTGTA-3'<br>3'-GTGTL̲ATGACLACAT-5' | 39 (2 eq Ag$^+$) | 1 | 47 |
| | 3 | | 5'-CACL̲TTALTGTL̲GTA-3'<br>3'-GTGL̲AATL̲ACAL̲CAT-5' | 45 (3 eq Ag$^+$) | | |
| | 3 | N,N'-bis-(sal cylidene)-ethylerdia mine (Salen) | 5'-CGGALGACLAGCG-3'<br>3'-GCCTL̲CTGL̲TCGC-5' | 42<br>83 (1 eq Cu$^{2+}$)<br>92 (2 eq Cu$^{2+}$) | 3 | 15d |
| Adjacent [ML$_n$] | 2 | Pyridine-2,6-dicarboxylate-Pyridine (Dipic-Py) | 5'-CACATTLL'TGTTGTA-3'<br>3'-GTGTAAL'LACAACAT-5' | 39 (15 eq Cu$^{2+}$) | 2 | 10,63,66 |
| | 4 | L = Dipic, L' = Py | 5'-CACATL'LLL'GTTGTA-3'<br>3'-GTGTALL'L'LCAACAT-5' | 40 (15 eq Cu$^{2+}$) | | |
| | 2 | Pyridine-2,6-dicarboxamide-Pyridine (Dipam-Py) | 5'-CACATTLL'TGTTGTA-3'<br>3'-GTGTAAL'LACAACAT-5' | 43 (15 eq Cu$^{2+}$) | 1 | 63 |
| | 4 | L = Dipam, L' = Py | 5'-CACATL'LLL'GTTGTA-3'<br>3'-GTGTALL'L'LCAACAT-5' | 47 (15 eq Cu$^{2+}$) | | |
| | 3 | 1,2,4-triazole (Triazole-Triazole) | d(A$_7$L$_3$T$_7$) | 40 (hairpin)<br>42 (1 eq Ag$^+$)<br>45 (2 eq Ag$^+$)<br>49 (3 eq Ag$^+$)<br>62 (5 eq Ag$^+$) | 3 | 14a |

*(continued)*

**Table 10.2** (*Continued*)

| Relative position of [ML$_n$] | #M sites | Ligands | Sequence | Tm (°C) | [dsDNA] (μM) | Ref. |
|---|---|---|---|---|---|---|
| 7 | | Bipyridine (Bpy·Bpy, in PNA) | H-GTAGLLCACT-Lys<br>Lys-CATCLLGTGA-H | 37<br>57/45 (1/2 eq Ni$^{2+}$)<br>50/45 (1/2 eq Co$^{2+}$)<br>42/37 (1/2 eq Cu$^{2+}$) | 5 | 16c |
| | | | Control: AT/TA | 67 | | |
| 8 | 3 | 6-(2'-Pyridyl)-purine (Pur$^P$·Pur$^P$) | 5'-CTTTCTL$_3$TCCCCT-3'<br>3'-GAAAGAL$_3$AGGGA-5' | 21<br>64 (1.3 eq Ni$^{2+}$) | 2.5 | 13b |
| 9 | 3 | Thymine (T·T) Salicylaldehyde Sal·Sal | 5'-CGGCCTLLLLTTTTLCGCGC-3'<br>3'-GCCGGALLLLTTTTLGCGCG-5' | 32 | 3.0 | 15c |
| 10 | 9 | 1,3-dideaza-2'-deoxyadenosine Dda·Dda | 5'-LTLTLTLTLTLTLTLTLT-3'*<br>5'-LTLTLTLTLTLTLTLTLT-3' | N.D.<br>82 (2 eq Ag$^+$) | 1 | 79 |
| | | | 5'-LLLLLLLLTTTTTTTTT-3'<br>5'-LLLLLLLLTTTTTTTTT-3' | N.D.<br>62 (2 eq Ag$^+$)<br>83 (2 eq Ag$^+$) | 1<br>3 | |

*The two strands may be parallel (shown) or antiparallel oriented.

duplex in the presence of $Ni^{2+}$ (Table 10.2, entry 8) suggested that $Ni^{2+}$ coordinates to the **Pur**$^P$ ligands and that the staking of the base pairs was not affected by the metal complexes. UV titrations of the **Bpy**-modified PNA duplexes showed that they coordinate only two transition metal ions. Also the thermal stability of the PNA duplexes in the presence of three equivalents of $Ni^{2+}$ was not higher than in the presence of two equivalents (Table 10.2, entry 7) [16c]. The difference in behavior of the DNA and PNA duplexes, each containing three adjacent pairs of bidentate ligands may be due to: (a) the fact that the larger **Pur**$^P$ ligand can form bis-ligand but not tris-ligand complexes, (b) the difference in duplex length (i.e., 15-bp DNA duplex vs 10-bp PNA duplex), and/or (c) the DNA backbone can act as a counteranion for the positively-charged complex, thus reducing the electrostatic repulsion between the adjacent $[ML_n]^{2+}$ complexes, while the PNA backbone is neutral.

Given the complexity of the coordination possibilities to duplexes containing multiple ligands, structural information obtained by X-ray crystallography or NMR spectroscopy can provide valuable information on the relative position of the ligands and metal complexes with respect to the duplexes. Several structural studies of DNA, PNA, and GNA duplexes containing ligands and/or metal complexes have been published in the last decade. Brotschi proposed a structural model for the DNA duplexes with consecutive **Bpy·Bpy** pairs in which the **Bpy** ligands form a zipper-like, interstrand stacking motif. The bipyridines have their distal rings involved in π-stacking interactions and the DNA backbone is stretched [70b]. The zipper model was confirmed by NMR studies of a 10-bp DNA duplex that contained one central pair of biphenyl groups [71]. In contrast to this model for structural organization of **Bpy** in DNA, the crystal structure of a 9-bp PNA duplex with a central **Bpy·Bpy** pair showed that the two **Bpy**s are bulged out of the PNA duplex and participate in intermolecular π-stacking interactions with each other. The PNA duplex "collapsed" to allow the non-modified base pairs adjacent to the **Bpy·Bpy** pair to participate in regular π-stacking interactions with each other. Interestingly, the effect of bulging out of the PNA duplex of the two **Bpy**s causes a ~53° bending of the duplex, which is significantly larger than the corresponding angle of the non-modified 8-bp PNA (~25°) and is similar to that observed for a DNA duplex containing a synthetic cis-syn-cyclobutane–pyrimidine dimer-like lesion in its complex with DNA photolyase after *in situ* repair [72].

The crystal structure of a DNA duplex with a palindromic sequence d(5'-$C_1G_2C_3G_4$**Dipic**$_5A_6T_7$**Py**$_8C_9G_{10}C_{11}G_{12}$-3') and containing two isolated **Dipic**-$Cu^{2+}$-**Py** complexes was reported in 2001 [66]. In the duplex, the $Cu^{2+}$ ions were coordinated by the **Dipic** and **Py** ligands in a square planar arrangement, with two additional donor atoms from adjacent base pairs, namely the O4' of the $T_7$ nucleotide and the O6 of $G_4$, weakly coordinated in axial positions. The metal-containing DNA duplex had an alternating purine–pyrimidine (APP) sequence, which is typical for Z-DNA. Indeed, the duplex adopted a Z conformation in both the crystal and the solution. The d(5'-$C_1G_2C_3G_4A_5$**Dipic**$_6$**Py**$_7T_8C_9G_{10}C_{11}G_{12}$-3') duplex that contained two **Dipic·Py** binding sites but did not have an APP sequence adopted a B-conformation in solution and in the presence of $Cu^{2+}$ ions. These structural results suggest that the metal complex "conforms" to the DNA structure dictated by the DNA sequence.

The structure of a palindromic DNA duplex containing three adjacent, central [**Triazole.Ag.Triazole**] alternative base pairs was reported in 2010 [45]. Each of the three metal complexes contained two coplanar **Triazole** ligands. This ligand arrangement was in agreement with an earlier prediction of the geometry of the metal complex within a DNA duplex based on DFT calculations [43]. The calculations showed that, although in the lowest energy configuration the duplex contains a complex in which the two ligands coordinated to the metal ion are not coplanar, the energy of the duplex containing a complex with coplanar ligands is close to the lowest energy one. Hence, when incorporated in the duplex, the complex could adopt a planar geometry due to stacking interactions. The overall structure of the duplex in solution showed only small deviations from a B DNA structure. Specifically, the twist angle of the Ag-containing alternative base pairs is lower than that of nucleobase pairs in B DNA, making this part of the DNA a little less twisted than "regular" B DNA. The base pair rise of the metal-containing alternative base pairs was larger than that of the nucleobase pairs in B DNA and than that expected in the case of Ag-Ag bonds between $Ag^+$ ions in adjacent [**Triazole**.Ag.**Triazole**] complexes.

Shionoya and Carell demonstrated the synthesis of DNA duplexes containing stretches of up to ten, adjacent [$ML_2$] complexes flanked by stretches of GC base pairs [15d,f,h,73]. The stretch of metal complexes were either homometallic, namely up to five [**CuH$_2$**] complexes or up to ten [**CuSalen**] or [**MnSalen**] compexes, or heterometallic, namely a combination of [**CuH$_2$**] and [**HgPy$_2$**] complexes or of [**CuSalen**] and [**HgT$_2$**] complexes. The [$ML^2$] stoichiometry of the complexes was confirmed by UV titrations and ESI mass spectrometry. The stoichiometry studies of DNA duplexes containing a mixture of ligands confirmed also the metal selectivity of the pairs of different ligands coexistent in the DNA [15f]. The CD spectra revealed that the right-handed, helical B DNA structure adopted by the salicylalde-hyde-containing DNA duplexes in solution is significantly affected by the formation of [**MSalen**] complexes; identification of the structure of the metal-containing duplexes based on CD spectroscopy only was not possible [15d].

The possibilities that exchange interactions between paramagnetic transition metal ions in nucleic acid duplexes could be rationally controlled and that interesting magnetic properties such as single molecule magnetism could be attained are appealing. Information about the interactions between the spins of the $Cu^{2+}$ ions in adjacent complexes was obtained by EPR spectroscopy. These studies showed that the spins of adjacent [**CuH$_2$**] complexes were ferromagnetically coupled [73]. Based on the magnitude of the dipolar coupling between the adjacent $Cu^{2+}$ ions, the $Cu^{2+}$-$Cu^{2+}$ distance was estimated to be ~3.2 Å. The electronic structure of the DNA duplex containing five [**CuH$_2$**] complexes has been studied using spin polarized DFT calculations [74]. The backbone of the DNA was not included in the calculation which considered stacks of [**CuH$_2$**] complexes. These calculations could not determine if the ferromagnetic or the antiferromagnetic state of the array of adjacent $Cu^{2+}$ at the core of the DNA is lower in energy but addressed issues of charge localization in the ferromagnetic state of the array. Specifically, the studies showed significant charge delocalization on the ligands. The spin density had a $\sigma$ antibonding character and was distributed over the d orbital of Cu and p orbitals of the four coordinated oxygens. The highest occupied energy levels were discrete and a conduction band was not formed. A more recent theoretical study of [**CuH$_2$**] complexes in which the backbone was included showed that the lowest energy structure of the complex has a plane reflection symmetry [75]. A stack of [**CuH$_2$**] complexes with this symmetry would behave as an insulating ferromagnet, which is in agreement with the experimental findings.

EPR spectra have been also obtained for a DNA duplex containing two adjacent [CuSa-len] complexes [15h]. These spectra showed antiferromagnetic coupling of the $S = 1/2$ spins of the two $Cu^{2+}$ ions and that the [CuSalen] complexes are situated at a distance of 3.7 Å. The difference between the exchange coupling between adjacent [CuH$_2$] (ferromagnetic) and [CuSalen] (antiferromagnetic) complexes was attributed to differences in the relative position of adjacent $Cu^{2+}$ complexes in the DNA duplexes [15h], as predicted by two earlier theoretical studies [76].

The information provided by the spectroscopic characterization and electronic structure calculations of nucleic acid duplexes with multiple transition metal complexes will be useful in the exploration of applications of these duplexes in molecular electronics.

## 10.5 Metal-Containing, Ligand-Modified Nucleic Acid Triplexes

Two metal complexes of the [ML$_3$] type have been incorporated in DNA duplexes. The first example is that of a complex of $Ag^+$ with monodentate, pyridine ligands. $Ag^+$, which can form three coordinate complexes, was used to create an [Ag(**Py**)$_3$]$^+$ complex within a homopurine/pyrimidine DNA triplex. Each of the three DNA strands of the triplex had a central **Py** [12b]. The formation of the triplex in the presence of $Ag^+$ was confirmed by Job plots. Binding of $Ag^+$ to the **Py**-containing triplex was inferred based on the observation that the melting temperature for the triplex to duplex transition increased in the presence of $Ag^+$ from 16 to 18 °C. In contrast, the addition of $Ag^+$ to a triplex without **Py** lowered the temperature at which the duplex–triplex transition takes place. More recently, phosphoramidites containing bidentate, 3-hydroxy-4-pyridone **H** ligands were used to create oligomers of 2–4 ligands [77]. The formation of 2–4 adjacent, octahedral complexes of $Fe^{3+}$ from triplexes of these oligomers has been inferred based on UV-vis titrations and MALDI-ToF mass spectrometry. These triplexes adopted a chiral structure with a handedness that depended on the number of metal complexes in the triplex.

## 10.6 Summary and Outlook

This chapter presented the results of research aimed at the combined use of transition metal coordination and nucleic acid hybridization to construct hybrid inorganic–nucleic acid supramolecular structures. The binding of transition metal ions to DNA duplexes is relatively weak compared to their binding to many polydentate ligands used in coordination chemistry. This property makes possible the site-specific incorporation of transition metal ions in nucleic acid structures that are modified with ligands. The research aimed at the specific metal ion incorporation into ligand-modified nucleic duplexes has confirmed several general design principles. The ligands need to have a high affinity for transition metal ions and to possess an extended aromatic surface that allows stacking with adjacent nucleobases. Four-coordinate metal complexes of the [2 + 2] or [3 + 1] type have been used in many cases as alternative base pairs and examples of metal complexes with different coordination number and geometry have also emerged.

The effect of metal base pairs on the properties of nucleic acid duplexes is complex. The observation of an increase in the thermal stability of ligand-modified, nucleic acid duplexes in the presence of transition metal ions is typically indicative of the formation of a nucleic acid duplex by a combination of hydrogen bonds between nucleobases and coordinative bonds between the metal and the ligands from both strands, as well as $\pi$-stacking. However, the higher strength of coordination bonds compared to that of hydrogen bonds does not always increase the stability of metal-containing duplexes because the effect of metal-containing alternative base pairs on the duplexes depends on the geometry and charge of the metal complex and on the chemical nature and sequence of the nucleic acid.

Achievement of the potential of hybrid metal ion–nucleic acid structures depends on the ability of researchers to synthesize ligand-containing, nucleic acid oligomers and to characterize the complexes formed by these oligomers with transition metal ions. The use of synthetic and characterization methods specific to the fields of coordination and nucleic acid chemistry creates a concomitant opportunity and challenge. For example, the spectroscopic methods used to investigate the stoichiometry and coordination geometry of the metal complexes attached to nucleic acids may require concentrations exceeding those typically used or practically achievable for nucleic acid solutions. Also, the new structural motifs brought about by the stereochemistry of the metal complexes may create alternative, isomer structures that are close in energy and need to be isolated and characterized. The incorporation in duplexes of one metal-containing alternative base pair has been investigated more extensively than that of multiple metal base pairs. In duplexes with several adjacent ligand pairs, the metal ions can adopt different coordination modes. Furthermore, in the cases of both duplexes with one and with more metal binding sites, the metal ions can coordinate not only ligands extraneous to the nucleic acid but also nucleobases or phosphate groups. Therefore, to understand the role of the metal ions in the duplex, one needs to determine the stoichiometry and structure of the metal complex formed within the duplexes, besides measuring the thermal stability of the duplexes. Molecular modeling, X-ray and NMR structural studies, and optical and magnetic resonance spectroscopic studies of the metal-containing duplexes can provide information useful both for the interpretation of the properties of the duplexes and for the rational design of metal-containing duplexes with specific physical or chemical properties.

An advantage of DNA over artificial, hybrid inorganic–nucleic acid structures is that DNA can be replicated and amplified. In the last few years, progress has been made towards the replication and amplification of nucleic acids that can coordinate metal ions. In 2010, Park *et al.* showed that a DNA polymerase can extend a primed template that had a TT or CC mismatch at the position corresponding to the $3'$ end of the primer only in the presence of $Hg^{2+}$ or $Ag^+$, respectively. The dependence of polymerase activity on $Hg^{2+}$ or $Ag^+$ was exploited to construct AND, OR, and PASS1 molecular logic gates [40a]. This study showed that the $T \cdot Hg^{2+} \cdot T$ and $C \cdot Ag^+ \cdot C$ complexes fit in the active site of the polymerase just like the natural nucleobase pairs do. Shortly before this study, Urata *et al.* showed that a DNA polymerase could incorporate in a primer DNA strand a T nucleobase opposite a T nucleobase in a DNA template strand in the presence of $Hg^{2+}$ [40b]. The DNA polymerase was able to continue with the full extension of the primer past the $T \cdot Hg^{2+} \cdot T$ metal-mediated base pair created in the template-extended primer duplex. The incorporation of the dTTP by the polymerase across the T depended specifically on $Hg^{2+}$ and could not take place in the presence of dATP.

In the last year, the incorporation of ligands different from the nucleobases C and T by polymerases in DNA has been also demonstrated. The strategy used by Kalachova *et al.*

was to synthesize triphosphates of adenine and cytosine that had ligands, specifically **Bpy** and Tpy, as side chains. These ligand-containing triphosphates have been incorporated in oligonucleotides by DNA polymerases [78]. Kaul *et al.* also showed that a polymerase can incorporate a **Salen-Cu$^{2+}$** complex in DNA [15i]. In the presence of ethylenediamine and Cu$^{2+}$, salicylic aldehyde was incorporated by the polymerase in a DNA primer across a salicylic aldehyde situated in the template strand. Crystal structures of the complex formed by the primer and the polymerase have shown that the salicylic aldehyde interacted with the polymerase in a manner similar to that in which the natural nucleobases interacted with the enzyme. The incorporation of the salicylic aldehyde into the primer did not affect the subsequent elongation of the primer with natural nucleobases, which also occurred in PCR. The PCR cycle time for incorporation of the metal-based alternative base pairs was longer than that for natural nucleobase pairs because the kinetics of formation of the metal complex in the polymerase was slower than that for nucleobase pairs. Given the demonstration that polymerases can incorporate metal-based alternative base pairs in DNA, the next challenge is to test the possibility of *in vivo* synthesis of inorganic–nucleic acid structures. It can be easily envisioned that the range of structures generated by a strategy that combines metal coordination and nucleic acid hybridization is broad in terms of both topology and function. Given the electronic structure and spectroscopic and chemical properties of transition metal ions, they can be used as intrinsic reporters of the structure or as functional elements. For example, metal ions situated at specific locations in a periodic DNA lattice can direct the organization of proteins for crystallization experiments. To date, DNA nanostructures have been made functional by exploiting structural changes induced in DNA by small molecules, DNA single strands, or DNA-binding proteins to create nanomechanical devices. Distinct from these applications, the catalytic properties of the transition metal ions integrated in DNA nanostructures together with the template effect exerted by the nucleic acid part of the structure could be used to catalyze reactions of DNA-attached chemical substrates. Finally, as metal coordination is orthogonal to hydrogen bonding, the former interactions could be used to expand the genetic code.

## Acknowledgement

This work was supported by NSF Grant No. CHE-0848725.

## Abbreviations

| | |
|---|---|
| Aeg | 2(N-Aminoethyl)glycine |
| AP | 2-Aminophenol |
| APP | Alternating purine–pyrimidine |
| Bp | Base pair |
| Bpy | 2, 2′-Bipyridine |
| CD | Circular dichroism |
| CT | Calf thymus |

| | |
|---|---|
| Cyclen | 1,4,7,10-Tetraazacyclododecane |
| Dda | 1,3-dideaza-20-deoxyadenosine |
| DFT | Density function theory |
| Dipic | Pyridine-2,6-dicarboxylate |
| Dipam | Pyridine-2.6-dicarboxamide |
| DPA | 2,2′-Dipicolylamine |
| Dpp | Diphenylphenanthroline |
| ds | Double stranded |
| en | Ethylenediamine |
| EPR | Electron paramagnetic resonance |
| ESI | Electron spray ionization |
| eq | Equivalent |
| GNA | Glycol nucleic acid |
| H | Hydroxypyridone |
| IR | Infrared |
| L | Ligand |
| LNA | Locked nucleic acid |
| M-DNA | Metal-containing DNA |
| MeBpy | Methyl bipyridine |
| Me-Dipam | Pyridine-2,6-(N-methyl-)dicarboxamide |
| NMR | Nuclear magnetic resonance |
| PEG | Polyethylene glycol |
| PNA | Peptide nucleic acid |
| 2PA | Bis-(2-pyridylmethyl)-amine |
| $Pur^P$ | 6-(2′-Pyridyl)-purine |
| Py | Pyridine |
| $Pyr^P$ | 4-(2′-Pyridyl)-pyrimidine |
| Q | 8-Hydroxyquinoline |
| Sal | Salicylaldehyde |
| Salen | N,N′-bis(salicylidene)ethylendiamine |
| SAMs | Self-assembled monolayers |
| Spy | 2,6-Bis(ethylthiomethyl)pyridine |
| ss | Single stranded |
| TAR | Trans-activation response |
| $T_m$ | Melting temperature |
| Tpy | Terpyridine |
| UNA | Unlocked nucleic acid. |

## References

1. (a) Derose, V.J., Burns, S., Kim, N.K., and Vogt, M. (2004) DNA and RNA as ligands. *Compr. Coord. Chem. II*, **8**, 787–813; (b)Martin, R.B. (1996) Dichotomy of metal ion binding to N1 and N7 of purines. *Met. Ions Biol. Syst.*, **32**, 61–89; (c)Martin, R.B. (1985) Nucleoside sites for transition metal ion binding. *Acc. Chem. Res.*, **18** (2), 32–38.
2. (a) Schoenknecht, T. and Diebler, H. (1993) Spectrophotometric and kinetic studies of the binding of nickel, cobalt, and magnesium to poly(dG-dC).poly(dG-dC). Determination of the

stoichiometry of the Ni2+-induced B -> Z transition. *J. Inorg. Biochem.*, **50** (4), 283–298; (b) Gueron, M., Demaret, J.P., and Filoche, M. (2000) A unified theory of the B-Z transition of DNA in high and low concentrations of multivalent ions. *Biophys. J.*, **78** (2), 1070–1083; (c) Van Steenwinkel, R., Campagnari, F., and Merlini, M. (1981) Interaction of manganese(2+) with DNA as studied by proton-relaxation enhancement of solvent water. *Biopolymers*, **20** (5), 915–923. (d) Zimmer, C., Luck, G., and Triebel, H. (1974) Conformation and reactivity of DNA. IV. Base binding ability of transition metal ions to native DNA and effect on helix conformation with special reference to DNA-Zn(II) complex. *Biopolymers*, **13** (3), 425–453; (e) Moldrheim, E., Andersen, B., Froystein, N.A., and Sletten, E. (1998) Interaction of manganese(II), cobalt(II) and nickel(II) with DNA oligomers studied by 1H NMR spectroscopy. *Inorg. Chim. Acta*, **273** (1, 2), 41–46; (f) Froeystein, N.A., Davis, J.T., Reid, B.R., and Sletten, E. (1993) Sequence-selective metal ion binding to DNA oligonucleotides. *Acta Chem. Scand.*, **47** (7), 649–657; (g) Vinje, J., Parkinson, J.A., Sadler, P.J. *et al.* (2003) Sequence-selective metalation of double-helical oligodeoxyribonucleotides with PtII, MnII, and ZnII ions. *Chem. Eur. J.*, **9** (7), 1620–1630; (h) Vinje, J. and Sletten, E. (2006) Internal versus terminal metalation of double-helical oligodeoxyribonucleotides. *Chem. Eur. J.*, **12** (3), 676–688; (i) Jia, X., Zon, G., and Marzilli, L.G. (1991) Multinuclear NMR investigation of zinc(2+) binding to a dodecamer oligodeoxyribonucleotide: insights from carbon-13 NMR spectroscopy. *Inorg. Chem.*, **30** (2), 228–239; (j) Pullman, A., Pullman, B., and Lavery, R. (1983) Molecular electrostatic potential versus field. Significance for DNA and its constituents. *Theochemistry*, **10**, 85–91; (k) Abrescia, N.G.A., Malinina, L., Fernandez, L.G. *et al.* (1999) Structure of the oligonucleotide d(CGTA-TATACG) as a site-specific complex with nickel ions. *Nucleic Acids Res.*, **27** (7), 1593–1599; (l) Abrescia, N.G.A., Tam, H.-D., and Subirana, J.A. (2002) Nickel guanine interactions in DNA: crystal structure of nickel-d[CGTGTACACG]2. *J. Biol. Inorg. Chem.*, **7** (1–2), 195–199; (m) Labiuk, S.L., Delbaere Louis, T.J., and Lee, J.S. (2003) Cobalt(II), nickel(II) and zinc(II) do not bind to intra helical N(7) guanine positions in the B-form crystal structure of d (GGCGCC). *J. Biol. Inorg. Chem.*, **8** (7), 715–720; (n) Soler-Lopez, M., Malinina, L., Tereshko, V. *et al.* (2002) Interaction of zinc ions with d(CGCAATTGCG) in a 2.9. ANG. resolution X-ray structure. *J. Biol. Inorg. Chem.*, **7** (4–5), 533–538; (o) Kagawa, T.F., Geierstanger, B.H., Wang, A.H.J., and Ho, P.S. (1991) Covalent modification of guanine bases in double-stranded DNA. The 1. 2-. ANG. Z-DNA structure of d(CGCGCG) in the presence of copper(II) chloride. *J. Biol. Chem.*, **266** (30), 20175–20184.

3. Braun, E., Eichen, Y., Sivan, U., and Ben-Yoseph, G. (1998) DNA-templated assembly and electrode attachment of a conducting silver wire. *Nature*, **391** (6669), 775–778.

4. Jayaraman, S., Tang, W., and Yongsunthon, R. (2011) Electrochemical Synthesis of M:DNA Nanohybrids. *J. Electrochem. Soc.*, **158**, K123–K126.

5. (a) Keren, K., Krueger, M., Gilad, R. *et al.* (2002) Sequence-specific molecular lithography on single DNA molecules. *Science*, **297** (5578), 72–75; (b)Keren, K., Berman, R.S., and Braun, E. (2004) Patterned DNA metallization by sequence-specific localization of a reducing agent. *Nano. Lett.*, **4** (2), 323–326.

6. (a) Aich, P., Labiuk, S.L., Tari, L.W. *et al.* (1999) M-DNA: A Complex Between Divalent Metal Ions and DNA which Behaves as a Molecular Wire. *J. Mol. Biol.*, **294** (2), 477–485; (b)Wettig, S.D., Li, C.-Z., Long, Y.-T. *et al.* (2003) M-DNA: a self-assembling molecular wire for nanoelectronics and biosensing. *Anal. Sci.*, **19** (1), 23–26; (c)Wettig, S.D., Wood, D.O., Aich, P., and Lee, J.S. (2005) M-DNA: A novel metal ion complex of DNA studied by fluorescence techniques. *J. Inorg. Biochem.*, **99** (11), 2093–2101; (d)Wood, D.O., Dinsmore, M.J., Bare, G.A., and Lee, J.S. (2002) M-DNA is stabilized in G. C tracts or by incorporation of 5-fluorouracil. *Nucleic Acids Res.*, **30** (10), 2244–2250; (e)Wood, D.O. and Lee, J.S. (2005) Investigation of pH-dependent DNA-metal ion interactions by surface plasmon resonance. *J. Inorg. Biochem.*, **99** (2), 566–574.

7. (a) Alexandre, S.S., Murta, B.J., Soler, J.M., and Zamora, F. (2011) Stability and electronic structure of M-DNA: Role of metal position. *Phys. Rev. B Condens. Mat. Mater. Phys.*, **84**, 045413/1–045413/7; (b)Brancolini, G. and Di, F.R. (2011) Combined effects of metal complexation and size expansion in the electronic structure of DNA base pairs. *J. Chem. Phys.*, **134**, 205102/1–205102/12; (c)Rubin, Y.V., Belous, L.F., and Yakuba, A.A. (2011) Electronic and molecular structure of M-DNA fragments. *J. Mol. Model.*, **17**, 997–1006.

8. Tanaka, K. and Shionoya, M. (1999) Synthesis of a novel nucleoside for alternative DNA base pairing through metal complexation. *J. Org. Chem.*, **64** (14), 5002–5003.

9. (a) Cao, H. Tanaka, K., and Shionoya, M. (2000) An alternative base-pairing of catechol-bearing nucleosides by borate formation. *Chem. Pharm. Bull. (Tokyo)*, **48** (11), 1745–1748; (b) Shionoya, M. and Tanaka, K. (2000) Synthetic incorporation of metal complexes into nucleic acids and peptides directed toward functionalized molecules. *Bull. Chem. Soc. Jpn*, **73** (9), 1945–1954; (c)Tanaka, K., Tasaka, M., Cao, H., and Shionoya, M. (2001) An approach to metal-assisted DNA base pairing: novel b-C-nucleosides with a 2-aminophenol or a catechol as the nucleobase. *Eur. J. Pharm. Sci.*, **13** (1), 77–83; (d)Tasaka, M., Tanaka, K., Shiro, M., and Shionoya, M. (2001) A palladium-mediated DNA base pair of a β-C-nucleoside possessing a 2-aminophenol as the nucleobase. *Supramol. Chem.*, **13** (6), 671–675.

10. Meggers, E., Holland, P.L., Tolman, W.B. *et al.* (2000) A novel copper-mediated DNA base pair. *J. Am. Chem. Soc.*, **122** (43), 10714–10715.

11. (a) Weizman, H. and Tor, Y. (2001) Oligo-ligandosides: a DNA mimetic approach to helicate formation. *Chem. Commun.*, **2001** (5), 453–454; (b)Weizman, H. and Tor, Y. (2001) 2,2′-Bipyridine ligandoside: a novel building block for modifying DNA with intra-duplex metal complexes. *J. Am. Chem. Soc.*, **123** (14), 3375–3376.

12. (a) Tanaka, K., Tengeiji, A., Kato, T. *et al.* (2002) Efficient incorporation of a copper hydroxypyridone base pair in DNA. *J. Am. Chem. Soc.*, **124** (42), 12494–12498; (b) Tanaka, K., Yamada, Y., and Shionoya, M. (2002) Formation of Silver(I)-mediated DNA duplex and triplex through an alternative base pair of pyridine nucleobases. *J. Am. Chem. Soc.*, **124** (30), 8802–8803.

13. (a) Switzer, C. and Shin, D. (2005) A pyrimidine-like nickel(II) DNA base pair. *Chem. Commun.*, **2005** (10), 1342–1344; (b) Switzer, C., Sinha, S., Kim, P.H., and Heuberger, B.D. (2005) A purine-like nickel(II) base pair for DNA. *Angew. Chem., Int. Ed.*, **44** (10), 1529–1532; (c) Shin, D. and Switzer, C. (2007) A metallo base-pair incorporating a terpyridyl-like motif: bipyridyl-pyrimidinone·Ag(I)·4-pyridine. *Chem. Commun.*, **2007** (42), 4401–4403; (d) Heuberger, B.D., Shin, D., and Switzer, C. (2008) Two Watson-Crick-like metallo base-pairs. *Org. Lett.*, **10** (6), 1091–1094.

14. (a) Boehme, D., Duepre, N., Megger, D.A., and Müller, J. (2007) Conformational change induced by metal-ion-binding to DNA containing the artificial 1,2,4-Triazole nucleoside. *Inorg. Chem.*, **46** (24), 10114–10119; (b) Polonius, F.-A. and Müller, J. (2007) An artificial base pair, mediated by hydrogen bonding and metal-ion binding. *Angew. Chem. Int. Ed.*, **46** (29), 5602–5604; (c) Megger, D.A. and Müller, J. (2010) Silver(I)-mediated cytosine self-pairing is preferred over hoogsteen-type base pairs with the artificial nucleobase 1,3-dideaza-6-nitropurine. *Nucleosides Nucleotides Nucleic Acids*, **29** (1), 27–38; (d) Megger, D.A., Fonseca Guerra, C., Bickelhaupt, F.M., and Müller, J. (2011) Silver(I)-mediated Hoogsteen-type base pairs. *J. Inorg. Biochem.*, **105** (11), 1398–1404; (e) Megger, D.A., Guerra, C.F., Hoffmann, J. *et al.* (2011) Contiguous metal-mediated base pairs comprising two AgI ions. *Chem. Eur. J.*, **17** (23), 6533–6544, S6533/1–S6533/21.

15. (a) Clever, G.H. Polborn, K., and Carell, T. (2005) A highly DNA-duplex-stabilizing metal-salen base pair. *Angew. Chem. Int. Ed.*, **44** (44), 7204–7208; (b) Clever, G.H., Soeltl, Y., Burks, H. *et al.* (2006) Metal-salen-base-pair complexes inside DNA: complexation overrides

sequence information. *Chem. Eur. J.*, **12** (34), 8708–8718; (c) Tanaka, K., Clever, G.H., Takezawa, Y. *et al.* (2006) Programmable self-assembly of metal ions inside artificial DNA duplexes. *Nat. Nanotechnol.*, **1** (3), 190–194; (d) Clever, G.H. and Carell, T. (2007) Controlled stacking of 10 transition-metal ions inside a DNA duplex. *Angew. Chem. Int. Ed.*, **46** (1/2), 250–253; (e) Clever, G.H., Kaul, C., and Carell, T. (2007) DNA–metal base pairs. *Angew. Chem. Int. Ed.*, **46** (33), 6226–6236; (f) Tanaka, K., Clever, G.H., Takezawa, Y. *et al.* (2007) Programmable self-assembly of metal ions inside artificial DNA duplexes. [Erratum to document cited in CA147:380005]. *Nat. Nanotechnol.*, **2** (1), 63; (g) Gaub, B.M., Kaul, C., Zimmermann, J.L. *et al.* (2009) Switching the mechanics of dsDNA by Cu salicylic aldehyde complexation. *Nanotechnology*, **20** (43), 434002/1–434002/8; (h) Clever, G.H., Reitmeier, S.J., Carell, T., and Schiemann, O. (2010) Antiferromagnetic coupling of stacked CuII-salen complexes in DNA. *Angew. Chem. Int. Ed.*, **49** (29), 4927–4929; (i) Kaul, C., Mueller, M., Wagner, M. *et al.* (2011) Reversible bond formation enables the replication and amplification of a cross-linking salen complex as an orthogonal base pair. *Nat. Chem.*, **3** (10), 794–800.

16. (a) Popescu, D.-L., Parolin, T.J., and Achim, C. (2003) Metal incorporation in modified PNA duplexes. *J. Am. Chem. Soc.*, **125** (21), 6354–6355; (b) Watson, R.M., Skorik, Y., Patra, G.K., and Achim, C. (2005) Influence of metal coordination on the mismatch tolerance of ligand-modified PNA duplexes. *J. Am. Chem. Soc.*, **127**, 14628–14639; (c) Franzini, R., Watson, R.M., Patra, G.K., and Achim, C. (2006) Metal binding to bipyridine-modified peptide nucleic acids. *Inorg. Chem.*, **45** (24), 9798–9811; (d) Bezer, S., Rapireddy, S., Skorik, Y.A., Ly, D.H., and Achim, C. (2011) Coordination-driven inversion of handedness in ligand-modified PNA. *Inorg. Chem.*, **50** (23), 11929–11937; (e) Ma, Z., Olechnowicz, F., Skorik, Y.A., and Achim, C. (2011) Metal binding to ligand-containing peptide nucleic acids. *Inorg. Chem.*, **50** (13), 6083–6092.

17. (a) Seubert, K., Guerra, C.F., Bickelhaupt, F.M., and Mueller, J. (2011) Chimeric GNA/DNA metal-mediated base pairs. *Chem. Commun.*, **47** (39), 11041–11043; (b) Zhang, L., Peritz, A., and Meggers, E. (2005) A simple glycol nucleic acid. *J. Am. Chem. Soc.*, **127** (12), 4174–4175.

18. (a) Babu, B.R., Hrdlicka, P.J., McKenzie, C.J., and Wengel, J. (2005) Optimized DNA targeting using N,N-bis(2-pyridylmethyl)-b-alanyl 2′-amino-LNA. *Chem. Commun.*, **2005** (13), 1705–1707; (b) Kalek, M., Madsen, A.S., and Wengel, J. (2007) Effective modulation of DNA duplex stability by reversible transition metal complex formation in the minor groove. *J. Am. Chem. Soc.*, **129** (30), 9392–9400.

19. Kuklenyik, Z. and Marzilli, L.G. (1996) Mercury(II) site-selective binding to a DNA hairpin. relationship of sequence-dependent intra- and interstrand crosslinking to the hairpin-duplex conformational transition. *Inorg. Chem.*, **35** (19), 5654–5662.

20. Nielsen, P.E., Egholm, M., Berg, R.H., and Buchardt, O. (1991) Sequence-selective recognition of DNA by strand displacement with a thymine-substituted polyamide. *Science*, **254**, 1497–1500.

21. Beck, F. (2002) Solid phase synthesis of PNA oligomers. *Methods Mol. Biol.*, **208**, 29–41.

22. (a) Bahal, R., Sahu, B., Rapireddy, S. *et al.* (2012) Sequence-unrestricted, Watson-crick recognition of double helical B-DNA by (R)-MiniPEG-γPNAs. *ChemBioChem*, **13**, 56–60; (b) Corradini, R., Sforza, S., Tedeschi, T. *et al.* (2007) Peptide nucleic acids with a structurally biased backbone: effects of conformational constraints and stereochemistry. *Curr. Top. Med. Chem.*, **7**, 681–775; (c) D'Costa, M., Kumar, V., and Ganesh, K. (1999) Aminoethylprolyl peptide nucleic acids (aepPNA): chiral PNA analogues that form highly stable DNA:aepPNA2 triplexes. *Org. Lett.*, **1** (10), 1513–1516; (d)Dragulescu-Andrasi, A., Rapireddy, S., Frezza, B. *et al.* (2006) A simple gamma-backbone modification preorganizes peptide nucleic acid into a helical structure. *J. Am. Chem. Soc.*, **128** (31), 10258–10267; (e) Ganesh, K.N. and Nielsen, P.E. (2000) Peptide nucleic acids analogs and derivatives. *Curr. Org. Chem.*, **4** (9), 931–943;

(f) Govindaraju, T., Kumar, V., and Ganesh, K. (2004) Synthesis and evaluation of (1S,2R/1R,2S)-aminocyclohexylglycyl PNAs as conformationally preorganized PNA analogues for DNA/RNA recognition. *J. Org. Chem.*, **69** (6), 1858–1865; (g) Govindaraju, T., Kumar, V., and Ganesh, K. (2004) (1S,2R/1R,2S)-cis-cyclopentyl PNAs (cpPNAs) as constrained PNA analogues: synthesis and evaluation of aeg-cpPNA chimera and stereopreferences in hybridization with DNA/RNA. *J. Org. Chem.*, **69** (17), 5725–5734; (h) Govindaraju, T., Kumar, V., and Ganesh, K. (2004) cis-Cyclopentyl PNA (cpPNA) as constrained chiral PNA analogues: stereochemical dependence of DNA/RNA hybridization. *Chem. Commun.*, **2004**, 860–861; (i) Govindaraju, T., Kumar, V., and Ganesh, K. (2005) (SR/RS)-cyclohexanyl PNAs: conformationally preorganized PNA analogues with unprecedented preference for duplex formation with RNA. *J. Am. Chem. Soc.*, **127** (12), 4144–4145; (j) Govindaraju, T., Madhuri, V., Kumar, V., and Ganesh, K. (2006) Cyclohexanyl peptide nucleic acids (chPNAs) for preferential RNA binding: effective tuning of dihedral angle beta in PNAs for DNA/RNA discrimination. *J. Org. Chem.*, **71** (1), 14–21; (k) Haaima, G., Hansen, H., Christensen, L. *et al.* (1997) Increased DNA binding and sequence discrimination of PNA oligomers containing 2,6-diaminopurine. *Nucleic Acids Res.*, **25** (22), 4639–4643; (l) Kumar, R., Singh, S.K., Koshkin, A.A. *et al.* (1998) The first analogs of LNA (locked nucleic acids): phosphorothioate-LNA and 2'-thio-LNA. *Bioorg. Med. Chem. Lett.*, **8** (16), 2219–2222; (m) Lagriffoule, P., Eriksson, M., Jensen, K.K. *et al.* (1997) Peptide nucleic acids with a conformationally constrained chiral cyclohexyl derived backbone. *Chem. Eur. J.*, **3** (6), 912–919; (n) Pokorski, J., Witschi, M., Purnell, B., and Appella, D. (2004) (S,S)-trans-cyclopentane-constrained peptide nucleic acids. a general backbone modification that improves binding affinity and sequence specificity. *J. Am. Chem. Soc.*, **126** (46), 15067–15073; (o) Rapireddy, S., He, G., Roy, S. *et al.* (2007) Strand invasion of mixed-sequence B-DNA by acridine-linked, gamma-peptide nucleic acid (gamma-PNA). *J. Am. Chem. Soc.*, **129** (50), 15596–15600.

23. Diederichsen, U. (1996) Pairing properties of alanyl peptide nucleic acids containing an amino acid backbone with alternating configuration. *Angew. Chem. Int. Ed.*, **35** (4), 445–448.

24. Nielsen, P.E. (ed.) (2004) *Peptide Nucleic Acids: Protocols and Applications*, 3nd edn, Horizon Bioscience, Wymondham, p. 318.

25. (a) Wolak, M.A., Balaeff, A., Gutmann, S. *et al.* (2011) Electronic structure of self-assembled peptide nucleic acid thin films. *J. Phys. Chem. C*, **115** (34), 17123–17135; (b) Paul, A., Watson, R.M., Lund, P. *et al.* (2008) Charge transfer through single-stranded peptide nucleic acid composed of thymine nucleotides. *J. Phys. Chem. C*, **112** (18), 7233–7240.

26. Wittung, P., Eriksson, M., Lyng, R. *et al.* (1995) Induced chirality in PNA-PNA duplexes. *J. Am. Chem. Soc.*, **117** (41), 10167–10173.

27. (a) Rasmussen, H., Kastrup, J.S., Nielsen, J.N. *et al.* (1997) Crystal structure of a peptide nucleic acid (PNA) duplex at 1.7 A resolution. *Nat. Struct. Biol.*, **4** (2), 98–101; (b) Yeh, J.I., Pohl, E., Truan, D. *et al.* (2010) The crystal structure of non-modified and bipyridine-modified PNA duplexes. *Chem. Eur. J.*, **16** (39), 11867–11875; (c) Yeh Joanne, I., Pohl, E., Truan, D. *et al.* (2011) The crystal structure of non-modified and bipyridine-modified PNA duplexes. [Erratum to document cited in CA154:046259]. *Chem. Eur. J.*, **17** (44), 12227.

28. (a) Sforza, S., Tedeschi, T., Corradini, R., and Marchelli, R. (2007) Induction of helical handedness and DNA binding properties of peptide nucleic acids (PNAs) with two stereogenic centers. *Eur. J. Org. Chem.*, **2007** (35), 5879–5885; (b) Sforza, S., Haaima, G., Marchelli, R., and Nielsen, P.E. (1999) Chiral peptide nucleic acids (PNAs). Helix handedness and DNA recognition. *Eur. J. Org. Chem.*, **1999** (1), 197–204; (c) Pensato, S., Saviano, M., Bianchi, N. *et al.* (2010) γ-Hydroxymethyl PNAs: Synthesis, interaction with DNA and inhibition of protein/DNA interactions. *Bioorg. Chem.*, **38** (5), 196–201.

29. He, W., Hatcher, E., Balaeff, A. *et al.* (2008) Solution structure of a peptide nucleic acid duplex from NMR data: Features and limitations. *J. Am. Chem. Soc.*, **130** (40), 13264–13273.

30. Sosniak, A.M., Gasser, G., and Metzler-Nolte, N. (2009) Thermal melting studies of alkyne- and ferrocene-containing PNA bioconjugates. *Org. Biomol. Chem.*, **7** (23), 4992–5000.

31. (a) Kise, K.J. Jr and Bowler, B.E. (2002) A Ruthenium(II) Tris(bipyridyl) Amino Acid: Synthesis and Direct Incorporation into an α-Helical Peptide by Solid-Phase Synthesis. *Inorg. Chem.*, **41** (2), 379–386; (b) Nickita, N., Gasser, G., Bond, A.M., and Spiccia, L. (2009) Synthesis, spectroscopic properties and electrochemical oxidation of ruii-polypyridyl complexes attached to a peptide nucleic acid monomer backbone. *Eur. J. Inorg. Chem.*, **2009**, 2179–2186.

32. (a) Singh, S.K., Nielsen, P., Koshkin, A.A., and Wengel, J. (1998) LNA (locked nucleic acids): synthesis and high-affinity nucleic acid recognition. *Chem. Commun.*, **1998** (4), 455–456; (b) Obika, S., Nanbu, D., Hari, Y. *et al.* (1998) Stability and structural features of the duplexes containing nucleoside analogs with a fixed N-type conformation, 2′-O,4′-C-methyleneribonucleosides. *Tetrahedron Lett.*, **39** (30), 5401–5404.

33. (a) Wengel, J., Petersen, M., Frieden, M., and Koch, T. (2004) Chemistry of locked nucleic acids (LNA): Design, synthesis, and bio-physical properties. *Lett. Pept. Sci.*, **10** (3/4), 237–253; (b) Doessing, H. and Vester, B. (2011) Locked and unlocked nucleosides in functional nucleic acids. *Molecules*, **16** (6), 4511–4526.

34. Nielsen, K.E., Rasmussen, J., Kumar, R. *et al.* (2004) NMR studies of fully modified locked nucleic acid (LNA) hybrids: solution structure of an LNA:RNA hybrid and characterization of an LNA:DNA hybrid. *Bioconjugate Chem.*, **15** (3), 449–457.

35. (a) Meggers, E. and Zhang, L. (2010) Synthesis and properties of the simplified nucleic acid glycol nucleic acid. *Acc. Chem. Res.*, **43** (8), 1092–1102; (b) Zhang, L. and Meggers, E. (2005) An extremely stable and orthogonal DNA base pair with a simplified three-carbon backbone. *J. Am. Chem. Soc.*, **127** (1), 74–75.

36. Schlegel, M.K., Essen, L.-O., and Meggers, E. (2010) Atomic resolution duplex structure of the simplified nucleic acid GNA. *Chem. Commun.*, **46** (7), 1094–1096.

37. Johnson, A.T., Schlegel, M.K., Meggers, E. *et al.* (2011) On the structure and dynamics of duplex GNA. *J. Org. Chem.*, **76** (19), 7964–7974.

38. (a) Shionoya, M. and Tanaka, K. (2004) Artificial metallo-DNA: a bio-inspired approach to metal array programming. *Curr. Opin. Chem. Biol.*, **8** (6), 592–597; (b) Wagenknecht, H.-A. (2003) Metal-mediated DNA base pairing and metal arrays in artificial DNA: Towards new nanodevices. *Angew. Chem. Int. Ed.*, **42** (28), 3204–3206.

39. (a) Henry, A.A. and Romesberg, F.E. (2003) Beyond A, C, G and T: augmenting nature's alphabet. *Curr. Opin. Chem. Biol.*, **7** (6), 727–733; (b) Kool, E.T. (2002) Replacing the nucleobases in DNA with designer molecules. *Acc. Chem. Res.*, **35** (11), 936–943.

40. (a) Park, K.S. Jung, C., and Park, H.G. (2010) "Illusionary" polymerase activity triggered by metal ions: use for molecular logic-gate operations. *Angew. Chem. Int. Ed.*, **49** (30), 9757–9760; (b) Urata, H., Yamaguchi, E., Funai, T. *et al.* (2010) Incorporation of thymine nucleotides by DNA polymerases through T·HgII-T base pairing. *Angew. Chem. Int. Ed.*, **49** (37), 6516–6519.

41. Brotschi, C. and Leumann, C.J. (2003) Transition metal ligands as novel DNA-base substitutes. *Nucleosides Nucleotides Nucleic Acids*, **22** (5/8), 1195–1197.

42. Megger, D., Megger, N., and Müller, J. (2012) Metal-mediated base pairs in nucleic acids with purine- and pyrimidine-derived nucleosides. *Metal Ions Life Sci.*, **10**, 295–317.

43. Müller, J., Boehme, D., Lax, P. *et al.* (2005) Metal ion coordination to azole nucleosides. *Chem. Eur. J.*, **11** (21), 6246–6253.

44. Boehme, D., Duepre, N., Megger, D.A., and Mueller, J. (2007) Conformational change induced by metal-ion-binding to DNA containing the artificial 1,2,4-Triazole nucleoside. *Inorg. Chem.*, **46** (24), 10114–10119.

45. Johannsen, S., Megger, N., Boehme, D. *et al.* (2010) Solution structure of a DNA double helix with consecutive metal-mediated base pairs. *Nat. Chem.*, **2** (3), 229–234.

46. Smith, A.P. and Fraser, C.L. (2004) Bipyridine ligands. *Compr. Coord. Chem. II*, **1**, 1–23.

47. Zimmermann, N., Meggers, E., and Schultz, P.G. (2002) A novel Silver(I)-Mediated DNA base pair. *J. Am. Chem. Soc.*, **124** (46), 13684–13685.

48. Constable, E.C. (1989) Homoleptic complexes of 2,2′-bipyridine. *Adv. Inorg. Chem.*, **34**, 1–63.

49. Brotschi, C., Haberli, A., and Leumann, C.J. (2001) A stable DNA duplex containing a non-hydrogen-bonding and non-shape-complementary base couple: Interstrand stacking as the stability determining factor. *Angew. Chem. Int. Ed.*, **40** (16), 3012–3014.

50. Franzini, R., Watson, R.M., Popescu, D.-L. *et al.* (2004) Metal-containing modified peptide nucleic acids. *Polymer Preprints*, **45** (1), 337–338.

51. Wierzbinski, E., de Leon, A., Davis, K.L. *et al.* (2012) Charge transfer through modified peptide nucleic acids. *Langmuir*, **28** (4), 1971–1981.

52. Milani, B., Anzilutti, A., Vicentini, L. *et al.* (1997) Bis-chelated Palladium(II) complexes with nitrogen-donor chelating ligands are efficient catalyst precursors for the CO/Styrene copolymerization reaction. *Organometallics*, **16** (23), 5064–5075.

53. Schlegel, M.K., Zhang, L., Pagano, N., and Meggers, E. (2009) Metal-mediated base pairing within the simplified nucleic acid GNA. *Org. Biomol. Chem.*, **7** (3), 476–482.

54. Tanaka, K., Tasaka, M., Cao, H., and Shionoya, M. (2002) Toward nano-assembly of metals through engineered DNAs. *Supramol. Chem.*, **14** (2–3), 255–261.

55. Zhan, Z.-Y.J. and Lynn, D.G. (1997) Chemical amplification through template-directed synthesis. *J. Am. Chem. Soc.*, **119** (50), 12420–12421.

56. Czlapinski, J.L. and Sheppard, T.L. (2004) Template-directed assembly of metallosalen-DNA hairpin conjugates. *ChemBioChem*, **5** (1), 127–129.

57. El-Jammal, A., Howell, P.L., Turner, M.A. *et al.* (1994) Copper complexation by 3-Hydroxy-pyridin-4-one Iron chelators: structural and iron competition studies. *J. Med. Chem.*, **37** (4), 461–466.

58. (a) Ahmed, S.I., Burgess, J., Fawcett, J. *et al.* (2000) The structures of bis-maltolato-zinc(II) and of bis(3-hydroxy-1,2-dimethyl-4-pyridinonato)zinc(II) and -lead(II). *Polyhedron*, **19** (2), 129–135; (b) Griffith, W.P. and Mostafa, S.I. (1992) Complexes of 3-hydroxypyridin-2-one and 1,2-dimethyl-3-hydroxypyridin-4-one with second and third row elements of groups 6, 7 and 8. *Polyhedron*, **11** (23), 2997–3005.

59. Schlegel, M.K., Essen, L.-O., and Meggers, E. (2008) Duplex structure of a minimal nucleic acid. *J. Am. Chem. Soc.*, **130** (26), 8158–8159.

60. Liu, S., Clever, G.H., Takezawa, Y. *et al.* (2011) Direct conductance measurement of individual metallo-DNA duplexes within single-molecule break junctions. *Angew. Chem. Int. Ed.*, **50** (38), 8886–8890, S8886/1–S8886/12.

61. (a) Heuberger, B.D. and Switzer, C. (2008) An alternative nucleobase code: characterization of purine-purine DNA double helices bearing guanine-isoguanine and diaminopurine 7-deaza-xanthine base pairs. *ChemBioChem*, **9** (17), 2779–2783; (b) Karlsen, K.K., Jensen, T.B., and Wengel, J. (2009) Synthesis of an unlocked nucleic acid terpyridine monomer and binding of divalent metal ion in nucleic acid duplexes. *J. Org. Chem.*, **74** (22), 8838–8841.

62. Yang, H., Rys, A.Z., McLaughlin, C.K., and Sleiman, H.F. (2009) Templated ligand environments for the selective incorporation of different metals into DNA. *Angew. Chem. Int. Ed.*, **48** (52), 9919–9923.

63. Zimmermann, N., Meggers, E., and Schultz, P.G. (2004) A second-generation copper(II)-mediated metallo-DNA-base pair. *Bioorg. Chem.*, **32** (1), 13–25.

64. (a) Conley, H.L. Jr and Martin, R.B. (1965) Cupric ion catalyzed hydrolyses of glycine ethyl ester, glycinamide, and picolinamide. *J. Phys. Chem.*, **69** (9), 2914–2923; (b) Nawata, Y., Iwasaki, H., and Saito, Y. (1967) Crystal structure of bis(pyridine-2-carboxamido)nickel(II) dihydrate. *Bull. Chem. Soc. Jpn*, **40** (3), 515–521.

65. Bonomo, R.P., Cucinotta, V., and Riggi, F. (1980) EPR solution studies of copper(II) mixed complexes containing iminodiacetic or pyridine-2,6-dicarboxylic acids and pyridine, ethylenediamine or diethylenetriamine. *J. Mol. Struct.*, **69**, 295–300.

66. Atwell, S., Meggers, E., Spraggon, G., and Schultz, P.G. (2001) Structure of a copper-mediated base pair in DNA. *J. Am. Chem. Soc.*, **123** (49), 12364–12367.

67. Mokhir, A., Stiebing, R., and Krämer, R. (2003) Peptide nucleic acid-metal complex conjugates: facile modulation of PNA-DNA duplex stability. *Bioorg. Med. Chem. Lett.*, **13** (8), 1399–1401.

68. Hrdlicka, P.J., Babu, B.R., Sorensen, M.D., and Wengel, J. (2004) Interstrand communication between 2′-N-(pyren-1-yl)methyl-2′-amino-LNA monomers in nucleic acid duplexes: directional control and signalling of full complementarity. *Chem. Commun.*, **2004** (13), 1478–1479.

69. Hirva, P., Nielsen, A., Bond, A.D., and McKenzie, C.J. (2010) Potential cross-linking transition metal complexes (M=Ni, Cu, Zn) in the ligand-modified LNA duplexes. *J. Phys. Chem. B*, **114** (36), 11942–11948.

70. (a) Brotschi, C. and Leumann, C.J. (2003) DNA with hydrophobic base substitutes: a stable, zipperlike recognition motif based on interstrand-stacking interactions. *Angew. Chem. Int. Ed.*, **42** (14), 1655–1658; (b) Brotschi, C., Mathis, G., and Leumann, C.J. (2005) Bipyridyl- and biphenyl-DNA: a recognition motif based on interstrand aromatic stacking. *Chem. Eur. J.*, **11** (6), 1911–1923.

71. Johar, Z., Zahn, A., Leumann, C.J., and Jaun, B. (2008) Solution structure of a DNA duplex containing a biphenyl pair. *Chem. Eur. J.*, **14** (4), 1080–1086.

72. Mees, A., Klar, T., Gnau, P. *et al.* (2004) Crystal structure of a photolyase bound to a CPD-Like DNA lesion after in situ repair. *Science*, **306** (5702), 1789–1793.

73. Tanaka, K., Tengeiji, A., Kato, T. *et al.* (2003) A discrete self-assembled metal array in artificial DNA. *Science*, **299** (5610), 1212–1213.

74. Zhang, H.Y., Calzolari, A., and Di Felice, R. (2005) On the magnetic alignment of metal ions in a DNA-Mimic Helix. *J. Phys. Chem. B*, **109**, 15345–15348.

75. Jishi, R.A. and Bragin, J. (2007) Symmetry selection in artificial DNA base pairs. *J. Phys. Chem. B*, **111** (19), 5357–5361.

76. (a) Nakanishi, Y., Kitagawa, Y., Shigeta, Y. *et al.* (2009) Theoretical studies on magnetic interactions between Cu(II) ions in hydroxypyridone nucleobases. *Polyhedron*, **28** (9–10), 1714–1717; (b) Mallajosyula, S.S. and Pati, S.K. (2009) Conformational Tuning of Magnetic Interactions in Metal-DNA Complexes. *Angew. Chem. Int. Ed.*, **48** (27), 4977–4981, S4977/1–S4977/8.

77. Takezawa, Y., Maeda, W., Tanaka, K., and Shinoya, M. (2009) Discrete self-assembly of iron (III) ions inside triple-stranded artificial DNA. *Angew. Chem. Int. Ed.*, **48** (6), 1081–1084.

78. Kalachova, L., Pohl, R., and Hocek, M. (2012) Synthesis of nucleoside mono- and triphosphates bearing oligo-pyridine ligands, their incorporation into DNA and complexation with transition metals. *Org. Biomol. Chem.*, **10** (1), 49–55.

79. Megger, D.A., Guerra, C.F., Hoffmann, J. *et al.* (2011) Contiguous metal-mediated base pairs comprising two AgI ions. *Chem. Eur. J.*, **17** (23), 6533–6544.

# 11

# Interaction of Biomimetic Oligomers with Metal Ions

*Galia Maayan*

*Schulich Faculty of Chemistry, Technion – Israel Institute of Technology, Israel*

Metal ions are key elements in both the structure and function of natural biopolymers, being employed in tasks spanning from structure stabilization to catalysis, light energy conversion and recognition. In the context of structure, metal ions can either stabilize an existing structural fold or impose a conformational constraint on an unstructured polypeptide while facilitating its folding. The role of metal coordination in the folding of natural biopolymers, as well as the direct correlation between structure and function, inspires the design of single-stranded biomimetic oligomers, namely peptidomimetics and their abiotic analogues, that fold into three-dimensional structures upon metal binding. Unlike helicates, in which metal ions template a helical structure via self-assembly, in biomimetic oligomers metal ions nucleate the formation of a three-dimensional structure in a controlled manner. Nucleation of a secondary structure is a great challenge because it requires that metal ion coordination will initiate a series of cooperative, non-covalent interactions in the molecular structure, leading to its folding. In this chapter, we will begin with an introduction of biomimetic oligomers and describe their folding behavior. We will show how metal ions can nucleate the folding of unfolded oligomers as well as stabilize the existing secondary structures of biomimetic oligomers. Emphasis will be given to both linear and cyclic systems in which absolute helicity can be achieved or enhanced through a rational design of metallofoldamers. Next we will give some examples of systems in which the folding induced by metal coordination leads to three-dimensional architectures different from the helical structure, and we will end the chapter with conclusions and an outlook for some future challenges.

*Metallofoldamers: Supramolecular Architectures from Helicates to Biomimetics*, First Edition.
Edited by Galia Maayan and Markus Albrecht.
© 2013 John Wiley & Sons, Ltd. Published 2013 by John Wiley & Sons, Ltd.

## 11.1 Introduction

Biomimetic oligomers are synthetic molecules akin to natural biopolymers, namely peptides, proteins and oligonucleotides. Both natural biopolymers and their artificial mimics are sequence-specific oligomers capable of folding into well defined three-dimensional structures in solution. One difference between them, however, is the identity of their backbone: while peptides and proteins are composed of $\alpha$-amino acids, biomimetic oligomers are assembled from non-natural monomers. There are mainly two groups of biomimtic oligomers: single-stranded (peptidomimetics and their abiotic analogues) and self-assembled multiple-stranded (nucleotidomimetics and their abiotic analogues). As the mimics of oligonucleotides that bind metal ions were discussed in the previous two chapters, the folding behavior of single-stranded biomimetic oligomers upon metal binding is the focus of this chapter. Such abiotic oligomers were the subject of numerous studies in recent years, and their synthesis, characterization and folding behavior were investigated. This research revealed that, similar to natural biopolymers, the conformation of folded biomimetic oligomers ("biomimetic foldamers") could be controlled by a variety of strategies involving non-covalent interactions [1]. These include the use of specific non-covalent forces, such as hydrogen bonding, donor–acceptor complexation, aromatic $\pi$-stacking and metal–ligand interactions, as well as non-specific van der Waals interactions and solvophobic effects [1–5].

Among these interactions, metal–ligand coordination represents an exciting opportunity to gain structure stability in a selective manner; as proteins often select a specific metal from the pool of metal ions that are present in the cellular environment, biomimetic oligomers can be designed to fold upon binding to a particular metal ion. It is now well established that the identity of the metal-binding ligands and their coordination mode, as well as side chain interactions, have a crucial role in governing metal binding and selectivity in proteins. The selective affinity of a protein to a specific metal ion determines its final three-dimensional structure, which eventually leads to the metalloprotein's unique function. Despite the well known relationship between the structure and function of metalloproteins, the association between metal coordination and protein folding, including the contribution of metal sites to structural stability, is still, in most cases, poorly understood. In addition, it is also not quite clear whether metal-binding sites in proteins are generally rigid or flexible, how well the protein can adjust to the coordination constraints of the incoming metal ion, or, on the contrary, whether the metal ion can obey the requirements of the protein matrix. Thus, the incentive behind the generation of synthetic single-stranded biomimetic metallofoldamers is twofold. First, producing biomimetic oligomers which incorporate metal-binding ligands varying in their type, their coordination capabilities and their position along the oligomer spine should enable a detailed investigation of the oligomer folding behavior upon metal complexation. This will shed light on the correlation between metal binding, folding and structure stability. Second, the new biomimetic metallofoldamers are anticipated to encompass unique functions, such as selective catalysis and sensing. If a biomimetic oligomer could fold upon the selective binding of one metal ion in the presence of other metals, for example, detection of the folding event will serve as a sensor for this specific metal.

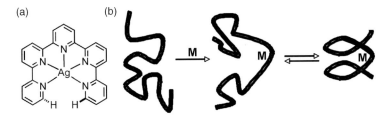

**Figure 11.1** *Classification of metallofoldamers: (a) templated helix-like structures and (b) nucleated secondary structures [6].*

The coordination of metal ions to single-stranded oligomers can either template a helical structure or nucleate its formation (Figure 11.1) [6]. J. Fox at the University of Delaware defined molecules that template an abiotic helix, such as the one represented in Figure 11.1a, as having an intrinsically helical metal coordination sphere. "Helicates" [7] are one example of molecules that template an abiotic helix, and they have been broadly discussed in earlier chapters. In contrast, molecules such as the one represented in Figure 11.1b do not have an inherently chiral metal coordination sphere, but instead, metal binding is likely to impose conformational restrictions which drive the folding of the biomimetic oligomer chain into a stable three-dimensional structure. This nucleation process resembles the folding of natural polypeptides and proteins in which metal coordination leads to a specific folding pathway by lowering the entropy of the unfolded state, thus speeding up the folding event [8].

Some metallofodamers are generated as a combination between a templated helical or other folded structure and a nucleated three-dimensional structure. In these cases, a metal ion will template a specific folding event and the resulted fold will then enable the nucleation of a stable structure by additional non-covalent interactions. This chapter will discuss such biomimetic metallofoldamers as well as abiotic single-stranded metallofoldamers, which formed solely upon nucleation by metal ion coordination. The discussion will not be limited to helical folding but rather will be further extended to metallofoldamers with various three-dimensional architectures.

## 11.2 Single-Stranded Oligomers in Which Metal Coordination Templates, or Templates and Nucleates the Formation of an Abiotic Helix

As shown in natural systems, an ordered structure within a biopolymer depends upon a combination of different physicochemical interactions [4]. Based on this notion, scientists began to develop abiotic oligomers in which metal coordination promotes various non-local interactions, leading to a controlled folded structure. The first example of an abiotic oligomer whose structure in solution was designed to involve both non-specific (solvophobic [9]) and specific (metal coordination) interactions, was reported by the group of Moore in 1999 [10]. The solution behavior of a meta-connected oligomer whose backbone consists of 12 non-polar phenylacetylene units was tested in the context of metal

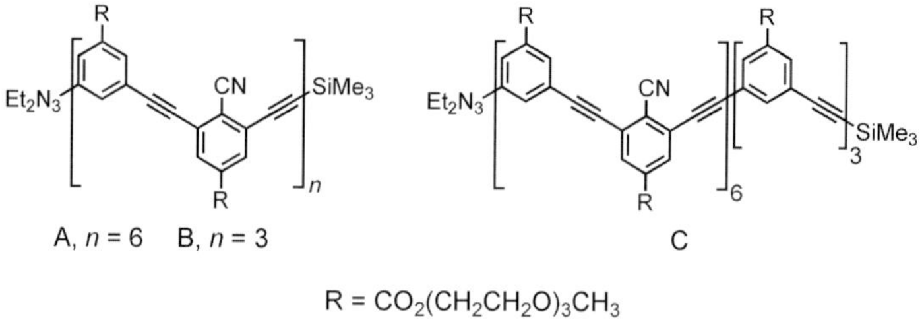

R = CO$_2$(CH$_2$CH$_2$O)$_3$CH$_3$

**Figure 11.2** *Abiotic oligomers for the examination of solvophobic and coordination interaction [11].*

binding. This oligomer contains six cyano groups located on alternating aromatic rings that are available for metal coordination (Figure 11.2, oligomer A).

In the helical conformation, this sequence places the six cyano groups into the interior of the tubular cavity, creating two trigonal planar coordination sites (Figure 11.3). The solvent of choice for metal-binding experiments was tetrahydrofuran, which does not cause a solvophobically driven helical structure in this system [11]. The metal selected was silver triflate (AgO$_3$SCF$_3$) because it can adopt a trigonal planar coordination geometry [12]. Changes in UV-vis spectra upon metal binding were indicative of a cisoid conformation of the diphenylacetylene units, consistent with a helical structure [11], which was further confirmed by $^1$H-NMR spectroscopy.

In addition, the UV titration spectra did not change after two equivalents of AgO$_3$SCF$_3$ were added, indicating that two Ag$^+$ ions were bound to each oligomer. The association constant of the overall reaction ($K_1K_2$) was estimated to be greater than $10^{12}$ M$^{-2}$. In order to further investigate the binding reaction, oligomers B and C (Figure 11.2), anticipated to bind one equivalent of AgO$_3$SCF$_3$, were synthesized and tested for metal coordination. UV-vis, $^1$H-NMR and ESI-MS spectra confirmed that only oligomer C binds to silver triflate, with an association constant of $K_1 = 2 \times 10^4$ M$^{-1}$. These results suggest that the binding of two equivalents of AgO$_3$SCF$_3$ in oligomer A is a cooperative process with $K_2 \gg K_1$ (Figure 11.3). Overall, this work demonstrates that folding is driven by a combination of solvophobic interactions that favor the helical structure and metal–ligand interactions. Hence, the oligomer can be modified to selectively bind metal ions as it templates the exterior turns of the helical structure, and consequently non-covalent interactions nucleate the formation of a central turn in the structure leading to a non-biological single-stranded helix.

While the above example involves linear oligomers, large macrocycles also can form helical complexes upon metal binding. Coordination to a metal ion forces the

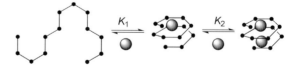

**Figure 11.3** *Representation of the metal-induced formation of helical structures as reported by the group of Moore. The metal ions (Ag$^+$) are shown as spheres [11].*

macrocycle to adopt a twisted conformation [13], in which its two halves create a double helical system. In rare cases, two diastereomeric structures of opposite helicity can be obtained for one compound by a thermodynamic inversion process. While helix inversion [14,15] between well defined and well characterized diastereomers is a biological phenomenon [16] found in natural systems, a similar process is not common in artificial systems. Following the idea that enantiomerically pure ligands will lead to metallofoldamers with a single-handed helical structure, Muller and Lisowski reported a chiral nonaazamacrocycle amine, which coordinates $Ln^{3+}$ ions to form enantiopure helical complexes (Figure 11.4) [17]. Moreover, helix inversion between the kinetic and thermodynamic binding products in the $Yb^{3+}$ complexes was also demonstrated. The nonaaza macrocycle **L** was prepared by the condensation of 2,6 diformylpyridine and trans-1,2-diaminocyclohexane [15a] leading to a $3 + 3$ macrocyclic Schiff base, which was then easily converted into the corresponding macrocyclic amine. The chiral macrocycle **L** was obtained in the enantiopure forms **L$_{RRRRRR}$** and **L$_{SSSSSS}$**, corresponding to an all-$R$ or all-$S$ configuration of the diaminocyclohexane carbon atoms, respectively [18]. Mixing ligand **L** with $Ln^{3+}$ precursors (Ln = Eu, Tb, Yb) resulted in the formation of metal complexes, as indicated by $^1H$-NMR spectroscopy, which were further isolated as enantiopure nitrate salts. The X-ray crystal structure of $(M)$-$[LnL_{RRRRRR}]^{3+}$ complexes revealed that they all adopt a unique type of geometry. Because the cavity radius of the "open" form of the ligand is too large to accommodate a single Ln ion, the macrocycle wraps around the cation in a helical fashion, leading to the generation of a left-handed M double helix.

The $^1H$-NMR spectra of the $(M)$-$[LnL_{RRRRRR}]$ complexes reflect their relatively high stability in solution. For instance, the $^1H$-NMR spectrum of a water solution of $(M)$-$[EuL_{RRRRRR}]^{3+}$ shows only traces of the $(P)$ complex after three weeks. The $(M)$-$[YbL_{RRRRRR}]^{3+}$ complex, however, is somewhat less stable because, in water, it gradually converts into the $(P)$ paramagnetic complex. After refluxing for 15 h, equilibrium is reached, with 95% conversion into the $(P)$-$[YbL_{RRRRRR}]^{3+}$ complex. The process can also be observed by CD spectroscopy, which reveals profound differences between the two forms. The inversion in helicity is explained by the notion that the less stable $(M)$-$[YbL_{RRRRRR}]^{3+}$ isomer is a kinetic product of the complexation of the free

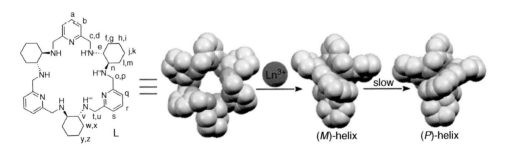

**Figure 11.4** *Schematic representation of macrocycle **L** (left), helix formation upon binding to a lanthanide ion and helix inversion, as demonstrated by the groups of Muller and Lisowski. Reprinted with permission from Ref. [17]. Copyright 2008 American Chemical Society.*

ligand (100% *ee*), while the $(P)$-$[YbL_{RRRRRR}]^{3+}$ isomer is a thermodynamic product (90% *ee*). In addition to the thermodynamic control over the absolute helicity, the inversion process is dependent also on the size of the $Ln^{3+}$ ion, for example, it is observed for solutions of the $[TbL]^{3+}$ complex, but not for $[EuL]^{3+}$.

## 11.3 Folded Oligomers in Which Metal Coordination Nucleates the Formation of an Abiotic Single-Stranded Helix

Nucleation of a stable three-dimensional structure by metal coordination requires careful design of biomimetic oligomers capable of non-covalent interactions upon metal binding. Studies of synthetic metallohelices, which use hydrogen bonding and π-stacking interactions predominantly in the formation of secondary structures in combination with metal–ligand coordination were pioneered in the mid1990s by the group of A.S. Borovik [19–21]. This group investigated the coordination of metal ions to the multidentate ligand 2,6-bis{[2-({2-acetylphenyl}-carbamoyl)phenyl]carbamoyl} pyridine, $H_2(1)$ (Figure 11.5). The ligand contains two aryl arrays that are held rigidly through hydrogen bonds and linked covalently to a pyridyl diamide tridentate chelate. In the solid state, $H_2(1)$ adopts a helical structure, which is mostly a result of the hydrogen bonds between the pyridyl nitrogen, the amide protons and the adjacent acyl oxygens. The ligand was designed such that, upon deprotonation of the pyridyl amides, a tridentate chelate is formed, enabling the binding of a metal ion via the pyridine, amides and the acyl oxygens. Due to the weak Lewis base character of these oxygen donors, it was also anticipated that their interactions with the metal ion would not depend on the geometric requirements of the ligand but rather on the stereochemical preference of the bonded metal ion. The design of this ligand was therefore aimed at answering two questions, the first being whether metal coordination could

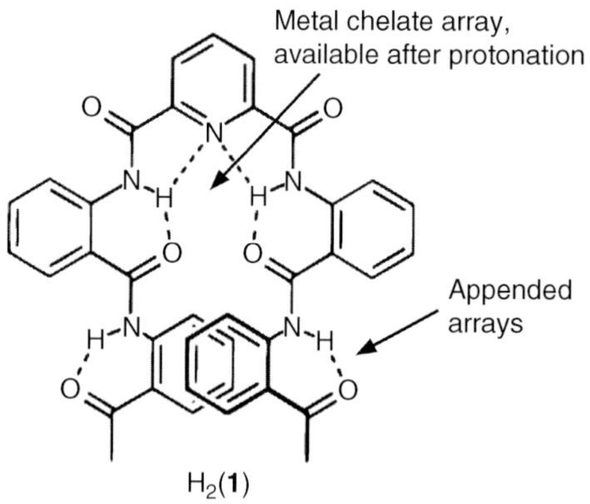

**Figure 11.5** *Structure of the multidentate ligand 2,6-bis((2-((2-acetylphenyl)-carbamoyl) phenyl)carbamoyl)pyridine, H₂(**1**). Reprinted with permission from Ref. [20] Copyright 1996 American Chemical Society.*

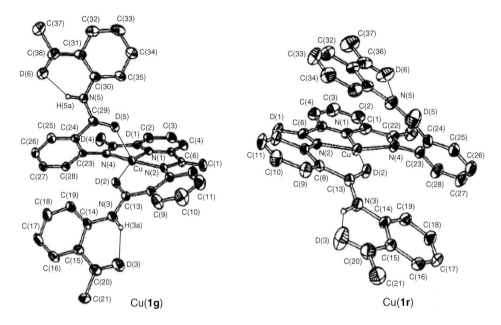

**Figure 11.6**   *X-Ray crystal structures of Cu(**1g**) (left) and Cu(**1r**) (right). The solvent molecules and non-amide hydrogens are removed for clarity. Atom key: open circles = carbon; circles with horizontal lines = nitrogen; circles with vertical lines = oxygen; and solid circles = copper. Reprinted with permission from Ref. [19]. Copyright 1996 American Chemical Society.*

nucleate a helical structure in solution and the second whether various metal ions would result in different three-dimensional architectures [19].

The combination between $Ni^{2+}$ and (**1**) [after its deprotonation to $H_2$(**1**)] afforded the complex Ni(**1**) which, according to its ¹H-NMR spectrum, exists in solution as a racemic mixture of left- and right-handed helices. The hydrogen bonds within the appended arrays are still present in the metal complex and participate in the stabilization of the metal-lohelix in solution. Subsequently, In order to assess whether coordination changes at the metal center could influence helicity, Cu(**1**) was synthesized. Crystallization of Cu(**1**) from solution afforded two structurally distinct isomers: a green complex having a distorted square pyramidal coordination geometry about the copper(II) ion, Cu(**1g**), and a tetra-coordinated red complex Cu(**1r**) (Figure 11.6). The differences in coordination numbers between these two clusters are due to the absence of the weakly coordinated amide–oxygen donor in Cu(**1r**). Both copper complexes nucleate the formation of helical structures; however, their different coordination geometries result in two distinct helices. In the case of Cu(**1g**), the additional Cu-O bond causes non-symmetrical Cu-O(amide) interactions between the arrays and the metal chelate, which results in a helical structure with a pronounced non-symmetrical twist and a microporous crystal lattice (Figure 11.7, left); the pores are formed via parallel aryl ring π-stacking between the axial appendages of the complex.

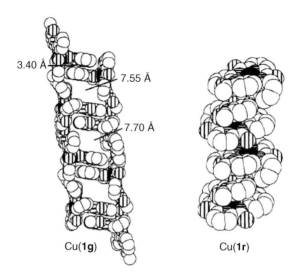

3.40 Å

7.55 Å

7.70 Å

Cu(**1g**)          Cu(**1r**)

**Figure 11.7**  *Space-filling representation of Cu(**1g**) (left) and Cu(**1r**) (right). The solvent molecules and non-amide hydrogens are removed for clarity. Reprinted with permission from Ref. [19]. Copyright 1996 American Chemical Society.*

In contrast, in the case of Cu(**1r**), the lack of one Cu-O bond leads to a realignment of the appended arrays and produces a more compact helix, which does not have a porous morphology (Figure 11.7, right); the helical molecules of Cu(**1r**) associate to form longer, extended helices in the solid state (Figure 11.3). Each extended helix consists of alternating right- and left-handed helical monomers that assemble through intermolecular π-stacking interactions. Although these results are only valid in the solid state [the structural properties of Cu(**1g**) and Cu(**1r**) are identical in solution], it is possible to conclude that metal coordination to $H_2(\mathbf{1})$ enables the nucleation of helical structures and the generation of new helices which can be altered upon variations in the metal ion precursor.

Studies on the complexes **Cu1** [including both Cu(**1g**) and Cu(**1r**)], as described above, suggested that the presence or lack of Cu-O amide bonds is a highly important factor for determining the final structure of the produced helices. To further investigate possibilities in which the helical structure could be modified, Borovik's group was interested in finding ways to control the formation of these Cu-O bonds. Thus, **Cu1** was treated with two different ligands, pyridine and lutidine, which are basic enough to substitute the oxygen amide groups [20]. Changes in the spectroscopic properties of **Cu1** upon binding to either pyridine, forming **Cu1**(py), or lutidine, forming **Cu1**(lut), implied that the coordination environment about the $Cu^{2+}$ centers has been altered. An X-ray diffraction study on **Cu1**(py) confirms that the binding of a single pyridine to copper causes a significant structural rearrangement. The two oxygen amide donors that were coordinated originally to the copper in **Cu1** are rotated away from the copper in **Cu1**(py) and no longer interact with the metal ion (Figure 11.8, left). Consequently, a chiral cleft about the exogenous pyridine is formed by the appended groups in $H_2(\mathbf{1})^{2-}$. An interesting characteristic of the **Cu1**(py) crystal structure is that all the complexes within the lattice have the same helical chirality.

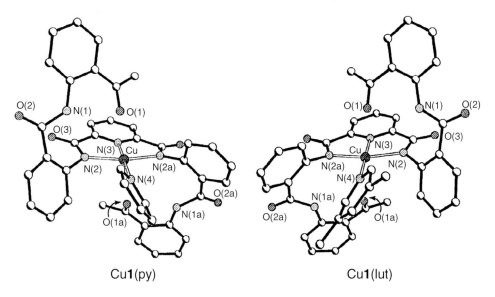

Cu1(py)                    Cu1(lut)

**Figure 11.8** *X-Ray crystal structures Cu1(py) (left) and Cu1(lut) (right). The solvent molecules and non-amide hydrogens are removed for clarity. Reprinted with permission from Ref. [20]. Copyright 1996 American Chemical Society.*

This is highly unexpected because $H_2(1)$ is achiral in solution. One possible explanation for the helicity of the crystals could be that spontaneous resolution occurs as individual crystals form in solution. Thus, a crystal of the other handedness has been isolated from a solution of Cu1(lut). It was characterized and found similar to that described above for Cu1(py), the major difference being the opposite helicity (Figure 11.8, right). This selective crystallization is dictated by the lattice architecture of the clusters in the solid state; stable crystals can be formed only if the helices within an array are perfectly aligned. This stabilization is only possible if individual metallohelices are of the same helicity. These results demonstrate how weak intra- and intermolecular interactions can be used to assemble chiral metallofoldamers.

The structural properties observed for Cu1(py) and Cu1(lut) appear to be present only in the crystalline state. Preliminary electronic absorbance and circular dichroism results suggest that the complexes racemize upon redissolution at room temperature. In order to overcome this problem, the Brovik group has synthesized derivatives of $H_2(1)$ in which the acetyl phenyl groups on the appended arms are replaced by enantiomerically pure groups, as depicted in Figure 11.9 [21]. The spectroscopic data obtained from the $^1$H-NMR spectra of all the new ligands, namely $H_2(2a)$, $H_2(2b)$ and $H_2(2c)$ indicate that the molecules maintain their hydrogen bonds in solution. These hydrogen-bonding interactions decrease the freedom of rotation of the two appended groups, resulting in a fairly rigid molecular structure. Moreover, the solid-state structure of crystalline $H_2(2a)$, as determined by X-ray diffraction methods, confirmed that, similar to $H_2(1)$, $H_2(2a)$ has a helical structure which is stabilized by networks of intramolecular hydrogen bonds.

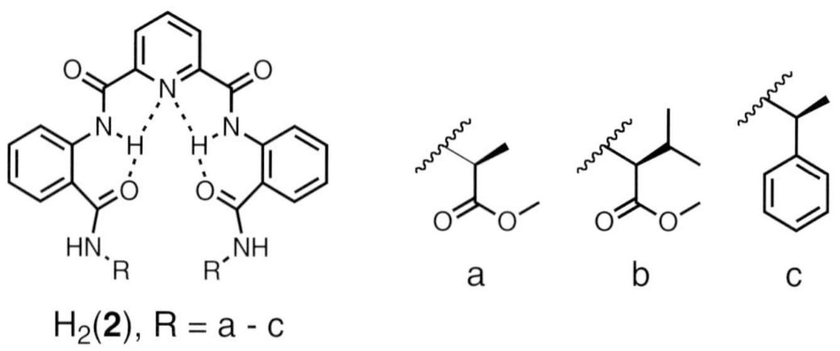

**Figure 11.9** *Structure of chiral 2,6-bis[(2-carbamoylphenyl) carbamoyl]pyridine ligands, H₂(2a-c) [21].*

Deprotonation of the pyridyl amides facilitated the ability of the $H_2(2)$ ligands to bind metal ions. The solid-state structures of these metal complexes was examined by an X-ray diffraction study conducted on $H_2(2b)$. This study revealed that a nickel complex is crystallized with two independent molecules, Ni(**2b′**) and Ni(**2b″**), in the asymmetric unit cell. Preliminary NMR studies suggest that the asymmetric arrangement of the appended arms found in the solid-state structure of Ni(**2b**) can be maintained in solution. Unfortunately, the Borovik group did not report chiral metal-helical compounds that maintain their structure in solution. Moreover, the molecules depicted in Figure 11.9 are capable of forming helical structures in solution as a result of their hydrogen-bonding network; hence, folding does not occur specifically upon metal binding. Metal coordination, as demonstrated in the above examples, can nucleate a helical structure but does not have a significant role in generating a new helix in solution.

The challenging task of nucleating a new helical structure was implemented by the group of Fox [6] using abiotic molecules based on salophen and salen ligands. Although these molecules do not adopt helical structures in the absence of metal ions, they were shown to fold into single-stranded helices in the presence of $Ni^{2+}$ or $Cu^{2+}$ (Figure 11.10).

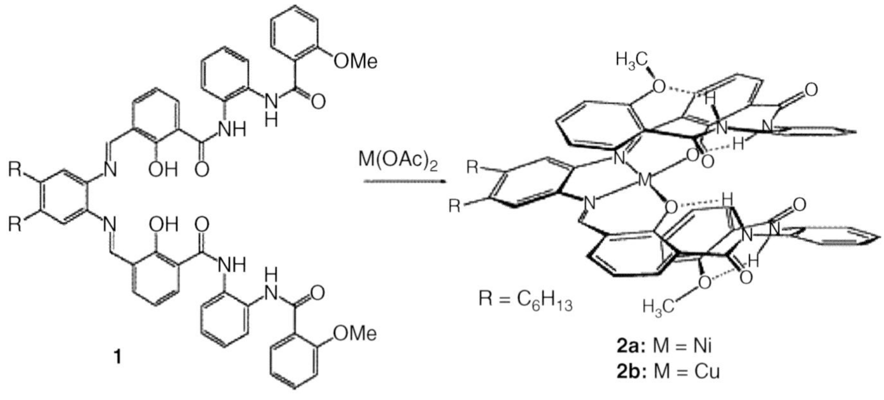

**Figure 11.10** *Abiotic oligomers for the examination of hydrogen bonding and coordination interaction. Reprinted with permission from Ref. [6]. Copyright 2005 American Chemical Society.*

The design of these ligands for the formation of single-stranded helical foldamers relies on the hypothesis that a combination between square planar metal complexes and a series of aromatic π-stacking interactions may result in a stable helix [22]. Ligand **1** was obtained in a five-step synthesis [23,24], with the advantage that the building block compounds can be prepared in multigram quantities and that the amide bond-forming reactions are straightforward. Metal complexes with $Ni^{2+}$ and $Cu^{2+}$ were prepared in good yields by mixing **1** with metal acetate precursors. These materials form helical structures in the solid state, as shown by X-ray diffraction studies, and are conserved in solution, as evident from NMR spectroscopy.

Circular dichroism (CD) analysis and optical rotation measurements revealed that the resolved crystalline metal complexes **2a** and **2b** (Figure 11.10) racemize quickly when dissolved at 5 °C. This observation implies that the secondary structure can reorganize easily and could potentially be used as a template for responsive materials. The secondary structure can also be altered as a consequence of an induced change in the coordination environment of the metallofoldamers. Indeed, electrochemical experiments have shown that structural reorganization occurs upon metal-centered reduction of $Cu^{2+}$-containing foldamers. When the reduction is carried out in the presence of coordinating ligands, it is proposed that apical binding of those ligands gives square pyramidal complexes. Semi-empirical (AM1) calculations support the idea that the helical structure can be disrupted by the reduction of $Cu^{2+}$ to $Cu^{+}$ with concomitant reorganization to a square pyramidal complex. This work is the first example of abiotic materials in which the metal coordination sphere is not inherently chiral but instead causes a series of cooperative, non-covalent interactions that ultimately result in a folded structure. The helical structure is therefore induced by metal coordination, which is required for folding and further reinforced by aromatic π-stacking interactions.

An alternative way to control the chiral environment of a metallofoldamer is via secondary sphere chirality [25], through which remote stereo-centers control the asymmetric environment about the metal [26]. In a subsequent study by the same laboratory [27], an enantiomerically pure oligomeric ligand **3** (Figure 11.11) was synthesized with the anticipation that it would fold into a discrete conformation upon metal binding, as the methyl groups positioned on the peripheral benzofuran rings would point to the outside of the helix [28]. A $Ni^{2+}$ complex was prepared accordingly, crystallized and analyzed by X-ray diffraction, showing that the helical structure is indeed controlled by the stereocenters at the periphery, however, with the peripheral carbonyls pointing to the interior of the helix. Solution NMR experiments at several temperatures (23 to −40 °C) revealed the existence of two species that undergo chemical exchange, with a barrier of $\sim 13\,kcal\,mol^{-1}$.

Further analysis by CD indicated that the complex is a mixture of helices **4(P)** and **4(M)** with opposite handedness (indicated by the spectrum of **4**, Figure 11.11).

In attempts to control the absolute helicity of the metallofoldamer, a different salophen (oligomeric ligand **6**, Figure 11.11) was designed to stabilize the (*M*)-helix by a three-center hydrogen bond [29] while destabilizing the (*P*)-helix through steric interference of the amide carbonyl by additional ester functions. Accordingly, ligand **6** was synthesized and mixed with $Ni(OAc)_2$ to obtain the corresponding Ni-complex (oligomer **5**, Figure 11.11). As predicted, X-ray diffraction analysis determines the folding of crystalline **5**

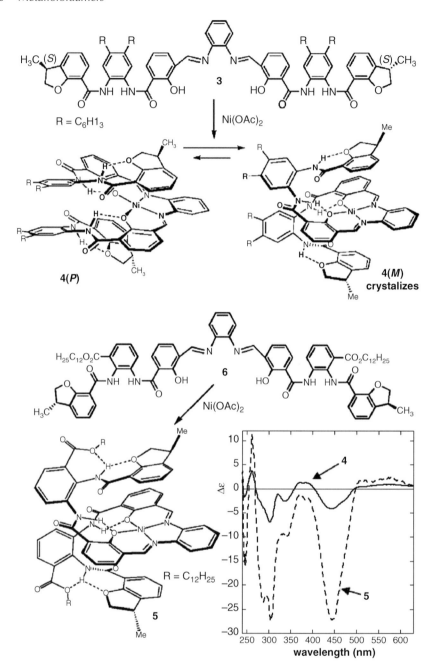

**Figure 11.11** *Synthesis and chiroptical properties of chiral metallofoldamers, as reported by the group of Fox. Reprinted with permission from Ref. [27]. Copyright 2006 American Chemical Society.*

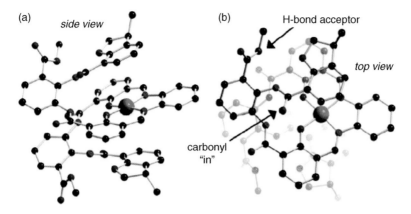

(a) side view

(b) H-bond acceptor

top view

carbonyl "in"

**Figure 11.12** *Molecular structures of a chiral foldamer from crystallographic coordinates. Reprinted with permission from Ref. [27]. Copyright 2006 American Chemical Society.*

into the "carbonyl-in" (*M*)-helix conformation (Figure 11.12). Evidence that the solution structure of **5** is helical was provided by CD and $^1$H-NMR measurements. Based on the results, the authors point out that either **5(***M***)** is the major conformation, or that the undetermined conformer also has *M*-helicity. Overall, this work demonstrates that control over the absolute sense of helicity can be effectively achieved by a combination of hydrogen bonding and steric interference arising from peripheral stereo-centers in a metal coordination derived foldamer.

The role of peripheral stereo-centers – chiral end groups – as directors of absolute helicity in salen- and salophen-based metallofoldamers was further investigated by the Fox group [30–32]. Initial studies showed that the chiral end group in metallofoldamers **4** and **5** are, in fact, the components responsible for the foldamer's absolute helicity. Even when *trans*-cyclohexane-1,2-diamine (TCDA), which is a common element of central chirality in salen catalysts, was embedded within oligomer **6**, the absolute helicity of its nickel complex was directed by the chirality of the end groups rather than the chirality of the central diamine group. This was proven by using an oligomer that contains both the TCDA group with an *S, S* chirality [bias for a (*P*)-helix] and end groups that are bias for a (*M*)-helix (oligomer **6TCDA**, Figure 11.13a) and demonstrating that the nickel complex **5TCDA** is a (*M*)-helix (Figure 11.13b) [30,31]. Moreover, the Fox group advocates that the ability of end groups to enforce absolute helicity does not necessarily require the incorporation of large and bulky groups; and they demonstrated that small chiral end groups could also control the absolute sense of helicity in salen and salophen metallofoldamers [32]. In their study, chiral 1-methylindan compounds were used as end groups and the new oligomers produced chiral metallofoldamers upon coordination to nickel ions. Spectroscopic studies in solution and NMR data, as well as an X-ray diffraction analysis performed on solid crystals of the metallofoldamers, revealed that these complexes adopt a helical structure with an absolute sense of helicity, which is dominated by the stereo-centers at the periphery [32]. Therefore, it can be concluded that the presence of chiral end groups, large or small, in metallofoldamers could play a crucial role in controlling their absolute helicity.

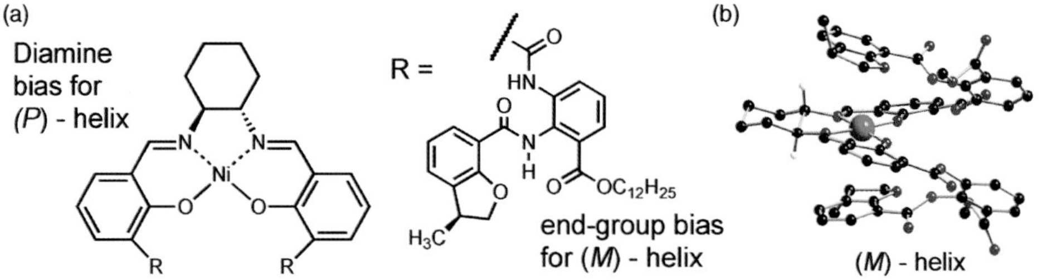

(a)

Diamine
bias for
*(P)* - helix

R =

end-group bias
for *(M)* - helix

H₃C

(b)

*(M)* - helix

**Figure 11.13** *Molecular structures of (a) **6TCDA** with two different chirality centers; the diamine and the end-group and (b) the metallofoldamer **5TCDA** from crystallographic coordinates. Reprinted with permission from Ref. [30]. Copyright 2007 American Chemical Society.*

Followed this work, the Fox group was interested to investigate other features that may influence the absolute helicity of salen metallofoldamers. One such factor is the nature of the internal chiral diamine. As discussed above and concluded from the work presented earlier in this chapter, *trans*-cyclohexane-1,2-diamine is only a weak director of absolute helicity in Ni-salen foldamers [30]. As *trans*-cyclohexane-1,2-diamine was the only chiral diamine studied, it was difficult to deduce a general conclusion regarding the influence of the diamine group on the helicity of metallofoldamers. Therefore, the Fox group wished to understand how the structure of different chiral diamines would bias absolute helicity in salen-based foldamers [33]. To this aim, three different chiral diamines, namely (1R, 2R)-cyclopentanediamine (**CP-DA**), (1S, 2S)-1,2-diphenylethylenediamine (**DPE-DA**) and (11R, 12R)-9,10-dihydro-9,10-ethanoanthracene-11,12 diamine (**DEA-DA**), were used to prepare new salen oligomers (Figure 11.14), and the properties of the Ni-salen foldamers

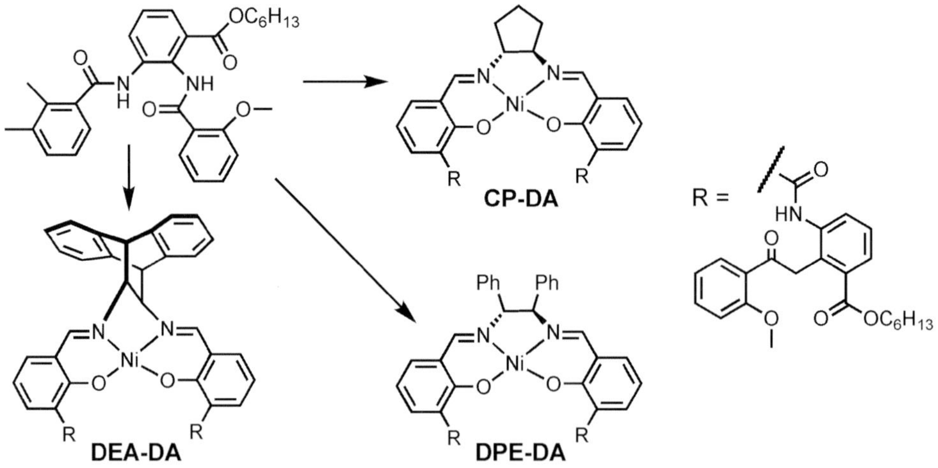

**CP-DA**

R =

**DEA-DA**

**DPE-DA**

**Figure 11.14** *Synthesis of Ni-salen foldamers from (1R, 2R)-cyclopentanediamine (**CP-DA**), (1S, 2S)-1,2-diphenylethylenediamine (**DPE-DA**) and (11R, 12R)-9,10-dihydro-9,10-ethanoanthracene-11,12 diamine (**DEA-DA**). These foldamers have achiral end groups and the diamines are the only source of central chirality [33].*

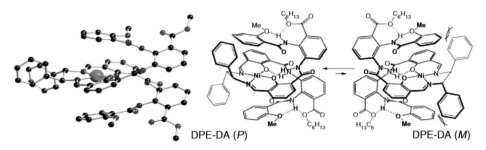

DPE-DA (*P*)                      DPE-DA (*M*)

**Figure 11.15** *X-Ray crystal structure of* **DPE-DA** *(left) and the rational for its preference for the (P)-helical conformer (right). [33] Reproduced by permission of The Royal Society of Chemistry.*

derived from those oligomers were compared both in solution and in the solid state. NMR studies in solution indicated that the metallofoldamers have substantially higher field chemical shifts relative to the salicylaldehyde starting material. These upfield chemical shifts are consistent with a helical structure in solution. CD spectra of the three metallofoldamers were also obtained and the intensity of their peaks was evaluated and compared to the CD peak intensity of the metallofoldamer **5TCDA**, which has positive dichroisms at 422 and 445 nm. It was found that the CD spectra of **CP-DA** and **DEA-DA** were very weak and there was almost no bias towards one helical conformation. In contrast, **DPE-DA** affords a CD spectrum which resembles that of **5TCDA**, with negative dichroisms at 428 and 456 nm and only 25% less intensity. This CD spectrum suggested that **DPE-DA** is composed mainly of a (*P*)-helical conformation, an assumption that was further confirmed by the crystal structure of **DPE-DA** that was determined and identified as the (*P*)-helix [33]. Subsequently, the salen metallofoldamers were studied by variable temperature NMR in order to assess the conformational ratios between the (*P*)- and (*M*)-helices. According to the obtained data, it was concluded that **DPE-DA** shows a large preference (6 : 1) for the (*P*)-helix, while **CP-DA** shows only a slight preference (2 : 1) for the (*M*)-helix and for **DEA-DA** the helical diastereomers are present in similar amounts. In order to rationalize the relatively large bias for the (*P*)-helical conformation in **DPE-DA** the authors propose that the (*M*)-helical conformation of **DPE-DA** suffers a steric interaction between the terminal anisoyl groups and the phenyl substituents from the diamine. These steric interactions are relieved in the (*P*)-helical conformation (Figure 11.15). Overall these studies illustrate how different internal chiral diamines affect the equilibrium between helical diastereomers in salen-based foldamers.

## 11.4 Folded Oligomers in Which Metal Coordination Enhances Secondary Structure and Leads to Higher-Order Architectures

One of the long-term goals in the development of functional folded materials is the creation of stable structures with protein-like properties. Despite recent advances in the stabilization of secondary structures upon metal coordination, the design of a sequence that can fold into a well defined tertiary structure in solution is still challenging. The focus of ongoing studies aiming at the generation of folded architectures by metal coordination is threefold: (i) enhanced stabilization of an existing secondary structure (e.g., helix),

(ii) generation of higher order architectures such as a two-helix bundle or (iii) both. Illustrative examples of both approaches are discussed in the following sections.

### 11.4.1 Metal Coordination in Folded Aromatic Amide Oligomers

Recently, hydrogen-bonding interactions have been used in combination with aromatic π-stacking interactions to induce folding in oligomer systems [1]. A common motif throughout this research is the use of aryl amides [3,34]. These foldamers form stable secondary structures in solution and solid state, therefore the incorporation of metal ions is anticipated to further stabilize and/or generate higher-ordered structures. The group of Parquette studied the impact of metal coordination on the conformational properties of dendrimers using small foldamers [35]. Their first efforts were to coordinate multiple $Cu^{2+}$ centers to pyridine-2,6-dicarboxamide dendrons that adopt compact helical conformations due to the syn-syn conformational preference of the pyridine-2,6-dicarboxamide repeat unit [36]. In these systems, the helical antipodes experience a highly dynamic equilibrium that interconverts the *M* and the *P* conformations rapidly with kinetic barriers too small to be measured by NMR. Coordination of $Cu^{2+}$, however, produced a kinetically stable, non-dynamic conformational state at room temperature. In a more recent work [37], this group reported the structural consequences of coordinating a different foldamer (2,6-bis{2-[(4S)-4,5-dihydro-1,3-oxazol-2-yl]phenyl}carbamoylpyridines, **7**) with divalent metal ions such as $Cu^{2+}$, $Ni^{2+}$ and $Zn^{2+}$ (Figure 11.16). As shown by X-ray crystallography

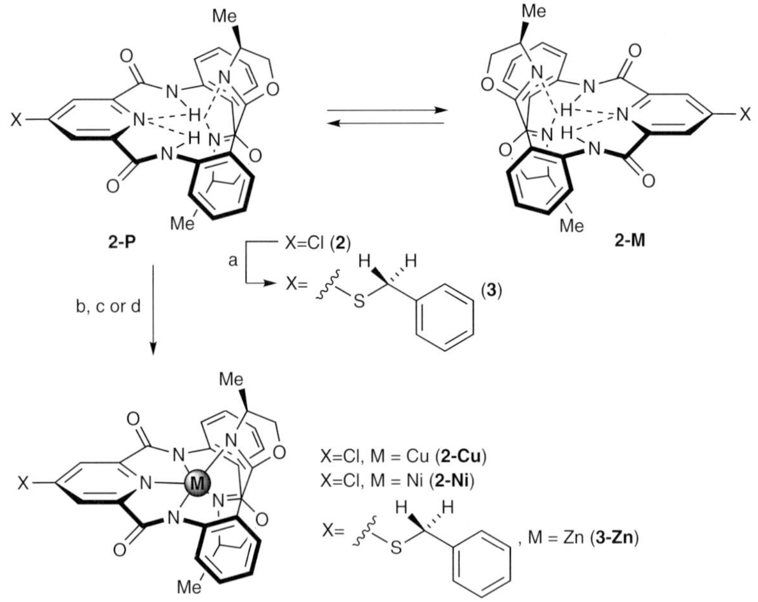

**Figure 11.16** *Synthesis of aromatic aryl amide foldamers and their metal complexation (reported by the Parquette lab.). (a) NaH, PhCH₂SH, THF, 72%. (b) Ni(OAc)₂, CH₂Cl₂-MeOH (9 : 1), (C₂H₅)₃N, 69% (for 7-Cu); (c) Cu(OTf)₂, CH₂Cl₂-MeOH (9 : 1), (C₂H₅)₃N, 87% (for 7-Ni); (d) (C₂H₅)Zn, toluene, (for 8-Zn). Reprinted with permission from Ref. [37]. Copyright 2006 American Chemical Society.*

analysis, a *P*-helical conformational preference was exhibited by all three of the complexes, identical to the preference observed for the parent (metal-free) dendron. Moreover, [1]H-NMR peak analysis indicated that metal coordination increases the helical inter-conversion barrier and thus the dynamic helicity of the dendron, making it conformationally "locked".

Nitschke and Huc reported the use of metal complexes as dynamic connection elements between oligomeric helical segments [38]. Specifically, $Cu^+$ and $Fe^{2+}$ complexes were applied to link and define the relative orientation of two helices mimicking a turn structure in proteins, but at an unconventional angle. This work focused on the aromatic oligoamides of 8-amino-2-quinoline carboxylic acid, which adopt particularly stable helical conformations in the solid state and in a wide variety of solvents [39]. Amine-functionalized tetramer **9** (Figure 11.17) adopts a helical conformation that spans over one and a half turns. The reaction of **9** with 6-methyl-2-formylpyridine and $CuBF_4$ produced a pseudotetrahedral $Cu^+$ complex **10**, as characterized by mass spectrometry, [1]H-NMR spectroscopy and X-ray crystallography. The conformation of **10** features several intrinsically chiral elements that are all expected to undergo dynamic exchange: the right (*P*) or left (*M*) handedness of the two helical segments and the $\Lambda$ or $\Delta$ configuration [40] of the metal complex. [1]H-NMR spectroscopy indicates a high degree of influence on the handedness of each helix by the configuration of the neighboring metal. Crystallographic investigations allowed the characterization of four out of the six possible forms of **10**.

As expected, the two 2-iminopyridine moieties formed a tetrahedral complex with $Cu^+$. This geometry dictated an unusual, perpendicular, orientation between the two helices (Figure 11.18). The unconventional 90° angle between two helices in **10** constitutes a unique motif that hints at the prospect of assembling large square structures comprised of helical oligomer "edges" bearing amine functions at both extremities linked by metal complexes at each "corner". Mixing **1** with 2-formylpyridine and $Fe(BF_4)_2$ in acetonitrile resulted in a mixture of compounds, but only one gave crystallography suitable crystals. X-Ray diffraction analysis provided the structure of a racemic *M*$\Delta$*M*/*P*$\Lambda$*P* $Fe^{2+}$ complex

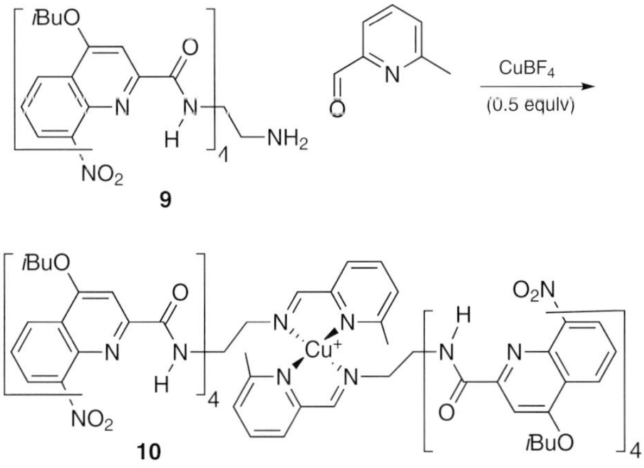

**Figure 11.17** *Equilibrium between **9** and tetrahedral $Cu^+$ complex **10**. Reprinted with permission from Ref. [38]. Copyright 2008 WILEY-VCH Verlag GmbH & Co. KGaA, Weinheim.*

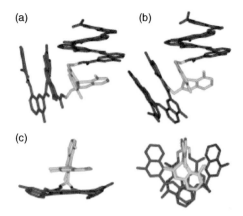

**Figure 11.18** *Crystal structures showing (a) the PΔM **10** and (b) MΛM **10**. (c) Top and side views of the overlay of fragments of the above complexes showing two Δ Cu$^{+1}$ complexes (in gray) and the first two quinoline residues of an M helix (in red) and of a P helix (in blue), as reported by the groups of Nitschke and Huc. Side chains, BF$_4^-$ ions, and included solvent molecules are omitted for clarity. Reprinted with permission from Ref. [40] (Nitschke's and Huc's groups). Copyright 2008 WILEY-VCH Verlag GmbH & Co. KGaA, Weinheim.*

bearing only two helix–iminopyridine ligands and two hydroxide counter ions bound directly to the metal center. In this case, the relative orientation of the two helices imparted by the octahedral geometry of Fe$^{2+}$ is almost parallel. The two hydroxide ligands that connected to Fe$^{2+}$ appear to play a role in this orientation as they seem to prevent the helices from folding back on the iminopyridine moieties. The structure of the complex formed upon binding to an Fe$^{2+}$ ion further validates metal-directed dynamic assembly as an efficient approach to generate helically folded, aromatic–amide oligomers and to precisely set their relative orientation.

### 11.4.2 Metal Coordination in Peptidomimetic Foldamers

Peptidomimetics are synthetic oligomers akin to natural peptides and polypeptides, which were designed to adopt three-dimensional structures by a controlled disposition of functional groups along their spine. Examples of metallo-peptidomimetics involve two types of oligomers: β-peptides and "peptoids" – a class of α-peptides. The secondary structure of both types has been well characterized [41], therefore metal coordination is anticipated to induce further stabilization and/or create higher-ordered structures. One advantage of peptidomimetics is their facile preparation achieved by efficient solid-phase synthesis [42]. This method can be automated, enabling the generation of several compounds in parallel. Moreover, it allows ambient temperatures, (typically 25 °C), faster synthesis and fewer purification steps.

β-Peptides are well characterized non-natural oligomers, which were developed within the last decade [41a]. Systematic studies on β-peptides were conducted in solution and in the solid state, revealing their potential to adopt well defined ordered conformations [41b,43]. Recently, Seebach [44] investigated the ability of Zn$^{2+}$ complexation to fortify and enforce β-peptide secondary structure towards the generation of artificial β-peptidic zinc fingers, a mimic of a natural motif found in proteins [45]. To this aim, a β-decapeptide, four β-octapeptides and a β-hexadecapeptide were designed and synthesized. For the first five β-peptides, the design was such that the peptides would: (i) fold to a *14*-helix (a helical secondary structure defined by 14-membered ring hydrogen bonds [43c]), a hairpin turn or neither and (ii) incorporate the cysteine and histidine side chains in strategic positions to allow the binding of $Zn^{2+}$ in order to stabilize or destabilize the intrinsic secondary structure of the peptide. The β-hexadecapeptide was designed to: (i) fold into a turn to which a *14*-helix is attached through a β-dipeptide spacer and (ii) contain two cysteine and two histidine side chains for $Zn^{2+}$ complexation in order to mimic a Zn finger motif. β-Peptides were generated by manual solid phase synthesis [46]. After cleavage from the resin, the β-peptides were purified by preparative HPLC and characterized by analytical HPLC, MS, NMR and CD measurements. Dramatic changes of the CD pattern take place when $ZnCl_2$ is added to aqueous solutions of the β-peptides, buffered at pH > 7 ($pK_a$ of histidine = 6.04). Although the CD spectra demonstrate that there are interactions between the β-peptides and $Zn^{2+}$ ions in solution, they do not provide any structural information for further qualitative assignments. Some CD spectra suggested the formation of 1:1 complexes, an observation that was further confirmed by electrospray mass spectrometry. $^1$H-NMR analysis in the absence or presence of $ZnCl_2$ indicates that the β-peptide, which is present as a *14*-helix in methanol, is forced into a hairpin-turn structure by Zn binding in water. In addition, the β-peptide with cystein and histidine residues positioned far apart from each other adopts a distorted turn structure in the presence of $Zn^{2+}$.

An alternative approach for mimicking natural metal-binding motifs is the use of "peptoids" – N-substituted glycine oligomers. Peptoids have emerged as intriguing mimics of polypeptides [47], particularly with respect to their ability to form well defined folded architectures [41b]. Moreover, many peptoid sequences exhibit a remarkable propensity for folding even at small oligomer chain lengths [48]. Peptoid oligomers can be synthesized efficiently by solid-phase methods, allowing the introduction of a variety of side chains (Figure 11.19) [42], therefore enabling the coordinated display of multiple chemical functionalities which can potentially emulate the active sites of proteins (e.g., metal-binding sites).

The group of Zuckermann described the introduction of a high affinity zinc binding function into a peptoid, which consists of two helices bound together by a short peptide coil, and demonstrated its folding into a two-helix bundle upon binding with a zinc ion (Figure 11.20) [49]. Each helix was designed to contain a bulky chiral side chain in two-thirds of the monomer positions, since these side chains are known to enforce helicity [48a]. The side chains that were used – (S)-N-(1-phenylethyl)glycine (Nspe), and (S)-N-

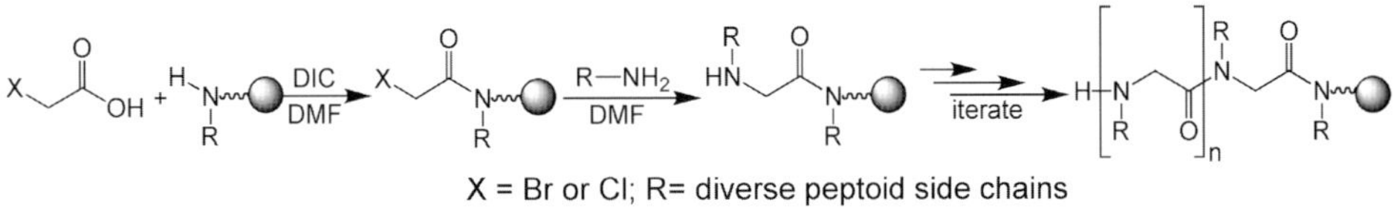

X = Br or Cl; R= diverse peptoid side chains

**Figure 11.19**  *Two-step solid-phase synthesis of peptoids.*

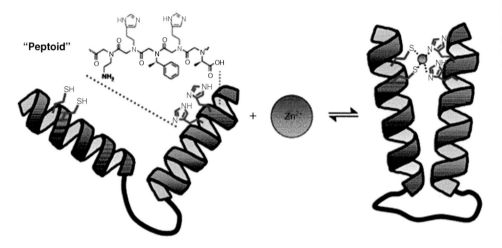

**Figure 11.20** *Schematic representation of peptoid two-helix bundle forming upon Zn²⁺ binding. Reprinted with permission from Ref. [49]. Copyright 2008 American Chemical Society.*

(1-carboxylethyl)glycine (Nsce) – impose steric hindrance on the backbone, generating a polyproline type I-like helix with three residues per turn. The chosen motif for the loop region was Gly-Pro-Gly-Gly, which has a propensity to form type II β-turns in proteins [50]. Borrowing from well understood zinc-binding motif $Cys_2His_2$ [45a,b], thiol and imidazole moieties were positioned within the peptoid such that both helices must align in close proximity to form a binding site. In order to measure the change of the distance between the two helical segments and to probe the binding of zinc, fluorescence resonance energy transfer (FRET) reporter groups were used [51]. Thus, a fluorescence donor (anthranilamide) was incorporated at one end of the peptoid and a quencher (nitrophenol) at the other end. FRET efficiencies were used to determine the folding of the peptoids and the affinity and selectivity of the zinc-binding peptoids. The first observation was that zinc binding increases the FRET efficiency in acetonitrile, indicating that zinc stabilizes the two-helix bundle by holding the two helical segments together. Moreover, in mixtures of two different peptoids, there was no change in the intermolecular FRET efficiencies upon the addition of zinc, proving that the two helix bundles are stabilized without inducing self-association. CD spectroscopy revealed no significant changes in the far-UV region upon zinc binding, indicating that the secondary structure of the individual peptoid helixes is unchanged. In addition, the position and number of zinc-binding residues, as well as the sequence and size of the loop that connects the two helices were systematically varied, followed by FRET efficiency measurements. The results suggest, for example, that a higher zinc-binding affinity was due to a longer loop region in the peptoid, probably because longer flexible linkers accommodate optimal zinc-coordination geometry. In addition, among various divalent metal ions (e.g., $Ca^{2+}$, $Mg^{2+}$, $Cu^{2+}$) the most significant change in the FRET efficiency was achieved by $Zn^{2+}$, with a binding affinity an order of magnitude higher than the other metal ions, a kind of selectivity that is also found in biological systems. This work is a prominent example of a tertiary structure formation governed by the docking of pre-organized helices.

Maayan *et al.* designed and synthesized helical peptoids bearing one or two multidentate ligands. Upon metal coordination, an enhancement of the secondary structure was demonstrated, either by stabilization of an existing helix or by the formation of a helical duplex. Moreover, the helical secondary structure environment induces chirality about the metal center and enforces the creation of a chiral metal complex from ligands which are not inherently chiral [52].

The introduction of multidentate metal-binding ligands as pendant groups in non-helical peptoid sequences was first demonstrated and presented in an earlier study by the same group [53]. For example, 8-hydroxy-2-quinolinemethylamine was prepared from the commercially available 8-hydroxy-2-quinolinecarbonitrile by a one-step hydrogenation procedure and incorporated within different peptoids without any need for protection of the hydroxyl group. In order to explore the influence of metal binding on the conformation of chiral helical peptoids, hydroxyquinoline ligands were incorporated to their scaffolds. These peptoid ligands were expected to bind divalent metal ions, producing tetra-coordinated metal species [54]. Specifically, $H_15$ and $H_26$ (Figure 11.21) were synthesized as model systems for a comparison of inter- and intramolecular metal complex formation, respectively.

The pentamer $H_15$, with one hydroxyquinoline site at the N-terminus, was expected to form a peptoid duplex upon metal binding (2 : 1 peptoid:metal). The hexamer $H_26$ contains two hydroxyquinoline ligands, endowing this oligomer with the capacity to form an

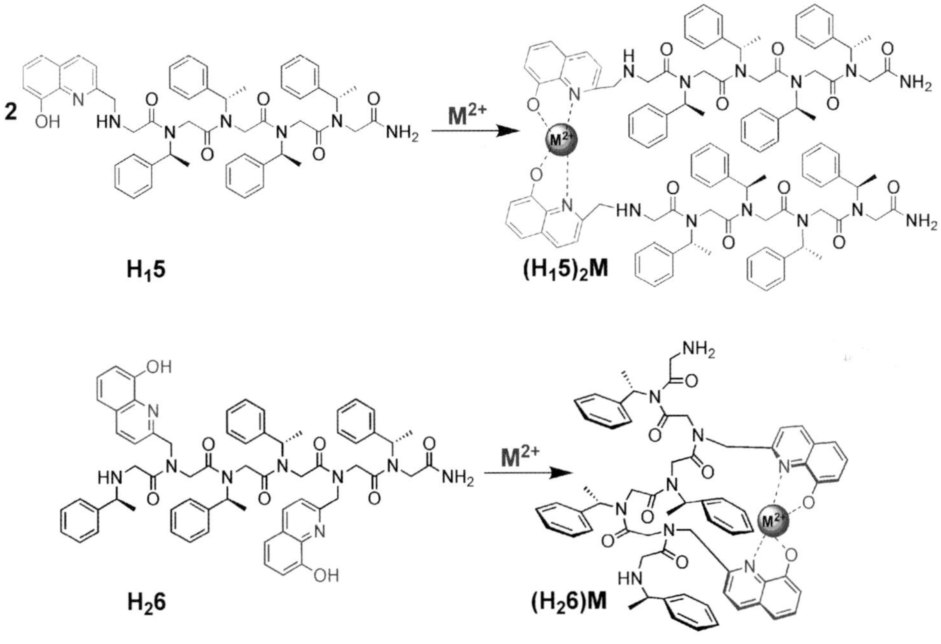

***Figure 11.21*** *Schematic representations of the peptoids $H_15$ and $H_26$ and their metal complexes [52].*

intramolecular 1 : 1 peptoid:metal complex (Figure 11.21). Positioning the ligands at $i$ and $i + 3$ in the sequence matches the pitch of the helix and is designed to orient these groups in proximity on the same face of the scaffold, separated by one helical turn.

Oligomers $H_1 5$ and $H_2 6$ were synthesized in good yields [52], and their ability to bind metal ions was evaluated by UV-vis spectroscopy (Figure 11.22, a and c). Job plots constructed from UV titration of the peptoids with either $Cu^{2+}$ or $Co^{2+}$ were consistent with a 2 : 1 $(H_1 5)_2 M$ (M= $Cu^{2+}$ or $Co^{2+}$) duplex and a 1 : 1 $(H_2 6)M$ intramolecular complex. The peptoid:metal ratio was corroborated further by mass spectrometry analysis. CD measurements revealed substantial changes upon metal complex formation (Figure 11.22, b and d). Solutions of the metal complexes exhibited increases in the magnitude of the CD signals near 200 and 220 nm, relative to the metal-free peptoids. In the case of $(H_1 5)_2 M$ this suggests an increase in conformational order, as the magnitude of the signal reflects the degree of helicity. The increase in the CD signal was more dramatic for $(H_2 6)M$, consistent with greater conformational constraint and enhanced secondary structure content due to

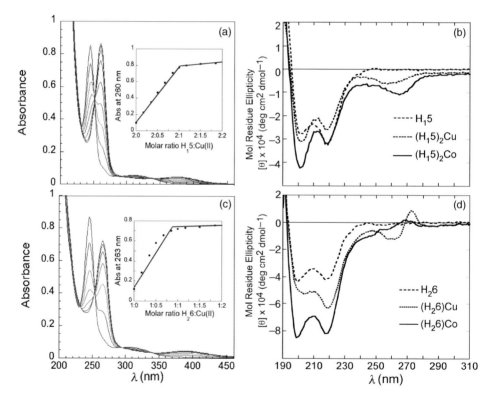

**Figure 11.22** *UV-vis spectra and Job plot for titration of (a) $H_1 5$ with $Cu^{2+}$ and (c) $H_2 6$ with $Cu^{2+}$. CD spectra for (B) $H_1 5$, $(H_1 5)_2 Cu$ and $(H_1 5)_2 Co$. CD spectra for (d) $H_2 6$, $(H_2 6)Cu$, and $(H_2 6)Co$ UV-vis spectra. (a) 54 $\mu$M peptoid in MeOH:$H_2 O$ (4 : 1) solution, (c) 40 $\mu$M peptoid in MeOH:$H_2 O$ (4 : 1) solution. Blue = free ligand, red= metal complex. CD spectra: 100 $\mu$M MeOH:$H_2 O$ (4 : 1) solutions [52].*

intramolecular metal complexation. Metal binding to $H_1 5$ and $H_2 6$ also produced new CD peaks between 240 and 280 nm, the region corresponding to the 8-hydroxy-quinoline $\pi-\pi*$ transition, which reflects the transmission of the stereogenic charac-ter of the peptoid scaffold to the metal center. These results indicate the reciprocating effects of metal binding – the chirality of the peptoid backbone estab-lishes an asymmetric environment about the metal center while metal complexation enhances the helical character of the backbone [55]. This synergistic interaction between helices and metal complexes, which has not been observed before in artifi-cial folded oligomers, holds potential for applications in asymmetric catalysis and material science.

The last example of metallopeptoids involves N-benzyloxyethyl cyclic oligomers of various sizes that bind alkali metal ions, a study conducted in the groups of Izzo and De Riccardis [56]. The synthesis of the linear N-benzloxyethyl glycine oligomers was accomplished both in solution [57] and through solid-phase methods. Head to tail macro-cyclizations of the linear compounds were achieved in the presence of different condens-ing agents, producing three cyclic peptides – trimer, tetramer and hexamer (Figure 11.23). The peptoids were characterized in solution by NMR spectroscopy and in the solid state by X-ray crystallography. The spectroscopic data of the cyclic trimer revealed a $C_3$-sym-metric all-*cis* "crown" conformation. In the case of the cyclic tetramer, a single crystal X-ray analysis demonstrated a *ctct* "chair" tetralactam core geometry [58]. The cyclic hexamer, in contrast, showed conformational disorder in solution, an observation that prompted metal complexation studies with this peptoid.

Indeed, stepwise addition of sodium picrate induced the formation of a new chemical species with a remarkably simplified NMR spectrum, suggesting the presence of an $S_6$-symmetry axis passing through the inner cavity of the sodium cation. Moreover,

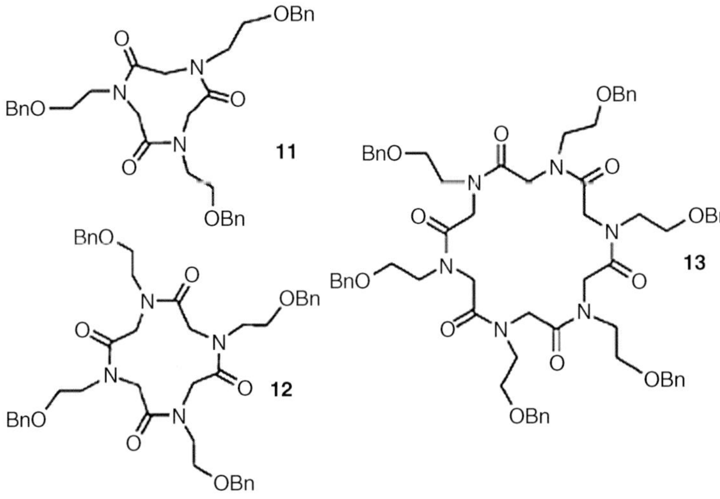

**Figure 11.23** *Izzo and De Riccardis's cyclic peptoids. [56] Reproduced by permission of The Royal Society of Chemistry.*

(a)  (b)

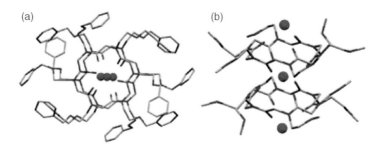

**Figure 11.24** *X-Ray crystal structure of 13₂·[Sr(Picr)₂]₃ complex. (a) Top view. (b) Side view. Hydrogen atoms and picrates have been omitted for clarity. [56] Reproduced by permission of The Royal Society of Chemistry.*

equilibrium NMR studies indicated that the electrostatic (ion–dipole) interactions stabilize this conformation in solution. The authors reported one successful attempt which produced needle-like crystals, suitable for X-ray structure analysis, of the cyclic hexamer as a 2:3 complex with strontium picrate (Figure 11.24). The analysis showed a unique peptoid bond configuration with the carbonyl groups alternately pointing toward the strontium cations and forcing the N-linked side chains to assume an alternate pseudo-equatorial arrangement. Although the authors do not discuss conformational control in cyclic peptoids upon metal coordination, this work was found to be interesting in the context of this chapter because, to date, it reports the first and only example of a metallopeptoid structure to be solved. Moreover, in a follow-up study, these laboratories reported cation transport carried by cyclic peptoids across a phospholipid membrane [59]. Two additional cyclic peptoids – an octamer and a decamer – were synthesized and characterized and their size-dependent selectivity for first group alkali metal cation transport was demonstrated.

## 11.5   Concluding Remarks

The selected examples of metallofoldamers discussed here illustrate the significant progress made in the design of foldamers that adopt well defined secondary structures upon metal–ligand coordination interactions in combination with a variety of other interactions, such as hydrogen bonding, aromatic π-stacking and solvophobic effects. Moreover, metallofoldamers have shown an impressive ability to form single-handed helical structures and other chiral architectures. The synergistic interaction between helices and metal complexes in metallopeptoids demonstrates the transfer of chiral information from a folded scaffold to an embedded metal center and, hence, has a potential for applications in asymmetric catalysis.

## References

1. Hill, D.J., Mio, M.J., Prince, R.B. *et al.* (2001) *Chem. Rev.*, **101**, 3893–4011.
2. Gellman, S.H. (1998) *Acc. Chem. Res.*, **31**, 173–180.

3. Huc, I. (2004) *Eur. J. Org. Chem.*, **2004**, 17–29.
4. Hecht, S. and Huc, I. (eds) (2007) *Foldamers: Structure, Properties, and Applications*, Wiley-VCH, Weinheim.
5. (a) Popescu, D.-L., Parolin, T.J., and Achim, C. (2003) *J. Am. Chem. Soc.*, **125**, 6354–6355; (b) Franzini, R.M., Watson, R.M., Patra, G.K. *et al.* (2006) *Inorg. Chem.*, **45**, 9798–9811; (c) Polonius, F.-A. and Müller, J. (2007) *Angew. Chem. Int. Ed.*, **46**, 5602–5604; (d) Clever, G. H., Kaul, C., and Carell, T. (2007) *Angew. Chem. Int. Ed.*, **46**, 6226–6236; (e) Müller, J. (2008) *Eur. J. Inorg. Chem.*, **2008**, 3749–3763; (f) Düpre, N., Welte, L., Gómez-Herrero, J. *et al.* (2009) *Inorg. Chim. Acta*, **362**, 985–992; (g) Takezawa, Y., Maeda, W., Tanaka, K., and Shionoya, M. (2009) *Angew. Chem. Int. Ed.*, **48**, 1081–1084.
6. Zhang, F., Bai, S., Yap, G.P.A. *et al.* (2005) *J. Am. Chem. Soc.*, **127**, 10590–10599.
7. (a) Lehn, J.-M., Rigault, A., Siegel, J. *et al.* (1987) *Proc. Natl Acad. Sci. USA*, **84**, 2565–2569; (b) Koert, U., Harding, M.M., and Lehn, J.-M. (1990) *Nature*, **346**, 339–342; (c) Piguet, C., Bernardinelli, G., and Hopfgartner, G. (1997) *Chem. Rev.*, **97**, 2005–2062; (d) Williams, A. (1997) *Chem. Eur. J.*, **3**, 15–19; (e) Ho, P.K.-K., Cheung, K.-K., Peng, S.-M., and Che, C.-M. (1996) *J. Am. Chem. Soc. Dalton Trans.*, **1996**, 1411–1417; (f) Mizutani, T., Yagi, S., Moringa, T. *et al.* (1999) *J. Am. Chem. Soc.*, **121**, 754–759; (g) Albrecht, M. (2001) *Chem. Rev.*, **101**, 3457–3497; (h) Stadler, A.-M., Kyritsakas, N., Graff, R., and Lehn, J.-M. (2006) *Chem. Eur. J.*, **12**, 4503–4522; (i) Xu, J. and Raymond, K.N. (2006) *Angew. Chem. Int. Ed.*, **45**, 6480–6485; (j) Leonard, J.P., Jensen, P., McCabe, T. *et al.* (2007) *J. Am. Chem. Soc.*, **129**, 10986–10987.
8. (a) Bushmarina, N.A. *et al.* (2006) *Protein Sci.*, **15**, 659–671; (b) Apiyo, D. and Wittung-Stafshede, P. (2002) *Protein Sci.*, **11**, 1129–1135.
9. Ben-Naim, A. (1971) *J. Phys. Chem.*, **54**, 1387–1404.
10. Prince, R.B., Okada, T., and Moore, J.S. (1999) *Angew. Chem. Int. Ed.*, **38**, 233–236.
11. Nelson, J.C., Saven, J.G., Moore, J.S., and Wolynes, P.G. (1997) *Science*, **277**, 1793–1796.
12. Venkataraman, D., Du, Y., Wilson, S. *et al.* (1997) *J. Chem. Ed.*, **74**, 915–918.
13. (a) Radecka-Paryzek, W., Patroniak, V., and Lisowski, J. (2005) *Coord. Chem. Rev.*, **249**, 2156–2175; (b) Hutin, M., Schalley, C.A., Bernadinelli, G., and Nitschke, J.R. (2006) *Chem. Eur. J.*, **12**, 4069–4076; (c) Houjou, H., Iwasaki, A., Ogihara, T. *et al.* (2003) *New J. Chem.*, **27**, 886–889; (d) Comba, P., Fath, A., Hambley, T.W. *et al.* (1998) *Inorg. Chem.*, **37**, 4389–4401; (e) Fenton, D.E., Matthews, R.W., McPartlin, M. *et al.* (1994) *J. Chem. Soc. Chem. Commun.*, **1994**, 1391–1392.
14. (a) Gregolinski, J. and Lisowski, J. (2006) *Angew. Chem. Int. Ed.*, **45**, 6122–6126; (b) Gregolinski, J., Slepokura, K., and Lisowski, J. (2007) *Inorg. Chem.*, **46**, 7923–7934.
15. (a) Zahn, S. and Canary, J.W. (2000) *Science*, **288**, 1404–1407; (b) Tang, H.Z., Novak, B.M., Ho, J., and Polavarapu, P.L. (2005) *Angew. Chem. Int. Ed.*, **44**, 7298–7301; (c) Vicario, J., Katoonia, N., Serrano Ramon, D. *et al.* (2006) *Nature*, **440**, 160–164; (d) Hembury, G.A., Borovkov, V.V., and Inoue, Y. (2008) *Chem. Rev.*, **108**, 1–73; (e) Miyake, H., Kamon, H., Miyahara, I. *et al.* (2008) *J. Am. Chem. Soc.*, **130**, 792–793.
16. Belmont, P., Constant, J.-F., and Demeunynck, M. (2001) *Chem. Soc. Rev.*, **30**, 70–80.
17. Gregolinski, J., Starynowicz, P., Hua, K.T. *et al.* (2008) *J. Am. Chem. Soc.*, **130**, 17761–17773.
18. (a) Gonzales-Alvarez, A., Alfonso, I., Lopez-Ortiz, F. *et al.* (2004) *Eur. J. Org. Chem.*, **2004**, 1117–1127; (b) Kuhnert, N., Rossignolo, G.M., and Lopez-Periago, A. (2003) *Org. Biomol. Chem.*, **1**, 1157–1170.
19. Kawamoto, T., Prakash, O., Ostrander, R. *et al.* (1995) *Inorg. Chem.*, **34**, 4294–4295.
20. Kawamoto, T., Hammes, B.S., Haggerty, B. *et al.* (1996) *J. Am. Chem. Soc.*, **118**, 285–286.
21. Yu, Q., Baroni, T.E., Liable-Sands, L. *et al.* (1998) *Tett. Let.*, **39**, 6831–6834.
22. Tanaka, M., Fujii, Y., Okawa, H. *et al.* (1987) *Chem. Lett.*, **1987**, 1673–1674.
23. Stanetty, P., Koller, H., and Pürstinger, G. (1990) *Monatsh. Chem.*, **121**, 883–891.
24. Toyota, E., Itoh, K., Sekizaki, H., and Tanizawa, K. (1996) *Bioorg. Chem.*, **24**, 150–158.

25. Borovik, A.S. (2005) *Acc. Chem. Res.*, **38**, 54–61.

26. (a) Moore, J.S. (1997) *Acc. Chem. Res.*, **30**, 402–413; (b) Mamula, O. and von Zelewsky, A. (2003) *Coord. Chem. Rev.*, **242**, 87–95.

27. Dong, Z., Karpowicz Jr., R.J., Bai, S. *et al.* (2006) *J. Am. Chem. Soc.*, **128**, 14242–14243.

28. (a) Sudhakar, A. and Katz, T.J. (1986) *J. Am. Chem. Soc.*, **108**, 179–181; (b) Katz, T.J., Sudhakar, A., Teasley, M.F. *et al.* (1993) *J. Am. Chem. Soc.*, **115**, 3182–3198.

29. Yuan, L., Sanford, A.R., Feng, W. *et al.* (2005) *J. Org. Chem.*, **70**, 10660–10669.

30. Dong, Z., Yap, G.P.A., and Fox, J.M. (2007) *J. Am. Chem. Soc.*, **129**, 11850–11853.

31. Fisher, L.A., Zhang, F., Yap, G.P.A., and Fox, J.M. (2010) *Inorg. Chim. Acta*, **364**, 259–260.

32. Dong, Z., Plampin, J.N. III, Yap, G.P.A., and Fox, J.M. (2010) *Org. Lett.*, **12**, 4002–4005.

33. Dong, Z., Bai, S., Yap, G.P.A., and Fox, J.M. (2011) *Chem. Commun.*, **47**, 3781–3783.

34. (a) Gong, B. (2008) *Acc. Chem. Res.*, **41**, 1376–1386; (b) Yi, H.-P., Wu, J., Ding, K.-L. *et al.* (2007) *J. Org. Chem.*, **72**, 870–877.

35. Rauckhorst, M.R., Wilson, P.J., Hatcher, S.A. *et al.* (2003) *Tetrahedron*, **59**, 3917–3923.

36. (a) Huang, B., Prantil, M.A., Gustafson, T.L., and Parquette, J.R. (2003) *J. Am. Chem. Soc.*, **125**, 14518–14530; (b) Huang, B. and Parquette, J.R. (2001) *J. Am. Chem. Soc.*, **123**, 2689–2690; (c) Recker, J., Tomcik, d.J., and Parquette, J.R. (2000) *J. Am. Chem. Soc.*, **122**, 10298–10307; (d) Gandhi, P., Huang, B., Gallucci, J.C., and Parquette, J.R. (2001) *Org. Lett.*, **3**, 3129–3132.

37. Preston, A.J., Gallucci, J.C., and Parquette, J.R. (2006) *Org. Lett.*, **8**, 5259–5262.

38. Delsuc, N., Hutin, M., Campbell, V.E. *et al.* (2008) *Chem. Eur. J.*, **14**, 7140–7143.

39. (a) Jiang, H., Léger, J.-M., and Huc, I. (2003) *J. Am. Chem. Soc.*, **125**, 3448–3449; (b) Jiang, H., Léger, J.-M., Dolain, C., and Guionneau, P. (2003) *Tetrahedron*, **59**, 8365–8374; (c) Dolain, C., Grélard, A., Laguerre, M. *et al.* (2005) *Chem. Eur. J.*, **11**, 6135–6144; (d) Gillies, E.R., Dolain, C., Léger, J.-M., and Huc, I. (2006) *J. Org. Chem.*, **71**, 7931–7939.

40. According to IUPAC recommendation the Λ/Δ nomenclature is restricted to tris(didentate)octahedral complexes. The axial chirality of a tetrahedral Cu[I] should in principle be described as *P*/*M*. However, to avoid any confusion with the *P*/*M* handedness of the oligomeric helices, the authors have used Λ = (*M*) and Δ = (*P*) to designate the absolute configuration of the Cu[I] complex.

41. (a) Seebach, D. and Cardiner, J. (2008) *Acc. Chem. Res.*, **41**, 1366–1375; (b) Yoo, B. and Kirshenbaum, K. (2008) *Curr. Opin. Chem. Biol.*, **12**, 714–723.

42. (a) Zuckermann, R.N., Kerr, J.M., Kent, S.B.H., and Moos, W.H. (1992) *J. Am. Chem. Soc.*, **114**, 10646–1647; (b) Murphy, J.E., Uno, T., Hamer, J.D. *et al.* (1998) *Proc. Natl Acad. Sci. USA*, **95**, 1517–1522; (c) Goodman, C.M., Choi, S., Shandler, S., and DeGrado, W.F. (2007) *Nat. Chem. Biol.*, **3**, 252–262.

43. (a) Chang, R.P., Gellman, S.H., and DeGrado, W.F. (2001) *Chem. Rev.*, **101**, 3219–3232; (b) Petersson, E.J. and Schepartz, A. (2008) *J. Am. Chem. Soc.*, **130**, 821–823; (c) Appella, D. H., Christianson, L.A., Karle, I.L. *et al.* (1996) *J. Am. Chem. Soc.*, **118**, 13071–13072.

44. Lelais, G., Seebach, D., Jaun, B. *et al.* (2006) *Helv. Chim. Acta*, **89**, 361–403.

45. (a) Coleman, J.E. (1992) *Annu. Rev. Biochem.*, **61**, 897–946; (b) Berg, J.M. and Godwin, H.A. (1997) *Annu. Rev. Biophys. Biomol. Struct.*, **26**, 357–371; (c) Cousins, R.J., Liuzzi, J.P., and Lichten, L.A. (2006) *J. Biol. Chem.*, **281**, 24085–24089.

46. Rossi, F., Lelais, G., and Seebach, D. (2003) *Helv. Chim. Acta*, **86**, 2653–2661.

47. (a) Kirshenbaum, K., Zuckermann, R.N., and Dill, K.A. (1999) *Curr. Opin. Struct. Biol.*, **9**, 530–535; (b) Barron, A.E. and Zuckermann, R.N. (1999) *Curr. Opin. Struct. Biol.*, **9**, 681–687.

48. (a) Kirshenbaum, K., Barron, A.E., Goldsmith, R.A. *et al.* (1998) *Proc. Natl Acad. Sci. USA*, **95**, 4303–4308; (b) Wu, C.W., Sanborn, T.J., Zuckermann, R.N., and Barron, A.E. (2001) *J. Am. Chem. Soc.*, **123**, 2958–2963; (c) Wu, C.W., Sanborn, T.J., Huang, K. *et al.* (2001) *J. Am. Chem. Soc.*, **123**, 6778–6784; (d) Wu, C.W., Kirshenbaum, K., Sanborn, T.J. *et al.*

(2003) *J. Am. Chem. Soc.*, **125**, 13525–13530; (e) Gorske, B.C., Bastian, B.L., Geske, G.D., and Blackwell, H.E. (2007) *J. Am. Chem. Soc.*, **129**, 8928–8929.

49. Lee, B.-C., Chu, T.K., Dill, K.A., and Zuckermann, R.N. (2008) *J. Am. Chem. Soc.*, **130**, 8847–8855.
50. Hutchinson, E.G. and Thornton, J.M. (1994) *Protein Sci.*, **3**, 2207–2216.
51. Lee, B.-C., Zuckermann, R.N., and Dill, K.A. (2005) *J. Am. Chem. Soc.*, **127**, 10999–11009.
52. Maayan, G., Ward, M.D., and Kirshenbaum, K. (2009) *Chem. Commun.*, **2009**, 56–58.
53. Maayan, G., Yoo, B., and Kirshenbaum, K. (2008) *Tetrahedron Lett.*, **49**, 335–338.
54. Nicolau, D.V. and Yoshikawa, S.J. (1998) *Mol. Graphics Model.*, **16**, 83–96.
55. (a) Gochin, M., Khorosheva, V., and Case, M.A. (2002) *J. Am. Chem. Soc.*, **124**, 11018–11028; (b) Case, M.A. and McLendon, G.L. (1998) *Chirality*, **10**, 35–40.
56. Maulucci, N., Izzo, I., Bifulco, G. *et al.* (2008) *Chem. Commun.*, **2008**, 3927–3929.
57. Abdel-Magid, A.F., Carson, K.G., Harris, B.D. *et al.* (1996) *J. Org. Chem.*, **61**, 3849–3862.
58. Pedersen, C.J. (1968) *Fed. Proc.*, **27**, 1305.
59. De Colaa, C., Licenb, S., Comegnaa, D. *et al.* (2009) *Org. Biomol. Chem.*, **7**, 2851–2854.

# 12

# Applications of Metallofoldamers

*Yan Zhao*

*Department of Chemistry, Iowa State University, USA*

Foldamers are synthetic chain molecules made of multiple repeat units. They mimic biopolymers such as proteins and nucleic acids in that their chains can adopt ordered, yet often tunable conformations [1–7]. Since metal ions are essential to the structure and function of metalloproteins in the biofoldamer world, chemists have been quick to create synthetic analogues in which metal ions interact with the foldamer backbone to modulate the latter's conformation. This chapter focuses on the applications of metallofoldamers. It starts with a general discussion on the creation of functions from structures and then illustrates the opportunities that metal ions provide to foldamer-based materials. The chapter covers three main applications of metallofoldamers: molecular recognition, sensing, and materials synthesis. Examples are chosen to illustrate the principles underlying the structure–function correlation and are by no means exhaustive. The chapter ends with some brief conclusions and an outlook for future challenges.

## 12.1  Introduction

Although synthetic chemists are good at and often spend most of their time designing and synthesizing molecular structures, it is the potential functions of these molecules that appeal to the majority of the public who are nonchemists. Even for many chemists, it is quite accurate to say that their primary research goals nowadays are the creation of new functions instead of structures. This is not to imply that fundamental studies of molecular structures are of little importance. Rather, chemists should strive to apply their structural understanding to create molecules and materials that can solve the practical problems

*Metallofoldamers: Supramolecular Architectures from Helicates to Biomimetics*, First Edition.
Edited by Galia Maayan and Markus Albrecht.
© 2013 John Wiley & Sons, Ltd. Published 2013 by John Wiley & Sons, Ltd.

faced by humanity. Such practices not only produce a positive return on the investment that society has made to chemistry research but also prompt chemists to seek a deeper understanding of how their molecules/materials behave in "real environments" as opposed to modeled environments such as a homogenous solution in a beaker.

In much of the last two decades of foldamer chemistry, chemists spent their time looking for different ways to generate synthetic analogues of biofoldamers [1–7]. It is extremely exciting to see that chemists, using their imagination and creativity, can prepare abiotic molecules that resemble proteins and nucleic acids in conformational order. If nature can use 20 amino acids and four nucleotide bases to create the exquisite and essentially limitless structures and functions with biofoldamers, what vast potential have chemists when the genetic and biological constraints are removed? Helices, single and multiple strands, turns, and sheets have all been realized in synthetic foldamers in this quest [1–7]. Foldamers with quaternary structures have been made via the bottom-up approach [8–10]. As the quest continues, however, chemists are no longer satisfied with mere new foldameric structures. Instead, the question of increasing importance becomes "what can the new foldamer do?" Numerous chemists have answered to such challenges and created many useful materials out of foldamers, including antimicrobial materials [11–15], protein surface-binders and inhibitors [16–19], vesicles [20], organogellators [20], and enantioselective catalysts [21].

It is natural for a curious reader to wonder at this point what benefits foldamers can offer as a class of materials, in terms of both structure and function. According to one advanced organic text, "conformations are the different shapes that a molecule can attain without breaking any covalent bonds [22]." These shapes, of course, are the results of many rotations around the $\sigma$ bonds within the molecule. IUPAC puts it another way: conformations are "the spatial arrangement of the atoms affording distinction between stereoisomers which can be interconverted by rotations about formally single bonds [23]." To a chemist, both definitions are illuminating, as the shape of a molecule can strongly influence its physical properties, and the spatial arrangement of the atoms in a molecule determines its three-dimensional distribution of functional groups and, in turn, the chemical and physical properties of the molecule. Clearly, if chemists can master the skill to control the conformation of a molecule, they can unlock the secrets in "taming" the molecule and potentially can regulate its physical and chemical behavior on demand. Moreover, since the interconversion between different conformers is often strongly influenced by their environment, conformational control serves as a rational way to design environmentally responsive materials. Cells rely on such responsiveness of proteins and nucleic acids to survive and adapt, whether to temperature, pH, nutrients, or specific signal molecules. Chemists collectively have just begun to design such functions through the approach of conformational control.

What do metal ions bring to this picture? We may catch a glimpse of the answer by looking into the metalloproteins, nature's examples of metallofoldamers. The earlier chapters of this book have extensive discussions on this topic. In brief, metal ions can bring profound influence to both the structure and the function of a foldamer. Metal ions tend to have their preferred coordination patterns. Because of the potentially strong metal–ligand interactions, metal ions may overwhelm other factors or at least play a significant role in determining the conformation of a foldamer. Moreover, metal ions can possess electronic, photonic, or catalytic properties not found in the organic backbone of a metallofoldamer and thus endow new functions to the molecule. In the meantime, the conformational preference of the foldamer chain can create unusual coordination motifs around the metal and allow it to behave in unconventional ways. Such interplay between the metal and the foldamer expands the repertoire of both and represents the most exciting aspect of metallofoldamers. In the following sections, we will illustrate how these interactions play out and benefit both partners with selected examples of synthetic metallofoldamers.

## 12.2 Metallofoldamers in Molecular Recognition

Moore and coworkers were among the first to design foldamers whose conformations depend on both the noncovalent interactions of the foldamer chain and the metal–ligand coordination [24]. The metallofoldamers were based on their well characterized *meta*-phenyleneethynylene (*m*PE) foldamer **1**. The foldamer contains an aromatic backbone that is solvated well by chloroform but poorly by acetonitrile. When the number of aromatic units reaches 8 to 10, the *m*PE oligomer adopts a helical conformation in acetonitrile, with the collapsed aromatic core surrounded by the solvophilic tri(ethyleneglycol) or Tg groups [25]. The foldamer can be viewed as a conformationally tunable mimic of macrocycle **2**. Because of the 120° bond angle in the *meta*-phenylene linkage, a folded *m*PE has six aryl units per helical turn. The helix has a tubular cavity in the middle which can bind hydrophobic molecules with complementary geometry (Figure 12.1) [26].

By equipping every other aryl group in the *m*PE oligomer with a cyano group, Moore turned **1** into a metal-binding foldamer (**3**). Unlike a multidentate metal-binding macrocycle such as an azacrown, **3** may adopt either a folded or unfolded conformation. The authors studied the binding of Ag$^+$ in tetrahydrofuran (THF), a solvent in which *m*PE oligomers adopt random coil conformations. Upon the addition of silver triflate (AgOTf), the control compound **1** ($n = 12$) showed little change in the $^1$H NMR or UV/Vis spectra but **3** ($n = 6$) displayed the typical transition to the *cisoid* conformation associated with the folded *m*PE [24]. The latter was found to bind two equivalents of silver ions as

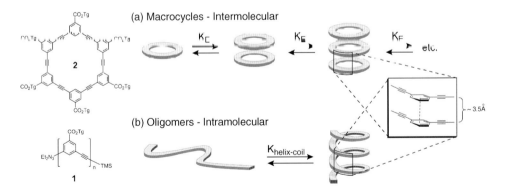

**Figure 12.1** *Relationship between the intermolecular aggregation of (a) mPE macrocycle **2** and the intramolecular folding of (b) linear mPE oligomer **1**. Reprinted with permission from [ref 25] Copyright 2006 American Chemical Society.*

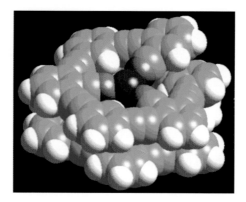

**Figure 12.2** *Space-filling model of* **3** *(n = 6) coordinating to two Ag⁺ ions. The triethylene-glycol side chains have been omitted for clarity. Reprinted with permission from [ref 24] Copyright 1999 WILEY-VCH Verlag GmbH & Co. KGaA, Weinheim.*

expected from the trigonal planar coordination of the metal (Figure 12.2). The solvopho-bic interactions of the *m*PE backbone were critical to the metal binding, as the shorter derivative, **3** ($n = 3$), was unable to bind $Ag^+$.

In a following work, Moore and coworkers demonstrated another intriguing application of metallofoldamers in molecular recognition [27]. Instead of putting the metal-binding ligands along the tubular cavity, they placed a single pyridyl group at the end of the *m*PE. Unlike a simple pyridyl ligand, however, the *m*PE-functionalized pyridyl (**4**) has the potential to fold and thus could benefit from the solvophobic and π-π interactions among the aromatic groups (Figure 12.3). The anticipation was that the entropic costs associated with the complexation could be offset by the stabilization energy in the newly formed, folded metal complex. The effect was similar to the chelating effect found in a bidentate such as ethylenediamine, except that the two pyridyl ligands in the $Pd^{2+}(4)_2$ complex were coupled by noncovalent instead of covalent bonds.

The folding and metal-binding of **4** were investigated by a number of techniques. The unfolded *m*PE oligomers had a peak centered at 303 nm. As the chain length increased, the peak at 303 nm gradually disappeared and a new peak at 289 nm appeared for the 2 : 1

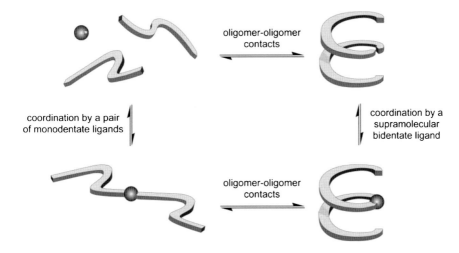

**Figure 12.3**   *Schematic representation of the concept of folding-based supramolecular chelation. Reprinted with permission from [ref 27] Copyright 2005 American Chemical Society.*

complex of **4** and $Pd^{2+}$ in acetonitrile, indicative of a transition from the unfolded to the folded conformation. The conformational change was also confirmed by $^1$H NMR spectroscopy. Isothermal calorimetry revealed that the overall free energy for the complexation had a nearly linear relationship with the chain length and became more negative as the chain lengthened. Most interestingly, the binding enthalpy was more favorable in **4** ($n-7$) than either the shorter ($n=5$) or the longer ($n-9$) ligand. The enthalpic gain from the pentamer (**4**, $n=5$) to the heptamer (**4**, $n=7$) was anticipated because more aryl–aryl contacts would occur in the folded complex when the chain became longer. For the pyridyl with nine $m$PE units (**4**, $n=9$), the total number of aryl groups was 10 and large enough for the ligand itself to fold in acetonitrile; thus, the net increase in aryl–aryl interactions was not as significant during complexation with the metal. In other words, the benefit of folding toward complexation was the largest when the metal–ligand complex was long enough to fold but the ligand itself was not. The concept was latter applied to the synthesis of supramolecular coordination polymers from bispyridyl-functionalized $m$PE oligomers [28].

In the above examples, metal ions were used to promote the folding. Ajayaghosh *et al.* reported the opposite effect of the metal ion; that is, unfolding a helix by metal–ligand interactions [29]. Similar to the $m$PE oligomers, **5** was a random coil in chloroform and a helix in acetonitrile. The folding was characterized by a number of techniques including solvent titration, fluorescence spectroscopy, and circular dichroism (CD). Because the folding of **5** was driven by solvophobic and $\pi$-$\pi$ interactions in acetonitrile, the hydrophobicity of the foldamer backbone was critical to the folded conformation. For the same reason, as the polarity of the aromatic backbone was increased by the coordination of the bispyridyl groups to zinc cations, the helix unfolded into a random coil. The effect was completely reversible. Once the metal ions were removed by a stronger ligand, EDTA, the extended chain spontaneously folded back to the helix.

**5**

An important feature of a metallofoldamer is the conformational tunability of the organic foldamer; most examples in the literature indeed used metal ions to modulate the conformation of the foldamer chain. Since many transition metal ions have different oxidation states and these different oxidation states tend to have their own preferred coordination motifs, electrochemistry may be used to control the conformation of the metallofoldamer by tuning the oxidation state of the metal. With such a goal in mind, Fox and coworkers designed ligand **6**, which contains a central square planar metal-binding salophen unit [30]. The rest of the foldamer backbone has hydrogen-bonding and aromatic groups. As previously described in Chapter 9, ligand **6** was designed to fold upon binding of metal ions. Indeed, the addition of Ni(II) and Cu(II) gave rise to the metallofoldamers **7a** and **7b**, respectively.

**6**                    (R = C$_6$H$_{13}$)                    **7a**: M = Ni; **7b**: M = Cu

The copper complex **7b** was found to go through reversible redox Cu(II)/Cu(I) changes by cyclic voltammetry but the reactions became irreversible when 5% *N,N*-4-(dimethylamino)-pyridine (DMAP) was added to the THF solution or when the redox was performed in acetonitrile, a more strongly coordinating solvent. The authors proposed that the Cu(I) complex initially generated was tetracoordinate but, in the presence of a stronger ligand (DMAP or acetonitrile), became pentacoordinate and nonhelical. The mechanism was supported by semiempirical (AM1) calculations. The Fox group also demonstrated in a following work that chiral end groups and steric interactions could be used to control the absolute helicity of a similar Ni complex (**8**) [31]. When the salophen group was replaced with a chiral salen, competition came into play as the chiral end groups and the central chiral diamine biased toward different helicity [32,33]. The chiral salen ligand, which has extensive applications in asymmetric catalysis, turned out as only a weak director for the absolute helicity [32].

**8(M)**          new H-bond          **8(P)**          new steric interaction

(a)

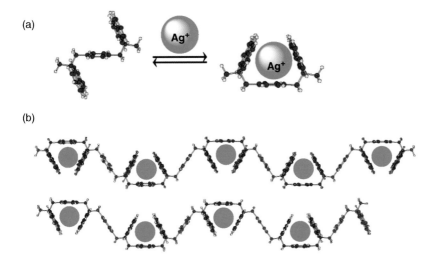

(b)

**Figure 12.4** *Calculated molecular structures of (a) unfolded and silver-bound folded conformation of a fluorene-p-xylene-fluorene unit and (b) representative examples of fluorene-p-xylene oligomers in their silver-bound conformation. Reprinted with permission from [ref 34] Copyright 2007 American Chemical Society.*

Most metal-binding ligands employ heteroatoms as the electron donors. Chebny and Rathore reported fluorene-*p*-xylene oligomers (**9**) with up to 10 fluorene groups in the main chain [34]. The fluorene-*p*-xylene-fluorene unit was found to form a receptor for Ag$^+$ with the $\pi$ electrons as the donor (Figure 12.4a). The binding was surprising strong ($K_a = 15\,000\,M^{-1}$) in a mixture of chloroform and methanol. The longer oligomers underwent similar conformational changes upon the binding of multiple silver cations (Figure 12.4b).

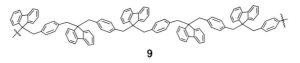

**9**

Instead of using transition metals, Ramakrishnan and coworkers synthesized polymers (**10**) whose folding was triggered by alkali metal ions [35,36]. The polymers contained alternate aromatic donors and acceptors, a feature similar to the *aeda*mers (*a*romatic *e*lectron *d*onor–*a*cceptor oligomers) reported by Iverson [37–39]. These aromatic units provided not only solvophobic and aromatic interactions to stabilize the stacked structure but also the spectroscopic handle in the UV-vis and NMR spectra to characterize the folded conformation. The intramolecular charge transfer (CT) complex formed between the electron donor and acceptor was found to increase upon the addition of methanol to a solution of the polymer in chloroform. Alkali metal ions also promoted CT complexation. The effect followed the order of K$^+ >$ Na$^+ >$ Li$^+$, consistent with the ability of the metal ions to complex with the hexa(ethylene oxide) tethers. The concept was also applied to foldable polymers responsive to ammonium salts [40–42].

**10**, n = 3, 4, 5

## 12.3   Metallofoldamers as Sensors for Metal Ions

Molecular recognition is the corner stone of supramolecular sensing. When a specific binding event is expressed in a readable signal, the host has the potential to become a sensor. Of course, to be truly qualified as a sensor, the supramolecular host needs to bind the target analyte with high specificity and affinity, preferentially in the presence of competing species.

Foldamer and nonfoldamer sensors are similar in that both possess a molecular recognition site and a signal-generating site. The latter is commonly a chromophore or fluorophore that undergoes spectroscopic changes upon binding with the analyte. What distinguishes a foldamer and a nonfoldamer sensor is the former's conformational change during the binding. As mentioned in the introduction of the chapter, changes in the physical and/or chemical properties often accompany the conformational change of a molecule, a feature immensely useful to the design of sensors.

An early example of a foldamer-based sensor was reported by Ajayaghosh and colleagues [43]. Molecule **11** has two squaraine dyes connected by a flexible oxyethylene tether. The unfolded conformer had its maximum absorption at 630 nm in acetonitrile. The addition of $Na^+$ and $K^+$ ions affected neither the absorption nor the emission of the compound but the addition of $Ca^{2+}$ turned the color of the solution from light blue to intense purple-blue. In the UV-vis spectrum, the peak at 630 nm decreased while a new peak at 552 nm appeared. The fluorescence of the compound ($\lambda_{max} = 652$ nm) was significantly quenched. Importantly, similar alkaline earth metals such as $Mg^{2+}$, $Sr^{2+}$, and $Ba^{2+}$ brought minimal changes to the absorption and emission spectra of the compound. The changes were proposed to occur as a result of the face-to-face stacked conformation. The arrangement of the dyes resembled the "H" aggregates of squaraine dyes, which tend to display blue-shifted absorptions compared to the monomers. The folding was found to be reversed by the addition of EDTA that removed the metal ion from the folded structure [44].

**11**

Molecule **11** is unable to fold on its own. Its metal-assisted folding was mainly exploited to generate the signal. If the sensor itself is able to fold, will its conformational property affect its ability to bind the metal? Zhao and coworkers addressed this question in a series of works, with their oligocholate foldamers (Figure 12.5).

The oligocholates were prepared from cholic acid [45], a widely used building block for supramolecular receptors [46–53]. Their folding is driven by an unusual type of solvophobic interactions [54–56]. When dissolved in a largely nonpolar solvent mixture, the

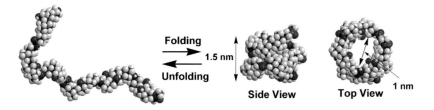

**Figure 12.5** *Space-filling molecular models of the unfolded and folded cholate hexamer. Reprinted with permission from [ref 55] Copyright 2007 American Chemical Society.*

hydrophilic faces of an oligocholate prefer to be solvated by the polar solvent but cannot do so in the unfolded conformation (Figure 12.6, left). To avoid the unfavorable hydrophilic–hydrophobic contact, the oligocholate folds into a helical structure, with the polar solvent microphase-separated from the bulk and placed in the internal nanocavity of the folded helix (Figure 12.6, right). Such an arrangement efficiently satisfies the needs of both the cholate hydrophilic faces to be solvated by polar solvent in a nonpolar mixture and the polar solvent to be located in a polar environment. Because folding requires phase separation of the polar solvent, formation of the helix is most favorable in a solvent mixture with marginal miscibility. Folding, for example, was more favorable in a 2 : 1 mixture of hexane and ethyl acetate (EA) than in a 1 : 1 mixture when the polar solvent was dimethyl sulfoxide (DMSO) or methanol [54]. This is because the polar solvent is miscible with EA but immiscible with hexane: as the amount of hexane increases in the ternary mixture, phase separation of the polar solvent becomes easier energetically.

The solvophobic folding of the parent oligocholates is extremely sensitive to solvent composition, with a few percentage change being typically sufficient to trigger complete unfolding [54–56]. Because the oligocholates are connected by amide bonds, their functionalization could be readily achieved by the incorporation of natural α-amino acids. Oligocholate **12** contains two methionines in the middle and a dansyl at the chain end [57]. Similar to the parent cholate hexamer, **12** could fold in <5% DMSO (or methanol) in 2 : 1 hexane/ethyl acetate [57].

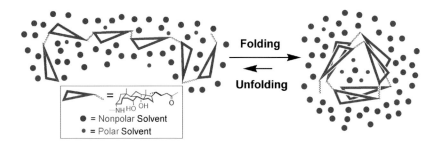

**Figure 12.6** *Preferential solvation of an oligocholate in a mixture of polar and nonpolar solvents. Reprinted with permission from [ref 55] Copyright 2007 American Chemical Society.*

**12**

The presence of two thioether groups in **12** allowed it to bind mercury ions. In 2 : 1 hexane/EA with 5% methanol, $Hg^{2+}$ quenched the dansyl emission efficiently and even 20 nM of $Hg^{2+}$ could be easily detected by the fluorescence spectroscopy. The binding was determined by the Job's plot to be 1 : 1. Unlike conventional sensors, however, the mercury binding took place in the folded conformation of **12**, as the methionines were too far apart in the unfolded conformer. Binding was extremely strong when the folded conformer was favored by the solvents and very weak when the unfolded form was favored. Experimentally, the binding constant between **12** and $Hg^{2+}$ could be tuned over five orders of magnitude by simple solvent changes [57].

The conformation of the sensor was the dominant factor controlling the binding affinity. Interestingly, despite its relatively flexible nature, this foldamer demonstrated good selectivity toward $Hg^{2+}$ in comparison to other divalent cations such as $Mg^{2+}$, $Zn^{2+}$, $Cu^{2+}$, $Co^{2+}$, $Ni^{2+}$, and even $Pb^{2+}$. Quite likely, in addition to the different binding affinity, the quenching effect of the metal ion was also important to the selectivity. Mercury is known, for example, to quench dansyl more strongly than most other metal ions [58].

Because of the large hydrophobic surface of an oligocholate, even multiple hydroxyl and amide groups are insufficient to give any water solubility to the foldamer. The Zhao group found that oligocholates could be solubilized by common surfactant micelles in water and the micelles greatly impacted the conformations of the foldamers [56,59,60]. Foldamer **12** could be included within the micelles of surfactants of many types, including the cationic cetyltrimethylammonium bromide (CTAB), anionic sodium dodecylsulfate (SDS), and nonionic Triton X-100 [61]. The environmentally sensitive emission of the dansyl group made it easy to study the conformation of the oligocholate in micelles. From the emission properties of the micelle-solubilized **12**, it was clear that nonionic Triton X-100 formed micelles with the most hydrophobic interior. The charge of the surfactant impacted the binding strongly, following the order: cationic micelle $\ll$ nonionic micelle $<$ anionic micelle. The different bindings were attributed to the electrostatic interactions between the positively charge $Hg^{2+}$ and the micelle [61].

The Zhao group later reported dicarboxylated oligocholate **13** [62]. The foldamer was functionalized with two pyrenyl groups at the chain ends. Due to its long fluorescence lifetime, pyrene could form excimers quite readily [63]. The folding of **13**, therefore, could be detected by the excimer formation at 470 nm in the fluorescence spectrum. The two carboxylic acid groups of **13** helped the folding significantly. Whereas the parent hexamer could not fold in the binary methanol/EA mixture at all, **13** could fold with up to 15% methanol. When **13** was titrated with zinc acetate, the excimer/monomer ratio underwent characteristic changes. In 5 and 10% methanol, the compound was already folded and the addition of $Zn^{2+}$ caused very little change to the excimer/monomer ratio. Although the addition of $Zn^{2+}$ enhanced the excimer formation above 15% methanol, the enhancement was the largest with 15% methanol.

It appeared that the most sensitive detection occurred when most of the foldamers just became unfolded.

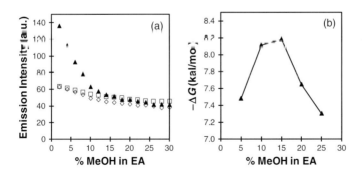

**13**

The principle of preorganization has brought great advancement to supramolecular chemistry over the last several decades. The idea originated from the Fischer's lock–key theory and was articulated by Cram [64]. In order to obtain high-affinity hosts, chemists normally resort to macrocyclic structures that require very little conformational changes to bind the guest, with the assumption that the cost of the conformational change during the binding needs to be paid from the binding interactions. Foldamer-based hosts, of course, deviate from this convention and, as described earlier, frequently experience large-scale conformational changes during the binding. Interestingly, although a huge number of rigid supramolecular receptors have been synthesized by chemists, the vast majority of them give very poor binding affinity in comparison to foldamer-based protein hosts [65]. Not surprisingly, some researchers in recent years began to question whether chemists, in an effort to rigidify supramolecular hosts, have moved away from some key elements of success [66–68].

Zhao and coworkers recently reported the naphthyl-dansyl functionalized dicarboxy-lated oligocholate **14** [69]. Similar to **13**, the compound had a cooperative conformational transition in methanol/EA mixtures and became fully unfolded in 15% methanol (Figure 12.7a). Fluorescence titration revealed that its binding affinity for $Zn^{2+}$ was highest in 15% methanol (Figure 12.7b). The results suggested that the conformational change

**Figure 12.7** *(a) Emission intensity of dansyl at 492 nm for **14** (▲), **15** (◇), and **16** (□) in MeOH/EA mixtures. The high intensity of **14** in <15% methanol came from the FRET in the folded conformation. (b) Binding free energy between **14** and Zn(OAc)₂ in MeOH/EA mixtures. Reprinted with permission from [ref 69] Copyright 2011 American Chemical Society.*

of the oligocholate and its guest binding were intimately related and seemed to benefit each other, unlike what the principle of preorganization assumes.

The authors then investigated a series of diamines, $H_2N(CH_2)_nNH_2$, which could only form weak carboxylate–ammonium ion pairs with **14** in the polar MeOH/EA mixtures. For diamines with intermediate chain length ($n = 6$ and 8), similar binding strengths were observed, once again peaking at 15% methanol. The results were completely unexpected for the hydrogen-bonded ion pairs, which are typically weakened by protic solvents such as methanol. The same trend was observed in the binding of the stronger ion-pairing diguanidine **19** and also in another oligocholate (**18**) that folded in a different solvent mixture [69].

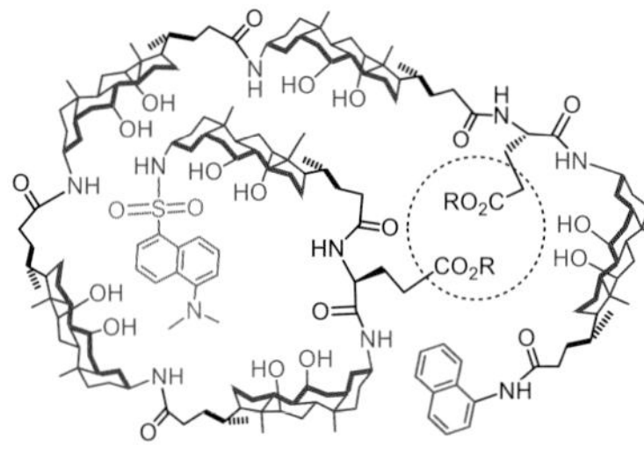

**14**, R = H; **15**, R = Me 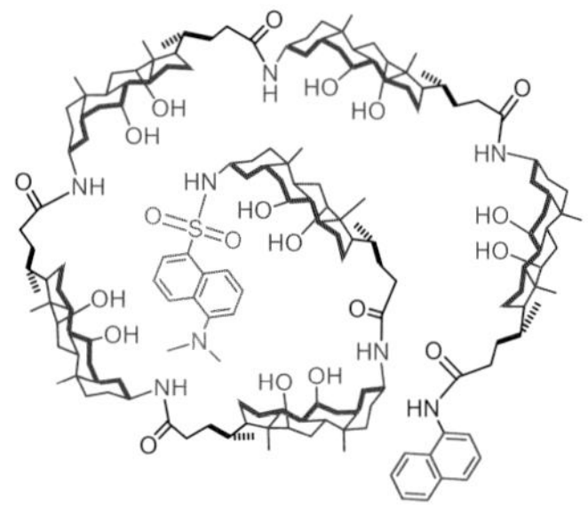          **16**

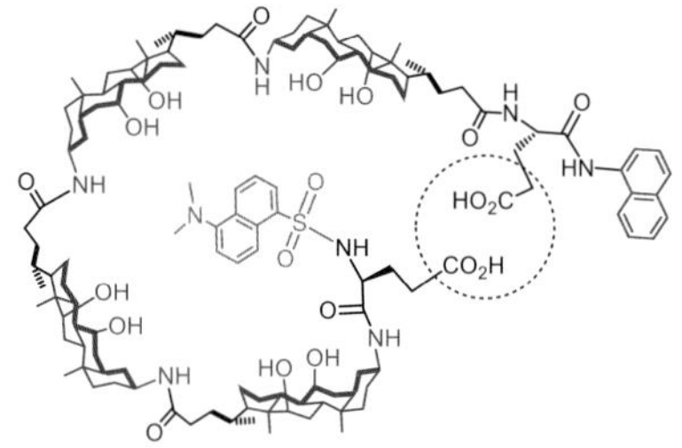

**17** 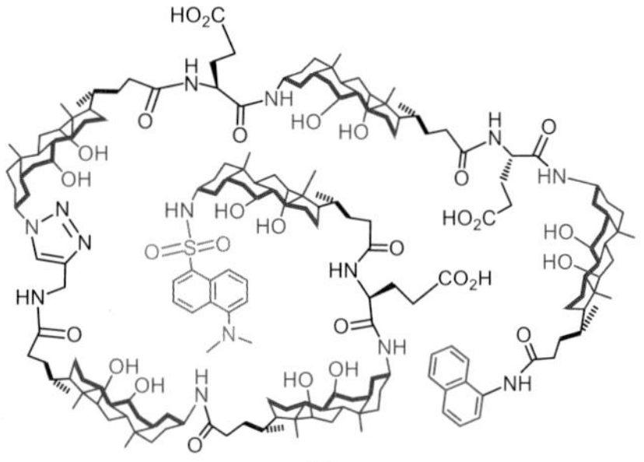          **18**

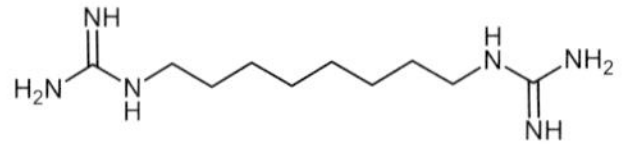

**19**

The authors attributed the unusual solvent effect to the positive cooperativity between the host's conformation and the host–guest binding. Essentially, strong interactions already exist between different segments of the host and with the solvents when **14** is folded; the guest-binding, under such a circumstance, brings little benefit to the *host*. When the host is far into the unfolded region, the guest-binding has to overcome an unfavorable unfolding–folding conformational change, which undermines the binding affinity [57]. When the host is unfolded but near the unfolding–folding transition, however, the conformational change that occurs during guest-binding could engage additional interactions within the host and with the solvents. Because these additional interactions also contribute to the overall binding equilibrium, the binding affinity is "magnified" by the synergism or positive cooperativity between the conformation of the host and host–guest binding interactions.

## 12.4 Metallofoldamers as Dynamic Materials

The above examples amply demonstrate that not only may a metal ion impact the conformation of a metallofoldamer but also the conformation of the foldamer can influence its binding with the metal ion. In addition to their applications as novel supramolecular hosts and sensors, metallofoldamers are very promising as environmentally responsive materials. Nature has long perfected its skills in creating responsive materials for highly sophisticated tasks through the conformational approach. An amazing example is kinesin, a motor protein that walks on microtubules to deliver various cargos within a cell [77]. Myosin, a related class of motor protein, is responsible for creating muscle contraction by walking on actin filaments [77]. Chemists are nowhere close to creating such intricately organized foldamer complexes that couple catalysis, conformational change, and binding to achieve desired mechanical work in the nanoworld. The following examples, nonetheless, demonstrate that chemists have been able to generate prescribed local motions in synthetic metallofoldamers with external stimuli.

In an effort to create motional dynamic devices (i.e., molecular or supramolecular structures that undergo motional or mechanical changes triggered by external stimuli), Lehn and colleagues synthesized alternatively joined pyridine–pyrimidine (py-pym) oligomers (**20**) [78]. These oligomers represent primitive molecular actuators whose motions are driven by ionic interactions. The work was built on the earlier observation that *meta*-connected oligoheterocycles folded into helical structures due to the preferred *transoid, transoid* conformation [79–82]. When Pb(OTf)$_2$ was added, the helix switched into a linear strand due to the *cisoid, cisoid* conformation induced by metal-binding (Figure 12.8).

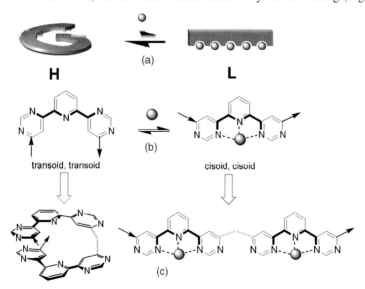

**Figure 12.8** (a) Ionic modulation of extension/contraction motions interconverting the helical, metal-free ligand **H** and the extended linear multinuclear complex **L**. (b, c) The transoid, transoid conformation of the helix and the metal-induced cisoid, cisoid conformation of the linear strand. Reprinted with permission from [ref 78] Copyright 2002, National Academy of Sciences, USA.

The transition was confirmed by X-ray crystallography, NMR spectroscopy, and electrospray mass spectrometry. A large change in the dimension of the molecular structure accompanied the metal-binding. As the helix unwound, the end-to-end distance of the oligomer could increase by as much as 500% (Figure 12.8). The dimensional change was completely reversible. Once the metal ions were removed by a stronger metal-binding azacryptand, the linear structure reversed to the helix. Addition of triflic acid protonated the azacryptand and released the metal ions to the py-pym oligomers, converting them to linear stands.

**20**

An impressive feature of biofoldamers is the enormous diversity in the structures and functions created out of a relatively small set of building blocks. Such diversity was demonstrated by Lehn and coworkers in **20**. As the number of repeat units decreased, the curvature of the unfolded conformer became smaller (Figure 12.9). When $n$ was reduced to 4, the curvature was small enough so that the lead ions on one strand of the linear strand could complex with other strands to form molecular grids (Figure 12.10). Depending on the amount of lead cations added, the compound was shown to form either a $[4 \times 4]$ molecular grid or a $[4 \# 4]$ double-cross-type architecture [83]. These are very different structures from the metal-free helices. The transition could be viewed as a two-dimensional motion triggered by the ionic signal.

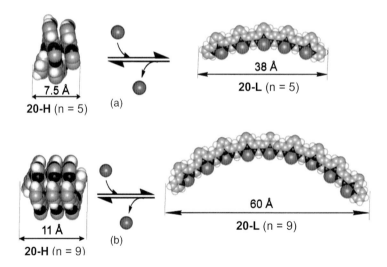

**Figure 12.9** *Molecular models of the helical and extended rack-type complexes of* **20** *(n = 5 and 9). Reprinted with permission from [ref 78] Copyright 2002, National Academy of Sciences, USA.*

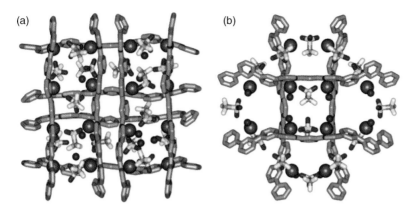

**Figure 12.10** *Crystal structures of (a) the [4 × 4] grid complex,* **20**$_8$–*(Pb*$^{2+}$*)*$_{16}$*, and (b) the double-cross complex,* **20**$_4$–*(Pb*$^{2+}$*)*$_{12}$*. Reprinted with permission from [ref 83] Copyright 2003 American Chemical Society.*

The organic framework may be modified to introduce additional functions to the oligo-heterocycles. When Lehn and colleagues replaced the pym groups with 1,8-naphthyridine (napy), $^1$H NMR spectroscopy indicated that the helical conformer was maintained in the free ligand [84]. The napy group enlarged the inner cavity of the folded helix and also strengthened the solvophobic and π-π interactions in the folded state. Remarkably, these molecules could aggregate along the helical axis to form longer helices. The aggregation was shown to be promoted by a decrease in the size of the aromatic side chain, an increase in the ionic strength of the solution, and, most interestingly, the binding of cationic metal ions in the interior of the helices.

21a, Ar =

21a, Ar =

21a, Ar =

Different metal ions often have different coordination motifs. The py-pym oligomers **22** and **23** both formed single-stranded helical structures in solution and in the solid state. The addition of AgOTf brought significant changes to the $^1$H NMR spectra of the compounds. With less than one equivalent silver ion, complex spectra were obtained, indicating the coexistence of many slowly exchanging species. At [oligomer]/[Ag$^+$] = 1 : 1, however, the spectra were simplified, consistent with the formation of highly symmetrical structures. The spectra did not change upon further increase of the silver ions. A combination of NMR spectroscopy, mass spectrometry, and X-ray crystallography established that a double-stranded helix was formed in the presence of Ag$^+$[85]. The interconversion

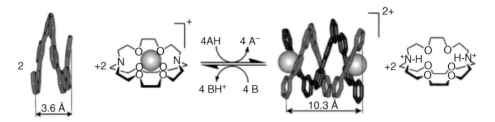

**Figure 12.11**   *Ionic modulation of reversible extension/contraction motional process upon single/double helix interconversion of 22. Reprinted with permission from [ref 85] Copyright 2003 WILEY-VCH Verlag GmbH & Co. KGaA, Weinheim.*

between the single- and double-stranded structures could be modulated once again by the addition of $Ag^+$ and the cryptand to remove the metal. As shown by the crystal structures, switching between the two forms was accompanied by a dimensional change of nearly 300% (Figure 12.11).

**22**, R = H; **23**, R = Ph

The structure of the py-pym oligomers was dominated by *transoid* conformation of the $\alpha,\alpha'$-connected aza-heterocycles. Once the heterocycles are connected by a hydrozone (hyz) unit, linear, bent, and helical structures all become possible, depending on the type and substitution of the aromatic heterocycles [86]. When these oligomers interact with metal ions, a variety of motions become possible. As shown in Figure 12.12, for three *meta*-attached pym units connected by hydrazones, the angle defined by the three hetero-cycles is 60° in the free ligand. Metal coordination bends the structure in opposite direc-tions in a "flapping" motion to produce a linear structure. For three *para*-linked pyrazine (pz) units, the free ligand is linear. Even though the metal coordination does not change the overall linear structure, the chain is rotated by 60° in a "twirling" motion. Since the information of preferred conformation is coded within the foldamer sequence, it is easy to image that combination of these "motional codons" will lead to materials with highly sophisticated molecular motions.

Schafmeister and colleagues prepared spiroladder oligomers based on bis-functional-ized amino acids such as **24–27** [87–90]. This approach was very different from those taken by most foldamer chemists. Instead of trying to control the conformation of flexible foldamer, they used the covalent framework and the inherent conformational pref-erence of the spirorings to generate rigid structures with predetermined 3-D structures. Molecule **28** contains two rigid molecular rods made by four bis(amino acids). The two 8-hydroxyquinoline end groups are known to form a 2 : 1 complex with $Cu^{2+}$. The two rods are connected by a flexible ornithine, which could act as a hinge as the molecule undergoes expansion and contraction upon reversible metal binding. The molecule indeed

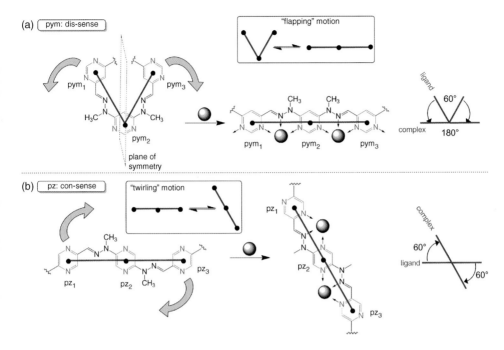

**Figure 12.12** *(a) Dis- and (b) con-sense in-plane displacements induced by cation-coordination to two sequences containing as central ring either (a) a pyrimidine (pym) or (b) a pyrazine (pz) group. Reprinted with permission from [ref 86] Copyright 2010 WILEY-VCH Verlag GmbH & Co. KGaA, Weinheim.*

bound one equivalent of $Cu^{2+}$, as determined by spectrometric titration [91]. The sedimentation coefficient for the metal-complexed species was found to be 18% larger than the metal-free oligomer, in line with the change in the hydrodynamic radii. This change compares favorably with the conformation-induced change in sedimentation coefficients for natural proteins such as Troponin C [92] and smooth-muscle myosin [93].

**24**

**25**

**26**

**27**

**28**

The above dynamic materials maintain their constitutions during their metal binding. When the foldamer chain contains reversible linkages such as imine bonds, a constitutional dynamic polymer library (CDPL) is created which may respond to external stimuli [94]. Lehn and coworkers combined two amines (**29** and **30**) and two aldehydes (**31** and **32**) to form four possible imine-linked polymers (**33–36**). Among the four polymers, only **35** could be stabilized by the CT interactions and the complexation between the oligoether tethers and alkali metal ions.

When an equimolar mixture of preformed **33** and **34** was allowed to equilibrate at room temperature, the **33/34/35/36** ratio was found to be $22:22:28:28$ by $^1$H NMR spectroscopy, suggesting that there was no preference for any structure under the experimental conditions. In the presence of NaOTf and LiOTf, the polymeric mixture shifted toward **35**. $^1$H NMR spectroscopy showed that the **33/34/35/36** ratio at equilibrium was $8:8:42:42$ and $7:7:43:43$, respectively, after the addition of the metal ions. The increased CT complexation also induced a color change of the solution, allowing visual detection of the changes [94].

Dynamic materials from metallofoldamers are not limited to solution. Lee and coworkers prepared a bipyridyl–Ag$^+$ complex whose conformation and self-assembly depended on the counterion in the solid state [95]. With nitrate as the anion (A$^-$), the silver complex **37** self-assembled into a 2-D hexagonal phase with a lattice constant $a = 29.6$ Å, but the helix seemed to be somewhat disordered. When the nitrate was replaced by $BF_4^-$, a similar hexagonal phase was obtained with $a = 31.3$ Å and a second periodicity of 9.6 Å existed in the direction of the stacked hexagonal columns. Calculations indicated that the compound formed helical structures with a pitch of five repeat units. The helical pores were filled with the $BF_4^-$ counterions. When the anion was triflate, X-ray diffraction and calculations were most consistent with hexagonal stacks of cyclic dimers linked by two silver ions (Figure 12.13). The even larger counterion, heptafluorobutyrate, prompted the transition to a lamellar structure with a layer thickness of 32.2 Å. It was proposed that the bulky anion could no longer be contained within the pores formed by the *cis*-oriented aromatic backbones and forced the unfolding of the metallofoldamer chain into a zigzag structure. Replacing the $Ag^+$ ions in **37** with $Pd^{2+}$ or $Cu^{2+}$ and a slight elongation of the side chain brought significant changes to the metallofoldamer in the solid state. The Pd(II)-based coordination polymer formed lamellar structures with the pyridyl ligands trans to each other in a square planar metal complex. The polymer based on $Cu^{2+}$ formed double helices through metal–chloride dimeric interactions [96].

**37**

$R' = $

Interestingly, **37** ($A^- = BF_4^-$) formed gels spontaneously above 2.5 wt% in water [97]. The addition of tetra-*n*-butylammonium fluoride ($Bu_4N^+F^-$) triggered depolymerization of the coordination polymers due to the strong electrostatic interactions between $Ag^+$ and $F^-$, turning the gel into a soluble mixture. The addition of $Bu_4N^+BF_4^-$ gelled the solution again. TEM analysis of the gels showed right-handed helical bundles of fibers with diameters of 6–30 nm. These fibers entangled to form a network. The emission of the gel was red-shifted and significantly weaker in comparison to that of the free ligand, supporting a helical structure similar to that formed in the solid state. The gel–sol transition could also be triggered by the addition of a larger ion, $C_2F_5CO_2^-$. CD, fluorescence, and $^{19}F$ NMR spectroscopy indicated that, instead of depolymerizaiton, the sol formation was caused by the unfolding of the coordination chains into a zigzag conformation.

The Lee laboratory also reported two phenanthrene-based metallofoldamers **38–39**. [98]. The fluorescence of the phenanthrene and the chiral side chain allowed them to characterize the coordination polymers by multiple techniques. In comparison to the ligands, the coordination polymers had the emission peak red-shifted by about 15 nm and significantly quenched. The CD spectra showed strong signals for the aromatic chromophores,

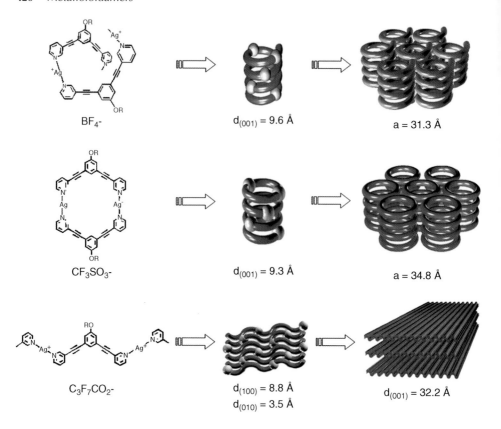

**Figure 12.13** *Schematic representation of the self-assembly of* **37** *and their subsequent organization in the solid state. Reprinted with permission from [ref 95] Copyright 2004 American Chemical Society.*

consistent with helical structures with preferred handedness. Both DLS and TEM showed the formation of cylindrical aggregates. In TEM, for example, **39** formed nanofibers with a diameter of $6.5 \pm 0.4$ nm and several microns in length.

**38**                    **39**

Most interestingly, both compounds displayed reversible changes in the UV-vis, fluorescence, and CD spectra upon heating and cooling. When heated to 40 °C, the UV absorptions for a 0.1 wt% aqueous solution of **39** red-shifted by about 20 nm. A similar red shift (about 10 nm) was observed for **38**. Heating also shifted the CD signals to the red and decreased their intensity. These changes were consistent with lower excition coupling within the aromatic stacks and a longer conjugation length. The cylindrical fibers were shown by TEM to remain at the higher temperature but the cross-section became 5.5 ± 0.4 nm or about 15% smaller in comparison to what was observed at the room temperature. Meanwhile, the emission intensity increased significantly during heating, suggesting a larger distance between the stacked aromatic groups in the helical fiber. These results provided strong support for a more expanded helical structure upon heating and were proposed to take place as the oligo(ethylene oxide) side chains became dehydrated at elevated temperatures. This study is a remarkable example of how conformational change in a foldamer can be exploited to produce novel environmentally responsive optical materials.

Schizophyllan (SPG) is a β-1,3-glucan that forms a triple helix in water that dissociates into a single chain (s-SPG) in DMSO [99–101]. These polysaccharides were found by Shinkai and coworkers to be excellent receptors for macromolecules [102]. SPG can form macromolecular complexes with certain polynucleotides such as poly(C), poly(A), poly(dA), or poly(dT) in water [103–105]. The binding involves hydrogen-bonding and hydrophobic interactions among two s-SPG chains and one polynucleotide. Because the complex spontaneously dissociates upon protonation of the nucleotide bases at pH < 6, the polymer is very promising for the endosomal delivery of polynucleotides [106]. SPG could also be modified at the glucose side chain to facilitate its membrane permeability. The resulting materials were shown to deliver antisense DNA to cells efficiently [107].

**40, SPG**

Other macromolecules that may be encapsulated within SPG helices include single walled carbon nanotubes [108] and conductive polymers [109]. SPG was shown to bind a water-soluble polythiophene and force it to adopt a more planar conformation. During binding, the conjugated polymer became chiral as it adopted a right-handed helical structure. The complex may be viewed as an insulated molecular wire wrapped by two polysaccharide chains [110].

The glucose side chain is important to the solubility of SPG. Once the side chain is removed, the β-1,3-glucan (i.e., curdlan; CUR, **41**), has low solubility in most solvents. By converting the primary alcohol into an azide, Shinkai and colleagues were able to use the alkyne–azide click reaction [111–113] to install metal-binding side chains onto the polysaccharide [114]. The polymer (**43**) retained many of the

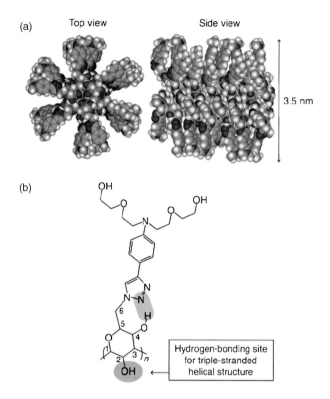

(a) Top view    Side view

3.5 nm

(b)

Hydrogen-bonding site
for triple-stranded
helical structure

**Figure 12.14** *(a) Computer-generated Corey–Pauling–Koltun (CPK) model for the right-handed, triple-stranded helical structure of CUR-DA. (b) Possible hydrogen-bonding mode between glucose in the main chain and the 4-anilino-1,2,3-triazole moiety. The diethylene glycol groups were substituted by methyl groups for clarity. Reprinted with permission from [ref 114] Copyright 2007 WILEY-VCH Verlag GmbH & Co. KGaA, Weinheim.*

conformational properties of SPG. It remained as a single-stranded random coil in DMSO, for example, and formed triple-stranded helices with >20% water in DMSO (Figure 12.14).

Normally, the triple-stranded helical structure of β-1,3-glucans is stable over a wide range of temperature, pH, and ionic strength [99–101]. For **43**, the CD signals displayed small changes at higher temperatures and with up to 60 mM of NaOH. The remarkable stability is beneficial to the protection of the encapsulated contents but problematic if one wants to release the contents. With the triazole groups in the structure, **43** could bind $Zn^{2+}$ and, interestingly, metal-binding was shown to dissociate the triple helix. The authors suggested that the $Zn^{2+}$ ions probably interfered with the hydrogen-bonding interactions of the triazole moieties with the polysaccharide backbone and triggered the dissociation. The process was completely reversible. When 1,4,8,11-tetraazacyclotetradecane (cyclam) was added to remove the zinc ions from the polymer, the characteristic absorption and CD signals for the triple helix resumed.

**41**, CUR

**42**, CUR-N$_3$  **43**, CUR-DA

## 12.5 Conclusions and Outlook

Much progress has been made in synthetic foldamers in the last two decades [1–7]. The incorporation of metal-binding sites on a foldamer backbone or at the side chains allows one to tune the conformation of a foldamer by potentially strong metal–ligand complexation. In the meantime, the conformational properties of a foldamer can greatly influence its coordination to the metal [115]. Such interplay opens up exciting opportunities for both the metal and the foldamer, enabling new applications in many areas, including molecular recognition, sensing, and material synthesis.

The examples described in this chapter point to a need to deepen our understanding of the interactions between metal ions and conformationally mobile foldamer chains. Depending on the exact role of the metal – as a connector of two foldamer fragments [27] or as a guest to bind with a multidentate foldamer host – the conformation of the foldamer may have very different effects on the metal-binding. The positive cooperativity found in oligocholates [69] represents the tip of the iceberg for the potential of foldamers as biomimetic supramolecular hosts. Chemists have always wanted to duplicate nature's exquisite ability to control structure, function, reactivity, and energy transfer with biofoldamers. Too many times have chemists failed to recreate the functions of proteins when they cut out "irrelevant" peptides and attempted to mimic enzyme active sites by rigid small molecule constructs. It is becoming clear in many cases that a wide variety of "conformational communication" exists between the active site and remote peptide chains and is essential to the function of the protein. A deeper understanding of cooperative conformational change is a fundamental challenge but is also the key to unlock nature's secrets in using weak noncovalent interactions to construct receptors with high affinity and specificity.

In comparison to the dynamic materials found in the biological world, what chemists have created so far is primitive. Nonetheless, as a Chinese proverb puts it, a journey of a thousand miles begins with a single step. Molecular recognition is at the heart of biology. Supramolecular interactions are critical to the performance of materials. Controlled mechanical movements formerly considered only possible with biofoldamers have already been realized by Lehn and coworkers in abiotic oligoheterocycles [78,83,85,86].

As chemists unravel the tricks in controlling functions using not only local structures but also remote features through conformational communication, they will enter a whole new world of cooperative molecules similar to proteins and other biofoldamers. The quest is already assured by nature to be fruitful and will lead to intelligent materials with medicinal, energy, and technological significance.

## References

1. Gellman, S.H. (1998) *Acc. Chem. Res.*, **31**, 173–180.
2. Hill, D.J., Mio, M.J., Prince, R.B. *et al.* (2001) *Chem. Rev.*, **101**, 3893–4012.
3. Hecht, S. and Huc, I. (eds) (2007) *Foldamers: Structure, Properties, and Applications*, Wiley-VCH, Weinheim.
4. Cubberley, M.S. and Iverson, B.L. (2001) *Curr. Opin. Chem. Biol.*, **5**, 650–653.
5. Kirshenbaum, K., Zuckermann, R.N., and Dill, K.A. (1999) *Curr. Opin. Struct. Biol.*, **9**, 530–535.
6. Goodman, C.M., Choi, S., Shandler, S., and DeGrado, W.F. (2007) *Nat. Chem. Biol.*, **3**, 252–262.
7. Bautista, A.D., Craig, C.J., Harker, E.A., and Schepartz, A. (2007) *Curr. Opin. Chem. Biol.*, **11**, 685–692.
8. Khakshoor, O., Demeler, B., and Nowick, J.S. (2007) *J. Am. Chem. Soc.*, **129**, 5558–5569.
9. Khakshoor, O., Lin, A.J., Korman, T.P. *et al.* (2010) *J. Am. Chem. Soc.*, **132**, 11622–11628.
10. Giuliano, M.W., Horne, W.S., and Gellman, S.H. (2009) *J. Am. Chem. Soc.*, **131**, 9860–9861.
11. Porter, E.A., Weisblum, B., and Gellman, S.H. (2002) *J. Am. Chem. Soc.*, **124**, 7324–7330.
12. Mowery, B.P., Lee, S.E., Kissounko, D.A. *et al.* (2007) *J. Am. Chem. Soc.*, **129**, 15474–15476.
13. Chongsiriwatana, N.P., Patch, J.A., Czyzewski, A.M. *et al.* (2008) *Proc. Natl Acad. Sci. USA*, **105**, 2794–2799.
14. Choi, S., Isaacs, A., Clements, D. *et al.* (2009) *Proc. Natl Acad. Sci. USA*, **106**, 6968–6973.
15. Tew, G.N., Scott, R.W., Klein, M.L., and Degrado, W.F. (2009) *Acc. Chem. Res.*, **43**, 30–39.
16. Rodriguez, J.M. and Hamilton, A.D. (2007) *Angew. Chem. Int. Ed.*, **46**, 8614–8617.
17. Wyrembak, P.N. and Hamilton, A.D. (2009) *J. Am. Chem. Soc.*, **131**, 4566–4567.
18. Horne, W.S., Johnson, L.M., Ketas, T.J. *et al.* (2009) *Proc. Natl Acad. Sci. USA*, **106**, 14751–14756.
19. Imamura, Y., Watanabe, N., Umezawa, N. *et al.* (2009) *J. Am. Chem. Soc.*, **131**, 7353–7359.
20. Cai, W., Wang, G.T., Xu, Y.X. *et al.* (2008) *J. Am. Chem. Soc.*, **130**, 6936–6937.
21. Maayan, G., Ward, M.D., and Kirshenbaum, K. (2009) *Proc. Natl Acad. Sci. USA*, **106**, 13679–13684.
22. Carey, F.A. and Sundberg, R.J. (2007) *Advanced Organic Chemistry*, 5th edn, Springer, New York, p. 142.
23. Moss, G.P. (1996) *Pure Appl. Chem.*, **68**, 2193–2222.
24. Prince, R.B., Okada, T., and Moore, J.S. (1999) *Angew. Chem. Int. Ed.*, **38**, 233–236.
25. Stone, M.T., Heemstra, J.M., and Moore, J.S. (2006) *Acc. Chem. Res.*, **39**, 11–20.
26. Tanatani, A., Hughes, T.S., and Moore, J.S. (2002) *Angew. Chem. Int. Ed.*, **41**, 325–328.
27. Stone, M.T. and Moore, J.S. (2005) *J. Am. Chem. Soc.*, **127**, 5928–5935.
28. Wackerly, J.W. and Moore, J.S. (2006) *Macromolecules*, **39**, 7269–7276.
29. Divya, K.P., Sreejith, S., Suresh, C.H., and Ajayaghosh, A. (2010) *Chem. Commun.*, **46**, 8392–8394.
30. Zhang, F., Bai, S., Yap, G.P. *et al.* (2005) *J. Am. Chem. Soc.*, **127**, 10590–10599.
31. Dong, Z., Karpowicz Jr, R.J., Bai, S. *et al.* (2006) *J. Am. Chem. Soc.*, **128**, 14242–14243.

32. Dong, Z., Yap, G.P., and Fox, J.M. (2007) *J. Am. Chem. Soc.*, **129**, 11850–11853.
33. Dong, Z.Z., Bai, S., Yap, G.P.A., and Fox, J.M. (2011) *Chem. Commun.*, **47**, 3781–3783.
34. Chebny, V.J. and Rathore, R. (2007) *J. Am. Chem. Soc.*, **129**, 8458–8465.
35. Ghosh, S. and Ramakrishnan, S. (2004) *Angew. Chem. Int. Ed.*, **43**, 3264–3268.
36. Ghosh, S. and Ramakrishnan, S. (2005) *Macromolecules*, **38**, 676–686.
37. Scott Lokey, R. and Iverson, B.L. (1995) *Nature*, **375**, 303–305.
38. Gabriel, G.J. and Iverson, B.L. (2002) *J. Am. Chem. Soc.*, **124**, 15174–15175.
39. Bradford, V.J. and Iverson, B.L. (2008) *J. Am. Chem. Soc.*, **130**, 1517–1524.
40. Ghosh, S. and Ramakrishnan, S. (2005) *Angew. Chem. Int. Ed.*, **44**, 5441–5447.
41. De, S., Koley, D., and Ramakrishnan, S. (2010) *Macromolecules*, **43**, 3183–3192.
42. Ramkumar, S.G. and Ramakrishnan, S. (2010) *Macromolecules*, **43**, 2307–2312.
43. Ajayaghosh, A., Arunkumar, E., and Daub, J. (2002) *Angew. Chem. Int. Ed.*, **41**, 1766–1769.
44. Arunkumar, E., Ajayaghosh, A., and Daub, J. (2005) *J. Am. Chem. Soc.*, **127**, 3156–3164.
45. Danielsson, H. and Sjövall, J. (1985) *Sterols and Bile Acids*, Elsevier, Amsterdam.
46. Davis, A.P. (1993) *Chem. Soc. Rev.*, **22**, 243–253.
47. Davis, A.P. and Joos, J.-B. (2003) *Coord. Chem. Rev.*, **240**, 143–156.
48. Li, Y.X. and Dias, J.R. (1997) *Chem. Rev.*, **97**, 283–304.
49. Maitra, U. (1996) *Curr. Sci.*, **71**, 617–624.
50. Virtanen, E. and Kolehmainen, E. (2004) *Eur. J. Org. Chem.*, **2004**, 3385–3399.
51. Janout, V., Lanier, M., and Regen, S.L. (1996) *J. Am. Chem. Soc.*, **118**, 1573–1574.
52. Bonar-Law, R.P. and Sanders, J.K.M. (1991) *J. Chem. Soc., Chem. Commun.*, **1991**, 574–577.
53. Bonar-Law, R.P., Mackay, L.G., and Sanders, J.K.M. (1993) *J. Chem. Soc. Chem. Commun.*, **1993**, 456–458.
54. Zhao, Y. and Zhong, Z. (2005) *J. Am. Chem. Soc.*, **127**, 17894–17901.
55. Zhao, Y., Zhong, Z., and Ryu, E.H. (2007) *J. Am. Chem. Soc.*, **129**, 218–225.
56. Cho, H. and Zhao, Y. (2010) *J. Am. Chem. Soc.*, **132**, 9890–9899.
57. Zhao, Y. and Zhong, Z. (2006) *J. Am. Chem. Soc.*, **128**, 9988–9989.
58. Métivier, R., Leray, I., and Valeur, B. (2004) *Chem. Eur. J.*, **10**, 4480–4490.
59. Zhong, Z. and Zhao, Y. (2008) *J. Org. Chem.*, **73**, 5498–5505.
60. Zhao, Y. (2009) *J. Org. Chem.*, **74**, 7470–7480.
61. Zhong, Z. and Zhao, Y. (2007) *Org. Lett.*, **9**, 2891–2894.
62. Zhao, Y. and Zhong, Z. (2006) *Org. Lett.*, **8**, 4715–4717.
63. Birks, J.B., Munro, I.H., and Dyson, D.J. (1963) *Proc. R. Soc. Lond. Ser. A*, **275**, 575–588.
64. Cram, D.J. (1986) *Angew. Chem. Int. Ed.*, **25**, 1039–1057.
65. Houk, K.N., Leach, A.G., Kim, S.P., and Zhang, X.Y. (2003) *Angew. Chem. Int. Ed.*, **42**, 4872–4897.
66. Williams, D.H., Stephens, E., O'Brien, D.P., and Zhou, M. (2004) *Angew. Chem. Int. Ed.*, **43**, 6596–6616.
67. Otto, S. (2006) *Dalton Trans.*, **2006**, 2861–2864.
68. Rodríguez-Docampo, Z., Pascu, S.I., Kubik, S., and Otto, S. (2006) *J. Am. Chem. Soc.*, **128**, 11206–11210.
69. Zhong, Z., Li, X., and Zhao, Y. (2011) *J. Am. Chem. Soc.*, **133**, 8862–8865.
70. Hunter, C.A. and Anderson, H.L. (2009) *Angew. Chem. Int. Ed.*, **48**, 7488–7499.
71. Dill, K.A. (1997) *J. Biol. Chem.*, **272**, 701–704.
72. Kraut, D.A., Carroll, K.S., and Herschlag, D. (2003) *Annu. Rev. Biochem.*, **72**, 517–571.
73. Koshland, D.E. (1995) *Angew. Chem. Int. Ed.*, **33**, 2375–2378.
74. Meskers, S., Ruysschaert, J.-M., and Goormaghtigh, E. (1999) *J. Am. Chem. Soc.*, **121**, 5115–5122.
75. Williams, D.H., Stephens, E., and Zhou, M. (2003) *J. Mol. Biol.*, **329**, 389–399.
76. Gonzalez, M., Bagatolli, L.A., Echabe, I. *et al.* (1997) *J. Biol. Chem.*, **272**, 11288–11294.

77. Vale, R.D. (2003) *Cell*, **112**, 467–480.
78. Barboiu, M. and Lehn, J.-M. (2002) *Proc. Natl Acad. Sci. USA*, **99**, 5201–5206.
79. Bassani, D.M., Lehn, J.-M., Baum, G., and Fenske, D. (1997) *Angew. Chem. Int. Ed.*, **36**, 1845–1847.
80. Bassani, D.M. and Lehn, J.-M. (1997) *Bull. Soc. Chim. Fr.*, **134**, 897–906.
81. Ohkita, M., Lehn, J.-M., Baum, G., and Fenske, D. (1999) *Chem. Eur. J.*, **5**, 3471–3481.
82. Howard, S.T. (1996) *J. Am. Chem. Soc.*, **118**, 10269–10274.
83. Barboiu, M., Vaughan, G., Graff, R., and Lehn, J.-M. (2003) *J. Am. Chem. Soc.*, **125**, 10257–10265.
84. Petitjean, A., Cuccia, L.A., Lehn, J.-M. *et al.* (2002) *Angew. Chem. Int. Ed.*, **41**, 1195–1198.
85. Barboiu, M., Vaughan, G., Kyritsakas, N., and Lehn, J.-M. (2003) *Chem. Eur. J.*, **9**, 763–769.
86. Stadler, A.M., Ramirez, J., and Lehn, J.M. (2010) *Chem. Eur. J.*, **16**, 5369–5378.
87. Levins, C.G. and Schafmeister, C.E. (2003) *J. Am. Chem. Soc.*, **125**, 4702–4703.
88. Levins, C.G. and Schafmeister, C.E. (2005) *J. Org. Chem.*, **70**, 9002–9008.
89. Pornsuwan, S., Bird, G., Schafmeister, C.E., and Saxena, S. (2006) *J. Am. Chem. Soc.*, **128**, 3876–3877.
90. Schafmeister, C.E., Brown, Z.Z., and Gupta, S. (2008) *Acc. Chem. Res.*, **41**, 1387–1398.
91. Schafmeister, C.E., Belasco, L.G., and Brown, P.H. (2008) *Chem. Eur. J.*, **14**, 6406–6412.
92. Kuo, I.C.Y. and Coffee, C.J. (1976) *J. Biol. Chem.*, **251**, 6315–6319.
93. Trybus, K.M., Huiatt, T.W., and Lowey, S. (1982) *Proc. Natl Acad. Sci. USA*, **79**, 6151–6155.
94. Fujii, S. and Lehn, J.-M. (2009) *Angew. Chem. Int. Ed.*, **48**, 7635–7638.
95. Kim, H.-J., Zin, W.-C., and Lee, M. (2004) *J. Am. Chem. Soc.*, **126**, 7009–7014.
96. Kim, H.-J., Lee, E., Kim, M.G. *et al.* (2008) *Chem. Eur. J.*, **14**, 3883–3888.
97. Kim, H.-J., Lee, J.-H., and Lee, M. (2005) *Angew. Chem. Int. Ed.*, **44**, 5810–5814.
98. Kim, H.-J., Lee, E., Park, H.-S., and Lee, M. (2007) *J. Am. Chem. Soc.*, **129**, 10994–10995.
99. Norisuye, T., Yanaki, T., and Fujita, H. (1980) *J. Polym. Sci.*, **18**, 547–558.
100. Yanaki, T., Norisuye, T., and Fujita, H. (1980) *Macromolecules*, **13**, 1462–1466.
101. Sato, T., Sakurai, K., Norisuye, T., and Fujita, H. (1983) *Polymer J.*, **15**, 87–96.
102. Sakurai, K., Uezu, K., Numata, M. *et al.* (2005) *Chem. Commun.*, **2005**, 4383–4398.
103. Sakurai, K. and Shinkai, S. (2000) *J. Am. Chem. Soc.*, **122**, 4520–4521.
104. Sakurai, K., Mizu, M., and Shinkai, S. (2001) *Biomacromolecules*, **2**, 641–650.
105. Bae, A.-H., Lee, S.-W., Ikeda, M. *et al.* (2004) *Carbohydr. Res.*, **339**, 251–258.
106. Sakurai, K., Iguchi, R., Mizu, M. *et al.* (2003) *Bioorg. Chem.*, **31**, 216–226.
107. Mizu, M., Koumoto, K., Anada, T. *et al.* (2004) *J. Am. Chem. Soc.*, **126**, 8372–8373.
108. Numata, M., Asai, M., Kaneko, K. *et al.* (2005) *J. Am. Chem. Soc.*, **127**, 5875–5884.
109. Numata, M., Hasegawa, T., Fujisawa, T. *et al.* (2004) *Org. Lett.*, **6**, 4447–4450.
110. Li, C., Numata, M., Bae, A.-H. *et al.* (2005) *J. Am. Chem. Soc.*, **127**, 4548–4549.
111. Rostovtsev, V.V., Green, L.G., Fokin, V.V., and Sharpless, K.B. (2002) *Angew. Chem. Int. Ed.*, **41**, 2596–2599.
112. Kolb, H.C., Finn, M.G., and Sharpless, K.B. (2001) *Angew. Chem. Int. Ed.*, **40**, 2004–2021.
113. Finn, M.G., Kolb, H.C., Fokin, V.V., and Sharpless, K.B. (2008) *Prog. Chem.*, **20**, 1–4.
114. Ikeda, M., Haraguchi, S., Numata, M., and Shinkai, S. (2007) *Chem. Asian J.*, **2**, 1290–1298.
115. Maayan, G. (2009) *Eur. J. Org. Chem.*, 5699–5710.

# Index

α-aminoxy acids  64–5
α-helices
  artificial DNA  303–4
  metallopeptides  276, 283, 292–7
α-lactalbumin  4–5
α-peptides  52, 60–3
α-synuclein  24
Aβ *see* β-amyloid
*ab initio* molecular orbital theory  59, 64
abiotic foldamers  52–3, 70–2
abiotic single-stranded helices  381–93
acridine orange quenching  217–19
AD *see* Alzheimer's disease
AFM *see* atomic force microscopy
aliphatic urea foldamers  63–4
allosteric cooperativity  101–2
ALS *see* amyotrophic lateral sclerosis
Alzheimer's disease (AD)  21, 28
amido foldamers  65–70
amidopyrrole oligomers  78–9
amino acid bridged dicatechol ligands  281
aminopeptidase  9–10
2-aminophenol  353–6
amyotrophic lateral sclerosis (ALS)  11
anion-induced organization  73–8
anthracene bridged dicatechol ligands  170
antibiotic metallopeptides  13–20
antibiotic salivary peptide  17
antiferromagnetism  325–6, 366
antiodontoblot microtubule  408
antiparallel configurations  281, 283,
  287–8, 298
aromatic aryl amide foldamers  394–6
aromatic bis-imine ligands  350–3, 362
aromatic interactions  70, 71–2
artificial DNA  303–32
  alternative base pairing systems  309–11

analytical characterization in solution
  316–17
applications  324–6
assembly and analysis of metal base
  pairs  315–18
automated oligonucleotide synthesis
  314–15
biological functions and beyond  305–6
conformational switching  324
design and synthesis of metal base
  pairs  311–15
DNA secondary structures  303–5
DNA technology  306–8
duplex stabilization  324
electron conductivity  308, 309–10
enzymatic oligonucleotide synthesis  315
future directions  326–7
hydrogen bonding  310
hydroxypyridone ligands  317–20, 324–6
imidazole, triazole and 1-deazaadenine-
  thymine base pairs  323
interactions of DNA with metal ions
  308–9
magnetism and electrical
  conductance  325–7
metal coordination modes  310–11
model studies  312
oligonucleotides as natural foldamers
  303–5
rational design  311–12
salen ligands  318–22, 325
sensor applications  325
shape complementarity  310
strategies for metal incorporation  315
synthesis of modified nucleosides
  312–14
synthetic metallofoldamers  318–23

*Metallofoldamers: Supramolecular Architectures from Helicates to Biomimetics*, First Edition.
Edited by Galia Maayan and Markus Albrecht.
© 2013 John Wiley & Sons, Ltd. Published 2013 by John Wiley & Sons, Ltd.

artificial DNA (*Continued*)
Watson–Crick base pairing 304–5,
309–11, 317
X-ray structure determination 317–18
arylamide oligomers 76–7, 394–6
asymmetric synthesis 52, 54–7
atomic force microscopy (AFM) 308
automated oligonucleotide synthesis 314–15
avidin 215–17

β-amyloid peptides 21–2, 28
β-carbon ligandosides 342
β-diketonate ligands 147–8
β-1,3-glucans 427–8
β-hairpins 298
β-peptides
biomimetic oligomers 396–7
unnatural foldamers 52, 53–9, 78, 80
β-sheets
artificial DNA 303–4
metallopeptides 276, 283, 297–8
unnatural foldamers 69–70
Bc-metal complexes 15–17
benzene-1,3,5-tricarboxylate anions 76–7
benzimidazole ligands 131–3
benzimidazolium cation encapsulation 174
bimetallic helicates
photophysical properties 207–9
thiolato donors 179–80, 182, 184–6
binaphthyl ligands 137–8
bioactivity of metallopeptides 290–2
biomimetic oligomers 379–405
abiotic helices 381–93
chiroptical properties 389–93
classification of metallofoldamers 380–1
folded oligomers 381, 384–402
higher-order architectures 393–402
metal coordination in folded aromatic amide
oligomers 394–6
nucleated secondary structures 379,
381–93
peptidomimetics 379, 380, 396–402
single-stranded oligomers 380, 381–4
bioprobes
breast cancer cells and tissues 215–17
cell penetration and staining 211–15
DNA analysis 217–19
lanthanoid helicates 210–19, 229
biotic foldamers 52–70
α-aminoxy acids 64–5

α-peptides 52, 60–3
aliphatic urea foldamers 63–4
amido foldamers 65–70
β-peptides 52, 53–9, 60–3, 78, 80
catalytic asymmetric synthesis 54–7
γ-peptides 52, 59–63
homogeneous foldamers 53
hybrid foldamers 60–3, 78
secondary and tertiary structures 55–63,
66–70, 78–80
biotinylation 215–17
bipyridine ligands
lanthanoid helicates 238–9
liquid-crystalline helicates 250, 252–7,
258, 262
metallopeptides 277–8, 286–7, 292–4,
297–8
molecular recognition 411
nucleic acids 350–3, 365
self-assembly of helicates 93–4, 115
structural aspects of helicates 140–3
Bipy–Schiff base metallohelicates 261–2
bis(amino acids) 422–3
bis(benzene-o-dithiolato) ligands 159–72
bis(benzimidazolepyridine) ligands 195, 205,
222–3
bis(bipyridyl) ligands 142–3
bis(catechol) ligands
coordination chemistry 166–72
metallopeptides 281–2, 290–2
structural aspects of helicates 127–8, 130–
1, 135, 148–50
bis-imine ligands 350–3, 365
bis(oxazoline) ligands 144–6
bis(pyridylimine) ligands 128
bleomycin 12
Blm-metal complexes 13–15
Born–Haber cycles 116–17
bovine spongiform encephalopathy (BSE) 22
breast cancer cells and tissues 215–17
bridging ligands
metallopeptides 281
structural aspects of helicates 125–6
thiolato donors 170
BSE *see* bovine spongiform encephalopathy

$Ca^{2+}$ binding 4–5, 7–8
cadmium regulation 6
cadmium sulfide nanoparticles 261
calcitonin 12

calmodulin   7–8
capsule formation
    dynamic materials applications   427–9
    thiolato donors   170–4, 176, 181
    unnatural foldamers   72–3, 77–80
carbon nanotubes (CNT)   326–7
carboxylate-containing siderophores   19–20
carboxypeptidases   8–9, 275
catalytic asymmetric synthesis   54–7
catechol-containing siderophores   19–20
CD *see* circular dichroism
CDPL *see* constitutional dynamic polymer
    library
cell wall biosynthesis   15–17
charge transfer (CT) complexes   413, 424
chelate cooperativity   102–3
chelate effect   94–5
chemical frustration   4
chiral lanthanoid helicates   236–9
chromium helicates   252
chromium-lanthanide helicates   227–8, 232–5,
    237–8
circular dichroism (CD)
    biomimetic oligomers   383–4, 389–93,
        397–8, 400–1
    biotic foldamers   66–7
    dynamic materials applications   425–9
    lanthanoid helicates   236–9
    metallopeptides   294, 296
    metalloproteins   5
    molecular recognition   411
    nucleic acids   343, 352, 355
    structural aspects of helicates   141
circular double-stranded helicates   129
circular helicates   100
circularly polarized luminescence
    (CPL)   236–9
*cis–trans* isomerism   65–6
CJD *see* Creutzfeldt–Jakob disease
CNT *see* carbon nanotubes
cobalt complexes   139–40, 166
cobalt regulation   7
coiled-coil structures   25–6
columnar mesophases   259–61, 264
complex anions   166–7, 169–71, 174–80
confocal luminescence microscopy
    213–15, 217
conformational communication   429
conformational switching   324
conformational tunability   412–13

con-sense in-plane displacements   422–3
constitutional dynamic polymer library
    (CDPL)   424
cooperativity
    liquid-crystalline helicates   255
    nucleic acids   352
    self-assembly of helicates   92, 100–4,
        108–10
    sensing applications   418, 429
copper catalysis   56
copper complexes
    artificial DNA   317–20, 324–6
    biomimetic oligomers   385–9, 394–6
    dynamic materials applications   425
    liquid-crystalline helicates   252–7, 258–60,
        262–6
    metallopeptides   277, 286–8, 297
    molecular recognition   412
    nucleic acids   336, 342, 351–9, 365–7, 369
    structural aspects of helicates   135–7,
        142–3, 145–6
Corey–Pauling–Koltun (CPK) model   428
Coulombic interactions   116–17
counterion template effects   129–32
CPK *see* Corey–Pauling–Koltun
CPL *see* circularly polarized luminescence
Creutzfeldt–Jakob disease (CJD)   22
cryptands   148, 420–2
crystallographic inversion centers   169–70,
    177, 180–1
CT *see* charge transfer
Curtius degradations   57
Cu,Zn-containing superoxide dismutase
    (SOD1)   10–12, 20, 28
cyanobacteria   2
cyclic β-amino acids   55
cyclic decapyrroles   76–8
cyclic hexaureas   76–7
cyclic peptoids   401–2
*trans*-cyclohexane-1,2-diamine (TCDA)
    391–2
cyclopentadienyl ligands   163–4
cytochrome-c   2, 3–4

d-f transitions   196–7, 205, 207–8, 220–36
dansyl ligands   416–17
1-deazaadenine-thymine base pairs   323
density functional theory (DFT)
    artificial DNA   325
    nucleic acids   353, 361, 366–7

deoxyribose nucleic acid (DNA)
   lanthanoid helicates   217–19
   metallopeptides   13–15
   metalloproteins   2–3
   nucleic acids   335–6, 342, 350–59, 365–7
   *see also* artificial DNA
design of foldamers   51
DFT *see* density functional theory
dianionic ligands   148–50
diastereoselectivity
   enantiomerically pure helicates from chiral
      ligands   139–50
   *meso*-helicates   135–6
   self-recognition   136–8
   structural aspects of helicates   135–51
differential scanning calorimetry (DSC)
   261–2
diffusion ordered spectroscopy (DOSY-
   NMR)   116
4-dimethylaminopyridine (DMAP)   72
dinuclear double-stranded helicates
   self-assembly   97, 116–17
   structural aspects   137, 146
   thiolato donors   165–7, 176–8
dinuclear metal centers   26–7
dinuclear triple-stranded helicates
   metallopeptides   284–5, 288–9
   photophysical properties   194, 198, 219–20,
      228–9
   self-assembly   111, 113–14, 118
   structural aspects   128, 137, 140
   thiolato donors   167–72, 178–81
dipam *see* pyridine-2,6-dicarboxamide
dipeptides   64, 284–5
diphenylphosphine-ethane dioxide
   ligands   240
dipic *see* pyridine-2,6-dicarboxylate
2,2′-dipicolylamine (DPA)   361
dipyrrolyl diketone boron complexes   74
direct counting method   99, 101
directional ligands   176–7
dis-sense in-plane displacements   422–3
1,4-disubstituted 1,2,3-triazoles   74–5
disulfide bonds   5
di(titanocene) complexes   163–4
ditopic bis(tridentate) ligands   198
ditopic organic receptors   102–3
DLS *see* dynamic light scattering
DMAP *see* 4-dimethylaminopyridine
DNA *see* artificial DNA; deoxyribose nucleic acid

DOSY-NMR *see* diffusion ordered
   spectroscopy
double helix structures   76
double-cross-type architectures   420–1
double-stranded helicates
   liquid-crystalline helicates   250, 253–4
   metallopeptides   280–2, 286–8
   self-assembly   97, 105–7, 109–12, 116–18
   structural aspects   129, 137–8, 141–6
   thiolato donors   165–7, 176–8
Down's syndrome   21
DPA *see* 2,2′-dipicolylamine
duplex stabilization   324
duramers   263–4
dynamic hybridization/dehybridization   308
dynamic light scattering (DLS)   343, 425
dynamic materials applications   407,
   419–28, 429

echistatin   292
EDTA *see* ethylene diamine tetraacetate
EF-hand domain   4, 7
effective molarity   97–9, 103–4
Eigen–Wilkins mechanism   105
electrical conductance   325–7
electron conductivity DNA   308, 309–10
electron paramagnetic resonance (EPR)
   spectroscopy
   metallopeptides   14
   nucleic acids   355, 360, 366–7
electron spin-echo envelope modulation
   (ESEEM) spectroscopy   14
electrostatic interactions   115–17
emission spectroscopy   200–1, 206, 213,
   220–1
enantioselective catalysis   408
energy transfer phenomena   208–10, 227–31,
   234–5
energy transfer upconversion (ETU)   234–5
enzymatic oligonucleotide synthesis   315
EPR *see* electron paramagnetic resonance
erbium helicates   204, 233–5
ESA *see* excited state absorption
ESEEM *see* electron spin-echo envelope
   modulation
ethidium bromide quenching   217–19
ethylene diamine tetraacetate (EDTA)   295
ETU *see* energy transfer upconversion
europium helicates   106–7, 113–16, 194–5,
   198–202, 206–30, 237–9

EXAFS *see* extended X-ray absorption fine structure
excited state absorption (ESA) 234–5
experimental matching 94
extended helical structures 239–40
extended rack-type complexes 420
extended X-ray absorption fine structure (EXAFS) 16

f-f transitions 196, 237
ferrichromes 19–20
ferritin 6
ferrocenyl ligands 278–9
ferromagnetism 325–6, 367
fibrillogenesis 21, 24
filament bundles 67–8
flapping motion 422–3
fluorene-*p*-xylene oligomers 413
fluorescence intensity 325
fluorescence resonance energy transfer (FRET) 398
fluorescence spectroscopy
  molecular recognition 411
  nucleic acids 343
  sensing applications 416–18
foldamers, definition 51–3
Fourier transform infrared (FTIR) spectroscopy 65–6, 253
FRET *see* fluorescence resonance energy transfer
FTIR *see* Fourier transform infrared

γ-peptides 52, 59–63
G-quadruplex structures 309
gadolinium helicates 134
gallium(III) complexes
  self-assembly of helicates 98–100
  structural aspects of helicates 128, 136, 148–9
glycol nucleic acids (GNA) 336, 338–40, 353–5, 357, 365
glycosylphosphatidylinositols 22
GNA *see* glycol nucleic acids
gramicidins peptide family 18–19
grid complexes 420–1

hairpin structures
  biotic foldamers 55–7
  metallopeptides 29
  nucleic acids 338, 343, 356–7

heavy metal regulation 6–7
HeLa cells 211–14
helical bundles 25–8
helicates
  allosteric cooperativity 101–2
  β-diketonate ligands 147–8
  β-peptides 55–9
  bipyridine ligands 140–3
  bis(benzene-*o*-dithiolato) ligands 159–72
  bis(catechol) ligands 127–8, 130–1, 135, 148–50
  chelate cooperativity 102–3
  chelate effect 94–5
  cooperativity 92, 100–4, 108–10
  coordination geometries for thiolato donors 159–60, 162–81
  dianionic ligands 148–50
  diastereoselectivity 135–51
  dinuclear double-stranded complexes 165–7, 176–8
  dinuclear triple-stranded complexes 167–72, 178–81
  enantiomerically pure helicates from chiral ligands 139–50
  future directions 150–1
  γ-peptides 59–60
  hexadentate N-donor ligands 144
  historical development of supramolecules 91–2
  hybrid foldamers 60–3, 78
  hydroxamic acid ligands 147
  interannular cooperativity 104
  intermolecular interactions 108–10
  kinetic aspects of multicomponent organization 92, 104–8
  *meso* helicates 135–6
  metallopeptides 279–88
  mixed benzene-*o*-dithiolato/catecholato ligands 174, 175, 178–81
  modelling intramolecular interactions 96–7
  mononuclear coordination complexes 93–6, 162–5
  non-covalently assembled ligand strands 150
  number of metal–ligand connections 99–100
  oxazoline ligands 144–6
  P-donor ligands 145–7
  peptide-helicates 281–8
  photophysical properties 193–248

helicates (*Continued*)
 polynuclear coordination complexes
  96–100
 2-pyridylimine ligands   144–5
 secondary structure and stabilizing
  interactions   118
 self-assembly   91–123, 135–41
 self-sorting effects   135–8, 150–1
 sequence selectivity   130–4
 site-binding model   110–12
 solvation energies and electrostatic
  interactions   115–17
 solvophobic interactions   72–3
 statistical factors in self-assembly   95–6,
  99, 101
 structural aspects   125–58
 structural dynamics   127–9
 subcomponent self-assembly
  reactions   181–6
 template effects   129–32
 terpyridine and quaterpyridine
  ligands   143–4
 theoretical models for effective
  molarity   97–9
 thermodynamic additive free energy
  model   112–15
 thermodynamic factors in self-
  assembly   92, 93–100, 110–15
 thiolato donors   159–92
 tripodal tris(benzene-*o*-dithiolato)
  ligands   172–6
 tris(benzene-*o*-dithiolato) ligands   159,
  162–76
 understanding the self-assembly
  process   108–17
 *see also* lanthanoid helicates; liquid-
  crystalline helicates
heme binding
 metallopeptides   26–7
 metalloproteins   2, 3–4
hemopexin   2
heteroarrays of transition metals   333–5
heterometallic helicates   196, 207–9, 221–36
heterostranded helicates   133, 135
hexadentate N-donor ligands   144
hexokinase   1
high performance liquid chromatography
  (HPLC)   315
histatin (Htn) peptide family   17
HIV protease   1

homochiral helicates   135–8
homogeneous foldamers   53
homometallic lanthanoid helicates   196,
  197–223
host–guest interactions   130, 132
HPLC *see* high performance liquid
  chromatography
Htn *see* histatin
Huntington's disease   20
hybrid foldamers   60–3, 78
hybrid inorganic–nucleic acid
  molecules   333–5, 337–8, 367–9
hydrogen bonding
 abiotic foldamers   70–1
 artificial DNA   310
 biomimetic oligomers   380, 388, 394
 biotic foldamers   58–63, 66
 dynamic materials applications   428
 metallopeptides   282–3, 298
 thiolato donors   165–6, 170–1
hydrophobic interactions   249, 295
hydroxamic acid ligands   147
hydroximate siderophores   19–20
hydroxypyridone ligands   317–20, 324–6
8-hydroxyquinoline ligands   353–6, 399–401,
  422–3

imidazole ligands   323, 342, 361
imino-bipyridine ligands   252–7, 258, 262
imino-phenanthroline ligands   252–7
imino-polypyridine ligands   251, 257–66
imino-terpyridine ligands   256, 259–61, 266
immunohistochemical luminescent
  assays   217
*in situ* polymerization   261–3
infrared (IR) spectroscopy   66, 68
interannular cooperativity   104
intermetallic interactions   106, 109–11,
  114, 117
intramolecular macrocyclization   96–8, 109–
  10, 114, 119
inversion centers   169–70, 177, 180–1
ionophores   18–20
IQ motif   8
IR *see* infrared
iron complexes
 artificial DNA   324
 biomimetic oligomers   395–6
 liquid-crystalline helicates   264
 metallopeptides   286–7, 290

nucleic acids    352, 359
  self-assembly of helicates    105–7
  structural aspects of helicates    128, 134, 147
iron-lanthanide helicates    228–30
iron–sulfur clusters    2
Irvin–Williams stability series    2
isothermal calorimetry (ITC)    354, 411
ITC *see* isothermal calorimetry

Job plots    400

kinesin    419

lanthanoid helicates
  advantages for photophysical
    properties    196–7
  biomimetic oligomers    383–4
  chiral    236–9
  control of f-metal ion properties by d-
    transition metal ions    228–35
  d-f transitions    196–7, 205, 207–8,
    220–36
  dinuclear triple-stranded helicates    194,
    198, 219–20
  energy transfer between lanthanoid
    ions    208–10, 227–31, 234–5
  extended helical structures    239–40
  heterometallic    196, 207–9, 221–36
  homometallic    196, 197–223
  luminescent bioprobes    210–19
  molecular upconversion    234–5
  nephelauxetic effect    200, 203–6, 224
  photophysical properties    193–248
  quantum yields    196, 198–204, 221,
    226–7, 235
  radiative lifetimes    196, 201–2, 203–6,
    229–33, 235
  research background    193–7
  self-assembly    106–7, 113–16, 118, 237–8
  sensitizing NIR-emitting lanthanoid
    ions    235–6
  site-symmetry analysis    204–5, 206–8
  switching luminescence on and off    228–9
  tri- and tetranuclear triple-stranded
    helicates    220–3
  triplet-state energy    198–203
  tuning lifetime of NIR-emitting lanthanoid
    ions    229–33
lanthanum helicates    209, 221–2
ligand-modified nucleosides    312–14

ligand-to-metal charge transfer (LMCT)
  220–1, 229
linear helicates    100
lipase catalysis    57
liquid-crystalline helicates    249–74
  applications    250, 257–8
  future directions    267–8
  imino-bipyridine ligands    252–7,
    258, 262
  imino-phenanthroline ligands    252–7
  imino-polypyridine ligands    251, 257–66
  mesomorphic materials    258–62, 264, 268
  mesophase packing    259–61, 264
  non-mesomorphic ligands    258–9, 262
  π-stacking interactions    249–50
  radical and *in situ* polymerization    261–3
  redox cycling    255–7
  self-organization    255–60
LMCT *see* ligand-to-metal charge transfer
locked nucleic acids (LNA)    336, 338–40,
  358–9, 361
loop/turn structures    276, 284, 288–92
luminescence dissymmetry factors    236–9
luminescence immunoassays    212–13
luminescence of lanthanoid helicates    196–7,
  202, 210–22, 228–33, 236–40
lyotropic metallomesogens    261

mAb *see* monoclonal antibodies
macrocyclization, intramolecular    96–8,
  109–10, 114, 119
magnetic properties
  artificial DNA    325–6
  lanthanoid helicates    228, 239–40
  nucleic acids    367
major histocompatibility complex
  (MHC)    78
MALDI-TOF spectroscopy    215
Mannich reactions    54–6
mass spectrometry (MS)
  artificial DNA    315
  biomimetic oligomers    382, 395
  dynamic materials applications    420–2
  liquid-crystalline helicates    252
  metallopeptides    281, 284
  nucleic acids    342, 366–7
  thiolato donors    172, 184
material synthesis applications    407,
  419–28
MCF-7 human breast cancer cells    215–17

mercury complexes
artificial DNA   310–11, 322
nucleic acids   342, 368–9
sensing applications   416
*meso*-complex anions   166–7, 169–71
*meso*-helicates   135–6
mesomorphic materials   258–62, 264, 268
mesophase packing   259–61, 264
*meta*-phenyleneethynylene (*m*PE)
foldamers   409–11
metal-containing, ligand-containing nucleic
acid duplexes   333–8, 340–69
[1+1] coordination   343–4
[2+2] coordination   344–7, 350–6, 359, 368
[3+1] coordination   347–9, 359–61, 368
[3+3] coordination   347, 357–9
alternative metal–ligand base pair with
different ligands   347, 359–61
alternative metal–ligand base pair with
identical ligands   342–59
chemical and physical characteristics
351–3, 361–5
design strategies   340–2
melting temperatures   344–9, 352–3, 361–4
multiple metal complex-containing
duplexes   361–7
N,N-coordinating ligands   343–53
O,N-coordinating ligands   353–6
O,O-coordinating ligands   356–7
metal-to-ligand charge transfer (MLCT)
223–5
metallomesogens *see* liquid-crystalline
helicates
metallopeptides   1, 12–29
α-helices   276, 283, 292–7
antibiotic metallopeptides   13–20
antibiotic salivary peptide   17
β-amyloid peptides   21–2, 28
β-sheets   276, 283, 297–8
Bc and cell wall biosynthesis   15–17
conformations and secondary
structures   276, 281–4, 288–99
examples of metal peptide conjugates
276–9
helicates   279–88
ionophores and siderophores   18–20
loop/turn structures   276, 284, 288–92
macrocycles   288–92
metal-assisted stabilization of peptide
microstructures   288–98

metal-triggered conformational
changes   25–8
metallo-Blm and DNA binding   13–15
N-terminal binding peptides and Ni-
SOD   24–5
as nature's metallofoldamers   22
neurodegenerative diseases   20–4
parallel and antiparallel configurations   281,
284–5, 287–8, 298
peptide-helicates   281–8
prion proteins and fragments   22–4, 28
self-assembly   287–8
sequential/directional ligands   285–8
metalloproteins   1–12, 28–9
α-lactalbumin and $Ca^{2+}$ binding   4–5
aminopeptidase and alternative catalysis
9–10
applications   408–9
calmodulin and $Ca^{2+}$ binding   7–8
carboxypeptidase A catalytic
mechanism   8–9
cytochrome-c and heme binding   2, 3–4
ligand-binding and conformation
changes   7–10
metal-triggered conformational changes
3–7
metallothionein and heavy metal
regulation   6–7
as nature's metallofoldamers   2–3
protein conformation, structure and
function   1–2
protein misfolding   10–12
superoxide dismutases   10–12
metallothionein   6–7
*N*-methylnicotinium ligands   140
MHC *see* major histocompatibility complex
micelles   416
mixed benzene-*o*-dithiolato/catecholato
ligands   159, 163, 176–81
MLCT *see* metal-to-ligand charge transfer
molecular cages   181
molecular grid complexes   420–1
molecular machines   308
molecular recognition   407, 409–13, 429
molecular scaffolds   250, 257
molecular upconversion   234–5
molten globular states   5
molybdenum complexes
metallopeptides   289, 291
thiolato donors   160, 182

monoclonal antibodies (mAb)   215–17
mononuclear coordination complexes   93–6,
   162–5
motor proteins   419
myosin   419, 423

*N*-methylnicotinium ligands   140
*N*-substituted oligoglycines *see* peptoids
N-terminal binding peptides   24–5
1,5-naphthalenediamido spacers   168–9
naphthyridine ligands   264–5
Nd:YAG lasers   229
near infrared (NIR) emissions   196, 204,
   229–36
neodymium helicates   204, 210, 233, 237–9
nephelauxetic effect   200, 203–6, 224
neurodegenerative diseases   20–4
Ni-containing superoxide dismutase
   (SOD1)   24–5
nickel complexes
   biomimetic oligomers   388–90, 392–5
   liquid-crystalline helicates   264
   molecular recognition   412
   nucleic acids   352–3, 356, 365
   thiolato donors   165–6, 183–6
NIR *see* near infrared
NMR *see* nuclear magnetic resonance
non-covalently assembled helicates   150
nuclear magnetic resonance (NMR)
   spectroscopy
   artificial DNA   313, 323
   biomimetic oligomers   382–5, 387–95, 397,
      401–2
   biotic foldamers   58, 64–6, 68
   dynamic materials applications   420–2,
      424–5
   helicates   92
   lanthanoid helicates   222, 231
   liquid-crystalline helicates   253
   metallopeptides   14–15, 21–3, 26,
      281–4, 294
   molecular recognition   409–11, 413
   nucleic acids   362, 368
   structural aspects of helicates of
      helicates   130, 141
   thiolato donors   167–8, 172–4, 176–9,
      184–5
nucleated secondary structures   379, 381–93
nucleic acids   333–78
   [1+1] coordination   343–4

[2+2] coordination   344–7, 350–7, 359, 368
[3+1] coordination   347–9, 359–61, 368
[3+3] coordination   347, 357–9
alternative metal–ligand base pair with
   different ligands   347, 359–61
alternative metal–ligand base pair with
   identical ligands   342–59
chemical and physical characteristics
   351–3, 361–5
design strategies   340–2
future directions   367–9
heteroarrays of transition metals   333–5
hybrid inorganic–nucleic acid
   molecules   333–5, 337–8, 367–9
melting temperatures   344–9, 352–3, 361–4
metal-containing, ligand-containing
   duplexes   333–8, 340–69
metal-containing, ligand-containing nucleic
   acid triplexes   367
metallopeptides   29
multiple metal complex-containing
   duplexes   361–7
N,N-coordinating ligands   343–53
O,N-coordinating ligands   353–6
O,O-coordinating ligands   356–7
synthetic analogues of DNA   336, 338–40,
   350–7

oligocholate foldamers   414–18, 429
oligoindoles   74
oligonucleotides
   alternative base pairing systems   310,
      318–19
   applications   324
   automated synthesis   314–15
   biological functions and beyond   305–6
   design and synthesis of metal base
      pairs   313–15
   DNA technology   306, 308
   enzymatic synthesis   315
   interactions with metal ions   308
   as natural foldamers   303–5
   synthetic metallofoldamers   320, 322
   X-ray structure determination   317–18
oligo(*m*-phenylene ethynylene)   72
oligopyridine ligands   251, 252, 267–8
optical microscopy   68–9
organogellators   408
oxazoline ligands   144–6
oxidative stress   21–3

π-stacking interactions
  biomimetic oligomers  380, 385–6, 389, 402
  dynamic materials applications  421
  liquid-crystalline helicates  249–50
  nucleic acids  336, 340–2, 365, 368
  unnatural foldamers  71–2
P-donor ligands  145–7
palladium complexes
  metallopeptides  288–90, 294, 296–7
  molecular recognition  410
  thiolato donors  184–5
PAM *see* phenylalanine amino mutase
parallel β-sheets  69–70
parallel configurations  281, 284–5, 287–8
Parkinson's disease  20, 24
PCR *see* polymerase chain reaction
PDI *see* protein disulfide isomerase
pentanuclear helicates  105, 108, 119
peptide nucleic acids (PNA)  51, 338–55, 357–9, 361, 365
peptide-helicates  281–8
  *see also* metallopeptides
peptoids  51
peptidomimetics  379, 380, 396–402
peptoids  396–402
phenylalanine amino mutase (PAM)
  catalysis  56
phosphoramidites  313–14, 367
phosphorescence  201
photoactivity  277
photophysical properties
  advantages of lanthanoid helicates  195–7
  chiral lanthanoid helicates  236–9
  control of f-metal ion properties by d-transition metal ions  228–35
  d-f transitions  196–7, 205, 207–8, 220–36
  dinuclear triple-stranded helicates  194, 198, 219–20
  energy transfer between lanthanoid ions  208–10, 227–31, 234–5
  extended helical structures  239–40
  heterometallic lanthanoid helicates  196, 207–9, 221–36
  homometallic lanthanoid helicates  196, 197–223
  lanthanoid helicates  193–248
  luminescent bioprobes  210–19
  molecular upconversion  234–5
  nephelauxetic effect  200, 203–6, 224

quantum yields  196, 198–204, 221, 226–7, 235
radiative lifetimes  196, 201–2, 203–6, 229–33, 235
research background  193–7
sensitizing NIR-emitting lanthanoid ions  235–6
site-symmetry analysis  204–5, 206–8
switching luminescence on and off  228–9
tri- and tetranuclear triple-stranded helicates  220–3
triplet-state energy  198–203
tuning lifetime of NIR-emitting lanthanoid ions  229–33
picrate ligands  401–2
pipyridyl ligands  75–6
plastocyanin  12–13
platinum complexes  296–7
pleated sheets  58
PNA *see* peptide nucleic acids
polarizability  205, 258–9
polymerase chain reaction (PCR)  306–7, 315
polymeric backbones  51
polynuclear coordination complexes  96–100
polynucleotides  427
polypeptides  80
polypyridine ligands  250–1, 257–66
polysaccharides  427–8
polytopic ligands
  lanthanoid helicates  194
  liquid-crystalline helicates  266–7
  nucleic acids  333–5
porphyrins  26–7
prion proteins  22–4, 28
process optimization  52
protein disulfide isomerase (PDI)  16
protein fibers  67–8
protein misfolding  10–12, 22–3
protein–protein interactions  2
protein surface-binders and inhibitors  408
proteins *see* metalloproteins
pyrazine ligands  422–3
pyrenyl ligands  416–17
pyridine-2,6-dicarboxamide (dipam)  360, 394
pyridine-2,6-dicarboxylate (dipic)  360–1, 365–6
pyridine-modified nucleic acids  343–5, 347–9, 357–9, 361–7
pyridine-pyridazine ligands  265

pyridine–pyrimidine oligomers  419–20, 422–3
2-pyridylimine ligands  144–5

quadruple-stranded helicates  128
quantum dots (QD)  325
quantum yields  196, 198–204, 221,
   226–7, 235
quaterpyridine ligands  143–4

radiative lifetimes  196, 201–2, 203–6,
   229–33, 235
radical polymerization  261–3
Ramachandran plots  283–4
randomization  51
reactive oxygen species (ROS)  21–3
redox cycling  255–7, 277
refractive index  204
RGD sequence  292
rhodium catalysis  56
rhodium-based metallopeptides  295–7
rhodotorulic acid  125–6
ribose nucleic acid (RNA)  304–5
ROS *see* reactive oxygen species
rotamers  57–8
rubredoxin  29
ruthenium catalysis  57
ruthenium-based metallopeptides  277–8, 294
ruthenium-containing nucleic acids  339–40
ruthenium-lanthanide helicates  232–3

salen ligands
   artificial DNA  318–22, 325
   biomimetic oligomers  391–3
   molecular recognition  412
   nucleic acids  350, 354–7, 366–7, 368
salophen ligands  389–91
SAM *see* self-assembled monolayers
samarium helicates  194–5, 203
scanning electron microscopy (SEM)  68–9
Scatchard plots  108–9
Schiff base condensations  182–3
schizophyllan (SPG)  427
screening  51
segetalins  290
self-assembled monolayers (SAM)  339
self-assembly
   allosteric cooperativity  101–2
   artificial DNA  305–7
   biomimetic oligomers  379
   biotic foldamers  63–4, 67, 78–80

chelate cooperativity  102–3
chelate effect  94–5
cooperativity  92, 100–4, 108–10
dynamic materials applications  424–5
helicates  91–123, 135–41
historical development of
   supramolecules  91–2
interannular cooperativity  104
intermolecular interactions  108–10
kinetic aspects of multicomponent
   organization  92, 104–8
lanthanoid helicates  106–7, 113–16, 118,
   237–8
liquid-crystalline helicates  249, 252,
   257, 267
metallopeptides  287–8
modelling intramolecular interactions  96–7
mononuclear coordination complexes  93–6
nucleic acids  333–5, 339
number of metal–ligand connections
   99–100
polynuclear coordination complexes
   96–100
secondary structure and stabilizing
   interactions  118
site-binding model  110–12
solvation energies and electrostatic
   interactions  115–17
statistical factors  95–6, 99, 101
theoretical models for effective
   molarity  97–9
thermodynamic additive free energy
   model  112–15
thermodynamic factors  92, 93–100, 110–15
thiolato donors  159, 171, 181–6
understanding the self-assembly
   process  108–17
self-discrimination  135
self-organization  255–60
self-recognition  135–8
self-sorting
   artificial DNA  305
   structural aspects of helicates  135–8, 150–1
SEM *see* scanning electron microscopy
sensing applications  407, 414–18, 429
sensor applications  325
sequence selectivity  130–4
sequence specificity  335–6
sequential/directional ligands  285–8
shape complementarity  310

siderophores 18–20
silica nanoparticles 217
silver complexes
  artificial DNA 310–11, 323–4
  biomimetic oligomers 382
  dynamic materials applications 421–2,
    424, 425
  molecular recognition 409–10, 413
  nucleic acids 335–6, 343, 358, 362, 366,
    368–9
  structural aspects of helicates 128–9, 144–6
single-stranded biomimetic oligomers 380,
  381–4
single-stranded DNA 306–8
single-walled carbon nanotubes
  (SWCNT) 326–7
site-binding model 110–12
site-symmetry analysis 204–5, 206–8
smectic mesophases 259–61
SOD *see* superoxide dismutases
sodium picrate 401–2
solid-state nuclear magnetic resonance
  (NMR) 21–3
solvation energies 115–17
solvophobic interactions
  abiotic foldamers 70, 71–3
  biomimetic oligomers 382, 402
  dynamic materials applications 421
  sensing applications 414–16
spectrophotometric titrations 254–5
SPGH *see* schizophyllan
spiroladder oligomers 422–3
squaraine dyes 414
stereochemical matching 93
stereogenic elements 135–7, 146–7
structural dynamics 127–9
subcomponent self-assembly reactions 181–6
*N*-substituted oligoglycines *see* peptoids
superoxide dismutases (SOD) 10–12, 20,
  24–5
surfactants 416
SWCNT *see* single-walled carbon nanotubes
symmetry numbers method 99, 101
synthetic analogues of DNA 336, 338–42,
  350–69

TCDA *see* trans-cyclohexane-1,2-diamine
TEM *see* transmission electron microscopy
template-controlled reactions 182–3
template effects 129–32
templated helix-like structures 381

terbium helicates 202–4, 209–10, 213,
  227, 240
terpyridine ligands
  liquid-crystalline helicates 256,
    259–61, 266
  metallopeptides 286–7, 290
  nucleic acids 357–9
  structural aspects of helicates 143–4
tetraguanidinium salts 75
tetrahedral clusters 130, 168–70, 172–3
tetrahedral helicates 100
tetranuclear clusters 148–9, 168–70, 172–3
tetranuclear triple-stranded helicates 114–15,
  220–3
tetrazole ligands 342, 361
tetrol ligands 148–9
thermodynamic additive free energy model
  (TAFEM) 112–15
thermotropic metallomesogens 261
thiolato donors
  bis(benzene-*o*-dithiolato) ligands 159–76
  coordination geometries 159–60, 162–81
  dinuclear double-stranded complexes 165–
    7, 176–8
  dinuclear triple-stranded complexes 167–
    72, 178–81
  helicates 159–92
  mixed benzene-*o*-dithiolato/catecholato
    ligands 159, 163, 176–81
  mononuclear coordination complexes
    162–5
  subcomponent self-assembly
    reactions 181–6
  tripodal tris(benzene-*o*-dithiolato)
    ligands 172–6
  tris(benzene-*o*-dithiolato) ligands 159,
    162–76
thiourea foldamers 63–4
time-resolved luminescence microscopy
  (TRLM) 211–12, 217
titanium complexes
  metallopeptides 281–5, 289
  structural aspects of helicates 131, 137,
    148–9
  thiolato donors 163–82
TP *see* trigonal-prismatic
*trans*-cyclohexane-1,2-diamine (TCDA)
  391–2
transcription factors 2–3, 6
transesterification catalysis 295
transmetallation reactions 182–6

transmission electron microscopy
(TEM) 426–7
triazole ligands
artificial DNA 323
dynamic materials applications 427–9
nucleic acids 343, 361
tridentate ligands 198, 357–9
tri(ethyleneglycol) 409–10
trigonal-prismatic (TP) coordination
geometry 159–60
trinuclear double-stranded helicates 105, 286–8
trinuclear triple-stranded helicates 133–4,
220–3, 228–9
tripeptide ligands 284–5
triple-stranded helicates
dynamic materials applications 427–9
liquid-crystalline helicates 250
metallopeptides 280, 284–5, 288–9
photophysical properties 194, 198, 202,
210–11, 219–23, 228–9
self-assembly 99, 105–7, 109, 111,
113–15, 118
structural aspects 128, 133–4, 137, 140,
147–9
thiolato donors 167–72, 178–81
triplet-state energy 198–203
tripodal tris(benzene-*o*-dithiolato)
ligands 172–6
tris(benzene-*o*-dithiolato) ligands 159, 162–76
tris(catecholato) ligands 164, 171, 173
tritopic ligands 234–5
TRLM *see* time-resolved luminescence
microscopy
Tröger's base scaffolds 137–8, 143
troponin 423
tungsten complexes 160
turcasarin 76–8
two-helix bundles 397–8

ultraviolet (UV) spectroscopy 343
unlocked nucleic acids (UNA) 338–40, 358–9
unnatural foldamers
abiotic foldamers 52–3, 70–2
anion-induced organization 73–8
applications 78–80
biotic foldamers 52–70
definitions 51–3
future directions 81
organization induced by external
agents 72–8
solvophobic interactions 70, 71–3

urea foldamers 63–4
urotensin II 291–2
UV-Vis spectroscopy
biomimetic oligomers 382, 400–402
dynamic materials applications 427
molecular recognition 409–10, 413
nucleic acids 367
sensing applications 414

vesicles 408
voltage-dependent $Ca^{2+}$ channels 8

Watson–Crick base pairing
artificial DNA 304–5, 309–11, 317
nucleic acids 335–6, 340, 355, 357

X-ray crystallography
artificial DNA 317–18
biomimetic oligomers 385, 387,
393–5, 401
dynamic materials applications 420–2
lanthanoid helicates 221–2
liquid-crystalline helicates 253–4, 264–6
metallopeptides 21–2
nucleic acids 343, 365, 368
structural aspects of helicates of
helicates 128–9, 130, 132–3, 148–50
unnatural foldamers 80
X-ray diffraction
biomimetic oligomers 387, 389–91, 395
biotic foldamers 58, 68–9
dynamic materials applications 425
liquid-crystalline helicates 262–4
metallopeptides 21
structural aspects of helicates 133
thiolato donors 177–9, 184

ytterbium helicates 203, 213, 232–3

zinc complexes
biomimetic oligomers 396–402
dynamic materials applications 427–9
metallopeptides 6–7, 275–6, 295, 298–9
molecular recognition 411
nucleic acids 352, 356
sensing applications 417–18
structural aspects of helicates 125–6, 127
thiolato donors 183–6
zinc finger proteins 2–3, 275–6, 299
zinc-lanthanide helicates 205–6, 223–6
zinc regulation 6–7